Pharmaceutical Inorganic Chemistry

Pharmaceutical Inorganic Chemistry

Bharath Shiromani

Prof. Dr. Kaza Somasekhara Rao
M.Sc, Ph.D

Professor of Inorganic Chemistry, (Retd)
Acharya Nagarjuna University – Nuzvid Campus,
Nuzvid, A.P – 521201, India.

Chennupati Venkata Suresh
M.Pharm (Ph.D)

Sri Siddhartha Pharmacy College
Nuzvid, A.P-521 201, India.

PharmaMed Press
An Imprint of **BSP Books Pvt. Ltd.**
4-4-309/316, Giriraj Lane,
Sultan Bazar, Hyderabad - 500 095.

Pharmaceutical Inorganic Chemistry
by **Prof. Dr. Kaza Somasekhara Rao and Chennupati Venkata Suresh**

Seventh Reprint 2026

Published by

PharmaMed Press
An Imprint of BSP Books Pvt. Ltd.
4-4-309/316, Giriraj Lane, Sultan Bazar, Hyderabad - 500 095.
Phone: 040-23445600, 23445688; Fax: 91+40-23445611
E-mail: info@pharmamedpress.com
www.pharmamedpress.com/pharmamedpress.net

Printed at
Repro India Limited.

Price: Rs. 795.00

ISBN: 978-81-7800-262-0 (Paperback)

Foreword

I am happy that Kaza Somasekhara Rao & Chennupati Venkata Suresh have undertaken the task of writing Text Book of Pharmaceutical Inorganic Chemistry. Although books on Pharmaceutical Inorganic chemistry have published in the past, the publication of this book brings up to date information in the field of Inorganic Chemistry. This book would be useful not only to the students of chemistry & Pharmacy but also to the teachers of Inorganic Chemistry.

The text book consists of Theory and Practical as two Major sections. Emphasis has been given to present the content in the light of modern developments in a very simple language and elegant style so that students who read the book should be able to follow the topics very easily without any difficulty. The book contains the information as prescribed by various universities of India for under graduate students and books such as these could help in bringing about uniformity not only in the standards of education, but also standards of teaching in pharmacy institutions.

Mr. Kaza Somasekhara Rao, who is an eminent personality in the field of chemistry, has rendered his valuable service to the profession of pharmacy by publishing this book, which should find a place in all Institutions providing pharmacy education.

Preface

This book of Pharmaceutical Inorganic Chemistry consists of Theory as well as Practical Parts. Every inorganic compound included in this book has been discussed under definition, preparation, tests for identity, tests for purity, assay method and uses. This caters the needs of Pharmacy and Chemistry students so far as the above subject. In Practical manual, we have discussed the qualitative, quantitative analysis, limit tests and some of the preparations. The authors covered all the topics by keeping in mind the syllabi of almost all Indian Universities. Thus, this book is helpful to the students definitely.

The authors thank Acharya Nagarjuna University, Nagarjuna Nagar, Guntur, A.P. and Sri Siddhartha Pharmacy College, Nuzvid for giving permission to write this book. The authors also acknowledge the co-operation extended by their well wishers and family members.

- Authors

Contents

CHAPTER 6

Pharmaceutical Aids

CHAPTER 12

Essential Trace Ions

CHAPTER 13

Gases and Vapours

CHAPTER 14

Radiopharmaceuticals

CHAPTER 15

Miscellaneous Pharmaceutical Agents

Part – II
Practical Lab Manual

4. Inorganic Qualitative Analysis

5. Inorganic Quantitative Analysis

7. Limit Tests

8. Preparation of Some Inorganic Compounds of Pharmaceutical Interest

Part - I

Theoretical Pharmaceutical Inorganic Chemistry

CHAPTER 1

Introduction

1.1 Inorganic Chemistry

The inorganic chemistry was originally meant for non-living chemistry. It was the part of chemistry that has arisen from the arts and recipes which deals with minerals and ores. This chemistry began with finding naturally occurring substances that had useful properties for e.g., flint (~5×10^5 years ago). The research continues in various fields of sciences like mineralogy, geology and later in other fields like pharmaceutical, bioinorganic enzymes, high temp-super conductors, metals etc. First chemical reaction was the discovery of fire, the most important early reactions were the reductions of metal oxides, carbonates and sulfides to metals. This chemistry mainly deals with the changes that can be effected in materials. There is a need to distinguish between superficial resemblances for e.g., between fool's gold (iron pyrites, FeS_2) and the element gold. Now we can see how far inorganic chemistry has developed from the days when it was meant for non-living to various other fields including living.

1.2 Inorganic Nomenclature

The complexity of inorganic compounds increases, there is difficulty in describing these compounds with simple names. The rules were issued by the International Commission on the Nomenclature of Inorganic Chemistry.

1.2 (a) Formulas and Names of Compounds

(i) In formulas the cation (electropositive constituent) will be placed first in the binary compounds.

$NaCl$, $MgSO_4$, $AlCl_3$ etc.

(ii) In the formulas of binary compounds between non-metals, the constituent that occurs in the sequence i.e., heavier non-metal first followed by the lower.

IF_7, NH_3, Cl_2O, OF_2 etc.

(iii) In naming of binary compound, the name of the electropositive constituent is not changed. But the electronegative constituent end in-ide (mono atomic or homo polyatomic). For e.g., HCl-hydrogen chloride, H_2S-hydrogen sulphide, NaCl-sodium chloride.

(iv) In naming the binary compounds, the name of the element standing later in the formula is end in – ide.

OF_2- oxygen difluoride, ClO_2- chlorine dioxide, SiC- silicon carbide,

(v) If the anion (electronegative constituent) is polyatomic, it should be named by the termination - ate, - ite, - ide

SO_4^{2-} -sulphate, NO_3^- -nitrate,

SO_3^{2-} -sulphite, NO_2^- -nitrite,

SO_2 -sulphur dioxide, NO -nitric oxide etc.

(vi) Some of the polyatomic compounds have central atom or characteristic atom. The atoms, radicals or molecules attached to central atom. Such polyatomic group is named as complex and the atom, radicals or molecules bound to the central atom are called ligands.

(vii) The molecular hydrides naming is naming the element followed by suffix – ane

PbH_4 -plumbane, GeH_4 -germane

SiH4 -silane, B_2H_6 -diborane etc.

(viii) There are exceptions to these rules.

PH_3 -phosphine, AsH_3 -arsine,

sbH_3 -stilbine, BiH_3 -bismuthane

1.2 (b) Name for Ions and Radicals

Radical is a group of atoms which occurs repeatedly in a number of different compounds.

(i) For mono atomic cations

Cu^+ Copper (I) ion, $Fe2^+$ iron (II) ion

I^+ Iodine (I) cation

(ii) For poly atomic cations

NO^+ nitrayl cation, NO_2^+ nitryl cation

(iii) For polyatomic cations derived by the addition of more photons end in-onium.

PH_4^+ - Phosphonium, As H_4^+ - Arsonium

NH_4^+ - Ammonium is exception for this.

(iv) Substituted ammonium ions derived form nitrogen bases with names ending in-amine to - ammonium.

NH_2-CH_2-CH_2-NH_2 -ethylene diamine.

OH NH_3^+ -hydroxyl ammonium

(v) The naming of monoatomic anions -end in - ide

H^- - hydride ion Cl^- - chloride ion

N_3^- - azide ion O^{2-} - oxide ion

S^{2-} - sulphide ion

(vi) The names of certain polyatomic anions -end in - ide

OH^- - hydroxide ion, NH^{2-} - imide ion

NH_2^- - amide ion CN^- - cyanide ion

HF_2^- -hydrogen difluride ion I_3^- -triiodide ion

(vii) The names of oxygen containing anions

NO_2^- - nitrite SO_3^{2-} - sulphite ClO_2^- - chlorite

$N_2O_2^-$ - hyponitrite $S_2O_6^{2-}$ – disulphite ClO^- hypochlorite

AsO_3^{3-} - arsenite $S_2O_4^{2-}$ – dithionite IO^- - hypoiodite

(viii) The names of certain neutral and cationic radicals containing oxygen or other chalcogens end in - yl.

OH - hydroxyl $PSCl_3$ - thiophosphoryl chloride

CO - carbonyl SO_2NH - sulphonyl imide

NO - nitrosyl SO - sulphynyl (thionyl)

PO - phosphoryl S_2O_5 - disuphynyl

ClO – chlorosyl SeO - selesninyl

ClO_2 - chloryl SeO_2 - selenonyl

ClO_3 - perchloryl CrO_2 - chromyl

$COCl_2$ - carbonyl chloride UO_2 - uranyl

1.2 (c) Acids

(i) Binary or psuedobinary compounds of hydrogen end in - ide

HCl - hydrogen chloride

HCN- hydrogen cyanide

H_2S - hydrogen sulphide

(ii) The oxo-acids name end with - ous or - ic.

$H_2N_2O_2$ - hyponitrous acid

$H_4P_2O_6$ - hypophosphic acid

HOBr - hypobromous acid

Prefix hypo - is used to denote lower oxidation state and per – used to denote higher oxidation state.

$HClO_4$ - perchloric acid

HIO_4 - periodic acid

Prefix ortho -, meta -, para - are used to distinguish acids differing in the 'content of water'.

H_3Bo_3 - orthoboric acid | $(HBo_2)_n$ - meta boric acid

H_4SiO_4 - orthosilicic acid | $(H_2SiO_3)_n$ - meta silicic acid

H_3PO_4 - ortho phosphoric acid | $(HPO_3)_n$ - meta phosphoric acid

H_6TeO_6- orthotelluric acid | $H_2S_2O_7$ - pyro sulphuric acid

$H_4P_2O_7$ - pyro phosphoric acid

The names of other oxo acids:

H_2CO_3 - carbonic acid

HOCN - cyanic acid

HNCO - isocyamic acid

HONC - fulminic acid

HNO_3 - nitric acid

HNO_2 - nitrous acid

$H_5P_3O_{10}$ - triphosphoric acid

$HClO_3$ - chloric acid

$HClO_2$ - chlorous acid

H_2SO_4 - sulphuric acid

H_2SO_5 - peroxomonosulphuric acid

$H_2S_2O_8$ - peroxodisulphuric acid

$H_2S_2O_3$ - thiosulphuric acid

$H_2S_2O_6$ - dithionic acid

$H_2S_2O_3$ - sulphurous acid

$H_2S_2O_4$ - dithonous acid

HOCN, HNCO and HONC are called chain compounds

1.2 (d) Elements, Atoms and Group of Atoms

(i) Identification of mass, charge and atomic number using indexes.

Mass, ionic charge, atomic numbers are indicated by means of:

Left upper index (superscript) – mass number

Left lower index (subscript) – atomic number

Right upper index (superscript) – ionic charge

Right lower index (subscript) – molecular formula

$^{32}_{16}S^{2+}_{4}$

(ii) *Isotopes*: Atomic number in same but mass number different

isotopes of hydrogen – ^1H, ^2H, ^3H named as portium, deuterium and tritium respectively.

The cations $^1H^+$, $^2H^+$, $^3H^+$ named as proton, deuteron triton respectively.

(iii) Name of an element or elementary substance of definite molecular structure or formula.

O_2 - oxygen, O_3 - ozone, P_4 - white phosphorous,

S_8 - α – sulphur, β - sulphur etc.

1.2 (e) Salts

(i) *salts*: A salt is a chemical compound consisting of a combination of cations and anions. H_3O^+ is normally called as acid (both salt and acid character)

(ii) Binary compound can be defined as that it contains only one kind of cation and one kind of anion and is still considered as salt.

NaCl - sodium chloride, LiH - lithium hydride

For polyatomic cations and / or anions enclosing marks should be used.

$Tl^{I}I_3$ - thallium (I) triiodide

$Tl^{III}I_3$ – thallium (III) triiodide

For salts containing acid hydrogen: Salts containing both a hydron which is replaceable and one or more metal cations are called acid salts.

$NaHCO_3$ - sodium hydrogen carbonate

LiH_2PO_4 - lithium dihodrogen phosphate

K_2HPO_4 - dipotassium hydrogen phosphate

$NaHSO_3$ - sodium bisulphite

1.2 (f) Prefixes or Affixes used in Nomenclature

(i) Multiplying affixes

Mono	1
Di	2
Tri	3
Tetra	4

Penta	5
Hexa	6
Hepta	7
Octa	8
Nona (ennea)	9
Deca, undeca (hendeca)	10
Dodeca	12, etc

These are used by direct joining without hyphens (-).

Other nomenclature used is:

bis (2), tris (3), tetrakis (4), pentakis (5) etc., also by direct joining without hyphens but usually with enclosing marks each whole expression to which prefix applies.

(ii) *structural affixes*

These affixes are italicized ad separated from the rest of the name by hyphens.

Cis	2 groups occupying adjacent positions
Trans	2 groups occupying opposite positions (i.e., in the polar position on a sphere)
Sym	Symmetric
Asym	Asymmetric
Closo	A cage or closed structure, especially a boron skeleton that is a polyhedron (a polyhedron having all triangular faces)
Nido	A nest like structure
Cyclo	A ring structure
Fac	3 groups occupying the corners of the same face of an octahedron
Hexahedro	8 atoms bond into a hexahedron (e.g. a cube)
Icasohedro	12 atoms bond into a triangular icosohedron
Octohedro	6 atoms bond into an octahedron
Dodecahedron	8 atoms bond into a dodecahedron with triangular faces
Quadro	4 atoms bond into a quadrangle (i.e., square)
Tetrahedro	4 atoms bond into a tetrahedron

Triangulo	3 atoms bond into a triangle
Triprismo	6 atoms bond into a triangular prism
Pentaprismo	10 atoms bond into a pentagonal prism
Hexaprismo	12 atoms bond into a hexagonal prism
antiprismo	8 atoms bond into a rectangular antiprism
Catena	A chain structure (used for linear polymeric substances)
η	Signifies that two or more contiguous atoms of the group are attached to metal
μ	Signifies that a group so designated bridges two centers of coordination
σ	Signifies one atom of group is attached to a metal.

(iii) *Substitutional Affixes:*

An affix such as 'ferra', 'nickela', 'sila', metalla' etc., can be used to indicate the formal substitution of a "heteroatom" in a hydrocarbon.

silabenzene-one silicon is substituted in place of carbon in benzene.

Ferra cyclopropane (metalla cycle)

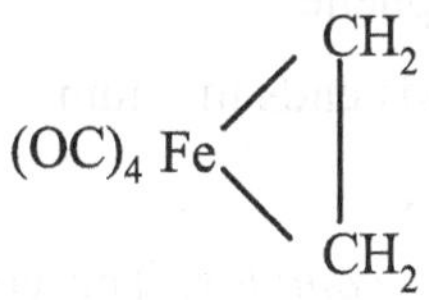

1.2 (g) Nomenclature of Complexes

To name the variety of coordination compounds in a more uniform fashion, Nomenclature committee of the International Union of Pure and Applied Chemistry (IUPAC) suggested the rules (1989).

(i) If a complex contains ions, the full name of the positive ion is given first followed by the full name of the negative ion. The central atom is named last though it is written first in the formula

$[Fe(H_2O)_6]^{2+}$ - Hexaaquairon (II) ion

$K_2[PtCl_6]$ - Potassium hexachloroplatinate (II)

$[Co(NH_3)_6]Cl_3$ - Hexammine cobalt (III) chloride

If the complex is neutral it is given one word name

$[Pt(NH_3)_2Cl_2]$ – Diammine dichloroplatinum (II)

(ii) In a complex ion, the order of presentation is – ligands, metal with oxidation state in Roman numerical.

$K_3 [Cr(CN)_6]$ - potassium hexa cyano chromate (III)

$[Cr(H_2O)_6]Cl_3$ - Hexaaquachlornium (III) chloride

(iii) *In writing the name of the complex:* The ligands are quoted in alphabetical order, regardless of their charge (followed by metal).

All negative (- ve) ligands end in - O

F^- - Fluoro	H^- -Hydrido	HS^- - mercapto
Cl^- - chloro	OH^- - Hydroxo	S^{2-} - thio
Br^- - Bromo	O^{2-} - Oxo	CN^- - cyano
I^- - Iodo	O_2^{2-} - peroxo	No_2^- - Nitro

Neutral ligands have no special endings.

NH_3 - Ammine	H_2O - Aquo	CO - carbonyl
NO- Nitrosyl	O_2 - dioxygen	N_2 - dinitrogen

Organic ligands are usually given their common names

CH_3 - Methyl C_6H_5 - phenyl

$(C_6H_5)_3$ P-triphenylphene

Positive (+ve) ligands ends in – ium

$N_2H_4^+$- hydrazinium, NO^+ - nitronoium

(iv) *In writing the formula of the complex*: The complex ion should be enclosed by square brackets. The metal is written first, then the ligands in the order – negative ligands, neutral ligands, and positive ligands (alphabetically according to first symbol within each group)

$[Co\ Cl_2(NH_3)_4]\ Cl$ – Tetraamminedichlorocobralt (III) chloride

$K_2 [Os\ Cl_5\ N]$ – potassium petachloro nitridoosmate (IV)

(v) *In a complex ion* – for the same ligands the prefixes di, tetra, penta, etc., are used and for the multidentate ligands (groups with complicated structure) the prefixes bis, tris, tetrakis, pentakis etc., and the ligand is enclosed in parentheses.

$K_2 [Cr\ (CN)_2\ O_2\ (O_2)\ NH_3]$ – potassium ammine dicyanodioxoperoxochromate (IV)

$[Co\ Cl_2\ (CH)_2]\ Cl$ - Dichlorobis (ethylene diammine) cobalt (III) chloride

$K_3 [Fe\ (ox)_3]$- Potassium tris (oxalato) ferrate (III)

(vi) *For Ambidentate ligands (ligands are attached by more than one atom):* The attached atom is specified.

$[Co(ONO)(NH_3)_5]Cl_2$ pentaammine nitro-O cobalt (III) chloride

$[Co(NO_2)(NH_3)_5]Cl_2$ pentaammine nitro-N cobalt (III) chloride

The other ambidenatate ligands are: –

M - S_2O_3	thiosulphato	- O, O
M - S_2O_3	thiosulphato	- O, S
M - SCN	thiocyanato	- S
M - SCN	thiocyanato	- N etc.

(vii) Bridged ligands are denoted by μ

μ – oxalato bis (triaquaoxalato titanium (III)

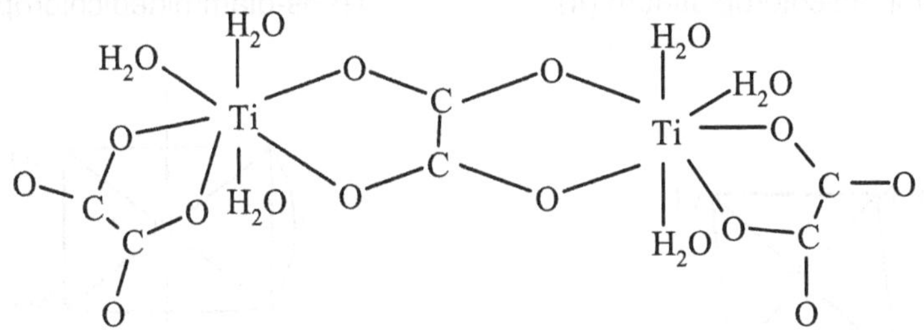

μ - amido bis (pentammine cobalt (III) nitrate

$[(NH_3)_5CoNH_2Co(NH_3)_5](NO_3)_5$

(viii) For metal to metal bonds, prefix bi – is used

sym-dichloro octakis (methylamine) biplatinum (II) chloride

For more than two metals

$Os_3(CO)_{12}$	Dodeca carbonyl-triangulotriosmium
$Cs_3[Re_3Cl_{12}]$	Cesium dodecachloro triangulotrirhenate
$[Nb_6Cl_{12}]^{2+}$	Dodeca- μ –chlorooctahedrohexaniobium (II) ion.

(ix) For geometrical isomers prefixes - cis, trans and number of coordinate positions.

Numbering of square planar octahedral complex structures

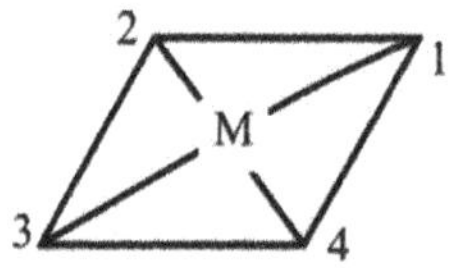

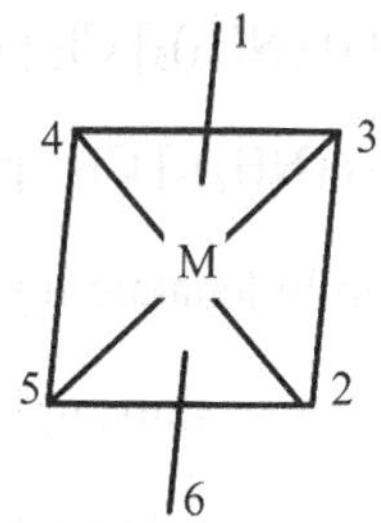

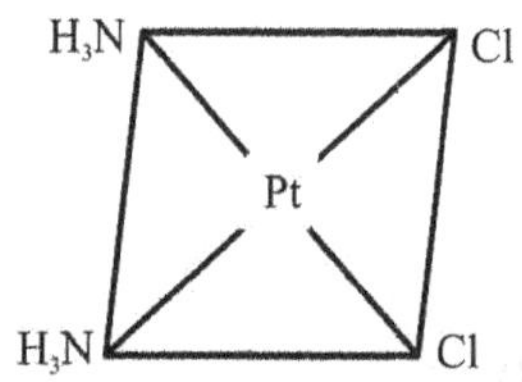

Cis-diamminedichloroplatinum (II)

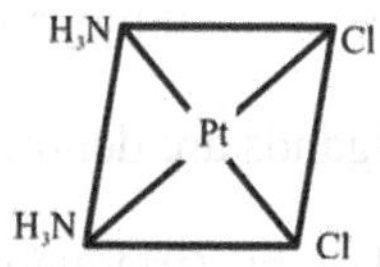

Trans-diamminedichloroplatinum (II)

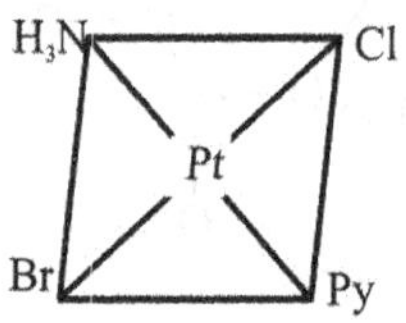

1-chloro-3-bromoamminepyridine platinum (II)

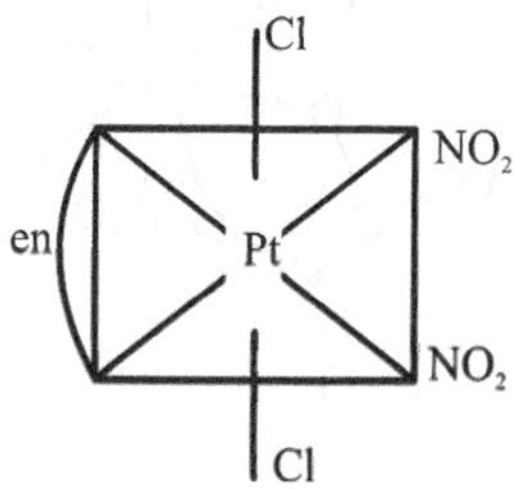

1,6-dichloro 2,3-dinitro (ethylene diammine) platinum (IV)

(ix) For optically active complex – represented by prefix D (dextro) and L (Levo) and if the ligand is optically active the prefix d – and l – are used

D-$[Co(en)_3]^{3+}$ D-$[Co(l-pn)_3]^{3+}$ D $[Co(d-pn)_3]^{3+}$

(x) For double salts

$AlK(SO_4)_2.12H_2O$

Aluminum potassium sulphate 12 water

$Fe(NH_4)_2(SO_4)_2.6H_2O$

Ferrous ammonium sulphate 6 water

1.2 (h) Chemical Reaction

A chemical reaction is a change of bonding between atoms. Bonding changes occur because of rearrangement or transfer of outer electrons about atoms. Chemical reactions are often accompanied by a change in energy.

Symbols used in chemical equation are:

(i)	+	mixed with
(ii)	$\rightleftharpoons$	Reaction favors right
(iii)	$\leftrightarrow$	Resonance
(iv)	$\rightarrow$	yields
(v)	$\nrightarrow$	No reaction
(vi)	$\downarrow$	precipitate
(vi)	$\uparrow$	gas evolves
(viii)	Δ	Heat of energy
(ix)	μ	light energy (in photoreactions)
(x)	aq	aqueous
(xi)	g	gas
(xii)	*l*	liquid
(xiii)	s	solid

1.3 Analytical Chemistry

Analytical chemistry is a measurement of science consisting of powerful ideas and methods that are useful in all fields of science and medicine. Analysis is:

(i) *Qualitative analysis:* It reveals that the identity of the elements and compounds in a sample.

(ii) *Quantitative analysis:* It indicates the amount of each substance in a sample.

Analytes are the components of a sample that are to be determined.

Analytical chemistry is applied throughout industry, medicine and in all the sciences. The relationship between analytical chemistry and other branches of chemistry and sciences is shown in Fig 1.1

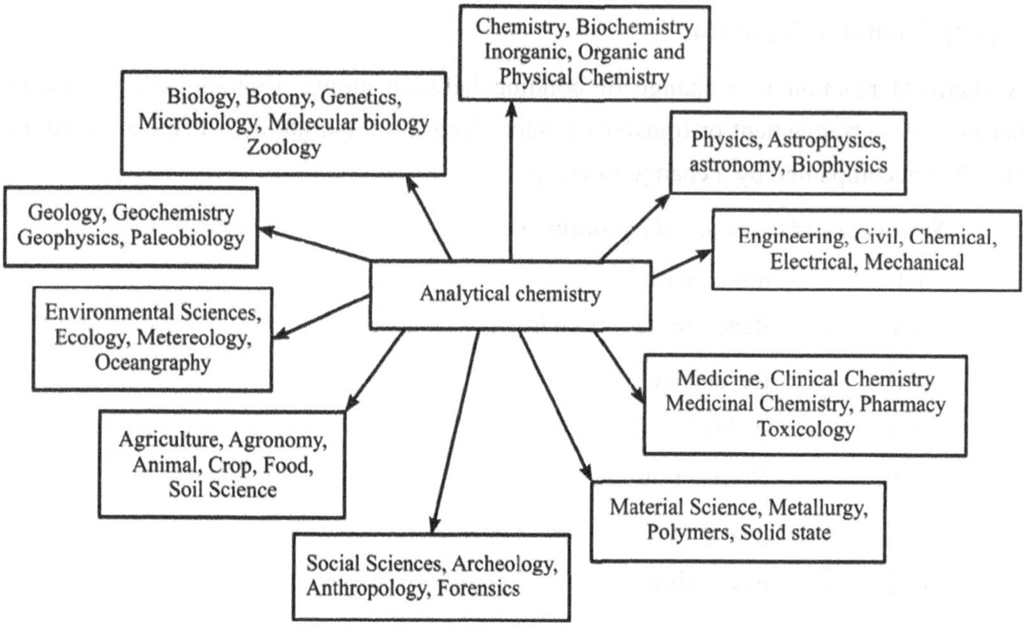

Fig. 1.1

1.4 Pharmaceutical Chemistry

1.4.1 Introduction

Chemistry is the defining science of pharmacy. To understand anything about a drug – the synthesis, the determination of its purity, the formulation into a medicine, the dose given, the absorption and distribution around the body, the molecular interaction of drug with its receptor, the metabolism of the drug and finally, the elimination of drug from the body - requires a thorough and comprehensive understanding of the chemical structure of the drug and how this chemical structure influences the properties and behavior of the drug in the body. For these reasons, chemistry is the most important of all the scientific disciplines contributing to the understanding of drugs and their actions in the body. A good understanding of the chemistry of drugs will allow the study of advanced topics such as drug design and medicinal chemistry, molecular pharmacology and novel drug delivery systems that are usually encountered in the later stages of a pharmacy or pharmaceutical science degree.

Pharmaceutical chemistry is a science that makes use of the general laws of chemistry to study drugs, i.e., their preparation, chemical nature, composition, structure, influence on an organism studies, the physical and chemical properties of drugs, quality control methods and their usage conditions etc.

Inorganic chemicals have been used in pharmacy and medicine for many reasons, ranging from therapeutic agents to nutritional supplements to pharmaceutical necessities. Inorganic pharmaceutical chemistry is a science that makes use of laws of chemistry to study inorganic substances as drugs. Inorganic drugs are majorly obtained from the natural source such as minerals. Examples include waters, inorganic chemicals, calcined inorganics, heavy metals etc. Although many inorganic chemicals are rarely used today, yet few inorganic chemicals are being still used in modern medicine.

1.4.2 Pharmacopeia

The word 'pharmacopoeia' is derived from the Greek words 'pharmakon' means drug or medicine and 'poeia' means to make. Pharmacopoeia, in its modern technical sense, is a book containing directions for the identification of samples and the preparation of compound medicines, and published by the authority of a government or a medical or pharmaceutical society. In a broader sense it is a reference work for pharmaceutical drug specifications. Most of the advanced countries have their own pharmacopoeia.

1.4.3 History of Pharmacopeia

In olden times, physicians used to prepare their own prescriptions. As the complexity of medical formulae increased, physicians delegated this laborious work to their assistants, called apothecaries. These apothecaries gradually become independent. They looked for guidance and knowledge to the books written by individuals who had achieved fame in medicine. The first known dated work of this kind published under civic authority appears to have been that of Nuremberg in 1542; a student named Valerius Cordus showed a collection of medical receipts, which he had selected from the writings of the most eminent medical authorities, to the physicians of the town, who urged him to print it for the benefit of the apothecaries, and obtained for his work the sanction of the senatus. The term *Pharmacopoeia* first appears as a distinct title in a work published at Basel in 1561 by Dr A. Foes, but does not appear to have come into general use until the beginning of the 17th century. The credit for producing the first national pharmacopoeia goes to England. Until 1617 such drugs and medicines as were in common use were sold in England by the apothecaries and grocers. In that year the apothecaries obtained a separate charter and it was enacted that no grocer should keep an apothecary's shop. The preparation of physicians' prescriptions was thus confined to the apothecaries, upon whom pressure was brought to bear to make them dispense accurately.

By the issue of a First Pharmacopoeia in May 1618 by the College of Physicians, and the wardens of the apothecaries received the powers to destroy all the compounds which they found unfaithfully prepared. Thus, the first authorized London Pharmacopoeia, was selected chiefly from the works of Mezue and Nicolaus de Salerno, but it was found to be so full of errors that the whole edition was cancelled, and a fresh edition was published in the following December. At this period the compounds employed in medicine were often heterogeneous mixtures, some of which contained from 20 to 70, or more, ingredients, while a large number of samples were used in consequence of the same substance being supposed to possess different qualities according to the source from which it was derived. During the 17th and 18th centuries more than 50 regional pharmacopoeias appeared.

1.4.4 Indian Pharmacopoeia

The pharmacopoeia originated as the Bengal Pharmacopoeia and General Conspectus of Medicinal Plants, 1844 and was known as the Bengal Pharmacopoeia. Its main focus was on indigenous drugs although it included some products imported from Europe. The first pharmacopoeia was published in 1868 under the authority of the Secretary of State for India in Council by a committee constituted in 1865. It contained standards for drugs official in the British Pharmacopoeia (BP) 1867 and a few selected indigenous drugs. A Supplement to the British pharmacopoeia came out in 1869 and remained in use till around 1885. With the publication of the British Pharmacopoeia in 1885 the then Government made this the sole authority on all matters relating to pharmacy. In 1900 the Indian and Colonial Addendum to the BP 1898 was published and an Indian edition of it with minor modifications came out in 1901. Until 1955 the British Pharmacopoeia remained our official guide on drugs standards. There started a move for a separate Indian Pharmacopoeia right after the publication of British Pharmacopoeia in 1914.

The BP Commission made provision for publication of Supplements to take care of local needs. In 1946 the Indian Pharmacopoeial List was prepared to serve as an Indian Supplement to BP 1932. After Independence an Indian Pharmacopoeia Committee, a permanent body was constituted in 1948 and it prepared the Pharmacopoeia of India (The Indian Pharmacopoeia) 1955. Subsequent editions were brought out.

The First Edition of the IP was published in 1955. It had laid down standards mainly for imported drug products and locally manufactured herbal products. In 1954 the original IP Committee had been reconstituted to bring out the second edition but managed to publish only a Supplement to the first edition in 1960.

In the meantime the country had made substantial progress in the manufacture of drug products and a few drug substances from imported intermediates.

The IP Committee was reconstituted again in 1962 with only one representative from the industry. It brought out the Second edition in 1967 with 885 monographs.

The revision of this edition was overdue and it was decided to bring out a Supplement with the help of a reconstituted Committee in 1972 which had representatives from different disciplines in the industry and the government. The Supplement came out in 1975. Simultaneously, work on the third edition had been initiated by the committee which had a wider representation. It had eminent physicians, pharmacologists, representatives of professional bodies, the regulatory agencies, the industry and independent professionals. It brought out the third edition in 1985. It had several changes in the contents and in the manner of presentation. It also published addenda to this edition in 1989 and 1991. A reconstituted committee published the fourth edition in 1996.

The latest edition of the Indian Pharmacopoeia entitled Indian Pharmacopoeia 2007 has been prepared by the Indian Pharmacopoeia Commission (IPC) in accordance with a plan and completed through the untiring efforts of its members and its Secretariat over a period of about two years. This is the fifth edition of the Indian Pharmacopoeia after independence. It supersedes the 1996 edition but any monograph of the earlier edition that does not figure in this edition continues to be official as stipulated in the second schedule of the drugs and Cosmetics Act, 1940. The Indian Pharmacopoeia 2007 is presented in three volumes. Volume I contains the general notices, preface, the structure of the IPC, acknowledgements, introduction, and the general chapters. Volume 2 deals with the general monographs on drug substances, dosage forms and pharmaceutical aids (A to M). Volume 3 Contains Monographs on drug substances, dosage forms and pharmaceutical aids (N to Z) followed by monographs on vaccines and immunizers for human use, herbs and herbal products, blood and blood related products, biotechnology products and veterinary products. The scope of the pharmacopoeia has been extended to include products of biotechnology, indigenous herbs and herbal products, viral vaccines and additional antiretroviral drugs and formulations, inclusive of commonly used fixed-dose combinations. To assure the health care professional's patients and consumers of the standards of drug in India the Secretary, Ministry of Health & Family Welfare, Shri Naresh Dayal who is also the Chairman of Indian Pharmacopoeia Commission released the Addendum 2008 to the Indian Pharmacopoeia (IP) 2007. The Addendum contains amendments to IP 2007 and adds 73 new monographs on different therapeutic groups representing synthetic, herbals and biological drugs. While the amendments to the IP 2007 will come into force with immediate effect, the new monographs will become effective from 1st July 2009.

A monograph in I.P. includes the following features:

Main title: Name of the substance.

1. *Synonyms:* The common names, if any.
2. *Chemical Formula:* Gives chemical structure of molecule
3. *Molecular Weight:* Calculated from molecular formula using standard atomic weights.
4. *Category:* It indicates drug's pharmacological activity.
5. *Doses:* Gives the average dose for adults.
6. *Description:* Statements on colour, odour and taste.
7. *Solubility:* Terms used to indicate different degrees of solubility are

Very soluble	Less than 1
Freely soluble	1 to 30
Soluble	10 to 30
Sparingly soluble	30 to 100
Slightly soluble	100 to 1,000
Very slightly soluble	1,000 to 10,000
Insoluble or practically insoluble	Greater than 10,000

8. *Standards:* I.P. prescribes standards of purity and strength in the monograph of almost all the official substances.
9. *Identification:* This includes some specific and some non-specific tests for identify of the substance.
10. *Test for purity:* I.P. prescribes for purity of almost all the official substances. These tests include melting point, boiling point, weight per ml, limit tests for chloride, sulphates, iron, heavy metals, lead and arsenic, specific optical rotation, sulphated ash, loss on drying, pH pf the solution, etc., as may applicable for the substance.
11. *Assay:* An assay gives in detail the analytical method for the substance in order to determine the % content of a particular chemical in the given test sample.
12. *Storage:* The substances and preparations of I.P. are to be stored under conditions that prevent contamination and deterioration.

    Terms used for indicating different temperature conditions are-

Term	Temperature range
Cold	2^0C to 8^0C (indicates storage in refrigerator)
Cool	8^0C to 25^0C
Room temperature	Temperature prevailing in working areas
Warm	30^0C to 40^0C
Excessive heat	Greater than 40^0C

1.4.5 United States Pharmacopeia

The United States Pharmacopeia (USP) is the official pharmacopeia of the United States, published dually with the National Formulary as the USP-NF. In 1817, Dr.Lyman Spadling of New York proposed a plan to the Medical society of the country at Newyork for publishing a National Pharmacopeia. The first edition of United States Pharmacopeia was compiled, edited and published on 15th December 1820 which was having 217 drugs in about 272 pages. On July 5, 1975, unification of USP and NF was announced. Convention, the N.F. has been published with U.S.P. in a single volume since 1980. Though U.S.P and N.F. were combined with in a single volume, they continued as two distinct official compendia. To set up the standards for all the drugs, the scope of both U.S.P and N.F. was changed drastically. The U.S.P was limited to drug substances and dosage forms and N.F. was limited to pharmaceutical ingredients. When a compound is both a drug and a pharmaceutical ingredient, it was included in the U.S.P with a cross reference from the N.F. to that U.S.P. monograph. In the single U.S.P-N.F. volume, a large amount of duplication of text was eliminated.

The edition of 1995 is U.S.P.23-N.F.18. This indicates that this is the 23rd edition of U.S.P and 18th edition of N.F. This new volume contains a total of 3410 monographs and 147 general chapters. There are 336 new monographs-320 in U.S.P and 16 in N.F., and 15 general chapters. The publication of U.S.P-N.F. is now undertaken by the United States

Pharmacopoeial Convention, Inc. Today the U.S.P-N.F. is recognized and used as a source of official drug standards by several countries throughout the world. The U.S.P.24-N.F.19 has been published in 1999 and gives drug standards that are official from January 2000.

The United States Pharmacopeial Convention (usually also called the USP) is the nonprofit organization that owns the trademark and copyright to the USP-NF and publishes it every year. Prescription and over-the-counter medicines and other health care products sold in the United States are required to follow the standards in the USP-NF. USP also sets standards for food ingredients and dietary supplements.

USP establishes written (documentary) and physical (reference) standards for medicines, food ingredients, dietary supplement products and ingredients. These standards are used by regulatory agencies and manufacturers to help to ensure that these products are of the appropriate identity, as well as strength, quality, purity, and consistency.

Prescription and over- the-counter medicines available in the United States must, by federal law, meet USP-NF public standards, where such standards exist. Many other countries use the USP-NF instead of issuing their own pharmacopeia, or to supplement their government pharmacopeia.

USP standards for food ingredients can be found in its Food Chemicals Codex (FCC). The FCC is a compendium of standards used internationally for the quality and purity of food ingredients like preservatives, flavorings, colorings and nutrients. While the FCC is recognized in law in countries like Australia, Canada and New Zealand, it currently does not have broad legal recognition in the United States. USP obtained the FCC from the Institute of Medicine in 2006. The IOM had published the first five editions of the FCC.

USP also conducts verification programs for dietary supplement products and ingredients. These are testing and audit programs. Products that meet the requirements of the program can display the USP Verified Dietary Supplement Mark on their labels. This is different from seeing the letters "USP" alone on a dietary supplement label, which means that the manufacturer is claiming to adhere to USP standards. USP does not test such products as it does with USP Verified products.

1.4.6 The British Pharmacopoeia

The British Pharmacopoeia (BP) is the official collection of standards for UK medicinal products and pharmaceutical substances. Produced by the British Pharmacopoeia Commission Secretariat of the Medicines and Healthcare products Regulatory Agency, the BP makes a valuable contribution to public health by setting publicly available standards for the quality of medicines.

The current edition of the British Pharmacopoeia comprises six volumes which contain nearly 3,000 monographs for drug substances, excipients and formulated preparation, together with supporting General Notices, Appendices (test methods, reagents etc) and Reference Spectra used in the practice of medicine, all comprehensively indexed and cross-referenced for easy reference. Items used exclusively in veterinary medicine in the UK are included in the BP (Veterinary).

Volumes I and II

- Medicinal Substances

Volume III

- Formulated Preparations
- Blood related Preparations
- Immunological Products
- Radiopharmaceutical Preparations
- Surgical Materials
- Homeopathic Preparations

Volume IV

- Appendices
- Infrared Reference Spectra
- Index

Volume V

- British Pharmacopoeia (Veterinary)

Volume VI

(CD-ROM version)

- British Pharmacopoeia
- British Pharmacopoeia (Veterinary)
- British Approved Names

The BP is available as a printed volume and electronically in both on-line and CD-ROM versions, the electronic products use sophisticated search techniques to locate information quickly. For example, pharmacists referring to a monograph can immediately link to other related substances and appendices referenced in the content by using 130,000+ hypertext links within the text.

CHAPTER 2

Atomic and Molecular Structures / Complexes

2.1 Introduction

The gradual development of modern atomic theory is one of the most fascinating stories in the history of scientific discovery. The idea of atom was originally developed by Leukippos with his atomic hypothesis around 500 B.C. and it was extended by Demokritos.

The hypothesis is "*The matter is made up of small, unalterable particles called atoms.*"

After 25 centuries later, John Dalton in 1803 postulated that:

(a) Matter is made up of extremely small particles called atoms,

(b) By chemical reaction atoms are indivisible and can neither be created nor destroyed,

(c) In an element all the atoms have same weight, size, shape, structure (different element atoms have different properties),

(d) Chemical combination between two or more atoms in stoichiometric ratios gives compounds.

The atoms are chemically indivisible particles of an element, can not exist freely and applies to elements only.

Berzelious stated that equal volume of all gases under the same conditions of temperature and pressure, contain equal number of atoms.

For *eg*: $H_2 + Cl_2 \rightarrow 2\ HCl$ .....(2.1)

Atomic number - It is the number protons

Atomic weight - The sum of the number of protons and neutrons

Isotope - Atomic number is same but atomic mass is different

2.2 Subatomic Particles

Atom possesses a complex structure consisting of fundamental particles (subatomic particles) like.

(a) ***Electron*:** This was the first subatomic particle recognized as a constituent of all atoms. J.J. Thomson is 1880 discovered in his experiment. From his experiment, he explains that the electrons are negatively (0.0005486 on physical scale) charged and the electron mass is 9.106×10^{-28} g or 1836 times less than the proton mass. The symbol is e^- negatively charged and the electron mass is 9.106×10^{-28} g (0.0005486 on physical scale) or 1836 times less than β-ray and charge is –1.

(b) ***Proton*:** The name proton (Greek meaning first) suggested by Ernest Ruther ford in 1920. It is having positive charge and its mass is 1.6725×10^{-24} g (1.008142 on physical scale). Symbol is H^1 or P^1 and charge is +1.

(c) ***Neutron***: Rutherford postulated that a proton might combine with an electron inside the nucleus to form a particle which would carry no charge but have some mass as that of proton and named it as neutrono.

For e.g.: Neutron was emitted in the reactions

$$_4Be^9 + {_2HC^4} \rightarrow {_6C^{12}} + {_0n^1} \qquad .....(2.2)$$

The symbol of neutron is $_0n^1$ and charge is zero. The mass is 1.008982 on physical scale.

(d) ***Positron*:** Positive electron was predicted by P.A.M Disac and from the study of cosmic rays C.O. Anderson discovered positron. The symbol is e^+ and charge is +1 and mass is 0.0005486 (on physical scale).

(e) ***Neutrono*:** It is a theoretical particle postulated by W.Pauli in 1931 to preserve the conservation of energy and angular momentum is beta decay.

$$\textit{Neutron} \rightarrow \text{Proton} + \text{electron} + \text{neutrono} \qquad .....(2.3)$$

The symbol is ν, charge is zero and mass is nearly zero (< 0.00002 on physical scale).

(f) ***Measons***: These are the particles required for the exchange of protons and neutrons into one another.

$$_1p^1 \rightarrow {}_0n^1 + \pi^+ \quad .....(2.4)$$

$$_1p^1 \rightarrow {}_1p^1 + \pi^0 \quad .....(2.5)$$

$$_1n^1 \rightarrow {}_1p^1 + \pi^- \quad .....(2.6)$$

The symbol is π, charge is +1, and –1 and mass is 0.151 on physical scale.

2.3 Models of Atoms

(a) ***Thomson's Model* (1904)**: The atom is a sphere of positive electricity in which electrons are embedded like raisins in bread pudding. The repulsion between electrons was balanced by attraction towards the centre of sphere.

Thompson explained the observed spectra of elements by assuming that electrons in an atom are not at rest but are vibrating about equilibrium position. Thus a vibrating electron would emit radiation. But this model could not explain the spectral lines obtained even in the simplest cases and only one spectral line is possible for H_2 in U.V. region.

(b) ***Rutherford Model* (1912)**: The atom consists of central nucleus of small dimension (diameter 10^{-8} cm) surrounded by electrons which are not at rest but are revolving round the nucleus in a closed path like the planets revolving round the Sun.

Atomic nucleus consists of protons (+ve charge) and most of the mass (neutrons). The model is shown in Fig. 2.1.

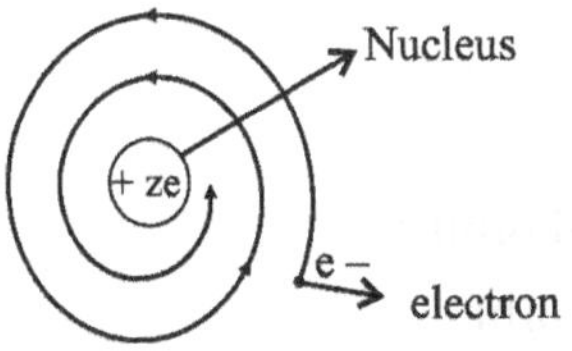

Fig. 2.1 Rutherford model of atom.

Drawback of this model is - when electron revolves around the nucleus, it will radiate out energy resulting in the loss of energy. This loss of energy makes the electrons to move slowly and consequently it will be moving in a spiral path and fall inside the nucleus and ultimately the atom collapse.

(c) ***Bohr's Model* (1913)**: Bohr applied the Planck's Theory i.e., the energy must be emitted or absorbed only in discrete amounts.

The Postulates of the Model are

(i) In the atom an electron can move, in certain specific (Fig. 2.2) orbits, without radiating any energy, around the nucleus

This orbital motion/rotation without emitting energy follows the Newtonian's law i.e., the force of attraction is equal to centrifugal force of attraction.

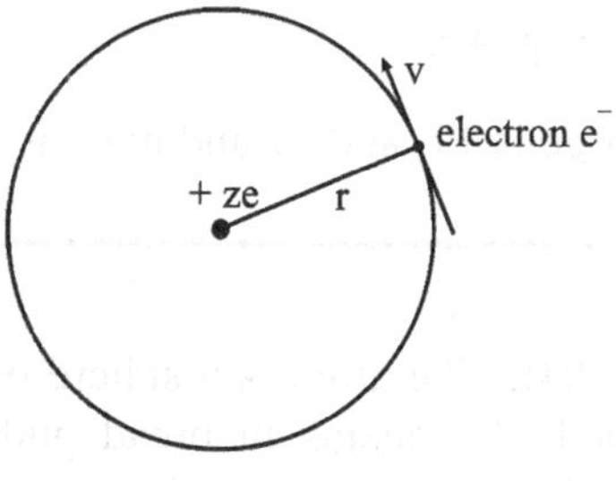

Fig. 2.2 Bohr atomic model.

(ii) The total angular momentum of the revolving electron is quantized which is an integral multiple of $h/2\pi$, where h is Planck's constant. Angular momentum is given by

$$mvr = n\frac{h}{2\pi} \qquad .....(2.7)$$

Where n = principal quantum number which is an integer 1, 2, 3 .....etc.,

This angular momentum limits the number of permissible orbits.

(iii) The energy is only radiated out when an electrons jumps from one stationary orbit to another. The frequency, ν of this radiation is given by the difference in energy between the initial and final orbits.

$$E_2 - E_1 = h\nu \qquad .....(2.8)$$

Where

E_1 = energy of initial orbit

E_2 = energy of final orbit

h = Planck's constant

ν = frequency

The energy of the revolving electron is negative because the electron is bound to the nucleus by attractive forces to that energy must be supplied to electron in order to separate it completely from the nucleus. The inner most orbital electron (n = 1) has the most negative value and decreases for n = 2, n = 3 … so on.

$$E = \frac{-2\pi^2 me^4 z^2}{n^2 n^2} \qquad .....(2.9)$$

Where m = mass of electron

e = charge of the electron

z = atomic number of atom

n = quantum number

if n = 1, and z = 1 for H atom

$E_1 = Z.18 \times 10^{-11}$ ergs

So $$h\nu = E_{n_1} - E_{n_2} = \frac{2\pi^2 me^4}{n^2}\left(\frac{1}{n_1^2} - \frac{1}{n_2^2}\right)z^2 \quad .....(2.10)$$

$$\nu = \frac{2\pi^2 me^4}{h^3}\left(\frac{1}{n_1^2} - \frac{1}{n_2^2}\right)z^2 \quad .....(2.11)$$

Since $c = \nu\,\lambda$ or $\frac{1}{\lambda} = \frac{\nu}{c}$

$$\frac{1}{\lambda} = \frac{2\pi^2 me^4}{c\,h^3}\left(\frac{1}{n_1^2} - \frac{1}{n_2^2}\right)z^2 \quad .....(2.12)$$

Since $\frac{2\pi^2 me^4}{c\,h^3}$ is constant which is known as Reidberg constant R,

then eq. 2.12 can be written as

$$\frac{1}{\lambda} = R\left(\frac{1}{n_1^2} - \frac{1}{n_2^2}\right)z^2 \quad .....(2.13)$$

where R = 109737.707 cm^{-1}

$\frac{1}{\lambda}$ is equal to wave number i.e., $\bar{\nu}$

Draw backs

- It cannot explain the spectra of other atoms except H atom.
- Each spectral line of hydrogen atom under high resolution consists of many closed lines.
- It failed to explain Zeeman effect (magnetic field) and Stark effect (electric field) i.e., the splitting of spectral lines.
- It failed to explain the distribution and arrangement of electrons in atoms.
- It used two theories namely the Planck's quantum theory and the laws of chemical mechanics which are opposed to each other.

Hydrogen atom emission spectrum consists five series namely – Lyman series (n_2 = 1), Balmer series (n_2 = 2), Paschen series (n_2 = 3), Brackette series (n_2 = 4) and Pfund series (n_2 = 5).

Later Sommerfied (1915) introduced the concept that the electrons are revolving around the nucleus in elliptical orbits to explain the multiplicity of spectral lines. These models gave four quantum numbers.

(d) ***Vector Atom Model* (1925):** This model of atom is extension of Bohr-Sommerfield model. Two concepts of this model are:

(i) ***Spinning Electron Hypothesis*:** Uhlenbeck and Goudsmith (1925) stated that the electron not only rotates about the nucleus but also about its own axis somewhat like the motion of planet is solar system i.e., earth moves around the sun and also about its own axis.

This concept give two angular momenta i.e., orbital angular momentum and spin angular momentum. Since the changed particle (electron) on rotation also gives orbital magnetic momentum and spin magnetic momentum.

The interaction between two angular momenta account for multiplicity of spectral lines and the interaction between two magnetic moments account for the fine structure.

(ii) ***Space Quantiration*:** In general the motion of electron in an atom is three dimensional so there are three degrees of freedom. The additional quantum number or third quantum number condition quantizes the orientation of elliptical orbit in three dimensional space and does not alter the size and shape of the orbits.

This atom model gives seven quantum numbers altogether

(e) ***Quantum Mechanical Atom Model*:**

Failure of Quantum theory is on the basis of

- Wave-particle dualistic nature of radiation.
- One quantum number does not cater the needs
- Introduction of new quantum number wherever it is necessary to explain the observation of spectrum.

The new Mathematical concept was introduced by taking into consideration the DeBroghe's Hypothesis $\gamma = h/p$ and Heisenberg's uncertainty principle in Quantum Mechanics to cover the draw backs arised in quantum theory.

The Quantum Mechanical Model of Atom can be explained by the following points.

(i) Electron moment represented by wave function is called as an orbital or the region of finding the electron is termed as an orbital.

(ii) Electron spends maximum time in its own orbit

(iii) Electron moment is in a haphazard (irregular) manner as shown in Fig. 2.3 i.e., like the insect moves around the electric bulb.

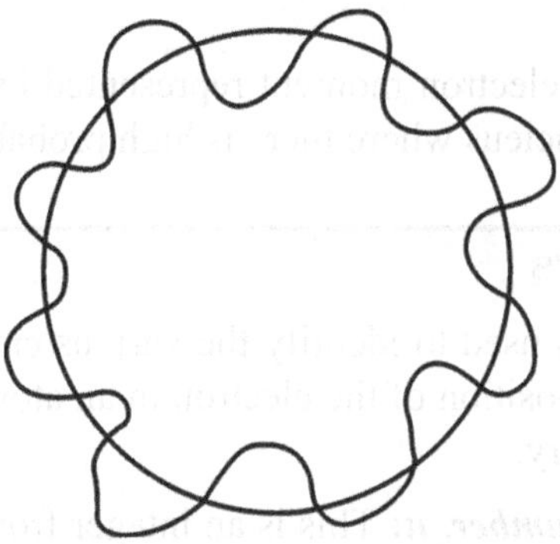

Fig. 2.3 Electron moment in a haphazard manner.

(iv) Electron is having penetration effect i.e., it spends sometime in the other orbits i.e., right from the nucleus, whereas in quantum theory it revolves only one circular orbit.

(v) In between orbits there are nodal surfaces i.e., zero electron density.

(vi) If electron moment is photographed, it looks like a cloud (Fig. 2.4).

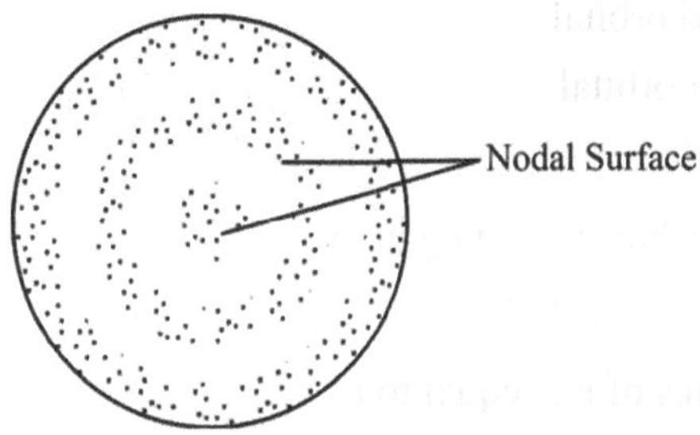

Fig. 2.4 Photograph of electron moment.

(vii) The Probability of finding the electron in atom is always real. Exact position of electron is not known so we can mention probability of the electron is an atom.

This Model is completely different from Bohr Model, where the electron is restricted to revolve around the nucleus in a fixed circular orbit. The Quantum Mechanical model is in accordance with Heisenberg's uncertainty principle (it is impossible to determine position and momentum of any particle/electron precisely and simultaneously).

Orbit and Orbital

The orbit is the circle around the nucleus with particular principal quantum number in which the electron revolves around the nucleus. The energies are different for different n values.

The orbital describes the electron moment represented by the wave function ψ or the region in space around the nucleus where there is high probability of finding the electron.

2.4 Quantum Numbers

The term quantum number is used to identify the various energy levels that are available to an electron. To know the position of the electron in an atom, four quantum numbers are required as per quantum theory.

(i) ***Principal Quantum Number, n***: This is an integer from 1, 2, 3 ...., n is never zero. It describes the size of the orbital depending on n value, the different shells K, L, M, N for n = 1, 2, 3, 4 respectively. The no of electrons present in each n is given by the formula $2n^2$. For e.g. n = 3 then $2 \times (3)^2 = 18$.

(ii) ***Azimuthal Quantum Number, l:*** This is an integer right from 0, 1, 2, ...n – 1. It describes the shape of the orbital i.e.,

$l = 0$ S orbital

$l = 1$ P orbital

$l = 2$ d orbital

$l = 3$ f orbital

The relation with n is

$l = n - 1$ values including zero

if n = 3 then $l = 0, 1, 2$

The number of values of n is equal to l.

(iii) ***Magnetic Quantum Number, m***: It describes the orientation of orbitals in presence of magnetic field. s, p, d, f orbitals contains suborbitals 1, 3, 5, 7. The relation between m and l is, $m = 2l + 1$ including zero. For e.g. if $l = 2$, $m = 2l + 1$ i.e., 5, then the m values are 2, 1, 0, –1, –2 = 5

(iv) ***Spin Quantum Number, s:*** It describes the spinning of the electrons on its own axis towards clockwise or anticlockwise. The values are half integers and have only two values i.e., $+\frac{1}{2}$ and $-\frac{1}{2}$.

These four numbers describes the position of electron in an atom.

Quantum Numbers Associated with Vector Atom Model

The Quantum Numbers are:

(i) Principal quantum number, n

(ii) Orbital angular momentum quantum number, l

(iii) Spin angular momentum quantum number, s

(iv) Total angular momentum quantum number, j – vectorial sum of orbital angular momentum and spin angular momentum.

(v) Orbital magnetic momentum quantum number, m_l.

(vi) Spin magnetic momentum quantum number, m_s.

(vii) Total magnetic momentum quantum number m_j – vectorial sum of m_l and m_s.

Quantum Numbers Associated with Quantum Mechanical Model

(i) ***Principal Quantum Number*****:** It is represented by n. It can have any integral value except zero. This arises from the solution for the radial part of wave function ψ and to the first approximation determines the energy of the orbital. It is quantitatively related to the distance from the nucleus to the most probable position of finding the electron. It describes the size of the orbital. This quantum number corresponds to K, L, M etc shells in the Bohr Model. n = 1, 2, 3, ……

(ii) ***Orbital Quantum Number*****:** Represented by l, all possible l values corresponding to n are given by

$$l = n - 1, \ldots .\ 0$$

Where ….. line = all integers necessary to complete the series

This quantum number arises from the solution of angular part ψ and refer, respectively to the magnitude and orientation of the angular momentum of an electron in a particular orbital.

$$\text{Angular momentum} = \sqrt{l(l+1)}\ \frac{h}{2\pi} \qquad \ldots..(2.14)$$

The quantum numbers l describes the shape of the orbital

s for $l = 0$

p for $l = 1$

d for $l = 2$

f for $l = 3$

(iii) ***Magnetic Quantum Number*****:** Represented by m_l for orbital. The possible m_l values corresponding to a given l are

$$m_l = +l,\ (l-1),\ \ldots .0 \ldots\ (-l+1),\ -l. \qquad \ldots..(2.15)$$

$= 2l + 1$ including zero.

The angular momentum is a vector quantity. When an external magnetic field is applied to system, the quantum number m_l can be employed to indicate the direction and magnitude for the component of projection of angular momentum vector relative to applied field. This quantum number also arises from the solution of the angular part of ψ. The principal component of this vector can only occupy certain quantized positions relative to the field.

For e.g., $l = 2$, the possible m_l values are equal to $2l + 1(2 \times 2+1) = 5$ and the values are +2, +1, 0, –1, –2.

Where +2 has lowest energy

The orientation of angular momentum vector for five d orbitals relative to the external field H, are shown in Fig. 2.5.

Where +2 has lowest energy

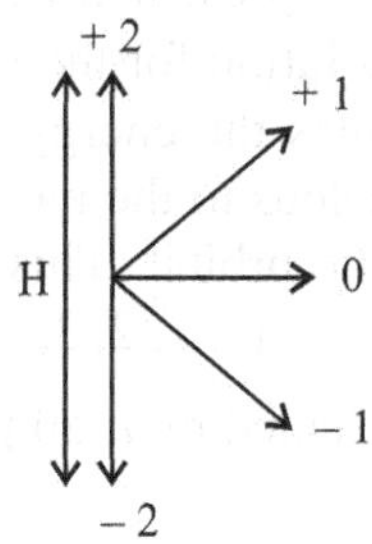

Fig. 2.5 Orientation of d Orbitals.

$$\text{The component of angular momentum} = m_l \frac{h}{2\pi} \qquad \text{.....(2.16)}$$

These three quantum numbers are solutions of Schrodinger equation.

(iv) ***Spin Quantum Number***: Spin magnetic moment quantum number is represented by m_s. Each orbital can accommodate two electrons and these are distinguished by m_s of either + 1/2 or –1/2. This quantum number is required to label uniquely all the electrons in an atom. This quantum number is associated with electron spin angular momentum. The spin angular momentum vector can be aligned parallel or anti parallel to the external field of direction and the corresponding values of the spin are

$$\text{Spin momentum} = +\frac{1}{2}\left(\frac{h}{2\pi}\right) \text{ and } -\frac{1}{2}\left(\frac{h}{2\pi}\right) \qquad \text{.....(2.17)}$$

Where $+\frac{1}{2}\left[\frac{h}{2\pi}\right]$ is slightly lower in energy

The quantum numbers and the orbitals are given in Table 2.1.

Table 2.1 The Quantum numbers and orbitals.

Quantum numbers			Orbital Name
n	*l*	m	
1	0	0	1s
2	0	0	2s
2	1	+1	$2p_x$
2	1	0	$2\ p_y$
2	1	–1	$2\ p_z$
3	0	0	3s
3	1	+1	$3p_x$
3	1	0	$3\ p_y$
3	1	–1	$3\ p_z$
3	2	+2	3 dxy
3	2	+1	3 dxz
3	2	0	3 dyz
3	2	–1	$3\ dx^2 – y^2$
3	2	–2	$3\ dz^2$
4	0	0	4s
4	1	+1	$4\ p_x$
4	1	0	$4p_y$
4	1	–1	$4p_z$
4	2	+2	4 dxy
4	2	+1	4 dxz
4	2	0	4 dyz
4	2	–1	$4\ dx^2 – y^2$
4	2	–2	$4\ dz^2$
4	3	+3	$4fx^3$
4	3	+2	$4fy^3$
4	3	+1	$4fz^3$
4	3	0	4 fxyz
4	3	–1	$4\ fx(y^2 –z^2)$
4	3	–2	$4fy\ (z^2 – x^2)$
4	3	–3	$4fz\ (x^2 – y^2)$

Quantum Numbers for Atom

(i) ***Quantum Number L:*** L govern the energy of particular atomic state and the energy difference of two states will be greater, greater will be the difference between L values.

$$L = |l_1 + l_2|, |l_1 + l_2 - 1|, |l_1 + l_2 - 2| .... |l_1 - 1| \quad .....(2.18)$$

For e.g.: $l_1 = 2$ and $l_2 = 1$

$$L = |2 + 1|, |2 + 1 - 1|, |2 + 1 - 2| ...... |2 - 1| = 3, 2, 1.$$

For many emission electrons $\Delta L = 0, \pm 1$ governs the allowed transitions.

(ii) ***Quantum Number S:*** Spin of the atom. The values of S are given by

$$S = \frac{N}{2}, \left(\frac{N}{2}-1\right), \left(\frac{N}{2}-2\right) \cdots \frac{1}{2} \text{ or } 0 \quad .....(2.19)$$

Where N = number of electrons

(iii) ***Quantum Number J:*** Total angular momentum of the atom. The values of J are given by

$$J = |L + S|, |L + S - 1|, |L + S - Z| ..... |L - S| \quad .....(2.20)$$

J is always positive if L > S then J = (2S + 1) values and

If L < S then J = (2L + 1) values.

J controls the energy values of L and S. The energy difference between the states of different J depends on the degree of interaction between L and S. Since the energies are different for different J values, it corresponds to different spectral terms, causing thereby multiplicity in the spectral terms to occur. Small J at the lowest of the spectrum.

2.5 Shapes of Orbitals

s orbital: Sharp is used for s orbital to describe certain lines in atomic spectra. For this if n = 1, l = 0, m = 0 then it is 1s orbital. The shape is spherical (Fig. 2.6).

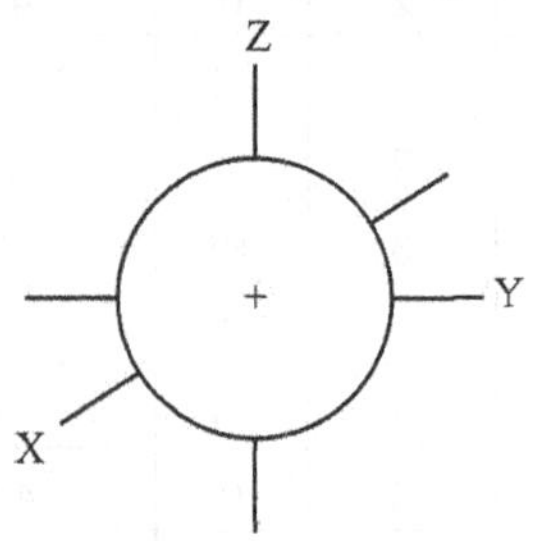

Fig. 2.6 The three dimensional surface representing the probability region of an s-orbital.

No node in this orbital and has the lowest energy.

***p orbital*:** Principal is used for p-orbital. For this if n = 2, l = 1, m = +1, 0, –1 then it is 2p orbital. The p-orbitals are three p_x, p_y, p_z corresponding m = $2l + 1$. The shape is dumbbell shape with two lobes and one node. Opposite lobes sign is different. The shape of p-orbital is shown in Fig. 2.7.

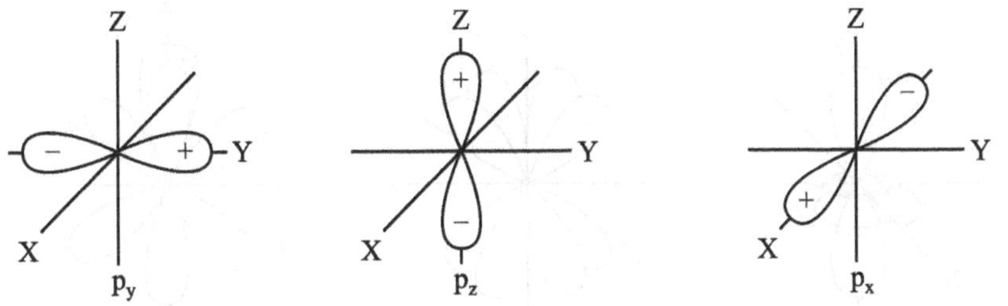

Fig. 2.7 Three dimensional representation of three p-orbitals.

The lobes of p_x align along x-axis, p_y align along y-axis p_z align along z axis. The energy of p > s since it is having one node.

***d-orbital*:** Diffuse is used for d-orbital. For this if n = 3, l = 2, m = +2, +1, 0, –1, –2 ($2l + 1 = 5$) then it is 3d orbital. So the d-orbitals are five, dxy, dyz, dxz, $dx^2–y^2$ dz^2. The shape is clove shape with four lobes and two nodes and the opposite lobes are having the same sign (Fig. 2.8).

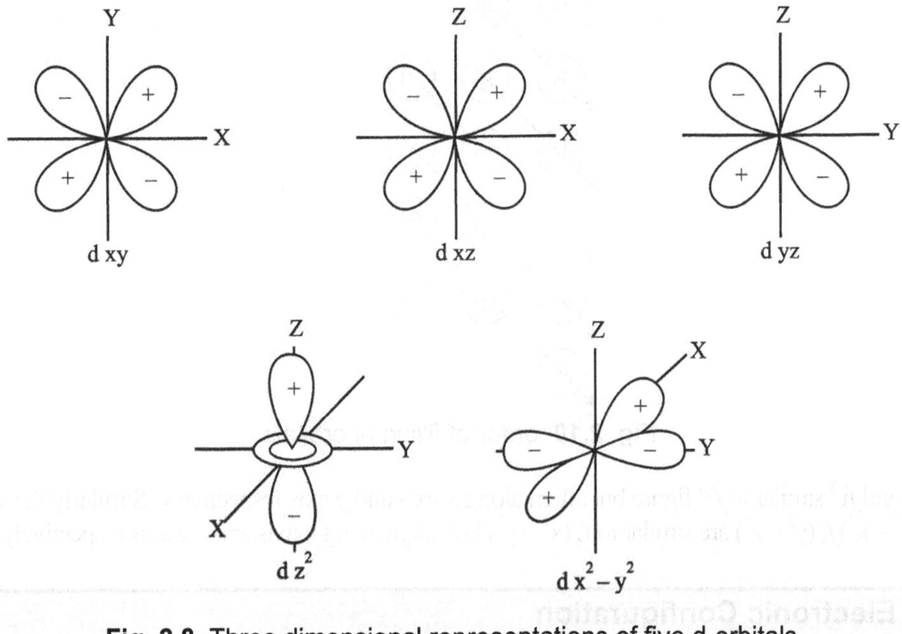

Fig. 2.8 Three dimensional representations of five d-orbitals.

The energy of d orbitals is greater than p since it is having 2 nodes.

f-orbital: Fundamental is used for f orbital. For this n = 4, *l* = 3, m = +3, +2, +1, 0, –1, –2, –3 (2*l* + 1 = 7), then it is 4f orbital. So f orbitals are seven f xyz, f x^3, f y^3, f z^3, fx(y^2 – z^2), f_y(z^2 –x^2), f_z(x^2 – y^2). The f orbitals have eight lobes and four nodes, opposite lobes are having opposite sign (Fig. 2.9).

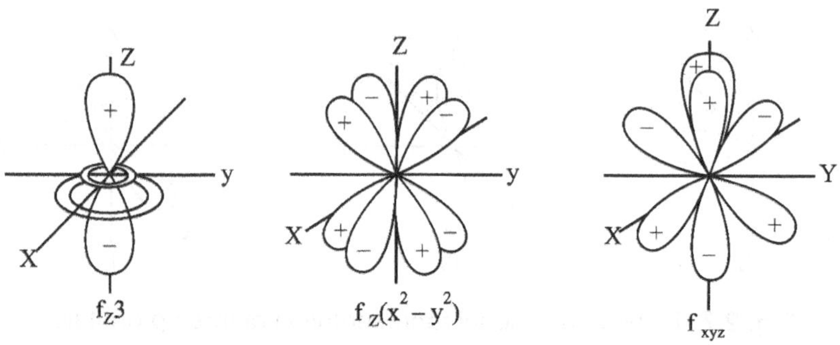

Fig. 2.9 Three directional representation of z component f orbital.

The increasing order of energy sequence of orbitals are as:

1s < 2s < 2p < 3s < 3p < 4s < 3d < 4p < 5 s < 4d < 5p < 6s < 4f < 5d < 6 p < 7s

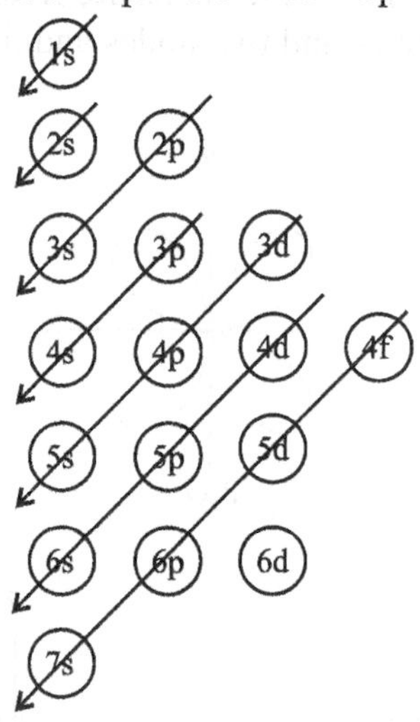

Fig. 2.10 order of filling of orbitals.

fx^3 and fy^3 similar to fz^3 figure but allign along x axis and y axis respectively. Similarly the shapes of f_y (x^2 – x^2) f_x (y^2 – z^2) are similar to f_z (x^2 – y^2) but allign along *y*-axis and *x*-axis respectively.

2.6 Electronic Configuration

In the atom filling of electrons follows regular pattern from one atom to the next atom. The rules in filling the electrons in any particular atom are:

(i) ***The Aufban Principle***: The method of determining the appropriate electron configuration of minimum energy (the ground state) makes use of Aufban principle or 'building up' of atoms one step at a time. Protons are added to the nucleus and the electrons are added to orbitals to build up the derived atom. It should be emphasized that this is only a formalism for arriving at the desired electron configuration, but an exceedingly useful one. In brief it can be said that the lower energy orbital will be filled first with electrons then higher energy orbital. The order of filling of orbitals is shown in Fig. 2.10.

For eg.:

H	$1s^1$
He	$1s^2$
Li	$1s^2\ 2s^1$
Be	$1s^2\ 2s^2$
B	$1s^2\ 2s^2\ 2p^1$
C	$1s^2\ 2s^2\ 2p^2$ etc

(ii) ***Hunds Rules***: In the ground state of any atom

- The electron enter first in the vacant orbital of lowest energy.
- Electrons occupy the orbitals singly with parallel spins as long as the same energy orbitals are available (degenerate orbitals) and then paired.

This is also known as **Hund's rule of maximum multiplicity**

For e.g.:

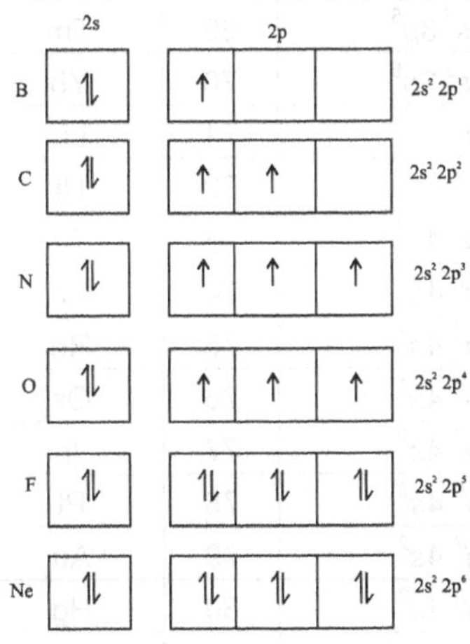

(iii) ***Pauli Exclusions Principle* (1925):** No two electrons in a single atom can have all their quantum numbers identical. All the electrons in any system distinguishable. That is in any orbital, the two electrons occupy with opposite spin (with + 1/2 and −1/2 value).

The electronic configuration of atoms are given in Table 2.2.

Table 2.2

Z	Element	Electron Configuration	Z	Element	Electron Configuration
1	H	$1s^1$	53	I	$[Kr]4d^{10}\,5s^2\,5p^5$
2	He	$1s^2$	54	Xe	$[Kr]4d^{10}\,5s^2\,5p^6$
3	Li	$[He]2s^1$	55	Cs	$[Xe]6s^1$
4	Be	$[He]2s^2$	56	Ba	$[Xe]6s^2$
5	B	$[He]2s^2\,2p^1$	57	La	$[Xe]5d^1\,6s^2$
6	C	$[He]2s^22p^2$	58	Ce	$[Xe]4f^{1}5d6s^2$
7	N	$[He]2s^22p^3$	59	Pr	$[Xe]4f^{3}6s^2$
8	O	$[He]2s^22p^4$	60	Nd	$[Xe]4f^{4}6s^2$
9	F	$[He]2s^22p^5$	61	Pm	$[Xe]4f^{5}6s^2$
10	Ne	$[He]2s^22p^6$	62	Sm	$[Xe]4f^{6}6s^2$
11	Na	$[Ne]3s^1$	63	Eu	$[Xe]4f^{7}6s^2$
12	Mg	$[Ne]3s^2$	64	Gd	$[Xe]4f^{7}\,5d^1\,6s^2$
13	Al	$[Ne]3s^2\,3p^1$	65	Tb	$[Xe]4f^{9}6s^2$
14	Si	$[Ne]3s^2\,3p^2$	66	Dy	$[Xe]4f^{10}6s^2$
15	P	$[Ne]3s^2\,3p^3$	67	Ho	$[Xe]4f^{11}6s^2$
16	S	$[Ne]3s^2\,3p^4$	68	Er	$[Xe]4f^{12}6s^2$
17	Cl	$[Ne]3s^2\,3p^5$	69	Tm	$[Xe]4f^{13}6s^2$
18	Ar	$[Ne]3s^2\,3p^6$	70	Yb	$[Xe]4f^{14}6s^2$
19	K	$[Ar]4s^1$	71	Lu	$[Xe]4f^{14}\,5d^1\,6s^2$
20	Ca	$[Ar]4s^2$	72	Hf	$[Xe]4f^{14}\,5d^2\,6s^2$
21	Sc	$[Ar]3d^1\,4s^2$	73	Ta	$[Xe]4f^{14}\,5d^3\,6s^2$
22	Ti	$[Ar]3d^2\,4s^2$	74	W	$[Xe]4f^{14}\,5d^4\,6s^2$
23	V	$[Ar]3d^3\,4s^2$	75	Re	$[Xe]4f^{14}\,5d^5\,6s^2$
24	Cr	$[Ar]3d^5\,4s^1$	76	Os	$[Xe]4f^{14}\,5d^6\,6s^2$
25	Mn	$[Ar]3d^5\,4s^2$	77	Ir	$[Xe]4f^{14}\,5d^7\,6s^2$
26	Fe	$[Ar]3d^6\,4s^2$	78	Pt	$[Xe]4f^{14}\,5d^9\,6s^1$
27	Co	$[Ar]3d^7\,4s^2$	79	Au	$[Xe]4f^{14}\,5d^{10}\,6s^1$
28	Ni	$[Ar]3d^8\,4s^2$	80	Hg	$[Xe]4f^{14}\,5d^{10}\,6s^2$
29	Cu	$[Ar]3d^{10}\,4s^1$	81	T1	$[Xe]4f^{14}\,5d^{10}\,6s^2\,6p^1$
30	Zn	$[Ar]3d^{10}\,4s^2$	82	Pb	$[Xe]4f^{14}\,5d^{10}\,6s^2\,6p^2$
31	Ga	$[Ar]3d^{10}\,4s^2\,4p^1$	83	Bi	$[Xe]4f^{14}\,5d^{10}\,6s^2\,6p^3$
32	Ge	$[Ar]3d^{10}\,4s^2\,4p^2$	84	Po	$[Xe]4f^{14}\,5d^{10}\,6s^2\,6p^4$

Table 2.2 *Contd...*

Z	Element	Electron Configuration	Z	Element	Electron Configuration
33	As	$[Ar]3d^{10}\,4s^2\,4p^3$	85	At	$[Xe]4f^{14}\,5d^{10}\,6s^2\,6p^5$
34	Se	$[Ar]3d^{10}\,4s^2\,4p^4$	86	Rn	$[Xe]4f^{14}\,5d^{10}\,6s^2\,6p^6$
35	Br	$[Ar]3d^{10}\,4s^2\,4p^5$	87	Fr	$[Rn]7s^1$
36	Kr	$[Ar]3d^{10}\,4s^2\,4p^6$	88	Ra	$[Rn]7s^2$
37	Rb	$[Kr]\,5s^1$	89	Ac	$[Rn]6d^1\,7s^2$
38	Sr	$[Kr]\,5s^2$	90	Th	$[Rn]6d^2\,7s^2$
39	Y	$[Kr]4d^1\,5s^2$	91	Pa	$[Rn]5f^{2}6d^1\,7s^2$
40	Zr	$[Kr]4d^2\,5s^2$	92	U	$[Rn]5f^{3}6d^1\,7s^2$
41	Nb	$[Kr]4d^4\,5s^1$	93	Np	$[Rn]5f^{4}6d^1\,7s^2$
42	Mo	$[Kr]4d^5\,5s^1$	94	Pu	$[Rn]5f^{6}7s^2$
43	Tc	$[Kr]4d^5\,5s^2$	95	Am	$[Rn]5f^{7}7s^2$
44	Ru	$[Kr]4d^7\,5s^1$	96	Cm	$[Rn]5f^{7}6d^1\,7s^2$
45	Rh	$[Kr]4d^8\,5s^1$	97	Bk	$[Rn]5f^{9}7s^2$
46	Pd	$[Kr]4d^{10}$	98	Cf	$[Rn]5f^{10}7s^2$
47	Ag	$[Kr]4d^{10}\,5s^1$	99	Es	$[Rn]5f^{11}7s^2$
48	Cd	$[Kr]4d^{10}\,5s^2$	100	Fm	$[Rn]5f^{12}7s^2$
49	In	$[Kr]4d^{10}\,5s^2\,5p^1$	101	Md[b]	$[Rn]5f^{13}7s^2$
50	Sn	$[Kr]4d^{10}\,5s^2\,5p^2$	102	No[b]	$[Rn]5f^{14}7s^2$
51	Sb	$[Kr]4d^{10}\,5s^2\,5p^3$	103	Lr[b]	$[Rn]5f^{14}6d7s^2$
52	Te	$[Kr]4d^{10}\,5s^2\,5p^4$	104	Rf[b]	$[Rn]5f^{14}6d^2\,7s^2$

2.7 Periodic Table

D.I. Mendeleeff (Russian scientist) and Lothar Mayer (German scientist) independently around 1868-1870 reported the periodic law and arranged the elements in the form of periodic table on the basis of atomic weights. As per the new periodic law, the periodic properties of elements can be represented as periodic functions of their atomic numbers. Present day long form is periodic table (Table 2.3) states that the chemical properties of an element are largely governed by electrons in the outer shell, and their arrangement. The elements with in a group should show similarities in physical and chemical properties because they have same outer electronic arrangement.

Table 2.3

New Designation / Original Designation

Non-Metals · Metals · Transition Metals · s-block · d-block · p-block · f-block

Key: Atomic # / Symbol / Atomic Mass

Period	1 IA (s-block)	2 IIA	3 IIIB	4 IVB	5 VB	6 VIB	7 VIIB	8 VIIIB	9 VIIIB	10 VIIIB	11 IB	12 IIB	13 IIIA	14 IVA	15 VA	16 VIA	17 VIIA	18 VIIIA (s-block)
1	1 H 1.0094																	2 He 4.00260
2	3 Li 6.941	4 Be 9.0122											5 B 10.81	6 C 12.011	7 N 14.007	8 O 15.999	9 F 18.998	10 Ne 20.179
3	11 Na 22.990	12 Mg 24.305											13 Al 26.982	14 Si 28.086	15 P 30.974	16 S 32.06	17 Cl 35.453	18 Ar 39.948
4	19 K 39.098	20 Ca 40.08	21 Sc 44.956	22 Ti 47.88	23 V 50.942	24 Cr 51.996	25 Mn 54.938	26 Fe 55.847	27 Co 58.933	28 Ni 58.69	29 Cu 63.546	30 Zn 65.39	31 Ga 69.72	32 Ge 72.59	33 As 74.922	34 Se 78.96	35 Br 79.904	36 Kr 83.80
5	37 Rb 85.468	38 Sr 87.62	39 Y 88.906	40 Zr 91.224	41 Nb 92.906	42 Mo 95.94	43 Tc (98)	44 Ru 101.07	45 Rh 102.91	46 Pd 106.42	47 Ag 107.87	48 Cd 112.41	49 In 114.82	50 Sn 118.71	51 Sb 121.75	52 Te 127.60	53 I 126.91	54 Xe 131.29
6	55 Cs 132.91	56 Ba 137.33	57 to 71	72 Hf 178.49	73 Ta 180.95	74 W 183.85	75 Re 186.21	76 Os 190.2	77 Ir 192.22	78 Pt 195.08	79 Au 196.97	80 Hg 200.59	81 Tl 204.38	82 Pb 207.2	83 Bi 208.98	84 Po (209)	85 At (210)	86 Rn (222)
7	87 Fr (223)	88 Ra 226.03	89 to 103	104 Unq (261)	105 Unp (262)	106 Unh (263)	107 Uns (262)	108 Uno (265)	109 Une (266)	110 Uun (267)								

(Mass Numbers in Parentheses are from the most stable of common isotopes.)

Phases: Solid, Liquid, Gas

Rare Earth Elements

	d-block	f-block													
Lanthanide Series	57 La 138.91	58 Ce 140.12	59 Pr 140.91	60 Nd 144.24	61 Pm (145)	62 Sm 150.36	63 Eu 151.96	64 Gd 157.25	65 Tb 158.93	66 Dy 162.50	67 Ho 164.93	68 Er 167.26	69 Tm 168.93	70 Yb 173.04	71 Lu 174.97
Actinide Series	89 Ac 227.03	90 Th 232.04	91 Pa 231.04	92 U 238.03	93 Np 237.05	94 Pu (244)	95 Am (243)	96 Cm (247)	97 Bk (247)	98 Cf (251)	99 Es (252)	100 Fm (257)	101 Md (258)	102 No (259)	103 Lr (260)

For e.g.,

I group elements

Na $\xrightarrow{-e^-}$ Na^+ one electron is lost

[Ne] $3s^1$ [Ne]

So, its valency is 1

II Group element

Mg $\xrightarrow{-2e^-}$ Mg^{2+} two electrons lost

[Ne] $3s^2$ [Ne] $3s^2$

So, its valency is 2

III Group element

Ga $\xrightarrow{-3e^-}$ Ga^{3+} three electrons lost

[Ar] $3d^{10}$ $4s^2$ $4p^1$ [Ar] $3d^{10}$

So its valency is 3

VII A Group

Cl $\xrightarrow{+e^-}$ Cl^-

[Ne] $3s^2$ $3p^5$ [Ne] $3s^2$ $3p^6$ or [Ar]

VIA Group

O $\xrightarrow{+2e^-}$ O^{2-}

[He] $2s^2$ $2p^4$ [He] $2s^2$ $2p^6$ or [Ne]

Ist group and IInd group elements are called as s-block elements, with outer ones electron, are known as alkali metals and two s electrons are known as alkaline eath metals.

p – block elements – III A to VII A group elements and are almost non-metals or post transition elements and these elements have outer incomplete p-orbitals and contain 6 elements in each period.

d - block elements – I B, II nd, B, III B to VIII B are called as d-block elements and these elements have incomplete (n – 1) d-orbitals and contains 10 elements.

f - block elements – These are also known as inner transition elements (Lanthanides and Actimides) occur in group III B. These elements have incomplete (n-2) f orbitals and contain 14 elements.

Zero group elements – These elements are called as inert gas elements with ns^2 np^6 configuration.

Finally the periodic table consists of 7 periods, 18 columns and 16 groups.

In the periods elements present are:

1st period consists 2 elements
2nd period consists 8 elements
3rd period consists 8 elements
4th period consists 18 elements
5th period consists 18 elements
6th period consists 32 elements
7th period consists 23 elements (Incomplete)

Periodic Properties

(i) ***Trends in Atomic Radii (Size of the Atom)*:** The size of radius of an atom is an important physical property of the elements which is a periodic function of atomic number.

The atomic radii of atoms decrease on moving from left to right in the period. As the atomic number increases across a period, the nuclear charge increases and the electrons are attracted more and more strongly towards the nucleus i.e., why the size or atomic radius decreases.

For eg.: The atoms of IInd period atomic radii are

Li – 1.23 Å, Be – 0.9 Å, B – 0.85 Å, C – 0.77 Å, N – 0.70 Å, 0 – 0.66 Å and F – 0.64 Å.

This shows the size is clearly decreases as atomic number increases

The atomic radii increases on descending a group in the periodic table. As we move down the group, the number of shells increases with increase in atomic number which is the reason for increase.

For e.g.: 1st group Li 1.25 Å Na –1.54 Å, K – 2.03 Å Rb – 2.16 Å Cs – 2.35 Å.

(ii) ***Ionisation Energy* (I.E):** The ionization energy or ionization potential is the minimum energy required to remove an electron from a free neutral gaseous atom in its ground state and it may be expressed in electron volts (ev) or kilo joules per gram atom ($kJ.g^{-1}$).

$$\text{Atom (g)} + \text{Energy} \rightarrow \text{positive ion (g)}^{+}\ \text{electron (g)} \qquad .....(2.21)$$

Successive ionization energies i.e., removal of one electron I_1, removal of second electron is the same atom I_2, third I_3 ....etc.

$$I_1 < I_2 < I_3 ....I_n$$

Trends: In the group as we move down, the ionization energy decreases. As the size increases, the distance between nucleus and electron increases, so the force of attraction decreases and I.E decreases.

For e.g.: Li – 5.39 ev, K – 4.34 ev, Cs – 3.90 ev.

In the period from left to right ionization energy increases.

For e.g., Li – 5.39 ev, C – 11.26 ev, O – 13.61 ev, F – 17.42 ev

(iii) ***Electron Affinity* (EA):** The electron affinity of an element may be defined as amount of energy which is released when an extra electron enters the valence orbital of an isolated neutral atom to form a negative ion.

$$\text{Atom (g)} + \text{electron (g)} \rightarrow \text{Negative ion (g)} + \text{energy} \quad .....(2.22)$$

Halogens have high electron affinity and chlorine has highest EA.

For e.g.: F – 3.45 ev, Cl – 3.61 ev, Br – 3.36, I – 3.06 ev. Electron affinity increases along the period and decreases down the group.

(iv) ***Electronegativity*:** Electronegativity is defined as the tendency of an atom to attract the shared electron pair or electrons towards itself in a covalent bond.

Trends – On moving down the group, the electronegativity decreases and on moving from left to right in period, the electro negativity increases i.e., increases the non-metallic character.

(v) ***Electropositivity (Metallic Character)*:** The tendency to lose electrons, if supplied energy, is called electropositivity.

$$M \rightarrow M^+ + e^- \quad .....(2.23)$$

The stronger this tendency, the more electropositive and more metallic an element.

Trends: It increases in the group as we move down and it decreases in the period if we move from left to right.

The electropositive scale and the electronegativity scale of all the elements is available in literature.

2.8 Molecules and Molecular Structure

Introduction

(a) The structure of molecules constitute the basis for their chemical behaviour. A good understanding of molecular structure is necessary to interpret differences in chemical behaviour of inorganic species. In the molecules, the nuclei of the constituent atoms at their most stable inter nuclear distances embedded in a matrix of electron. Most of the electrons are in atomic orbitals surrounding the individual nuclei, and the remainder value electrons are in more generalized multinuclear molecular orbitals. In combination of atoms into molecules, there are certain forces are involved.

(b) *Chemical Bonding and Bond Type*: The atoms approach each other, the repulsive forces occur to push the atoms away are:

- electron-electron repulsion between valence electrons of the approached atoms,
- nuclear repulsion between neighbouring nuclei.

When the proper balance between these forces exists and the energy of the resulting system is lower than the sum of the energies of the isolated atoms.

Chemical bond exists between two atoms when the bonding force between them is of such strength as to lead to an aggregate of sufficient stability to warrant their consideration as an independent molecular species.

(c) Types of Bonds: The bonding types are:

(i) *Ionic bond*: Occur between electropositive and electronegative elements due to coulumbic forces of attraction.

e.g.: $Na^+ + Cl^- \rightarrow Na\,Cl$

$Mg^{2+} + SO_4^{2-} \rightarrow Mg\,SO_4$

(ii) *Covalent bond*: Occur in homodiatomic, hetero diatomic and also between multi atom molecules. This occur due to sharing of electron pair equally or unequally between two atoms.

e.g.: H_2, Cl_2, I_2 (homodiatomic molecules)

HCl, CO, NO (heterodiatomic molecules)

The valence bond theory and molecular orbital theory explain this bond.

(iii) *Metallic bond*: Occur in between electropositive atoms (mostly in between metals).

e.g: Li, Cu, Zn etc. Band theory can explain this bond.

Fig. 2.11 is the triangle illustrating transitions between ionic, covalent and metallic bonds.

(iv) *Hydrogen bond*: Occur between hydrogen and two more electronegative ions.

e.g.: O – H....), F – H ..... F, N – H ......N.

(v) *Vander waals forces*: These are weak forces occur between molecules and are responsible to condence neutral molecules from gaseous state to liquid and liquid to solid.

(d) ***Bond Length*:** The equilibrium distance between the centres of two nuclei of two bonded atoms is known as bond length. It is expressed in Angstorm units (Å i.e., 10^{-8} cm) or nanometers (nm i.e., 10^{-9} m) or picometes (pm i.e., 10^{-10} cm).

e.g., $H-H = 0.74$ Å, $H-Cl = 1.27$ Å, $C-H = 1.09$ Å,

$C=N = 1.09$ Å $C-) = 1.42$ Å, $C=0 = 1.1.24$ Å.

Bond length decreases from single bond to double bond and double bond to triple bond. The magnitude of bond length provides the idea about nature of the bond.

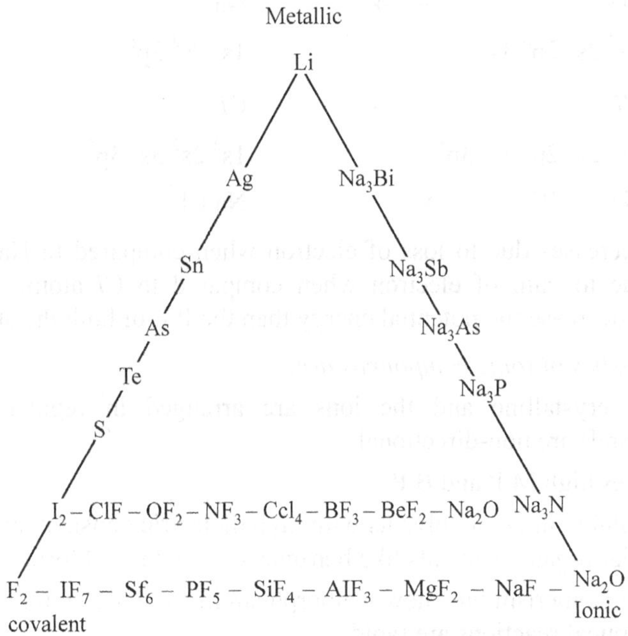

Fig. 2.11 The triangle.

(e) ***Bond Angle*:** Bond angle may be defined as the internal angle between those orbitals which have bonded electron pairs in valency shell of the central atom in a molecule.

e.g.: O = C = O is linear and angle 180°, in BF_3 the angle is 120°, CH_4 – 109°. 28' etc.

(f) ***Bond Energy*:** Band energy is defined as the energy required to break a mole of bond and separate the bonded atoms in a gaseous state. It is also a measure of bond strength and is expressed in kJ mol^{-1} or k Cal. mol^{-1} or electron volts.

For eg: H_2 (g) → 2H(g), bond energy is 435 kJ mol^{-1} for poly atomic molecule the bond energy is not equal to bond dissociation energy because the energy necessary to rupture a given bond will depend to some extent on the nature of reminder molecule.

2.9 Ionic Bond / Electrovalent Bond / Polar Bond

The ionic bond is formed between two ions of opposite charge by electrostatic force of attraction. It involves the transfer of certain number of electrons to another dissimilar atom which has tendency to gain electrons, so that both acquire stable inert gas configuration.

$$Na \xrightarrow{-e} Na^+ \quad .....(2.24)$$
$$1s^2\ 2s^2\ 2p^6\ 3s^1 \qquad 1s^2\ 2s^2\ 2p^6$$

$$Cl \xrightarrow{+e} Cl^-$$
$$1s^2\ 2s^2\ 2p^6\ 3s^2\ 3p^5 \qquad 1s^2\ 2s^2\ 3s^2\ 3p^6 \quad .....(2.25)$$

$$Na^+ + Cl^- \rightarrow Na\ Cl \quad .....(2.26)$$

Cation size decreases due to loss of electron when compared to Na atom and anion size increases due to gain of electron when compared to *Cl* atom. This electrostatic attraction always decrease the potential energy than the P.E of both the atoms.

The characteristics of Ionic compounds are:

(i) usually crystalline and the ions are arranged in regular way in lattice, compounds are non-directional.

(ii) possesses high M.P and B.P

(iii) quite soluble in polar solvents with high dielectric constant and are insoluble in non-polar organic solvents like benzene, carbon tetra chloride etc.

(iv) have no isomerism but show isomorphism in similar electronic compounds and the chemical reactions are rapid.

Born-Hober (1919) designed a cycle for the formation of ionic crystal on the basis of some thermodynamic data.

$$Q = S + I + \frac{D}{2} + E + U_0 \quad .....(2.27)$$

where

Q = heat of enthalpy for the formation of compound, it is –ve

S = heat of sublimation of Na solid atom to Na(g) atom

I = Ionisation energy for Na(g) to Na^+(g)

$\frac{D}{2}$ = Dissociation energy for Cl_2 molecule to Cl (g)

E = Electron affinity for the formulation Cl(g) to Cl^-(g). It is –ve

U_0 = The energy involved when one gram molecule of the crystal is to form the gaseous ion which is –ve.

For example for formation of NaCl

$$Q = 109 + 496 + 121 - 365 - 781 = -420 \text{ kJ mol}^{-1}$$

Q and lattice energy must be negative to form ionic compound.

In an ionic structure each ion is surrounded by a certain number of ions of the opposite sign. This number is called coordination number of the ions. For example every Na^+ is surrounded by six evenly spaced Cl^- ions and every Cl^- by six evenly spaced Na^+ ions. Based on the arrangement of cations and anions, ionic solids have various arrangements like cubic-close packed, hexagonal close packed, body centered cubic etc.

2.10 Polarisation

When two ions approach each other closely, the cation attracts the electron cloud of the anion and at the same time repels the anion nucleus. This process is called polarization or distortion (Fig. 2.12).

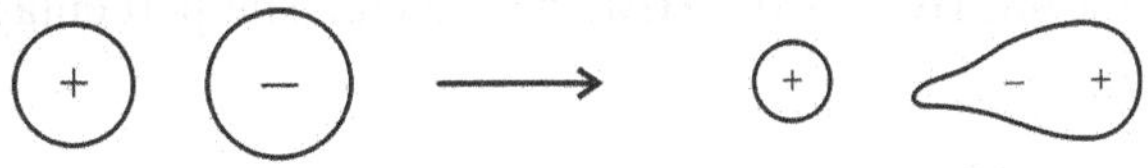

Fig. 2.12 Schematic representation of the polarization of ions.

The effect of polarization upon the passage from ionic to covalent bonding was described by K. Fajans. His description aids in the conceptual development of this process, and is based on the ionic charge, ionic radius and valence shell structure.

The Fajans rules are:

- A sizable charge on a cation, particularly if the radius is small, will have a pronounced polarizing effect on other entities in comparison to less highly charged ions.

  The greater distortion effect on an anion like O^{2-}, F^- decreases in the order $Al^{3+} > Mg^{2+} > Na^+$

- The ionic potential (cationic charge / radius) improves the polarizing power. For cations having same charge increases with the decrease of ionic radii.

  For eg: $Be^{2+} > Mg^{2+} > Ca^{2+} > Sr^{2+} > Ba^{2+}$

- A sizable charge on an anion, particularly if the radius is large, will be easily polarized. This is due to the fact that the electrons are at a relatively great distance from the positively charged nucleus and can be influenced more readily by neighbouring small cation. *For e.g.*: H^+ (small cations) and I^- (large cation) in HI.
- The cation have noninert gas configuration seems to have greater polarising power on anion than do those possessing inert gas configuration. This is may be due to incomplete shielding on nuclear charge in the non-inert gas configuration permitting more of the excess nuclear charge to penetrate the electron shell.

  For eg: The cations of alkali metals have inert gas configuration ($ns^2\ np^6$) cannot polarize. But Ag (1), Cu (1) have $(n-1)d^{10}$ configuration, they can polarize because d- electrons cannot screen the outer most electron.

2.11 Partial Ionic Character in Covalent Bond

The sharing of electrons approaches equality in covalent bond (non-polar bond). but deviates to a greater or lower extent. This deviation is called partial ionic character of a covalent bond.

Pure covalent compound has zero dipole movement (μ expressed Debye 1D = 10^{-18} electrostatic unit e.u.). But no covalent bond has μ = 0.

For eg:, μ of HF = 1.98, HCl =1.03, HBr. = 0.79 etc. The percentage of ionic character can be determined by

$$\%\ \text{ionic character} = 100\ \mu / e.r \quad .....(2.28)$$

where μ = dipole movement in Debye units

e = electronic charge

r = bond lengths

Hammy and Smith gave another equation based on electronegetivities.

$$\%\ \text{ionic character} = 16\ (x_A - x_B) - 3.5\ (x_A - x_B)^2 \quad .....(2.29)$$

where x_A = electronegetivity of atom A

x_B = electonegativity of atom B

2.12 Covalent Bond

The atoms combine to attain stability. Lewis states that the nonionic linkages amount to sharing of electron pairs accordance with **Octet Rule**.

Single bond - if one pair of electrons shared

For e.g.,

H – F – H $:\ddot{\underset{\cdot\cdot}{H}}:$, H_2 – H**:** H

The thick ones are shared pairs.

Double bond - if two pairs of electrons shared,

For e.g.; C_2H_4

Triple bond - if three pairs of electrons share.

For e.g., C_2H_2 HC:::CH

The Various Theories of Covalent Bond are:

(a) Heitler London Theory (Valence Bond Theory)

The pairing and the resultant neutralization of opposite electron spin leads to covalent bond.

An atom in order to enter into chemical combination, it must possess one or more unpaired electrons (valence electrons in atomic orbitals, AO's), which can pair with those in another atom through canceling the electron spins. The number of unpaired electrons determines the valency. The electron pairs are localized between the two bonded atoms. This Theory is also known as atomic orbital theory.

Draw backs:

(i) The assumption that the electron in a shared pair come from different atom is contrary.

(ii) Coordinate linkage ruled out

(iii) The spin theory fail to account for maximum covalency such as in PF_5, Perbromates etc.

(iv) Para magnetism in O_2 molecule cannot explain.

(b) Hund-Mulliken (Molecular orbital theory)

This theory employs molecular orbitals (MO's) rather than AO's. The valence electrons are considered to be associated with all the nuclei in the molecule. Thus AO's from different atoms must be combined to produce MO's and the atomic nuclei being separated by fixed equilibrium distances. The advantage of this theory is that it does not require electron spins and is applicable for many simple molecules.

(c) Pauling – Slatter Theory (Hybridization Theory)

It is the extension of Heitler-London theory. It account the directional characteristics of covalent bonds. In this AO's are hybridised to give hybrid orbitals. *The phenomenon of mixing or linear combination of pure atomic orbitals (i.e orbitals of similar energies belonging to the same atom or ion) in such a way as to form new atomic orbitals of equal energy is known as hybridisation and the new orbitals formed are known as hybrid orbitals.*

The no. of hybrid orbitals produced is equal to the no. of orbitals mixed. The hybrid bonds are stronger than the single non-hybridised bonds of comparable energy. The type of hybridization indicates the geometry of molecules. Once the hybrid orbital has been used to build a hybrid it is then no longer available to hold electron in its pure form.

Types of Hybrisation

(a) sp type-one s and on p orbital overlap

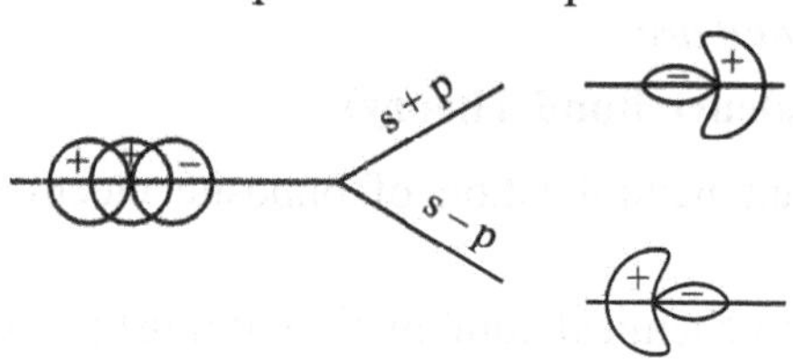

s : p = 1:1
Eg BeF_2
Bond angel =180°
Shape = linear

Fig. 2.13

(b) sp^2 type – one s and two p orbitals overlap

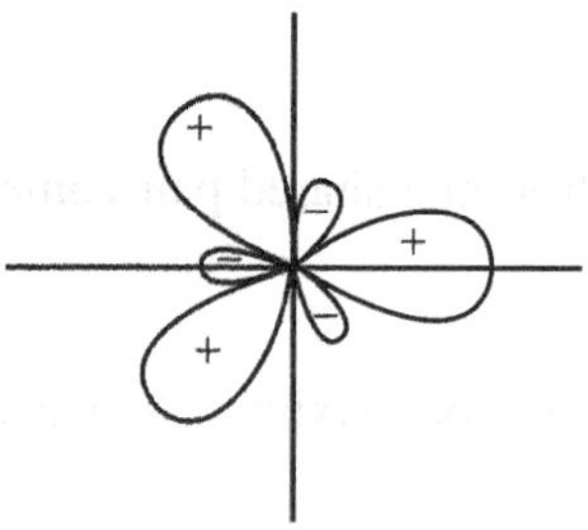

s:p = 1:2
es; Bf_3
Bond angle = 120°
shape = trigonal planer

Fig. 2.14

(c) sp^3 type-one s and 3 p orbitals overlap
s : p = 1: 3
e.g., CH_4
Bond angle = 109 °.28 '
Shape = tetrahedral.

Table 2.4

No of hybridization	Type of hybridization	Shape	Angle	Shape Figure
2	SP	Linear	180°	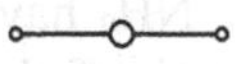
3	SP^2	Trigonal Planar	120°	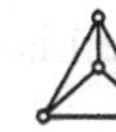
4	SP^3or $d_{\epsilon}^3\ S$	Tetrahedral	$140^{\circ}\ 28'$	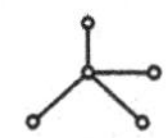
4	$dx^2 - y^2\ SP^2$	Square planar	90°	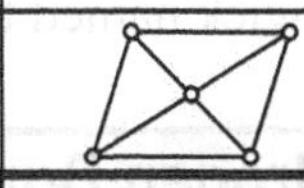
5	$d_2 2\ SP^3$	Trigonal bi pyramid	90° &120°	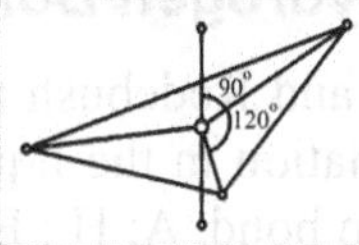
5	$d_x^2 - y^2\ SP^3$	Square pyramid	90°	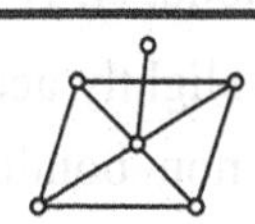
6	$d_r^2\ SP^3$	Octahedral	90°	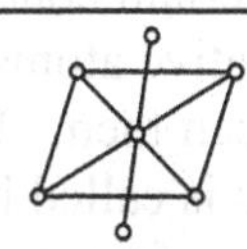
7	$d^3\ SP^3$	Pentagonal bi pyramid	90° & 72°	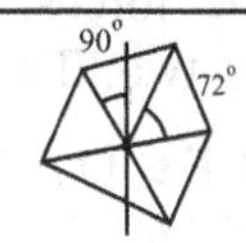

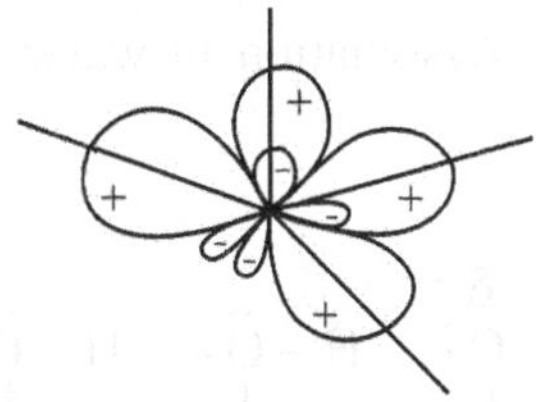

Fig. 2.15

The energy of hybridization is of order of magnitude of bond energies and can thus be important in determining the structure of molecules.

(d) Lowry theory / coordinate covalent bond / semi polar bond theory

The two electrons required to form bond are donated by one atom, which results in unequality in charge distribution i.e., semi polar bond.

E.g., NH_3 having eight electrons on Nitrogen i.e. 3 – bond pair electrons and one lone pair of electrons, BH_3 having 6 electrons on boron i.e. 3 – bond pair electrons. Outer shell is not octet. So boron accepts pair of electrons from nitrogen in NH_3.

$$H_3N \rightarrow BH_3$$
$$R_3N^+ \rightarrow O^{2-}$$

Sidgwick named this as coordinate bond later.

2.13 Hydrogen Bonding

Latimer and Rodebush (1920) introduced the term hydrogen bond to describe the nature of association in the liquid state of water and HF etc. The necessity for the formation of hydrogen bond, A: H : B, is

- a slightly acetic hydrogen.
- a non–bonding electron pair

Under certain conditions an atom of hydrogen is attracted simultaneously to two more electro negative atoms, instead of one, so it may be considered to be acting as a bridge or bond between them. This type of bond in which a hydrogen atom is associated with two other atoms is called hydrogen bond. The bond with one atom is covalent and the other is due to ionic forces. It is attracting only two atoms, so the coordination number of hydrogen is two. The bond strength of the bonds will increase with increasing electro negativity (F > O > N > Cl). In ice hydrogen bond $\overset{1000}{O}-H.......\overset{1.76}{O}$. So the bond length is larger i.e. 2.76 A° (weak bond) and dissociation energy is 3–6 k.cal. mol $^{-1}$. Bond energies ranges from 2 to 10 k.cal. mol $^{-1}$ i.e., very weak when compared ~ 100 k. cal. mol^{-1} for covalent or ionic bond. Association in water molecule and HCN is shown in Fig 2.16.

(i) H_2O

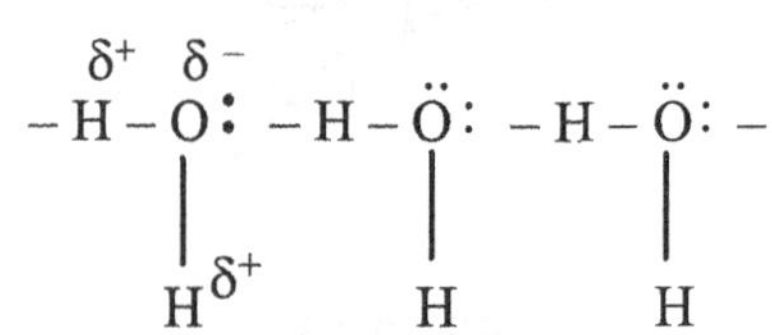

(ii) HCN $\quad H-C\equiv N.....H-C\equiv N.......H-C\equiv N.$

Fig 2.16

Importance

Hydrogen bonding plays an important role in solution formation and in water of crystalisation. It is important in interactions between complex molecules, and in secondary structures of proteins. It is also a secondary bind force in drug–receptor interactions.

2.14 Vander Waal's Forces (Non–Valence Cohesive Forces)

The cohesive (attractive) forces which cause the neutral molecules of a covalent compound or a polyatomic element or the atoms of a mono atomic non – metallic element condense into the liquid or solid state are collectively known as vander Waal's forces. The forces occur mostly in non–polar molecules and are dipolar in nature. They are

(i) *Ion-dipole forces (Debye forces in 1920):*

The dipolemovement $\mu = q^{\pm} r'$ .....(2.30)

where $q^{\pm}$ = opposite charges.

r' = distance between two charges.

These occur where a dipole is placed in an electric field which is the result of presence of ion, the dipole will orient itself so that the attractive end directed towards the ion. This effect is known as induction effect. Ion dipole forces are important in solution of ionic compounds in polar solvents. e.g., Na F in H_2O form solvated species such as $Na(OH_2)_x^+$ and $F(OH_2)_y^-$.

Another e.g., is $Co(NH_3)_6^{3+}$

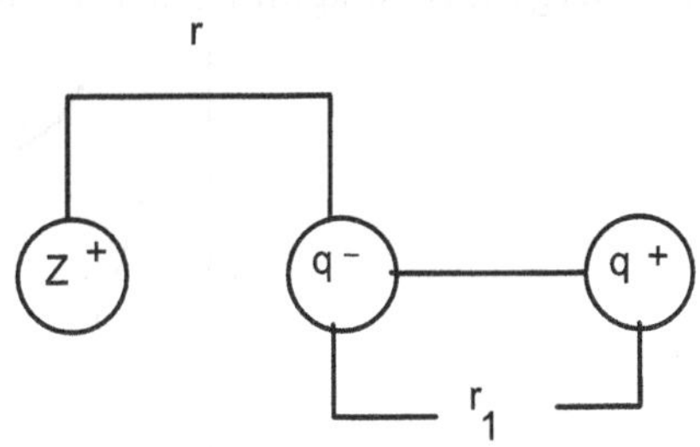

(ii) *Dipole –Dipole interactions (Keesom Forces 1912):*

The interaction of two dipoles i.e. either head-to-tail arrangement of dipoles

These are directional and are responsible for the association and structure of polar liquids

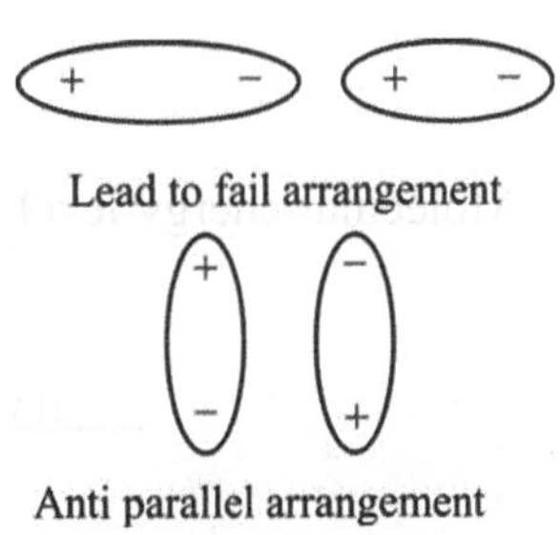

Fig. 2.18

(iii) *Induced dipole interactions (London Forces 1930)*:

If a charged particle, such as ion, is introduced into the neighbourhood of an uncharged, non-polar molecule (e.g., I_2, noble gas atom Xe etc.), it will distort the electron cloud of the atom or molecule in much the same way that a charged cation can distort the electron cloud of a large and soft anion. This effect is known as dispersion effect. The importance is limited to solutions of ionic or polar compounds in non-polar solvents.

All these forces are weaker when compared to ionic or covalent bond and these occur at very short distances.

2.15 Molecular Orbitals

The atomic orbitals of different atoms of almost same energy can overlap to give molecular orbitals. If the AO's overlap with the same sign, result in bonding MO's and with the opposite sign result in anti bonding MO's. If there are no suitable energy orbitals in the other atom, the AO's remain as non-bonding orbitals.

Bonding MO: The increased electron density in between the two nuclei in the orbital and is therefore called as bonding MO. The energy of BO's is lower than original AO's.

Antibonding MO: The electron density is zero in between the two nuclei and the energy of ABO's is higher than original AO's.

The energy level diagram is shown in Fig. 2.19.

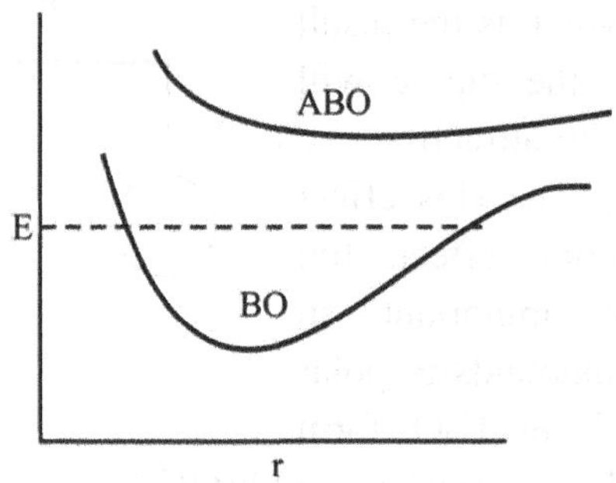

Fig. 2.19

Molecular energy level diagram for H_2 molecule is shown in Fig.2.20

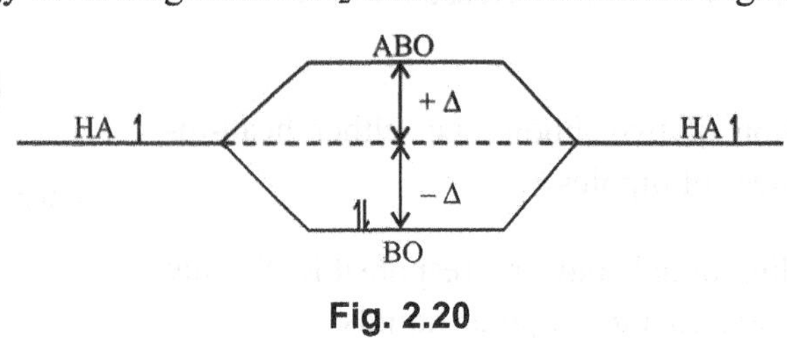

Fig. 2.20

By occupying the electron in BO, the molecule gains energy i.e.$2 \times -\Delta = -2\,\Delta$ which corresponds to bonding energy. The system is stabilized in this way and the bond is formed. The filling of electrons in MO's follows the same three rules as in electronic configuration of atom. i.e. Aufbua Principle, Hunds rule and Pauli exclusion principle. The bond order i.e., whether single bond, double bond... etc., can be obtained from the equation

$$\text{Bond order} = \frac{\text{Number of BO electron} - \text{Number of ABO electrons}}{2} \quad .....(2.31)$$

For eg:, of H_2 molecules Bond order $= \frac{2-0}{2} = 1$ so there is single bond in between 2H atoms.

MO's are as in AO's described by four quantum numbers. In MO's n, *l*, s are same but m is replaced by λ. If $\lambda=0$ then σ bond, $\lambda = \pm 1$ then π bond and $\lambda = \pm 2$ the δ (delta) bond.

σ bond accommodate maximum 2 electrons similar to s orbital in AO, π bond accommodate maximum 4 electrons, δ bond accommodate 4 electrons.

(i) ***Shape of σ Bond:*** s – s, s – p_x, p_x - p_x AO's overlap along the axis joining nuclei produces σ MO's. σ BO's are gerade and σ* ABO's (ABO's are marked with star) are ungerade.

e.g : s – s overlapping.

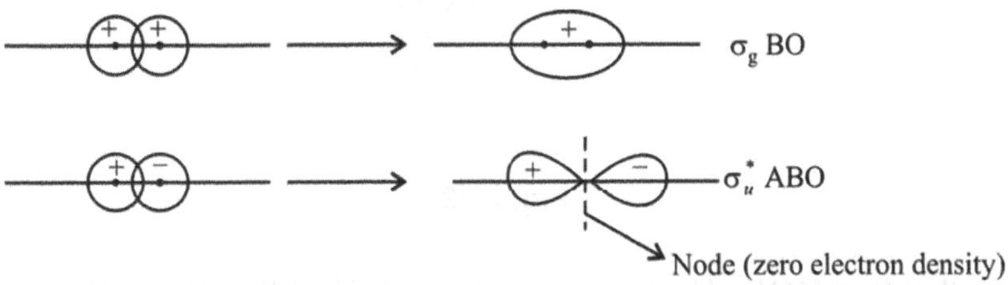

Fig. 2.21

On rotating about 180° on the axis, the sign is not inverted then gerade and if sign is inverted then ungerade. The signs are not the charges and only represent the direction of wave function.

(ii) ***Shape of π-Bond:*** p_y – p_y, p_z – p_z, p_y – d_{xy} AO's overlap laterally to form π MO's. The π bonds are having short bond length. π BO's are ungerade and π* ABO's are gerade.

e.g: $P_z - P_z$ overlapping

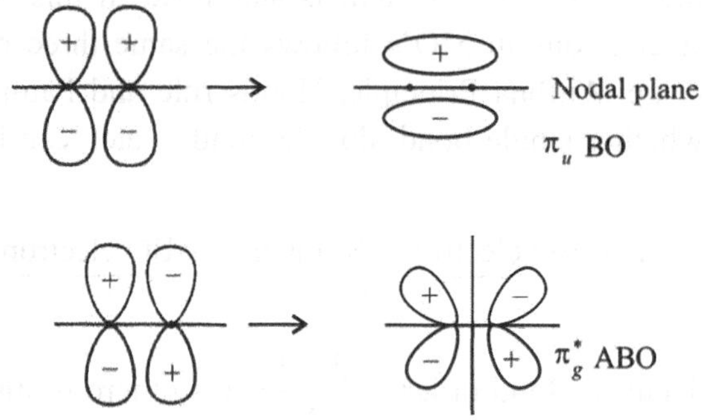

Fig. 2.22

(iii) ***Shape of δ (delta) Bond:*** Two δ AO's overlap sideways to form δ and δ^* MO's. δ BO's are gerade and δ^* ABO's are ungerade.

e.g: $dx^2 - y^2 - dx^2 - y^2$ overlapping

Fig. 2.23

(iv) ***Non-Bonding MO's*:** The symmetry of two AO's different i.e., rotation about the axis changes the sign of one and there is no overall change in energy and this neutralism is called as non-bonding.

e.g: $s - p_2$ overlapping remains non-bonding MO's

Fig. 2.24

2.16 Stereochemistry of Inorganic Molecules

The shapes of molecules depend on the arrangement of valencies within the molecule is called the geometry of the molecules. The theories regarding the geometry of molecules are :

(a) ***Helferich Rules:*** These rules are empirical and used to predict for simple molecules.

(i) AX_2: In this A is the central atom and if A has no unshared electrons, the shape is linear. Eg., CO_2,N_2O, N_3^- etc. If A has unshared pair of electrons (lone pair) the shape is angular. e.g H_2O, H_2S.

(ii) AX_3 : If A has no unshaped pair of electrons, the shape is trigonal planar with angle 120° . e.g BH_3, NO_3^- . If A has unshared pair of electrons, the shape is pyramidal with A at apex. E.g., NH_3, SO_3^{2-} ClO_3^-

(iii) AX_4: If A has no unshared pair of electrons, the shape is tetrahedral with angle 109°. 28' e.g., CH_4 NH_4^+. In some cases square planar e.g., Ni(11), Pd (11).

(b) ***Valence Shell Electron Pair Repulsion Theory (VSEPR Theory):***

This Theory was outlined by Sidgwick and Powell (1940) and was popularized by Gillepse and Nyholm. This theory is useful in predicting the structures of non-transition element molecules.

The geometry of the molecule depends upon the no. of bonding and non-bonding electron pairs (lone pairs) on the central atom (in the valence shell) which arrange themselves in such a way that there is minimum repulsion between them so that the molecule has minimum energy and maximum stability. Since there can be only one orientation of orbitals corresponds to minimum energy, a molecule of a given substance has invariably a definite shape i.e. a definite geometry.

This theory considers

- indistinguishability of electron
- the operation of columbic force ($1/r^2$)
- the operation of Pauli repulsive forces (like spin electrons far apart as much as possible) i.e., $1/r^m$.

The spatial difference between bonding pairs and lone pairs are shown in Fig.2.25

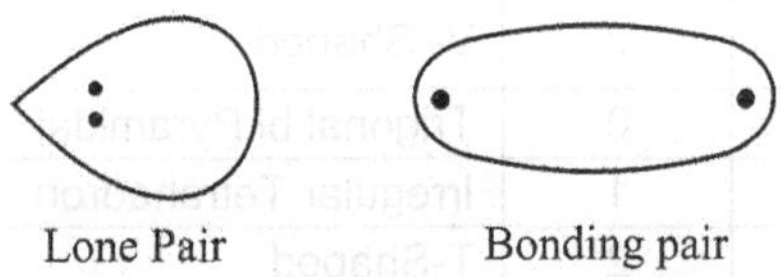

Fig. 2.25

The lone pair concerned to the one atom only and occupy larger volume and bond pair spread in between two atoms. The repulsion in between lone pair (L.P) and bond pair (B.P.) causes the differences in angles and shapes of the molecules. The repulsions increases in this order

B.P – B.P.< B.P. – L.P. < L.P. – L.P.

On this basis we can explain why the angle decreases from CH_4 to H_2O

CH_4	>	NH_3	>	H_2O
109°. 28'		107°.3'		104°.5'
All are B.P.		3B.P. and 1L.P.		2 B.P. and 2 L.P.

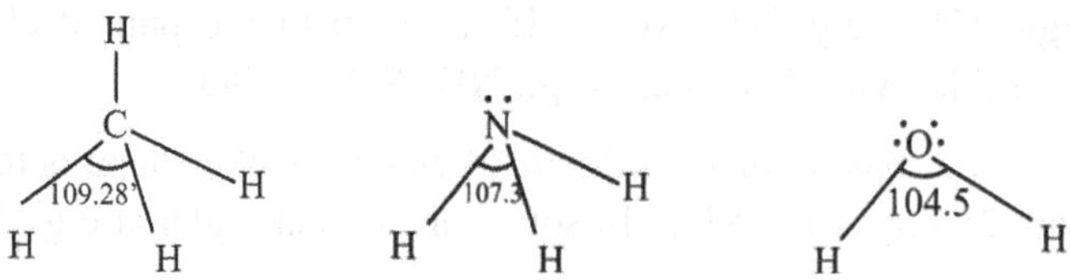

Fig. 2.26

So the shape of the molecule is effected by the presence of B.P. and L.P. number and repulsions on the central atom valence shell. Depending on the B.P. and L.P., and other rules as per VSEPR Theory, the geometries of non transmission elements are given in Table 2.5.

Table 2.5

S.No.	No. of electron Pair	No. of B.P.	No. of L.P.	Shape	Eg.,
1	2	2	0	Linear	$BeCl_2$, $HgCl_2$, Zn I_2, CO_2
2	3	3	0	Trigonal Planner	BCl_3, NO_3
3	3	2	1	V- Shaped	O_3, NO_2^-, Sn Cl_2
4	4	4	0	Tetrahedral	CH_4, ClO_4^-, Al_2Cl_6
5	4	3	1	Pyramidal	NF_3, H_3O^+, ClO_3^-
6	4	2	2	V- Shaped	H_2O, SCl_2, ClO_2^-
7	5	5	0	Trigonal bi Pyramidal	PCl_5, $PF_3(CH_3)_2$
8	5	4	1	Irregular Tetrahedron	SF_4, R_2TeCl_2
9	5	3	2	T-Shaped	ClF_3, $C_6G_5ICl_2$
10	5	2	3	Linear	ICl_2^-, XeF_2

Table 2.5 *Contd...*

S.No.	No. of electron Pair	No. of B.P.	No. of L.P.	Shape	Eg.,
11	6	6	0	Octahedral	SF_6, PCl_6^-, S_2F_{10}
12	6	5	1	Square pyramidal	BrF_5, $XeOF_4$
13	6	4	2	Square Planer	ICl_4^-, XeF_4
14	7	7	0	Pentagonal Bipyramid	IF_7^-
15	7	6	1	Irregular octahedron	$SbBr_6$, $Xe F_6$

(c) ***Hybridisation Theory***: This is another aspect to determine the geometry of molecule was discussed in 2.

2.17 Resonance

The term resonance refers to the fact that many molecules can not be represented by any single valence bond structure. But must be considered as intermediate between two or more structures, canonical structures, such molecules are called resonance hybrids. The molecule is not well represented by one structure during part of its life time, by another during another part, rather the structure is unique. But is most conveniently described by the mathematical expedient called resonance.

For e.g: CO_3 cannot be represented by any one of the structures 1 to 3 alone but is intermediate between all of them

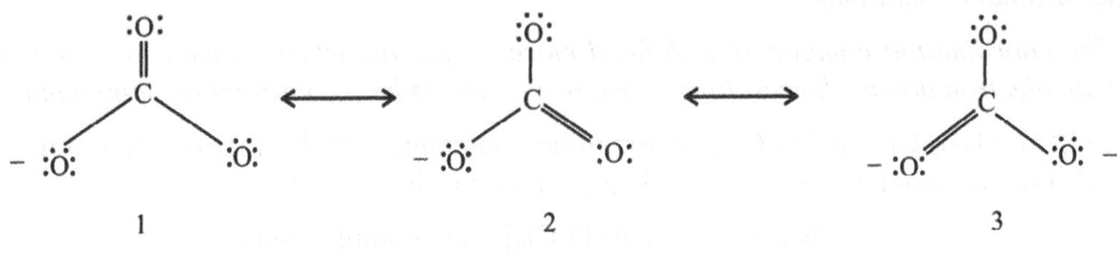

Fig. 2.27

The 4th bond i.e. π bond is not localized in any one of three bonds, but rather is delocalized overall three bonds. The bond order is 11/3. These 1, 2, 3 structures are cannonical structures and from one structure to another, the symbol to be given is double headed arrow. The other eg., O_3, C_6H_6, CO_2 etc. which can have resonance. The classic difficulty over the concept of resonance is the disposition regard the molecule as oscillating from one to another of the assigned forms.

If Ψ_1, Ψ_2, Ψ_3 .... are the various wave functions of the possible structures of a molecule contributing to its resonance hybrid (i.e. actual structure), the important ones will be those for which C is large (among possible structures) in the following equation,

$$\Psi = N(C_1\Psi_1 + C_2\Psi_2 + C_3\Psi_3 + ...) \qquad .....(2.32)$$

where Ψ = Wave function of actual structure of the molecule.

Resonance Energy: One of the most important point connected with resonance is that resonance hybrid is a more stable structure than any of the structures contributing to it. A molecule for which resonance is important is more stable than predicted from any one of the canonical structures. *The discrepancy between calculated and observed heat of formation is called resonance energy.* The greater the no. of significant contributing structures, the greater the resonance energy. For eg. CO the Δ_{cal} = 175 k.cal.mol $^{-1}$ and Δ_{obser} = 256 k.cal. mol^{-1}. So the resonance energy is 256 – 175 = 81 k. cal. mol^{-1}. Similarly for CO_2 the resonance energy is 383 – 346 = 37 k cal. mol^{-1}

2.18 Complexes

The coordination chemistry is largely transition metal chemistry. The first found complexe or coordination compound in 1753 is Prussian blue i.e. potassium hexacyanoferrate (III). In that century others found are K_2PtCl_6(1760-65), $[Co(NH_3)_6]$ Cl_3 (1798).

Any symmetrical arrangement of atoms about a central atom, in construction of a molecule, may be designated as coordination compound. A complex ion may be defined as an electrically charged radical or group of atoms such as $[Co(NH_3)_6]^{3+}$, *which has been formed by the direct union of simple ions or by the direct union of a simple central ion with neutral molecules and it is incapable of dissociation into their component parts under ordinary conditions.*

The coordination compound is defined more simply as, when a metal ion combines with an electron donar, the resulting substance is said to be a coordination compound.

E.g. $[Co(NH_3)_6]Cl_3$, $K_2[Pt\ Cl_6]$ coordination compounds. In the first example nitrogen in NH_3 is donar atom and in the second eg., chloride is donar atom.

$[Co(NH_3)]^{3+}$ and $[Pt\ Cl_6]^{2-}$ are complex ions.

In the complex ion, the neutral molecules or ions are attached to the central metal ion. They are named as ligands (ligand Latin name Ligare means to bind). The ligands may be anionic (Cl^-, SCN^-, NO_3^- ,CN^-), neutral (NH_3, Co, H_2O) or less frequently by cationic ($N_2H_5^+$). *The number of ligands (donar atoms) bond directly to the central metal ion defines the coordination number.* The coordination numbers vary from 2 to 9.

The presence of complex ions in solution can usually be demonstrated by a difference in chemical and /or physical properties between the complex ion and its component parts. To represent the complex ion it is put within the square brackets.

The nomenclature IUPAC different for coordination compounds from ordinary compounds and these compounds show structural isomerism and stereoisomerism.

The formation of these compounds i.e. even though the valency is satisfied in the formation of compounds why the metal ion still combining with other molecules is not known till 1900.

Alfred Werner interpreted the bonding in such type of compounds for which he was awarded Noble prize.

Werner's Theory: According to Werner

(i) Even when, to judge by the valence number, the combining power of metal is exhausted, they still possesses in most cases the power of combining further.

(ii) The metal ion possesses two types of valencies – Primary valency and the secondary valency.

(iii) There are two spheres of attraction- outer or isonisation sphere satisfied by primary valence and inner or coordination sphere satisfied by auxiliary or secondary valence

(iv) To distinguish between two spheres of attraction, Werner introduced square brackets[ ] for the inner sphere.

(v) Metal ion has fixed number of secondary valence which he named as coordination number.

(vi) The groups attached to the central ion through secondary valence are arranged in such a way to give spatial arrangement.

On the basis of his theory Werner assigned the formula to $CoCl_3$ $6NH_3$ as $[Co(NH_3)_6]Cl_3$, where Cl is satisfied by primary valence (outer sphere) and NH_3 by secondary valence. The complex ion $[Co(NH_3)_6]^3$ and the coordination number of Co^{3+} is 6. The structural formula is shown in Fig.2.28.

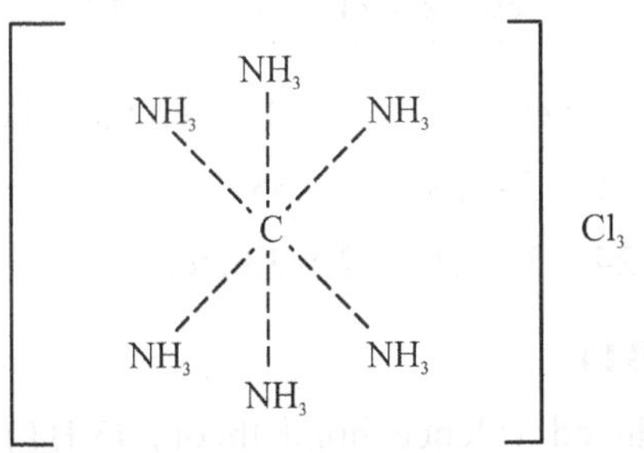

Fig. 2.28

Similarly for other compounds on the basis of experimental evidences like conductivity, freezing points, gravimetry, colour he had given the formulas.

$COCl_3$ $5NH_3$ as $[CO(NH_3)_5Cl]Cl_2$

$CoCl_3$ $4NH_3$ as $[Co(NH_3)_4Cl_2]Cl$. It also shown cis-trans isomerism

$CoCl_3$ $3NH_3$ as $[Co(NH_3)_3 Cl_3]$

There is no theoretical justification for Werners two types of valence till the development of electronic theory of valence and also it does not explain why the coordination no. is fixed for metals and what is the bonding type of secondary valence.

Sidgwick's Theory

N.G. Sidgwick solved this problem to some extent. The bond between central ion and the ligand is the coordinate bond (semipolar bond) in which a pair of electrons are donated by the ligands (M←L). Central ion is Lewis acid and ligand is Lewis base. He also introduced the concept of Effective Atomic Number (EAN) rule for knowing the coordination no. of central ion i.e. it is four or six etc. The metal ion in a compound further reacts with ligands to attain stability. The sum of electrons on the metal ion plus the electrons donated from the ligands was called the effective atomic number. The complexes obeying the EAN rule are given in table 2.6

Table 2.6

S.No	Complex	No. of electrons on M^{nt}	No. Electrons from ligands	EAN
1	$[Co(NH_3)_6]^{3+}$	27 – 3 = 24	12	24 + 12 = 36 (Kr)
2	$[Pt(NH_3)_6]^{4+}$	78 – 4 = 74	12	74 + 12 86 (Rn)
3	$[C(NH_3)_4]^{2+}$	48 – 2 = 46	8	46 + 8 = 54 (Xe)
4	$[Fe(CN)_6]^{4-}$	26 – 2 = 24	12	24 + 12 = 36 (Kr)

Some complexes may not obey this EAN rule even then they are stable.

For e.g. $[Fe(CN)_6]^{3-}$ 26 – 3 = 23 +12 = 35

$[Cr(NH_3)_6]^{3+}$ 24 – 3 = 21 + 12 = 33 etc.

Valence Bond Theory (VBT)

- Later Pauling introduced valence bond theory (VBT) to explain the bonding in complexes. The assumptions are:
- the central metal atom must make available a no of orbitals equal to its coordination number for the formation of σ covalent bonds,
- the original central metal AO's hybridise to form a new set of equivalent bonding orbitals with definite directional characteristics (responsible for shape and geometry of complexes).

A covalent σ bond arises from the overlap of a vacant metal orbital and a filled σ orbital of the donar atom of the ligand with lone pair of electrons., Later VBT introduced the concept of back-bonding i.e. π bonding (M→L) which strengthens the σ bond became of charge distribution between metal and ligand.

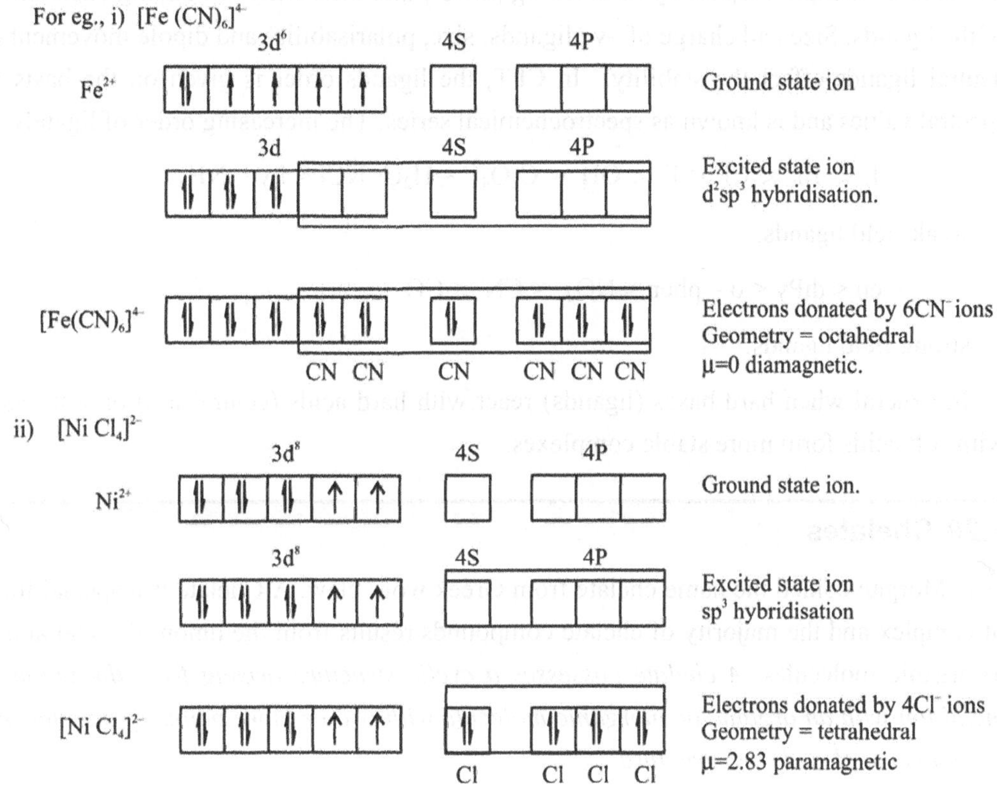

Fig. 2.29

Depending on the ligand type in the d orbitals, the pairing of electrons or no pairing of electrons takes place. So ligands are categorized as weak field or high spin or outer orbital complexes, $[Ni\ Cl_4]^{2-}$ and as strong field or low spin or inner orbital complexes, $[Fe(CN)_6]^{4-}$.

Later crystal field theory (CFT), Molecular orbital theory (MOT) and Ligand field Theory (LF) came from time to time to explain the bonding in complexes in a more reasonable manner.

2.19 Ligands

Ligands are generally anions or neutral molecules and rarely cations. Nature of the ligands will effect the stability of complexes. Stability of the complex increases with the Lewis base strength (capability of donating power) and increasing π bonding capabilities of the ligands. Size and charge of –ve ligands; size, polarisability and dipole movement of neutral ligands effect the stability. In CFT, the ligands order is given on the basis of spectral values and is known as spectrochemical series. The increasing order of ligands is

$$I^- < Br^- < Cl^- < F^- < OH^- < C_2O_4^{2-} \sim H_2O < NC_5 < Py \sim NH_3$$

weak field ligands.

$$< en < diPy < o-phen < NO_2^- < CN^- < CO$$

strong field ligands.

In general when hard bases (ligands) react with hard acids (central ion) or soft bases with soft acids form more stable complexes.

2.20 Chelates

G.T. Morgan coined the name chelate from Greek word claw. A Chelate is a special form of complex and the majority of chelate compounds results from the union of metal atoms to organic molecules. *A chelate possesses a cyclic structure arising from the union of metal ion with an organic or inorganic molecule with two or more points of attachments to produce a closed ring structure.*

If the ligands attached to the metal ion by donating only one pair of electrons, then it is named as monodentate ligand. If same ligand donate two lone pairs of electrons, then it is named as bidentate (two-toothed) ligand, three lone pair of electrons as tridentate (three-toothed), four lone pairs of electrons as quadridentate or tetradentate (four-toothed) and so on. Multi dentate lingands (chelates) form more stable complexes, than the monodentate lingands provided the donating atom is same in both. For e.g, β-value of $[Ni(NH_3)_4]^{2+}$ is 7.79, where as $[Ni(en)_2]^{2+}$ is 14.5 and $[Ni(penten)]^{2+}$ is 19.3, where pentene is tetrakis (aminoethyl) ethylene diammine (tetradentate). Some common polydentate ligands are listed in Table 2.7

Table 2.7 Common Polydentate Ligands.

S.No	Dentate	No. of membered rings	Example
1.	Bidentate	4	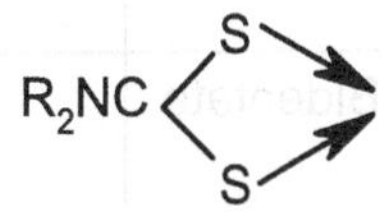 dithiocarbomate
2.	Bidentate	4	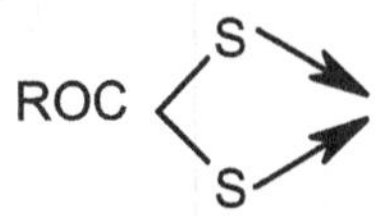 Xanthates
3.	Bidentate	4	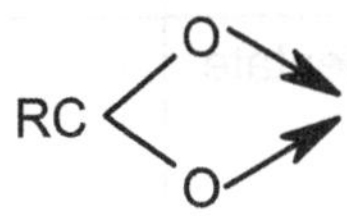 carboxylate
4.	Bidentate	5	Ethylenediamine, (en) 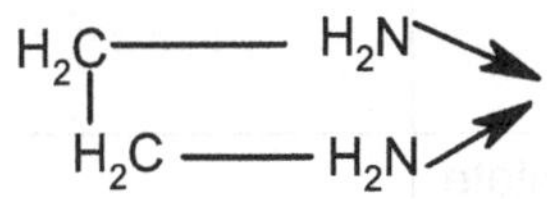
5.	Bidentate	5	2, 2-bipyridine, bipy 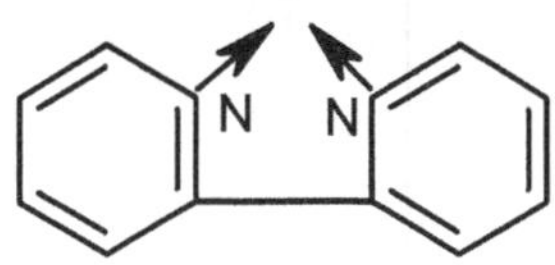
6.	Bidentate	5	1, 10-phenanthroline, (phen) 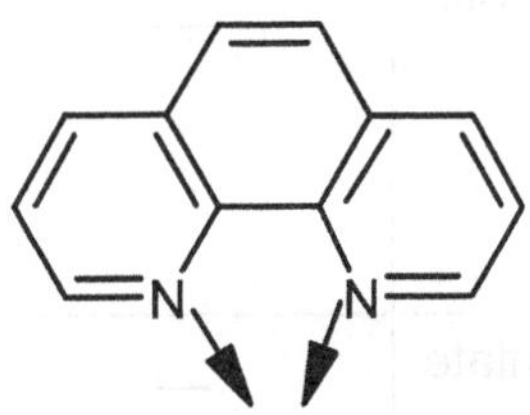
7.	Bidentate	5	0-Phenelene bis-(dimethylarsine), (diars) 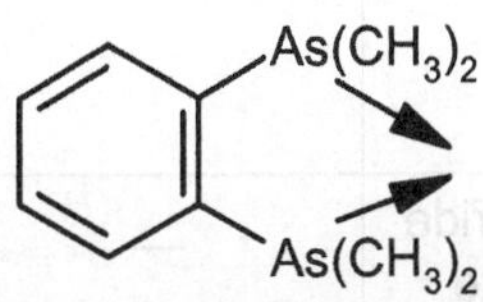

8.	Bidentate	5	1, 2-bis-(diphenylphosphino) ethane, (diphos) 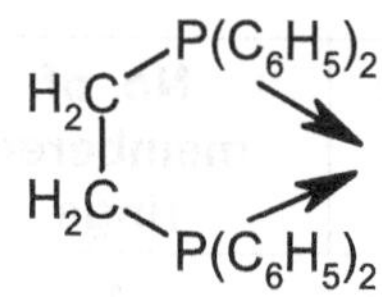
9.	Bidentate	6	Acetylacetonate, (acac) 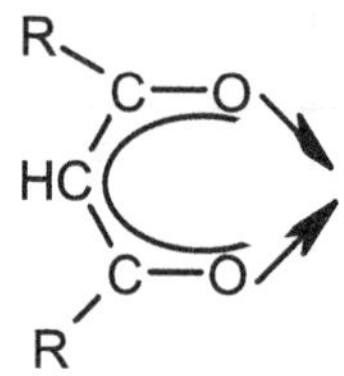
10.	Bidentate	6	Salicylaldiminato, (sal) 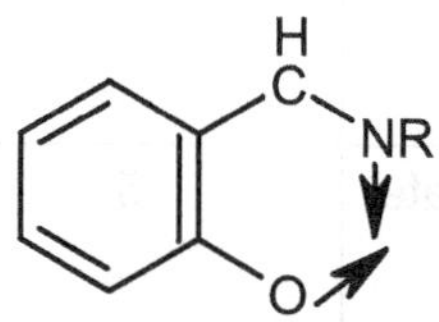
11.	Tridentate	—	Terpyridine, (terpy) 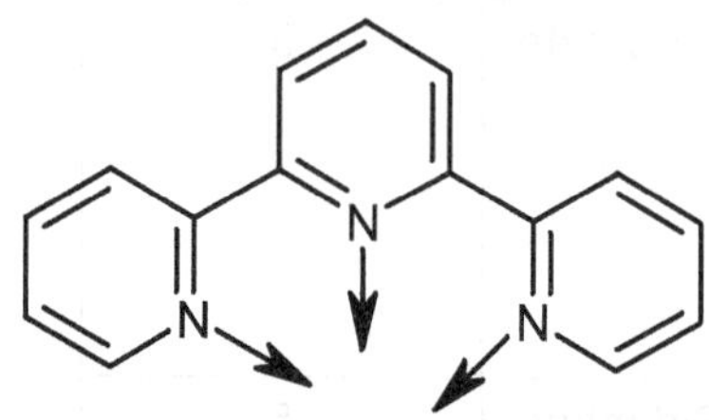
12.	Tridentate	—	Diethylenetriammine, (dien) 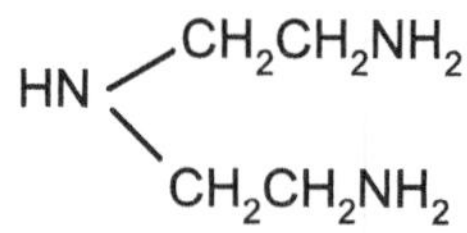
13.	Tridentate	—	bis-(3-dimethylarsinylpropyl methylarsine, (triars) 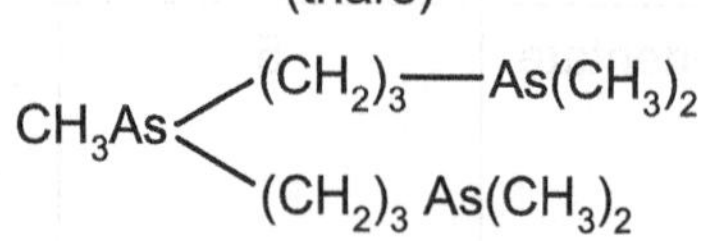
14.	Quadridentate	—	Trithylenetetrammine, (trien) $H_2N(CH_2)_2NH(CH_2)_2NH_2$

15.	Quadridentate	—	Schiff base derived from acac. 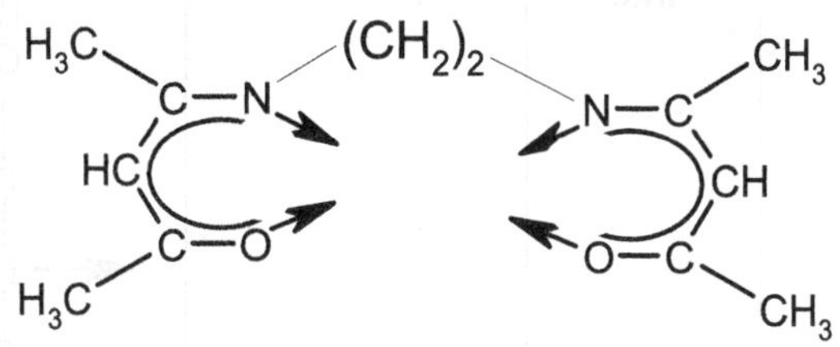
16.	Quadridentate	—	Porphyrins 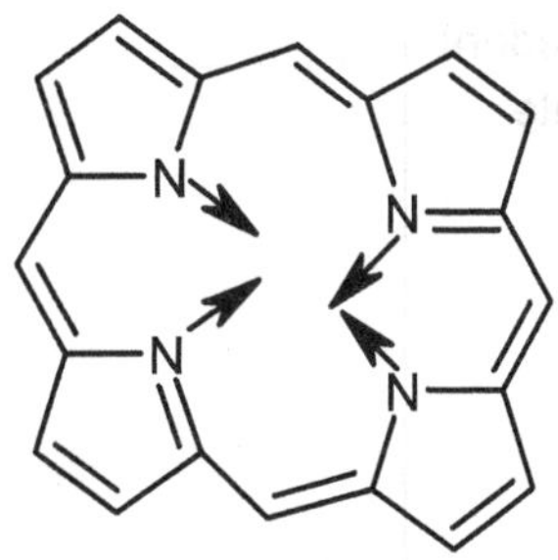
17.	Quadridentate	—	Tetrapyridyl 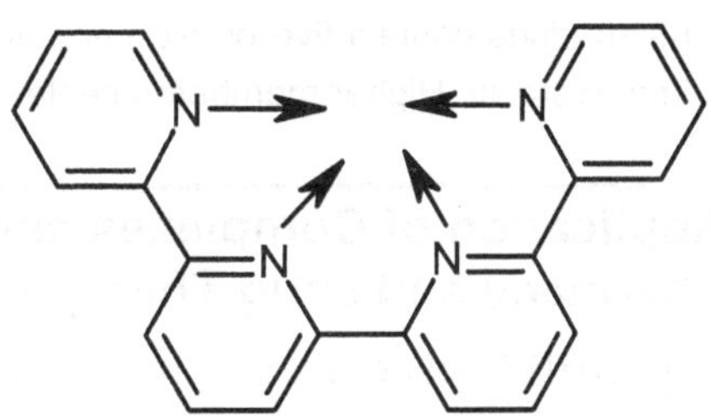
18.	Quadridentate	—	Tris (o-diphenylarsenephenyl)-arsine, (QAS) 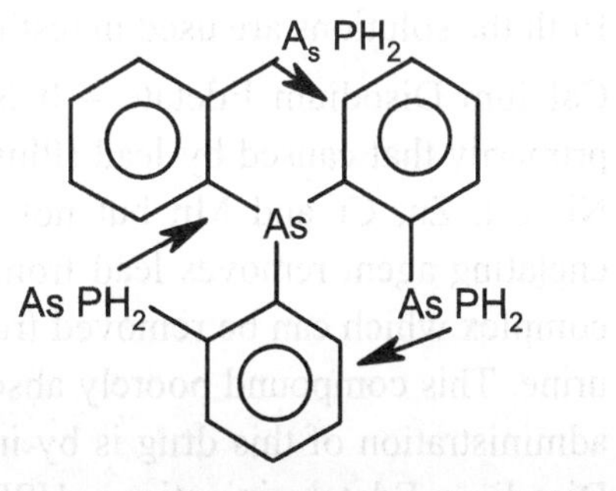

19.	Pentadentate	—	Ethylenediaminetetraacetic acid, (EDTAH3-) 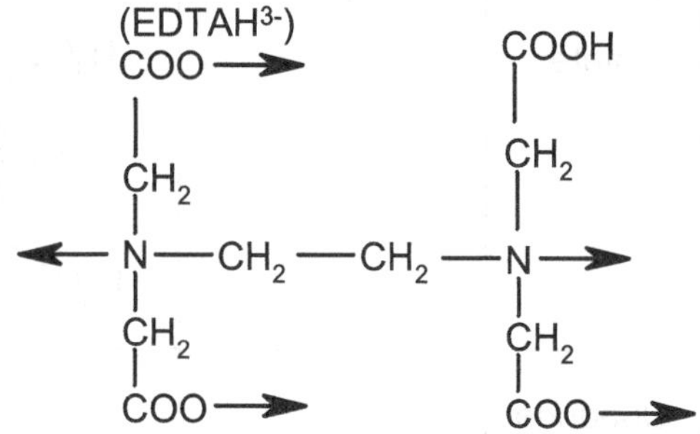
20.	Hexadentate	—	ethylenediaminetetracetate (EDTAH4-) 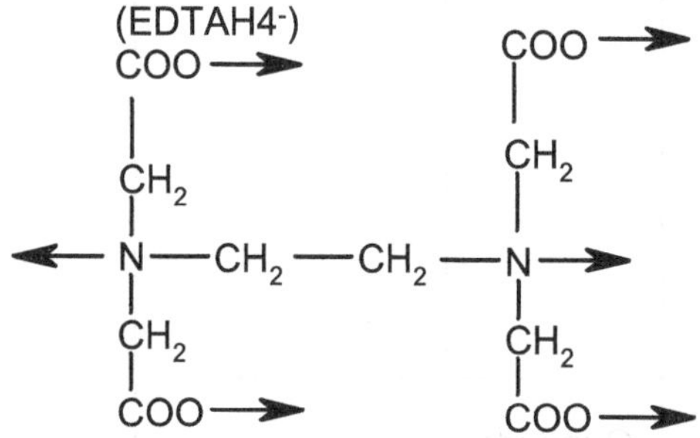

Usually chelate rings contain five or six members including the metal ion which results in reduced strain and more stable. Higher membered chelate rings are less stable

2.21 Application of Complexes and Complexing Agents in Pharmacy and Drug Therapy

(i) Copper(II) Chelates are employed for identifying reducing substances (e.g. sugars)

Benedicts solution – Copper(II) is chelated with citric acid.

Fehlings solution - Copper(II) is chelated with tartaric acid.

Both the solutions are used in testing sugar.

(ii) Calcium Disodium Edetate – It is used in the treatment heavy metal poisoning, primarily that caused by lead (Plumbism). Also used in the poisonings due to Cu, Ni, Cd, Zn, Cr and Mn but not useful in the toxicity of Hg, As and Au. This chelating agent removes lead from the tissues by forming an inactive soluble lead complex which can be removed from the circulation by the kidneys and excreted in urine. This compound poorely absorb lead from gastrointestinal track. The route of administration of this drug is by intravenous (I.V.) injection. The official Calcium Disodium Edetate injections, USP XVIII contains 180 mg-220 mg, in each ml.

Intramuscular injection (IM) administration is employed sometimes in diagnosis of metal poisoning. Excretion of Pb in the urine (500 μg L/24 hours or greater) is indicative of toxicity.

The Calcium Disodium Edetate. USP XVIII (Calcium Disodium Versenate) The name is calcium disodium ethylenediamine tetraacetate in the complex and the formula is $C_{10}H_{12}CaN_2Na_2O_8$ x H_2O and anhydrous complex mol. wt is 374.28. It is a white crystalline powder and stable in air. Freely soluble in water and the pH of aqueous solution is 6.5 to 8.0.

The Structure is

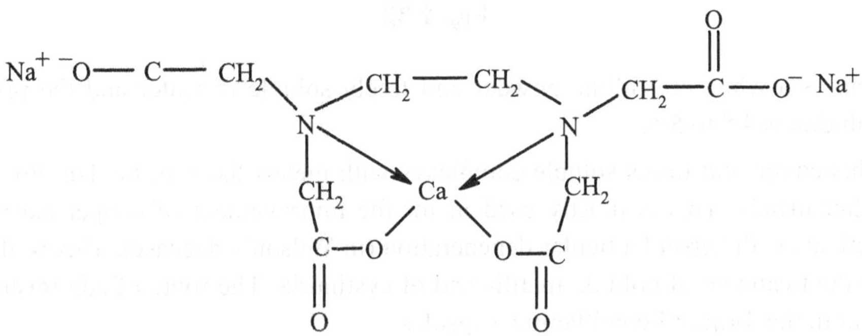

Fig. 2.30

(iii) Dimercaprol USP XVIII (2,3 – Dimercapto – 1- propanol: Britishanti lewisite, (BAL). The formula is $C_3H_8OS_2$ and the mol. wt. is 124.22. This is a chelating agent and is colourless liquid and is soluble in water, alcohol and benzoate,

The structure is

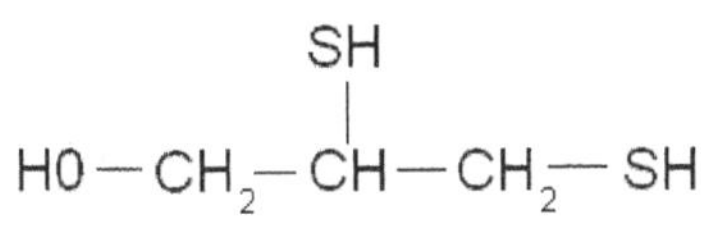

Fig. 2.31

This is used in the treatment of arsenic poisoning from other sources as an effective neutralising agent. This is also used for Hg and Au poisoning. It forms stable mercaptide of metals which are excreted in urine.

The administrative route is by intramuscular injection i.e. a solution of dimercaprol (100 mg in each ml) in a mixture of benzyl benzyte and vegetable oil.

(iv) Pencillanine, USP XVIII. This is a chelating agent, the name is β, β. Dimethyl cystine or D-(–) –3- Mercaptovaline. The formula is $C_5 H_{11} NO_2S$ and the mol. wt is 149.21.

The structure is

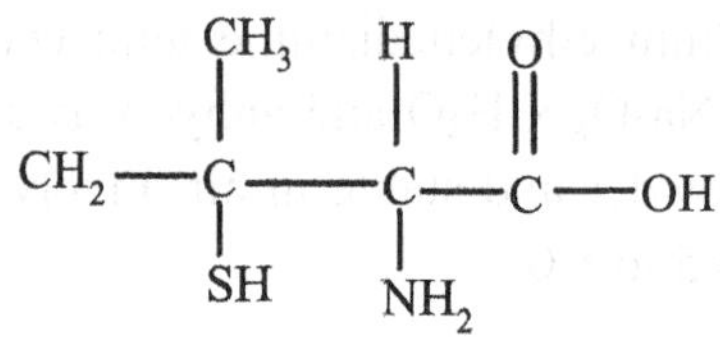

Fig. 2.32

This is a white crystalline powder and freely soluble in water and the pH of aq. solution is 4.5 to 5.5.

This compound forms soluble complexes with metals like Cu, Fe, Hg, Pb, Au and other metals. This is mostly used in for the improvement of copper excretion in patients with hepatolenticular degeneration (in Wilson's decease). This is also used in the treatment of gold dermatitis and of cystinuria. The route of administration is oral in the form of Penicillamine capsules

(v) Disodium Edetate USP XVIII. This is a chelating agent and the name is disodium ethylenediamine tetraacetate, the formula is $C_{10}H_{14}N_2Na_2O_82H_2O$ and the mol. wt. is 372.24. The structure is

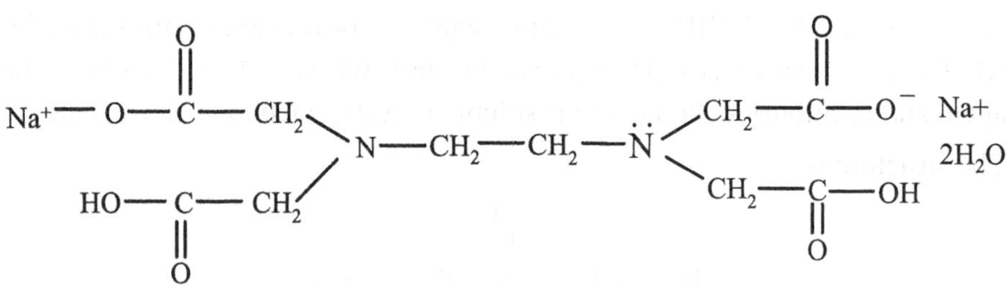

Fig.2.33

This compound is white crystalline powder, is soluble in water and the pH of aq. solution is 4.0 to 6.0. The compound is used in the treatment of occlusive vascular disease, cordiac arrhythmias. The administrate route is intravenous injection.

(vi) Complexalism of some chelating agents plays an important role in analylitical chemistry for the determination of methods by titration.

(vii) Excess in take of iron cause various problems known as siderosin. The Desferrioxamine B (Deferoxamine Mesylate) is the methylsulfuric acid salt. Deferoxamine is produced naturally by streptomyces pilosus as a Fe(III) Complex. After chemical removal of the iron, the merylate salt is the chelating agent. It is water soluble and form octahedral complexes with Fe(III) ions. The structure is

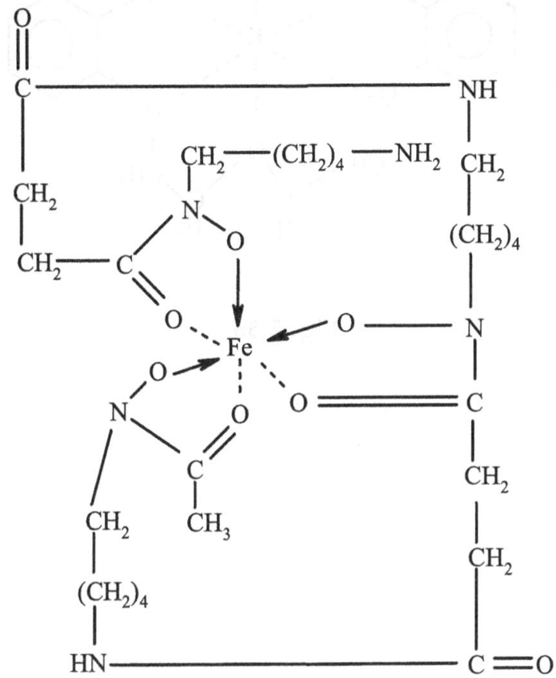

Fig. 2.34

This drug is used for African siderosis. The administration is through either IM or IV.

(viii) Vitamins B-12 (Cyanocobalamin). This compound is the complex and the chelateing agent is corrin group and the metal ion is cobalt(III). This is the commercial therapeutic product and is employed for pernicious anemia.

(ix) Macrocyclic complexes formed by polyethers with alkali metals and they are named as crown ethers. Poly ethers of this type act as ion carriers inside living cells to transport ions across cell membranes and thus maintain the balance between Na^+ and K^+ inside and outside cells.

E.g. K^+ (dibebzo – 18- crown 6). The structure is

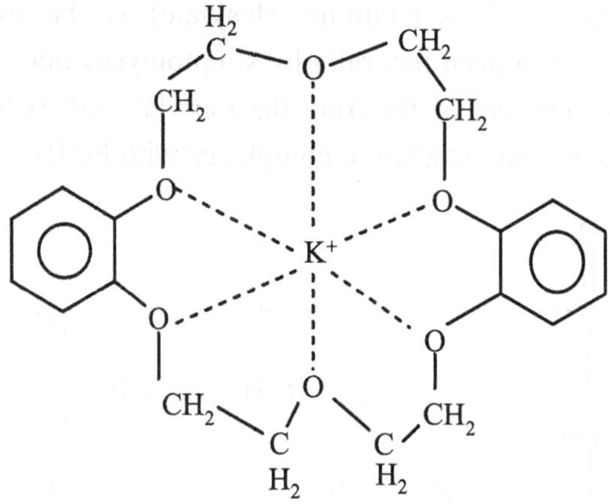

Fig. 2.35

CHAPTER 3

Solutions

3.1 Introduction

A solution can be defined as a homogeneous mixture of two or more substances, the composition of which may vary.

It consists of two components, solute and solvent.

Solvent – The substance which is present in larger proportion.

Solute – The substance which is present in smaller proportion.

Depending on the number of constituents, a solution may be referred to as :

1. Binary solutions (Two components)
2. Teretiary solution (three components)
3. Quaternary solution (four components) etc.

3.2 Types of Solutions

A solution may exist in solid, liquid or gaseous state. Depending on the physical state of the solute and solvent, there are nine types of solutions. Different types are given in Table 3.1

Table 3.1

S. No.	Solute	Solvent	Solution	Example
1	Gas	Gas	Gases in gases	Mixture of Gases (air)
2	Gas	Liquid	Gases in liquids	Acreated water (CO_2 in water under pressure)
3	Gas	Solid	Gases in solids	Gas adsorbed by metals or minerals H_2 in palladium

Table 3.1 *Contd....*

S. No.	Solute	Solvent	Solution	Example
4	Liquid	Gas	Liquids in gases	Moist air
5	Liquid	Liquid	Liquids in liquids	Alcohol in water
6	Liquid	Solid	Liquids in solids	Mercury in Zinc (Zinc amalgam)
7	Solid	Gas	Solids in gases	Camphor in air
8	Solid	Liquid	Solids in liquids	Salt in water
9	Solid	Solid	Solids in solids	Alloys Brass, Bronze etc.,

A solution is always in the same physical state as the solvent. The solutions of solids in liquids and liquids in liquids are important.

Aqueous solutions — Solutions of substances in water solvent.

Non-aqueous solutions — Solutions of substances in other solvents.

Amalgam — Solutions metals in mercury, E.g., Sodium amalgam, Zinc amalgam etc.

True solution: When alcohol is added to water, we get at equilibrium, a single homogeneous single phase system composed of two different molecular species which are uniformly and intimately mixed with each other. Such a system constitutes a true solution.

3.3 Expression of Concentration of Solutions

The concentration of a solution, depending on the relative amounts of solute and solvent, could be expressed by using different terms.

1. ***Molarity (M):*** *Molarity is defined as the number of moles (gram-molecular weight) of the solute present in $1dm^3$ (1 L or 10^{-3} m^3) of the solution.*

   For e.g., 0.1M means when 0.1 mole of a solute present in one cubic decimeter of the solution

   $$Molarity = \frac{Number\ of\ moles\ of\ the\ solute}{Volume\ of\ the\ solution\ in\ dm^3} \quad .....(3.1)$$

   Volume of the solution should be expressed in dm^3 for expressing the concentration in terms of molarity.

   40 g of NaOH dissolved in 1000 ml of solution, then it is 1M 0.4 g of NaOH dissolved in 1000 ml of solution, then it is 0.1M.

Problem 1: Calculate the molarity of NaOH if 375 ml of solution contains 5.50 g of NaOH.

Ans: No. of milli moles of the solute $= \frac{5.5}{0.04} = 137.45$

$$\therefore M = \frac{No.\ of\ millimoles\ of\ solute}{Volume\ of\ the\ solution\ in\ ml} = \frac{137.45}{375} = 0.3667$$

2. ***Molality (m):*** Molality is defined as number of moles (gram-molecules) of a solute present in kilogram of solvent. When one mole of a solute is dissolved in one kilogram of water, the concentration of the solution is one molal (1 m)

$$Molality(m) = \frac{No.\ of\ moles\ of\ the\ solute}{Mass\ of\ the\ solvent\ in\ kg} \quad .....(3.2)$$

3. ***Normality (N):*** The number of gram-equivalents (equivalent weight) of a solute present in 1 dm^3 of the solution is called its normality.

   A one normal solution contains one gram equivalent of a solute in 1 dm^3 (one litre) of solution and is denoted by 1N.

   The equivalent weight depends on the nature of reactions.

   For e.g., the equivalent weight of monobasic acids like HCl, HNO_3, CH_3COOH etc. is identical with its molecular weight.

   The equivalent weight of dibasic acids like H_2SO_4, $H_2C_2O_4$ is ½ of mol. wt. Equivalent weight of acid is that weight of it which contains one gram of replaceable hydrogen i.e. 1.008 gm of H and that of base is that weight of it which contains one replaceable hydroxyl group i.e. 17.008 gm of ionisable OH^- is equivalent to 1.008 gm of H. The normalities of common acids and ammonia are given in Table 3.2

Table 3.2 Normalities of common acids.

S. No.	Name	Normality	Sp. gr	% by wt.	Vol. required to prepare 1N Solution (ml)
1	Hydrochloric acid	11.3	1.18	35	89
2	Nitric acid	16.0	1.42	70	63
3	Sulphuric acid	36.0	1.84	96	28
4	Phoshoric acid	41.1	1.69	85	23
5	Acetic acid	17.4	1.05	99.5	58
6	Aq. amomonia	14.3	0.90	27	71

For salts formed from reaction of acid with base, the equivalent weight depends on the reaction

$$Na_2\ CO_3 + 2HCl \rightarrow H_2CO_3 + 2NaCl. \quad .....(3.3)$$

Sodium carbonate reaction with two molecules of HCl to form 2 molecules of NaCl, hence the equivalent weight is ½ mol. wt. for Na_2CO_3.

Equivalent weight of oxidation reduction reactions: The equivalent weight of an oxidizing or reducing agent is most simply defined as that weight of the reagent which reacts with or contains1.008 gm of available H or 8.00 gm of available oxygen. The partial ionic equations also used in the calculation of the equivalent weights of oxidizing and reducing agents.

For e.g.,

$KMnO_4$ (acid): ionic equation is

$$MnO_4^- + 8H^+ + 5e^- \rightleftharpoons Mn^{2+} + 4H_2O \quad .....(3.4)$$

$$\text{Eq. wt} = \frac{\text{Formula weight}}{5} \quad .....(3.5)$$

$K_2Cr_2O_7$: ionic equation is

$$Cr^2O_7^{2-} + 14H^+ + 6e^- \rightleftharpoons 2Cr^{3+} + 7H_2O \quad .....(3.6)$$

$$\text{Eq. wt} = \frac{\text{Formula weight}}{6} \quad .....(3.7)$$

H_2S: ionic equation is

$$H_2S \rightleftharpoons 2H^+ + S + 2e^- \quad .....(3.8)$$

$$Eq.wt = \frac{\text{Formula weight}}{2} \quad .....(3.9)$$

$H_2C_2O_4$: ionic equation is

$$C_2O_4^{2-} \rightleftharpoons 2CO_2 + 2e^- \quad .....(3.10)$$

$$\text{Eq. wt.} = \frac{\text{Formula weight}}{2} \quad .....(3.11)$$

Milliequivalent weight is one thousandth of the equivalent weight. The gram milliequivalent weight is the milliequivalent weight expressed in grams. A normal solution containing 1 gm eq.wt ml^{-1}. Hence the number of gram milliequivalent weights present in a solution can be found from

Number of gram-milliequivalent weight

$$= \text{number of millilitres} \times \text{normality} \quad .....(3.12)$$

For e.g., 2.00 ml of 6.00N HCl contains

$$\text{no. of meq.wt} = 2 \times 6 = 12$$

4. ***Mole fraction (x):*** The mole fraction of a solute in a solution is the ratio of the number of moles of a solute to the total no. of moles of the solute and solvent in a solution.

For e.g., If n_2 mole of a solute is dissolved in n_1 mole of a solvent, the mole fraction of the solvent and the solute are given by

$$\text{Mole fraction of the solvent } x_1 = \frac{n_1}{n_1 + n_2} \qquad .....(3.13)$$

$$\text{Mole fraction of the solute } x_2 = \frac{n_2}{n_1 + n_2} \qquad .....(3.14)$$

The sum of the mole fraction of all the components in a solution is always equal to one.

5. ***Percentage:*** In terms of percentage, the concentration of a solution may be expressed in four different ways

   10 ml of alcohol present in 100 ml of solution = 10 % (v /v)
   10 g NaCl present in 100 ml of solution = 10 % (w/v)
   10 ml of alcohol present in 100 g of solution = 10 % (v/w)
   10 g of NaCl present in 100 g of solution = 10 % (w/w)

Unless specified, a 10 % solution may always taken as weight by weight (w/w)

6. ***Parts per million (ppm):*** When a solute present in the solution is in very minute amounts, the concentration is usually expressed in parts per million.

For e.g., The amount of oxygen dissolved in sea water is 5.8 g in 10^6 (1 million) of sea water. It means 5.8 parts of oxygen are present in one million parts of sea water. Hence the concentration of oxygen in sea water is 5.8 ppm.

$$1 \text{ ppm} = \frac{\text{Mass of solute(g)}}{\text{Total mass of solution (g)}} \times 10^6 \qquad .....(3.15)$$

Parts per billion (ppb)

$$1\text{ppb} = \frac{\text{Mass of solute (g)}}{\text{Total mass of solution}} \times 10^9 \qquad .....(3.16)$$

Presently instead of ppm, mg per litre is using i.e. milligrams of solute present in one litre of solution.

This ppm or mg per litre is mostly used in expressing pollutants in the environment. All concentration units, except mole fraction vary with temperature.

7. ***Specific Gravity:*** The specific gravity (density) of the solution of a single solute is a measure of the concentration of the solute in solution (percentage by weight composition).

For e.g.,: HCl specific gravity 1.12 means 23.8 g of hydrogen chloride in 100 g of solution

Conc. HNO_3 Sp.gr. 1.42 i.e. 70 % HNO_3

Conc. H_2SO_4 Sp gr 1.84 i.e. 95.5 % H_2SO_4

Sp. Gr × % of solute present = concentration of solute present in 1 ml of solution.

$$1.84 \times 0.955 \simeq 1.76 \text{ g } H_2SO_4 \text{ ml}^{-1}$$

So Molarity is 18M and Normality is 36 N

The calculation is difficult and specific gravities of common reagents are found in the hand books.

Problem 2: What volume of 0.6380M KOH solution will neutralize 430.0 ml of 0.4000M sulfuric acid.

Ans.: 1 mole of H_2SO_4 = 2 moles of KOH

430 ml of 0.4000M solution contains $= \frac{430}{1000} \times 0.4000$

$= 0.1720$ moles of H_2SO_4

0.1720 mole of H_2SO_4 = 0.3440 mole of KOH

1 ml of KOH solution contains 0.0006380 mole KOH

$\therefore$ Volume of KOH required $= \frac{0.3440}{0.0006380} = 539.2$ ml

Problem 3: What is the approximate molarity and normality of a 13.0 % solution of H_2SO_4? To what volume should 100 ml of the acid be diluted in order to prepare a 1.50N Solution?

Ans.: Specific gravity of H_2SO_4 acid is = 1.090

1 litre weighs 1090 g and contains = 1090 × 0.130

= 142 g H_2SO_4

1 mole of H_2SO_4 = 98.08 g

Molarity of solution $= \frac{142}{98.08} = 1.45$M

1 g equivalent H_2SO_4 = 98.08/2 = 49.04 g

Normality of solution $= \frac{142}{49.04} = 2.90$N

100 ml contains 290 meq. H_2SO_4

After dilution, x ml of 1.50N contains 290 meq.

$x \times 1.50 = 290$

Volume of acid diluted (x) = 193 ml

i.e., the dilution is made by adding 93 ml of water.

Problem 4: A solution contains 3.30 g of $Na_2CO_3.\ 10H_2O$ in each 15 ml. What is its normality and molarity?

Ans: Mol. wt of $Na_2CO_3.\ 10H_2O = 286$

$$\text{Eq. wt. of } Na_2CO_3.\ 10H_2O = \frac{286}{2} = 143$$

meq. Wt =0 143.

$$\text{Solution contains } Na_2CO_3.\ 10H_2O = \frac{3.30}{15.0} = 0.220 \text{ g ml}^{-1}$$

A normal solution contain = 0.143 g ml^{-1}

$$\text{Normality} = \frac{0.220}{0.143} = 1.54\text{N}$$

$$\text{Molarity} = \frac{0.220}{0.286} = 0.\ 77\text{M}$$

3.4 Solutions of Solids in Liquids (Solubility)

The solubility of a substance is defined as the maximum weight of the substance which will dissolve in 100 g of solvent at a given temperature or the volume of solvent required to dissolve 1 g of the compound.

A saturated solution is one which has dissolved all the solute it is capable of holding at a given temperature to form a homogeneous mixture. The concentration of the solute in this solution is termed as solubility of solute in the solvent at that temperature.

For E.g., **Saline solution** – NaCl in water

Boric acid solution – Boric acid in water.

Saturated solution – ionic product = solubility product

Unsaturated solution – ionic product > solubility product

Super saturated solution – ionic product < solubility product.

Why a solid dissolves in a liquid cannot be said definitely but it depends on the temperature, pressure, particle size of the solute.

The Methods used for the determination of solubility of sparingly soluble salts are:

(a) ***Conductivity method:*** The solubility may readily be obtained from conductivity measurements when a saturated solution of the salt is so dilute that complete ionization may be assumed.

The equivalent conductivity $\wedge$ at V dilution is given by

$$\wedge = KV \quad .....(3.17)$$

where K = specific conductivity using Kohlrausch's law, the solubility can be determined.

(b) ***Radioactive method (Paneth):*** The solubility can be determined by adding radioactive isotope to the sparingly soluble salt, precipitation, and measuring radioactivity

(c) Electromotive force (EMF) method.

(d) ***Colorimetric method:*** The principle of this method is to compare the depth of the colour of a solution with that of some standard solution.

Theoritical principles of Reaction in Solution

Many of the reactions take place in solution. The solvent is most commonly water but other liquids may also be used. So general knowledge, of the conditions which exist in solutions and the factors which influence chemical reaction are to learn.

(a) ***Chemical equilibrium:*** The equilibrium mixture remains constant provided the temperature remains constant. The composition of a given equilibrium system can be altered by changing conditions like – temperature, pressure and concentration of components.

(i) *Lechatlier-Braun Principle:* Any alteration in the factors which determine equilibrium causes the equilibrium to become displaced in such a way as to oppose as far as possible, the effect of alteration.

From this theorem it can be concluded for the reaction

$$2NH_3 \rightleftharpoons N_2 + 3H_2 \quad .....(3.18)$$

- at a given pressure or volume an increase in temperature increases the degree of dissociation of ammonia,
- at constant temperature an increase in pressure will diminish the degree of dissociation,
- an increase in concentration of H_2 will diminish that of N_2 and increase in concentration of NH_3,

This theorem is applicable only when the state of equilibrium is reached.

(ii) *Law of mass action (Guldberg and Waage 1967):* The driving force (which determines the rate of chemical reaction) or the rate of chemical reaction, proportional to the active masses of the reacting substances which present at that time. Active mass is the molar concentration of the reacting substances.

Consider a reversible reaction

$$\underset{\text{Reactants}}{L + M} \rightleftharpoons \underset{\text{Reaction products (resultants)}}{G + H} \qquad .....(3.19)$$

Velocity with which L and M react is

$$V_1 = K_1 \times [L] \times [M] \qquad .....(3.20)$$

Where K_1 = constant (velocity coefficient)

Square brackets denote molecular concentration.

For reverse reaction:

$$V_2 = K_2 \times [G] \times [H] \qquad .....(3.21)$$

At equilibrium $V_1 = V_2$

$$K_1 \times [L] \times [M] = K_2 \times [G] \times [H] \qquad .....(3.22)$$

$$K = K_1 / K_2 = \frac{[G]\ [H]}{[L]\ [M]} \qquad .....(3.23)$$

Where K is equilibrium constant of the reaction at a given temperature.

This is derived on the basis of Kinetic theory.

(b) *Factors affecting chemical reaction solution are:*

(i) ***Nature of solvent:*** The solvent is polar or non-polar solvent depends on permanent dipole moment, μ and dielectric constant of the solvent. The electrostatic (dipole-dipole interaction), hydrogen bonding are responsible. Polar solvents have high dielectric constants.

Dipole movement: The dipole movement, μ is defined as the product of the electric charge and the distance between the two average centres of +ve and –ve charges

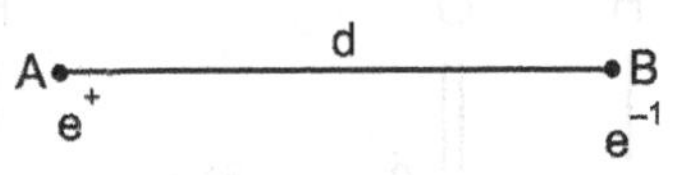

$$\mu = d \times e \qquad .....(3.24)$$

Where d = distance between A and B

e = is equal to both electrical charges.

The dipole moment usually expressed in Debye units, D.

$$D = 10^{-18} \text{ esu-cm} \qquad .....(3.25)$$

The substances which show finite dipole moment are polar, while others are non-polar.

Dielectric constant: This is defined as the ratio of the strength of the electric field E_o, between two charged plates separated by vacuum, for a given potential difference and that of the electric field E when the substance in question occupies the space between the plates at the same potential difference.

$$\frac{E_o}{E} = \in > 1 \qquad .....(3.26)$$

Dielectric constant varies with the first power of the number of molecules in unit volume and with the square of the dipole moment. The value indicates the average degree of polymerization.

For e.g., the dielectric constants $\in$ of:

Water = 78.5

Benzene = 2.27

Ethyl alcohol = 24.3

Chloroform = 4.8

Glycerol = 42.5

Polar solvents have high dielectric constants when compared to non-polar solvents.

The inter molecular forces in non-polar solvents are due to weak forces or non-valence cohesive forces or vander Waal's forces. vander Waal's forces are dipolar in nature.

Dipole-Dipole forces (Keesom forces 1912): These will occur between polar molecules.

For e.g.,

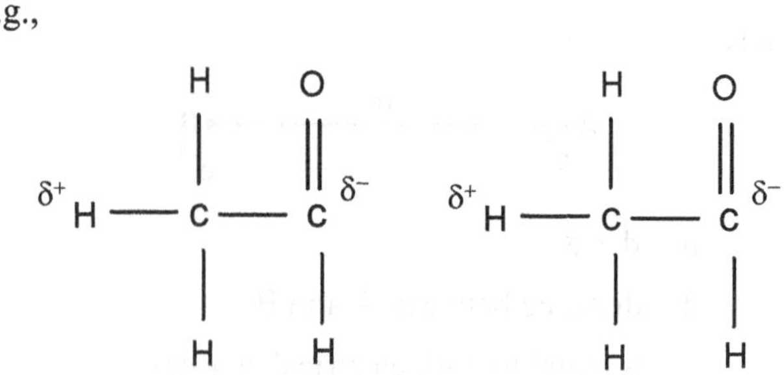

Ion-dipole forces (Debye forces 1920): These forces are important in solution of ionic compounds in polar solvents. For e.g., NaF in water gives solvated species such as $\overset{+}{Na}(OH_2)x$ and $F(OH_2)_y^-$. The forces are temperature independent.

For e.g.,

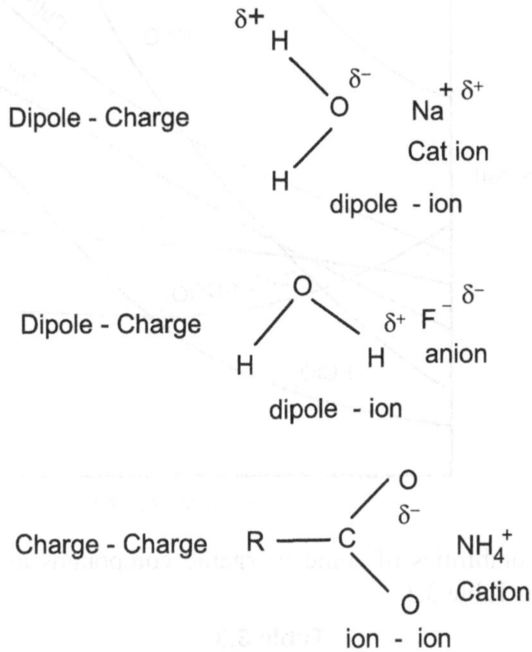

Induced dipole interactions (London forces): The importance of these is limited to solutions of ionic or polar compounds in non-polar solvents.

For e.g,

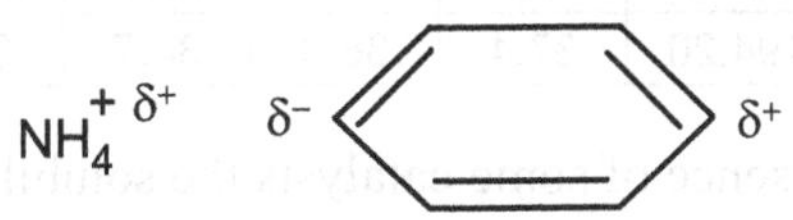

KI is iodine solution

$$K^+ + I^- + I_2 \rightleftharpoons K^+ + I_3^- \quad .....(3.27)$$

Induced dipole-Induced dipole interaction: There are responsible for the solubility of hydrocarbons in other hydrocarbons. These are responsible for the interactions between π orbitals in aromatic rings and are considered to be secondary binding forces between organic drug molecules.

(ii) ***Temperature:*** Solubility of a solid in liquids varies with temperature. The plot of solubility against temperature is called solubility curve (Fig 3.1).

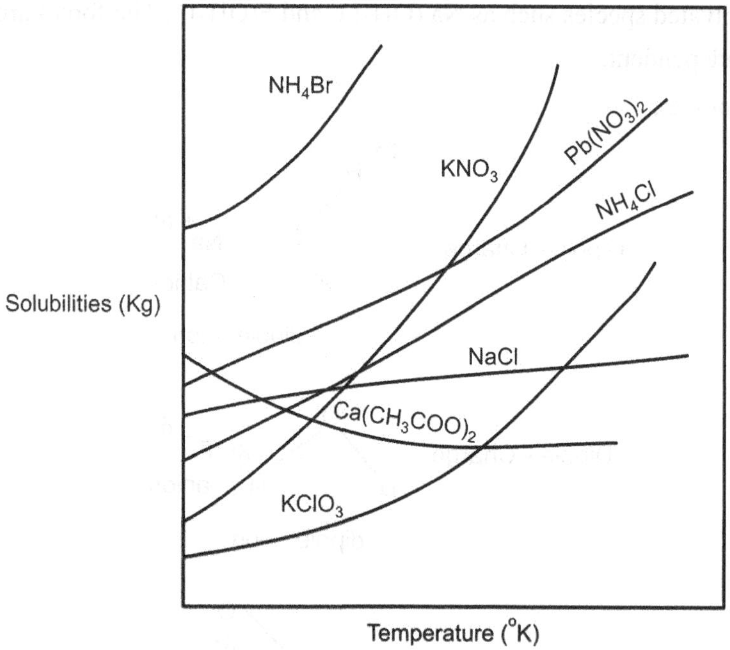

For e.g:, Solubilities of some inorganic compounds at various temperatures are given in Table 3.3

Table 3.3

	Mol.Wt.	0°	10°	20°	30°	40°	100°
NaCl	58.46	35.6	35.7	35.8	36.6	36.0	39.1
$AgNO_3$	169.89	55.6	63.3	69.5	74.0	-	90.0
NH_4NO_3	80.05	118.3	-	192.4	241.8	297.0	871.0
$Ca(C_2H_3O_2)_2 2H_2O$	194.20	37.4	36.0	34.7	33.8	38.2	29.7

(iii) ***Catalysis:*** In presence of some catalysts the solubility increases.

(iv) ***Solubility product (S):*** Solubility product is the product of the total molecular concentrations of the ion at constant temperature.

Consider the electrolyte

$$AB \rightleftharpoons A^+ + B^- \quad .....(3.28)$$

$$S_{AB} = [A^+] \times [B^-] \quad .....(3.29)$$

(Solubility product of AB)

This concept is applicable for sparingly soluble salts i.e. those of which the solubility is less than 0.01g moll^{-1}. The solubility product is the ultimate value which is attained by the ionic conc. product when equilibrium has been established between the solid phase of a difficultly soluble salt and the solution.

For e.g., solubility products of

$$AgCl = 1.7 \times 10^{-10}$$

$$BaSO_4 = 9.9 \times 10^{-11}$$

$$CaC_2O_4 = 2.3 \times 10^{-9}$$

$$Fe(OH)_3 = 4.0 \times 10^{-38}$$

(v) ***Common ion effect:*** The solubility of a salt in the presence of a relatively large amount of the common ion decreases.

MA salt and M^+A is another salt containing common ion.

$$[M^+] \times [A^-] = S_{MA} \quad .....(3.30)$$

$$[A^-] = S_{MA} / [M^+] \quad .....(3.31)$$

For e.g., the solubility of AgCl is decreased 100 times on adding 0.001M NaCl and 1000 times 0.01M NaCl. Here Cl^- is the common ion. Similar results obtained with 0.001M Ag NO_3 and 0.01M Ag NO_3 (Ag^+ is common ion).

(vi) pH or hydrogen ion exponent

(vii) Hydrolysis of salts

(viii) Complex ions and complexation

(c) ***Buffer solutions:*** The resistance of a solution changes in pH upon addition of small amounts of acid or alkali is termed as buffer action. A solution which possesses such properties is known as buffer solution. Further discussion is given in chapter 5.

3.5 Solutions of Liquids in Liquids

Binary liquid solutions i.e. containing two liquids can be classified as

(i) ***Completely miscible solutions:*** The two liquids *p* and *q* are completely miscible in all proportions

e.g., water and ethanol

Toluene and benzene.

(ii) ***Partially miscible solutions*:** The two liquids *p* and *q* are only partially miscible.

e.g., Water and Phenol.

(iii) ***Immiscible solutions:*** The liquids *p* and *q* are not at all miscible or completely immiscible

e.g., Water and carbon tetrachloride.

The solutions again classified into ideal solutions and non-ideal solutions.

(i) ***Ideal solutions:*** *An ideal solution may be defined as one in which the various pure constituents involved do not experience any modification of properties beyond that of dilution.* The molecules have similar structure, polarity, intramolecular attraction are alike. No heat is evolved or absorbed, $\Delta H_{mixing} = 0$, the partial vapour pressure of one component is directly proportional to the fraction of molecules of that component i.e. ideal solution obey Raoult's law. The $\Delta V_{mix} = 0$

Ideal liquids mixtures examples are:

- ethylene bromide and ethylene chloride
- n- hexane and n-heptane
- benzene and toluene
- carbon tetra chloride and silicon tetra chloride etc.

(ii) ***Non-ideal Solutions:*** *A non-ideal solution may be defined as the one in which a considerable change like in volume and enthalpy is experienced by the constituents.* The non-ideal solution does not obey Raoult's law and the $\Delta H_{max} \neq 0$, $\Delta V_{mix} \neq 0$.

The non-ideal solutions mixture examples which show positive deviations from ideal solution are:

- water and ethyl alcohol
- dimethyle ketone and ethyl alcohol
- heptane and ethyl alcohol
- ethyl alcohol and n – hexane.

If A and B are the two liquids in a mixture P_A and P_B are their partial pressures and X_A and X_B are the mole fractions and their measured vapour pressures are greater than expected (not obey Raoult's law) $P_A > P_A^0 X_A$ and $P_B > P_B^0 X_B$ such solutions exhibit positive deviation.

The non-ideal solution mixture example which show negative deviation from ideal solution are:

- Water and hydrochloric acid
- Water and sulphuric acid
- Dimethyl ketone and chloroform.

If $P_A < P_A^0 X_A$ and $P_B < P_B^0 X_B$, such a system exhibit negative deviation.

3.6 Colligative Properties

A colligative property can be defined as any property which depends only on the number of particles in solution i.e. the molecular concentration of the solute in the solution and not in anyway on the nature of solute

These properties provide valuable methods for the determination of the molecular weight of the dissolved substance (non-volatile, non-electrolyte) and for evolution of certain important thermodynamic quantities.

(i) Osmotic pressure; (ii) Lowering of vapour pressure;
(iii) Elevation of boiling point; (iv) Depression of freezing point;
(v) Plasmolysis

(i) Osmotic Pressure: ***Osmosis:*** If two aqueous solutions of different concentrations or when a solution is separated from pure solvent by a semi-permeable membrane, permeable to solvent but impermeable to solute, the solvent (if aqueous, pure water) will tend to pass spontaneously through the membrane is called Osmosis (Greek meaning is push) and is first reported by Abbe Nollet. This spontaneous process will continue until the solvent concentration on both sides of the membrane is equal.

Osmotic pressure: The extra pressure applied to a solution in order just to prevent the flow of solvent through the membrane from the pure solvent is known as osmotic pressure. The most common method for measuring osmotic pressure of solutions is Berkeley and Hartley.

The Osmotic pressure of a solution is not a pressure produced by the solution, but it is measure of some real difference, expressed in pressure units, in the nature of pure solvent and solution and the semi-permeable membrane is merely the artifice by which this difference is made manifest.

Accordingly to van't Hoff equation osmotic pressure π is given by

$$\pi V = n RT \qquad \text{.....(3.32)}$$

Where π = Osmotic pressure

n = no. of moles of solute

V = litres of solution

R = gas constant

T = absolute temperature

Problem 5: Calculate the Osmotic pressure of 1 % solution of cane sugar at 0°C

Ans: n = 1, V = 34.2 (1 mole of sugar = 342 g)

so 342 × 100 = 34200 i.e. 34.2 litres

$$T = 273.15 \quad R = 0.08122$$

$$\pi = \text{(Osmotic pressure)} = \frac{1 \times 0.08122 \times 273.15}{34.2} = 0.649 \text{ atmosphere.}$$

According to van't Hoff's equation, the osmotic pressure should be colligative property of the solution i.e. at any given temperature it should depend only on the concentration and not at all on the nature of solute. This is applicable for only dilute solutions.

Osmometer (Nollet's) apparatus)

Animal membrane is Pig's bladder (semi-permeable membrane). The animal membrane is stretched across the mouth of an inverted thistle funnel which is then partly filled with a strong solution of cane sugar and dipped into a beaker of distilled water (Fig.3.2)

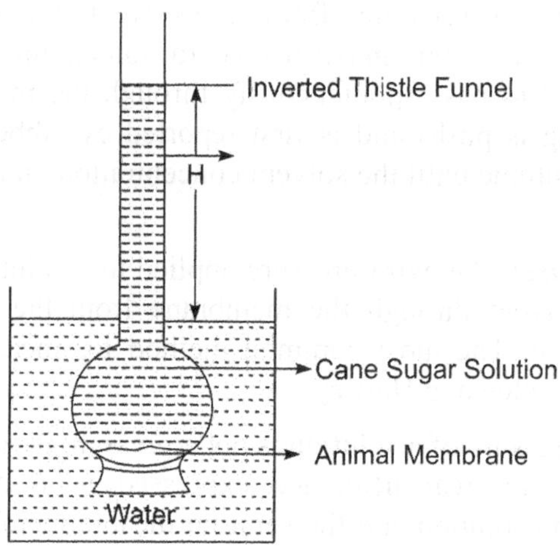

Fig. 3.2

Osmotic properties are important to the mechanism of action of drugs such as osmotic diuretics, saline cathartics etc. These are also important in the maintenance of fluid balance across biological membranes.

(ii) Lowering of Vapour Pressure: The vapour pressure of a liquid is lowered when a non-volatile substance is dissolved in it and the lowering or decrease is proportional to the amount of solute added.

Raoult's law: *It states that the relative lowering of vapour pressure is equal to the ratio of the number of molecules of the solute and the total no. of molecules in the solvation i.e. molar fraction of the solute.*

$$\frac{p^o - p}{p} = \frac{n_2}{n_1 + n_2} = N_2 \qquad .....(3.33)$$

Where n_1 = no. of moles of solvent

n_2 = no. of moles of solute

N_1 = mole fraction of solvent

N_2 = mole fraction of solute

P^o = vapour pressure of pure solvent at a particular temperature

P = vapour pressure of the above solvent in solution at same temperature

By subtracting unity from both from both sides of the eq. 3.33,

$$\frac{p}{p^o} = 1 - N_2 \qquad .....(3.34)$$

The relative lowering of vapour pressure is not dependent on the nature of the solute and solvent but it depends on the concentration of the solute. So the relative lowering of vapour pressure of the solvent is a colligative property.

We can write the eq. 3.33 in another way

$$\frac{p^o - p}{p} = \frac{W_2 / M_2}{(W_1 / M_1) + (W_2 / M_2)} \qquad .....(3.35)$$

$$\frac{p^o - p}{p} = \frac{W_2}{W_1} \times \frac{M_1}{M_2} \qquad .....(3.36)$$

Where W_1 = wt of solvent in g.

W_2 = wt of solute in g

M_1 = molecular wt of solvent

M_2 = molecular wt of solute

From this it is possible to calculate the molecular weight of a dissolved substance.

Problem 6: Calculate the molecular weight of sugar. The lowering of vapour pressure by adding 29.0358 g of sugar to 1000 g of water is 0.00705 mm. The vapour of water at that temp is 4.62 mm.

Ans:

$$\frac{p^o - p}{p} = \frac{0.00705}{4.62}$$

$W_2 = 29.0358$

$W_1 = 1000$

$M_1 = 18$

$$\frac{0.00705}{4.62} = \frac{29.0358}{1000} \times \frac{18}{M_2}$$

$$M_2 = 342$$

Dynamic and static methods used for the determining the lowering of vapour pressure.

(iii) Elevation of boiling point: The temperature at which the vapour pressure of a liquid becomes equal to external pressure acting upon the surface of the liquid is called the boiling point of the liquid.

The boiling of a solution containing a non-volatile solute is higher than that of pure solvent. *The difference between the boiling points or pure solvent and the solution at any constant pressure is known as the elevation or boiling point and is proportional to the concentration of the solute.*

The elevation of boiling point curves of solvent and solute in shown in Fig. 3.3

ABC – Represents the vapour pressure temperature curve of the pure solvent

DFEH – Represents the vapour pressure temperature curve of solution.

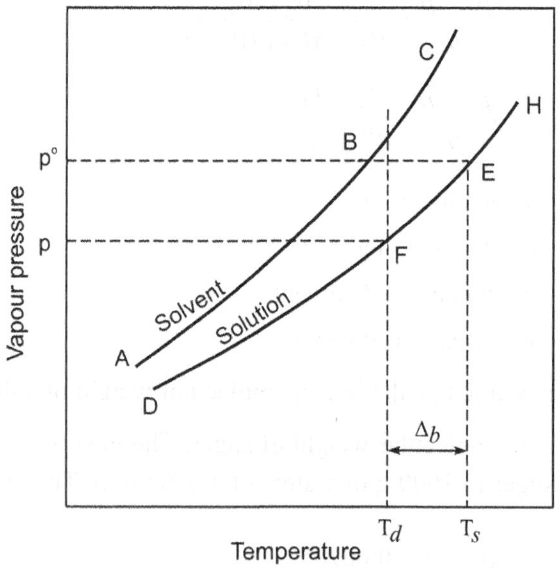

Fig 3.3 Elevating of boiling point of solvent, T_s = boiling point of solution.

According to the definition of the boiling point, to reach B.P at some external pressure P^o, say 1 atmosphere, the solvent and solution must be heated to temperatures at which their respective vapour pressures becomes equal to the given external pressure.

But $T_S > T_l$ from the figure. The elevation of boiling point Δ b is

$$\Delta T_b = T_S - T_l \quad .....(3.37)$$

Lowering of vapour pressure is proportional to $p^o - p/p^o$. BE is elevation of boiling point and BF is relative lowering of vapour pressure. The ratio of BE/BF will be constant for these solutions. The relative lowering is equal to N_2, the mole fraction of the solute

$$\Delta T_b = K_b N_2 \quad .....(3.38)$$

Where Kb = proportionality constant

$$\Delta T_b = \frac{RT_l^2}{1000H_{vap}} m \quad .....(3.39)$$

$$K_b = \frac{RT_l^2}{1000H_{vap}} m \quad .....(3.40)$$

Where K_b is constant for each solvent and depends only on the boiling point and heat of vaporization of the solvent.

$$\Delta T_b = K_b m \quad .....(3.41)$$

Where m is molality = $\frac{W_2}{W_1 M_2} \times 1000$

On combining eq. 3.39 and 3.41

$$\Delta T_b = K_b \frac{1000\ W_2}{M_2 W_1} \quad .....(3.42)$$

$$\therefore\ M_2 \text{ (the molecular weight of solute)} = = K_b \frac{1000\ W_2}{\Delta T_b W_1} \quad .....(3.43)$$

From the elevation of boiling point, it is possible to calculate the molecular weight of solute.

Problem 7: 0.56 g of camphor to 16.0 g of ethyl alcohol. Molar elevation constant of ethyl alcohol is 1.2°C. Elevation of boiling point is 0.278°C. Calculate mol. wt. of camphor.

Ans: $K_b = 1.2°C$, $W_1 = 16$ g, $W_2 = 0.56$ g, $\Delta T_b = 0.278°C$

$$M_2 = \frac{1.2 \times 1000 \times 0.56}{0.278 \times 16} = 151.1$$

The elevation of boiling point for dilute solutions is directly proportional to the mole fraction of the solute only, and is not dependent on the nature of the solute. Hence it is a colligative property.

(iv) Depression of freezing point: *The freezing point of a solution is defined as the temperature at which the solid solvent and the solution exist in equilibrium. The freezing point of solution is always lower than that of pure solvent. The depression of freezing point is proportional to the concentration of the solution. It is a direct consequence of lowering of vapour pressure.* The freezing point curves are shown in Fig. 3.4.

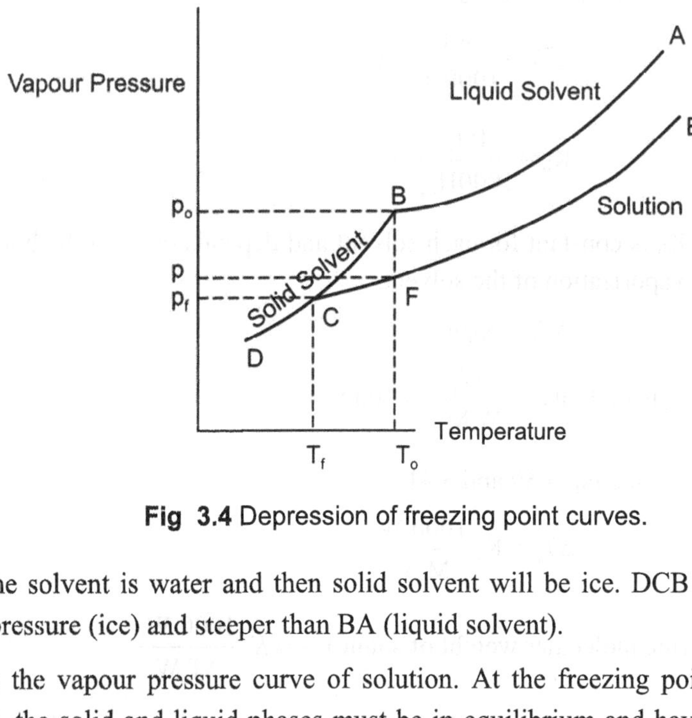

Fig 3.4 Depression of freezing point curves.

Here the solvent is water and then solid solvent will be ice. DCB is sublimation curve pressure (ice) and steeper than BA (liquid solvent).

EFC is the vapour pressure curve of solution. At the freezing point of the pure solvent, the solid and liquid phases must be in equilibrium and have same vapour pressure.

T_o = Freezing point of pure solvent,

T_f = Freezing point of solution

Depression of freezing point

$$\Delta T_f = T_o - T_f \quad \text{.....(3.44)}$$

$$\Delta T_f = K_f N_2 \quad \text{.....(3.45)}$$

Where K_f = proportionality constant.

N_2 = mole fraction of solute.

$\Delta T_f = K_f m$

Where m = molality

$$\text{molecular weight of solute } M_2 = K_f \frac{1000\, W_2}{\Delta T_f W_1} \quad .....(3.46)$$

The depression of freezing point for dilute solution is directly proportional to the mole fraction of solute only and is not dependent on the nature of solvent (colligative property).

Problem 8: For a naphthalene in benzene, 0.142 g of naphthalene to 20.25 g of benzene observed, $\Delta T_f = 0.284°C$ and $K_f = 5.12$. Calculate molecular weight of naphthalene.

Ans:
$$M = \frac{5.12 \times 1000 \times 0.142}{0.284 \times 20.25} = 126$$

(v) **Plasmolysis:** Devaries employed plant cells for measuring roughly and comparing the osmotic pressure of solutions of certain substances. The cells are bounded by more or less firm walls lined with a membrane permeable to water but impermeable to substance dissolved in cell-sap (tone) e.g. certain potassium salts, Glucose etc. Now if such a cell is immersed in solution of lower osmotic pressure than that of the solution inside the cell, no appreciable change will be observed in the size of the cell on account of rigidity or cell walls, the water is not allowed to flow from outside to inside. But when cell is immersed in a solution of higher osmotic pressure than that of cell-sap, the cell membrane will partially collapse as there is diffusion of water from the interior of the cell to surrounding solution. This phenomenon is known as plasmolysis. By bringing cells of the same kind into solutions of different concentrations, he determined the concentration at which the plasmolysis ceases or is just detectable. Such solutions are called isotonic, ie. Solutions have the same osmotic pressure. All isotonic solutions have the same molal concentration. In other words, isotonic solutions have osmotic pressures equal to the osmotic pressure of intracellular fluid i.e. $\pi_{Solu} = \pi_{Cell}$. These solutions can be applied to tissues also or injected without causing damage to cells though osmotic effects. If $\pi_{Solu} > \pi_{Cell}$, the solutions are known as hypertonic. There is cell shrinkage since water leaves the intracellular compartment (occurs in red blood cells).

If $\pi_{solu} < \pi_{cell}$, the solutions are known as hypotonic. There is cell swelling, distention and finally rupture since the cell will imbibe water (hemolysis in red blood cells)

These solutions are sometimes used to advantage to electrolyte therapy.

CHAPTER 4

Treatment of Analytical Data

4.1 Introduction

The function of the analyst is to obtain a result as near to true value as possible by the correct application of analytical procedure employed so the presentation of Analytical data of the measurements is of utmost important. In this chapter it will be explained some of the terms employed and to out line statistical procedures which may be applied to the analytical results.

4.2 Divisions of Analytical Chemistry

Analytical Chemistry is ordinarily divided into qualitative analysis and quantitative analysis.

In qualitative analysis a compound or mixture is analysed to determine what constituents or components present.

In quantitative analysis a compound or mixture is analysed to determine the proportion in which the constituents and components present.

Calculation in qualitative analysis is not much important but in quantitative analysis is more extensive and is based on numerical data obtained from careful measurements of masses and volumes of chemical substances.

4.3 Sampling

The objective of sampling is to collect a portion of material small enough in volume to be transported convenient and handled in the laboratory while still accurately representing the material being sampled.

A sample is a portion of material selected which possesses essential characteristics of the bulk of the material.

A sampling unit may be defined as a minimum sized package in the consignment of material taken from any bulk sample.

A gross sample is one which is prepared by mixing various increments of samples.

A sub sample is smallest portion of the main sample just like a gross sample.

The analyte is the amount of sample taken for analysis. An ideal sample would be identical in all its intensive properties to the bulk of the material from which it is removed.

Precautions:

The sampling program defines the proportion of the whole to which the test results apply. Account must be taken of the variability of the whole with respect to time, area, depth and in some cases rate of flow. The samples do not deteriorate or become contaminated before it reaches the laboratory. For this purpose various precautions to be taken while sampling.

Types of samples

(a) ***Grab or catch Samples:*** A sample collected at a particular time and place can represent only the composition of the source at that time and place. When a source is known to be fairly constant in composition over a considerable period of time or over substantial distances in all directions, then the sample said to represent a longer time period or a larger volume or both, then the specific point at which it was collected. In that case some sources may be represented by quite well by single grab samples e.g., waste water streams, surface water. When a source is known to vary with time, number of grab samples to be collected.

(b) ***Composite Samples:*** The term composite sample refers to a mixture of grab sample collected at the same sampling point at different times. The time composite samples are useful only for determining components that can be demonstrated to remain unchanged under the conditions of sample collection and presentation.

(c) ***Integrated Samples:*** It is best to analyse the mixtures of grab sample collected from different points simultaneously or as nearly so as possible. Such mixtures are called integrated samples.

Methods of Sampling:

(a) ***Manual Sampling:*** It involves no equipment but may be unduly costly and time consuming for routine or large scale sampling programmes.

(b) ***Automatic Sampling:*** Automatic samples are being used increasingly. Various devices are available but no one is universally ideal.

There are various techniques are there for sampling solids, liquids and air.

Collection of sample is the most important aspect before analysis and treatment of analytical data.

4.4 Processing Data

The data must be presented either in tables or in graphs with a clear mention of the units used. Particular case should be taken in selection scale for graphs or histograms. If the data collected are to be interpreted and some inference is sought or a pattern is to be studied or the validity of the data is to be tested or the interrelationship of various data is to be studied. Statistics comes as a handy tool. The statistics includes.

(i) ***Descriptive Statistics:*** Presentation of data

(ii) ***Interfacial Statistics:*** The different statistical methods used

The different Samplings are

(i) ***Random Sampling:*** This procedure is one in which each portion of the whole is given an equal chance of appearing in the sample.

(ii) ***Aliasing (Biased sampling):*** A cyclic variation in quantity would fall into phase and lead to biased sampling.

(iii) ***Stratified Sampling*** This procedure should lead to estimates of central value and dispersion that are as accurate as possible and is unbiased estimate.

The various types of data and statistical methods to be applied are subjected to many question are presented in Table 4.1

Table 4.1 Various types of data, Questions and statistical methods

S. No	Types of Data	Method	Question/Remark
1.	Observation on one variable	Mean, variance, standard deviation mean deviation	Most representative value. How much variation?
2.	One sample and one variable	Standard score	How typical is observation of the sample it comes from?

Table 4.1 *Contd...*

S. No	Types of Data	Method	Question/Remark
3.	2 Sample and one variable	Significance text, t-test and F-test	Do samples have significantly different means
4.	One sample and two variable	Correlation and regression	How does one variable respond to changing other? Strength of association between variables.
5.	4 Sample, 2 variables	Multiple correlation and regression	Relation between dependent and several independent variable

Probability: It is applied to any subjects which involves chance or random happenings. It ranges from zero (absolute in possibility) to one (complete certainty).

$$P = \frac{h}{n} \qquad .....(4.1)$$

Where h = an event depends on no. of times it occur

n = total number of possible ways

For e.g., If we roll the six sided dies 100 times and observe that number 6 comes 15 times, then the empirical probability $P = \frac{15}{100} = 0.15$

Degree of freedom ν: These are no. of values in a set data which are free to vary

$$\nu = \frac{A+B+C}{3} = 9$$

Average: Average is calculated to find out a typical representative of all the observation of a variable.

$$\text{Arithmetic mean } \bar{X} = \frac{\text{sum of are values}}{\text{no.of total values}} = \frac{x}{n} \qquad .....(4.2)$$

4.5 Classification of Errors

In the filed of science, numerical data and numerical results obtained in analysis are subject to errors. The independent measurements of the same quantity, even when made under apparently identical conditions, usually differ to some extent.

Errors can be classed as

(a) Systematic or Determinate errors,

(b) Random error or indeterminate errors

(a) Systematic or Determinate errors

Systematic or Determinate errors are errors that persist in a definite way and to a fixed degree from one determination to another and are such nature that their magnitudes can be determined and their effects eliminated or at least greatly reduced.

These errors further classed as:

(i) ***Operational or personnel errors:*** An example of which is the error caused by consistently establishing a colour change too late.

These are due to factors for which the individual analysts are responsible and are not connected with the method or procedure and form part of personal equation of an observer. These errors arise from the constitutional inability of an individual to make certain observation accurately. Thus some persons are unable to judge colour changes sharply in visual titration, which may result in a slight overstepping of the end point. The errors are mostly physical in nature and occur when proper analytical technique is not followed.

For e.g., Mechanical loss of materials in various steps of analysis, allowing hygroscopic materials to absorb moisture before weighing, use of impure chemicals, no proper washing and drying of precipitates etc.

(ii) ***Instrumental and reagent errors:*** An example of which is the error caused by the use of a balance with arms of unequal lengths.

These arise due to faulty construction of balances, the use of uncelebrated weights and glass ware, other instruments etc.

(iii) ***Methodic errors:*** An example of which is the error caused by the presence of a foreign substance in a weighed precipitate.

These originate from incorrect sampling and from incompleteness of a reaction.

(iv) ***Additive or proportional errors:*** The proportional errors may arise from an impurity in a standard substance, which leads to an incorrect value for the morality of a standard solution. Other proportional errors may not vary linearly with the amount of the constituent, but will at least exhibit an increase with the amount of constituent Present.

For e.g., Cu (11) determination in the alloy (Cu20% and 0.2% Fe) by iodometric method $2Cu^{2+} + 4I^- \rightarrow 2CuI + I_2$

The results of copper determination is shown in Fig 4.1 for a range of sample sizes.

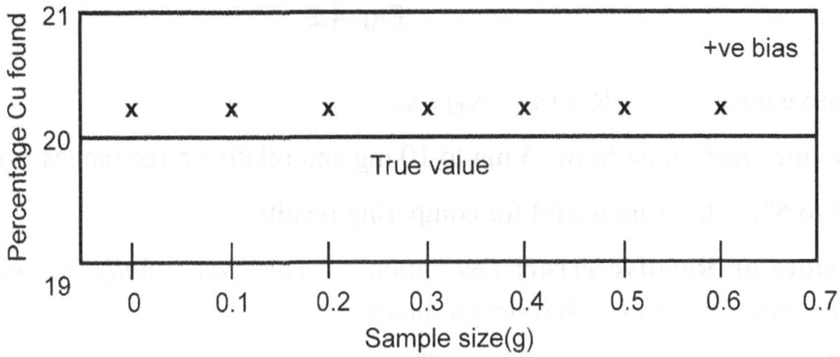

Fig. 4.1 Effect of proportional error in determination of copper (II) in presence of iron

The same absolute error + 0.2% or relative error 1% occur.

The additive error absolute value is independent of the amount of constituent presented determination. Examples are loss in weight of a crucible in which a precipitate is ignited and errors in weights.

The absolute error E_A is a measurement or results X_M is given by

$$E_A = X_M - X_T \qquad \text{.....(4.3)}$$

where X_T = true or accepted value

Absolute and relative errors in analysis of 200 mg aspirin standard, has been analysed a no. of times and is shown in Fig 4.2.

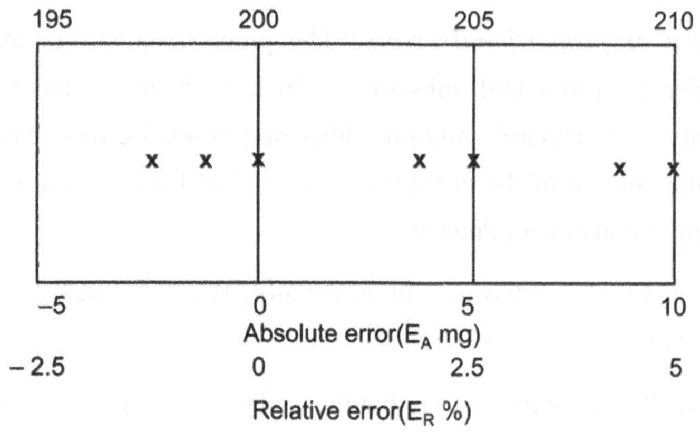

Fig. 4.2

Relative error $ER = (x_M - x_T) / x_T$ .....(4.3a)

Absolute error range from –5 mg to 10 mg and relative error ranges from –2.5 to 5%. There are useful for comparing results.

(v) Absolute or Relative error: *Determinate errors can usually be corrected for by calibration or by other experimental means.*

(b) Random or indeterminate errors: There are errors that are more or less beyond the control of the observer and that have signs and magnitude determined solely by chance.

These may be caused by the factors like

- Fluctuations in temperature and pressure
- Inability of the observer to estimate correctly fractional parts of marked divisions
- General fatigue of the eye.

They are characterized by the fact that positive and negative errors are equally likely to occur. The arithmetical mean of the numerical results of a series of similar observations subject only to random errors can be taken as the most provable value.

Number of measurements are made under same prescribed conditions and represented graphically. The frequency occurrence of experimental value is plotted as a function of the error or deviation from the average mean. The mathematical model that best satisfies such a distribution of random errors is called normal or Gaussian distribution. This bell shaped curve is shown in Fig. 4.3 i.e., symmetrical about the mean.

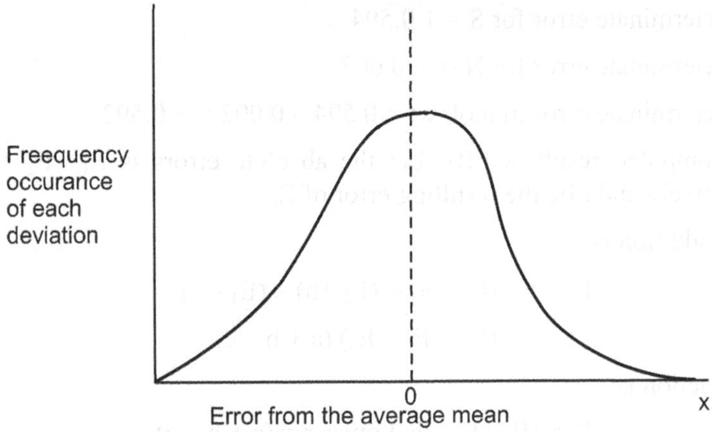

Fig. 4.3 Normal or Gaussian distribution curve.

The curve satisfied the equation

$$\frac{1}{\sigma\sqrt{2\pi}} e^{\frac{-(x-\mu)^2}{2\sigma}} \qquad \text{.....(4.4)}$$

Where σ = Standard deviation

μ = mean deviation

s = population mean

x = population standard deviation

Mathematical calculation of errors

Accumulated errors (propagation of Errors): Errors associated with energy measurement made in an analytical procedure and there will be aggregated in the final calculated result. The accumulation or propagation of errors is treated similarly for both determinate and random errors.

Determinate errors can be either +ve or –ve, so the overall errors is calculated using one or two alternative expression. They are

(i) Addition and substraction determinate error

The overall absolute determinate error

$$E_r = E_1 + E_2 + E_3 \qquad \text{.....(4.5)}$$

For e.g.: Mol. Wt g. AgSCN = 165. 954

If the at. wt. of S is taken as 32.660 instead of 32.066 and N is taken an 14.005 instead 14.007.

The determinate error for S = + 0.594

The determinate error for N = – 0.002

The determinate error in mol. wt = 0.594 – 0.002 = + 0.592

The computed result is ER_2. Let the absolute errors is $E_1 + E_2 + E_3$ be a, b, c respectively and r be the resulting error of E_r.

Then addition is

$$E_r + r = (E_1 + a) + (E_2 + b) - (E_3 - c) \quad .....(4.6)$$

$$= (E_1 + E_2 - E_3)\,(a + b - c)$$

Substraction is

$$E_r = (E_1 + E_2 - E_3) \text{ gives } r = (a + b - c) \quad .....(4.7)$$

These errors are (the absolute errors) directly transmitted into results

(ii) Multiplication and Division Determinate Error

The overall relative determinate Error

$$E_{TR} = E_{1R} + E_{2R} + E_{3R} \ldots. \quad .....(4.8)$$

Where $E_{1R} + E_{2R}$ = relative determinate error in the individual measurements taking sign into account.

Let $$E_r = \frac{E_1E_2}{E_3} \quad .....(4.9)$$

Actual measurement are $(E_1 + a)$ $(E_2 + b)$ as $(E_3 + c)$

Then $$E_r + r = \frac{(E_1 + a)(E_2 + a)}{(E_3 + c)} = \frac{E_1E_2 + aE_2 + bE_1 + ab}{(E_3 + c)} \quad .....(4.10)$$

Relative error $$\frac{r}{E_r} = \frac{a\ E_2E_3 + bE_1E_3 - cE_1E_2}{E_1E_2(E_3 + c)} \quad .....(4.11)$$

These errors (the relative errors) are transmitted into result.

4.6 Minimization of Errors

Systematic error can often be reduced by following one of the methods given below.

- Calibration of apparatus and application of correction.
- Running a blank determination: By carrying out the experiment under identical condition without the sample.

- Running a control determination: By carrying out the experiments under identical conditions for the standard sample and unknown sample.
- Use if independent methods of analysis: The accuracy of a result may be established by using different methods. For e.g., Fe(III) by gravimetric method and by titrimetric method. If the results obtained are concordant, it is probable that the values are correct within small limits of error.
- Running parallel determinations for the same sample solutions.
- Standard addition: A known amount of the constituent being determined is added to the sample which is then analyzed. The difference between the analytical results for sample with and with out the added constituent gives recovery of the amount of the added constituent.
- Isotopic dilution: A known amount of the element being determined containing a radioactive isotope, is mixed with the sample and the element is isolated in a pure form. The radioactivity of the isolated material is measured and compared with that of the added element.

4.7 Accuracy

The accuracy of numerical value is the degree of agreement between it and the true value or most probable value and it expresses the correctness of measurement

Since the true value is never known except within certain limits, the accuracy of a value is never known except within those limits. There are two method of determining accuracy for Analytical methods

(i) ***Absolute Method:*** A synthetic sample containing known amounts is question is used. These substances are primarily standards, may be available commercially or they may be prepared by analysts. The test of accuracy of the method under consideration is carried out by taking varying amounts of the constituent and proceed according to specified methods. Because of the determinate errors, the amount of constituent must be varied and in the procedure may be a function of the amount used.

The difference between two numerical values can be expressed as the absolute difference or as the relative difference (expressed in parts per thousand)

Problem 1: The absolute difference between the values 2.431 (True value) and 2.410 (obtained value) is 0.021. Find the relative difference.

Ans: The relative difference is $= \dfrac{0.021}{2.410} \times 1000$

$= 8.7$ parts per thousand (ppt)

(ii) ***Comparative method:*** Sometimes it is not possible prepare the desired samples synthetically for e.g., in the case of minerals, alloys or ores etc. The constituents can be determined by supposedly accurate methods of analysis for e.g., spectrophotometer, spectrographic, titrimetric etc. This comparative method involving, i.e., agreement between any two methods of analysis, secondary standards, is not satisfactory from theoretical point of view but is very useful in applied analysis.

4.8 Precision

Precision is defined as the concordance of a series of measurement of the same quantity or it is a measure of variability or dispersion within a set of replicated values or results obtained under the same prescribed conditions

Precision expresses the reproducibility of a measurement and it always accompanies accuracy.

There precision and accuracy of the numerical data can be explained using the plots of titration data as shown in Fig. 4.3

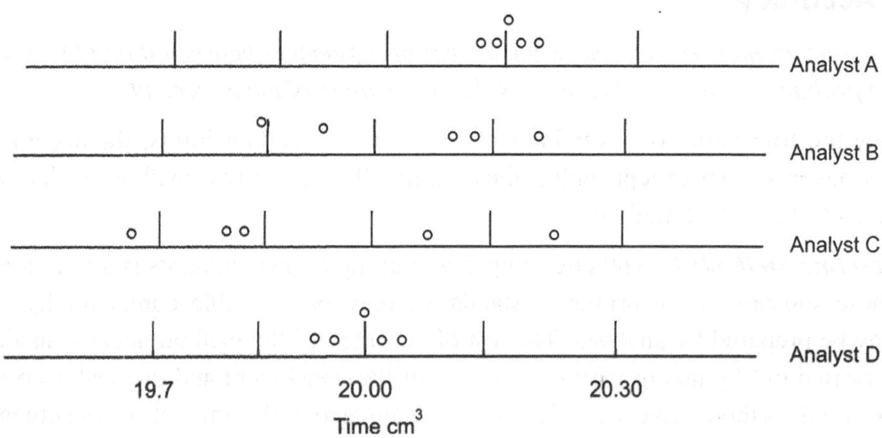

Fig. 4.4 Plot of Titration data to distinguish accuracy and precision

We can summarise the results analysis as follows

The average values for analyst B and D are very close to 20.00 cm^3. These two sets are therefore said to have good accuracy.

- The average value for analysts A and C and well above below 20.00 cm^3 respectively. These are said to have poor accuracy.

- The five titre values of for analysts A and D are very close to one another within each set. These two sets therefore both show good precision.
- The five titre values analysts B and C are spread widely within each set. These two sets show poor precision.

Therefore the confidence in the analytical procedure and the result is greater when good precision can be there (D).

4.9 Precision Measures

In a series of independent determination of a given quantity, if determinate errors have been effectively eliminated or corrected for, the average, or mean, the numerical values obtained can be taken as the most probable value of the series, and a measure of the degree of precision of this mean value can be considered as a measure of the limiting degree to which the result is likely to differ from the unknown true value. It is therefore a measure of the reliability of the result. For e.g., suppose the following nine values are obtained

41.62, 41.76, 41.53, 41.60, 41.47, 41.71, 41.60, 41.57, 41.64,

(i) Average: Average is calculated to find out a typical representative of all observations of a variable

Arithmatic mean $$\overline{X} = \frac{\text{Sum of all values}}{\text{number of total values}} = \frac{x}{n} \quad .....(4.12)$$

$$= \frac{374.64}{9} = 41.627$$

(ii) Mean duration x_i: The difference between any one of the values and this mean is the deviation of that value from the mean

In the case of above results mean deviation (with out regard to sign):

0.007, 0.133, 0.027, 0.157, 0.083, 0.027, 0.013, 0.097, and 0.083

(iii) Average deviation, d: The average deviation (d) of a single measurement is the mean of deviations of all the individual measurements.

$$d = \frac{\Sigma X_i}{n} \quad .....(4.13)$$

Where ΣX_i = sum of individual deviations from the mean.

For the above results d = 0.070 and represents the amount by which average independent measurement of the ionics is likely to differ from the most the probable value.

(iv) Average deviation of the mean, D – It is numerically equal to the average deviation of a single measurement divided by square root of the number of measurements made.

$$D = \frac{d}{\sqrt{n}} = \frac{0.070}{\sqrt{9}} \quad .....(4.14)$$

For the above results $D = \frac{0.070}{\sqrt{9}} = 0.023.$

(v) Standard deviation of single measurement, S: It is often used as a precision measure and is considered to be more reliable measure than average deviation.

$$S = \frac{\sqrt{\Sigma(x_i)^2}}{n-1} \quad .....(4.15)$$

where n – 1 Σ = summation = the number of degrees of freedom.

For the above results.

$$S = \frac{\sqrt{(0.007)^2 + (0.133)^2 + (0.027)^2 + ...}}{8}$$

$$= 0.091$$

(vi) Standard deviation, S: This number serves to give an indication of the reliability of the mean.

$$S = \frac{S}{\sqrt{n}} \quad .....(4.16)$$

$$S = \frac{0.091}{\sqrt{9}} = 0.030$$

The true value falls within the limits 41.627 – 0.030 and 41.627 + 0.030.

The most widely used measure of precision and is a parameter of normal error or gaussian curve, Fig. 4.4 shows two curves for frequency distribution of two theoretical sets of data, each having an infinite number of values and is known as a statistical population.

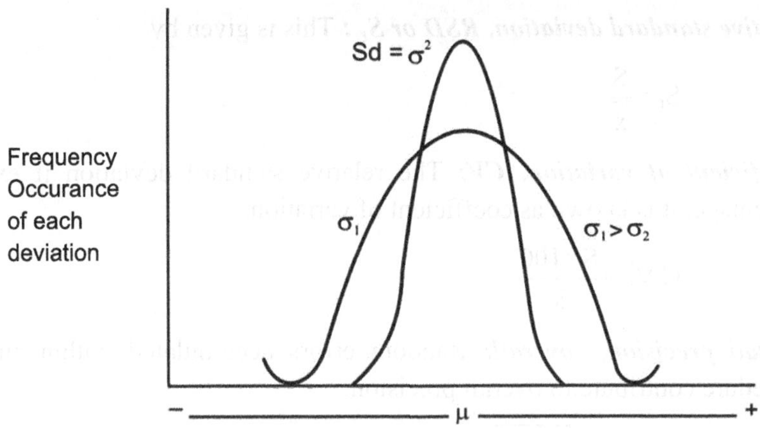

Fig 4.5 Goussian curve for the frequency distribution of two statistical populations with different spreads

The maximum in each curve corresponds to population mean, which for these examples the same value of μ. The spread of values for the two sets is quite different and this is reflected in the half widths of the two curves at the prints of inflection, which is the population standard deviation. The precision of the σ_2 is better than σ_1. The abscission values are +ve and –ve deviations from the mean.

Population standard deviation

$$\sigma = \frac{\sqrt{\Sigma(x-\mu)^2}}{x} \quad .....(4.17)$$

Where x = represents individual value in the population

n = total no. of values

μ = population mean.

The estimated standard deviation S in given by the equation 4.15.

(vii) Variance, S^2: The square of the standard deviation is called variance or it is the arithmetical mean of the squares of the deviations of the values from the mean of the data

$$S^2 = \frac{\sqrt{\Sigma(x_i)^2}}{n} \quad .....(4.18)$$

If n – 1 degrees of freedom are used for small number of values i.e. < 30

$$S^2 = \frac{\sqrt{\Sigma(x_i)^2}}{n-1} \quad .....(4.19)$$

(viii) ***Relative standard deviation, RSD or S_r :*** This is given by

$$S_r = \frac{S}{\overline{X}} \qquad .....(4.20)$$

(ix) ***Coefficient of variation, CV:*** The relative standard deviation if expressed in percentage, it is known as coefficient of variation.

$$C.V. = \frac{S \times 100}{\overline{X}} \qquad .....(4.21)$$

(x) ***Overall precision, ^{S}overall:*** Random errors accumulated within an analytical procedure contribute to overall precision.

$$S_{overall} = \sqrt{(S_1)^2 + (S_2)^2 + (S_3)^2 + ...} \qquad .....(4.22)$$

$$= \sqrt{(0.30)^2 + (0.30)^2} = 0.042 \text{ cm}^3$$

The overall relative precision in given by

$$S_{roverall} = \sqrt{Sr_1^2 + Sr_2^2 + Sr_3^2 + ...} \qquad .....(4.22)$$

(xi) ***Confidence interval:*** When small number of observations is made, the value of standard deviation S, does not by itself give a measure of how close the sample mean $\overline{x}$ might be to the true mean. To estimate the range within which the true mean may be formed it is possible to determine confidence interval, known as confidence limits.

Confidence limits of μ, for n replicable measurements is

$$\mu = \overline{x} \pm \frac{ts}{\sqrt{n}} \qquad .....(4.23)$$

where t = parameter depends on no. of degrees of freedom ν and confidence level required.

Problem 2: The mean $\overline{x}$ of four determination of the copper content of a sample of an alloy is 8.27% and S = 0.17%. Calculate 95% confidence limit for the true value.

Ans:

$$95\% \text{ C.L. for } \mu = 8.27 \pm \frac{3.18 \times 0.17}{\sqrt{4}}$$

$$= 8.27 \pm 0.27\%$$

The true value of copper lies in the range 8.00 to 8.54 percent.

4.10 Significant Figures

The precision of a numerical value is best indicated by the number of significant figures used in expressing that value.

A number is an expression of quantity.

A figure or digit is any one of the characters 01,2,3... which, alone or in combination, serves to express a number.

A **Significant figure** is a digit that denotes the amount of the quantity in the place it would stand. For e.g., in the case of 243, the figures signify that there are two hundreds, four tens and three units and are therefore all significant.

The character 0 is used in two ways.

It may be used as a significant figure or it may be used merely to locate the decimal point.

For e.g., The weight of crucible in 10.603g. In this all five figures are significant including zeros.

If the weight of the crucible in 10.710 meaning that the weight as measured is nearer to 10.710 than to 10.709 or 10.611, both zeros are significant.

If weight of ash is 0.00004 g, here zeros are not significant but serve to show merely that the figure 4 belongs to fifth place to the right of decimal point. The same is true for 45600 in, when signifying the distance between two given points as measured by instruments that are accurate to **three figures** only. The above value can be written as 4.65×10^5 in.

The following rules are applicable in analytical chemistry.

Rule 1 : Retain as many significant figures in a result and in data in general as will give only one uncertain figure.

For e.g., the value 25.34 (burette reading) contains the proper no. of significant figures and for the digit 4 is obtained by estimating an ungraduated scale division and is doubtless uncertain.

Rule 2: In rejecting superfluous and inaccurate figures increase by 1, the last figure retained if the following rejected figure is 5 or over.

For e.g., in the number 17.378, if the last figure is rejected, the near value is 17.38.

Rule 3 : In adding or subtracting, a number of quantities extend the significant figure in each term and in the sum or difference only to the point corresponding to that uncertain figure occurring farthest to the left relative to the decimal point.

For e.g., the sum of three terms, 0. 0211, 35.75, 2.05671, on the assumption that the last figure in each is uncertain, then

$$\begin{array}{r} 0.02 \\ 35.75 \\ 2.06 \\ \hline 37.83 \end{array}$$

Rule 4 : In multiplication or division, the percentage precision of the least precise factor entering into the computation. Hence in this retain as many significant figures in each factor having the largest percentage deviation.

For e.g., the product of $0.0211 \times 35.75 \times 2.05671 = 1.551$.

For 0.0211 the percentage deviation $= \dfrac{1}{211} \times 100 = 0.47$

35.75 percentage deviation $= \dfrac{1}{35.75} = 0.23$

2.05671 percentage deviation $= \dfrac{1}{2.05671} \times 100 = 0.0005$

The first term has largest percentage deviation, therefore governs the number of significant figure that can be properly retained in the product, for the product cannot have a precision greater that 0.47%. The last figure in the product as expressed with true significant figures above is doubtful. So poorer number of significant figures retained.

Rule 5 : Computation involving a precision not greater than 0.25% can be made with a 10 in-slide rule.

Rule 6 : In carrying out the operating of multiplication or division by the use of logarithms, retain as many figures in the mantissa of the logarithm of each factor.

4.11 Comparison of Results

The comparison of values obtained from a set of results with either the true value or other sets of data makes it possible to determine whether the analytical procedure has been accurate and/or precise, or it is superior to another method.

The significance tests are used to compare individual values or sets of values for significant differences.

Significance test: It involve a comparison between a calculated experimental factor and a tabulated factor determined by the number of values in the set(s) of experimental data and a selected probability level that the conclusion is correct. They are use for.

- To check individual values in a set of data for the presence of determinate errors.
- To compare precision of two or more sets of data using their variances.
- To compare the means of two or more sets of data with one another or with known values to establish levels of accuracy.

The tests are based on:

(i) **Null hypotheses:** An assumption that there is no significant difference between the values being compared.

For e.g., if a test indicates that this hypothesis is correct and that there is no significant difference between two values at 95% probability level. It also allows the possibility that there is significant difference at the5% level. Separate tabular values for one significance test factors have been complied as.

(a) **One tailed test:** It is used either to either to establish whether one experimental value is significantly greater than the other or the other way around.

(b) **Two tailed test:** It is used to establish whether there is a significant difference between the two values being compared, whether one is higher or lower than the other, not being specified. This is mostly used.

(c) **Outlier:** A measurement or result that appears to differ significantly from other in the same set of replicates is described as an outlier.

There are tests for comparing results. This method of test requires knowledge of what is known as the numbers of degrees of freedom. In statistical terms this is the number of independent values necessary to determine the statistical quantity the test are:

(i) Q-test (Dixon's Q. Test):

This Q test is used to determine whether to reject or retain the suspected outlier.

$Q_{expt.}$ = [Suspected value-nearest value]/(largest-smallest) .....(4.24)

$Q_{expt.}$ is then compared with Q_{tab} (tabulated value) at a selected level or probability for a test of n values (table 4.2)

Table 4.2 Critical values of Q at the 95% level for two tailed test

S.No	Sample size	Critical Value
1	4	0.831
2	5	0.717
3	6	0.621
4	7	0.570

If $Q_{expt.} < Q_{tab}$, then as per the null hypotheses, that there is no significant difference between the suspect value and the other values in the set is accepted and the suspect value is retained for further data processing.

If $Q_{expt.} > Q_{tab}$ the suspect value is rejected. The rejected value should be used for remaining calculation.

Problem 3: The four replicate values for the determination are 0.403, 0.410, 0.401, 0.380 µg dm^3

Ans: Inspection data says 0.380 µg dm^3 is a possible outlier.

$Q_{expt.} = |0.380 - 0.401| / |0.401 - 0.381| = 0.70.$

Q_{tab} = 0.83 for four values at 95% probability level.

As $Q_{expt.} < Q_{tab}$ 0.380 µg dm^3 should be retained.

(ii) F-test (variance ratio test): The F-test enables the precision of two sets of data to be compared using their variances.

This is used to compare the precisions of two sets of data, for example; the results of two different analytical methods or the results of from two different laboratories.

$$F = \frac{S_A^2}{S_B^2} \quad .....(4.25)$$

The larger value of S always used as the numerator so that $F \geq 1$.

If $F_{expt} < F_{tab}$: there is no significant difference between two variances and hence between the precision of two sets of data, is accepted.

If $F_{expt} > F_{tab}$, there is significant difference between the precision of two sets of data.

Problem 4: The standard deviation for one set of 10 determinations $S_A = 0.210$, and the standard deviation from another 12 determinations $S_B = 0.641$. Find out if there was significant difference between the precision of there two sets of results.

Ans:
$$F = \frac{(0.641)^2}{(0.210)^2} = \frac{0.411}{0.044} = 9.4$$

For Probability

$$P = 0.10 \quad 0.05 \quad 0.01$$
$$F = 2.28 \quad 2.91 \quad 4.71$$

2.28 corresponds to 10% probability, 2.91 corresponds to 5% probability and 4.71 corresponds t 1% probability.

Under these conditions there is less than one chance in 100 that these precisions are similar. The F value turned out to be less than 2.28 then lit would have been possible to say that there was no significant difference between the precisions, at the 10 % level.

Student t – test: This is a test used for small samples. The purpose of this test is to compare the mean from a sample with some standard value and to express some level of confidence in the significance of comparison. It is also used to test the difference between the means of two sets of data $\bar{X}_A$ and $\bar{X}_B$. The value of t is obtained from equation

$$t = \frac{(\bar{X} - \mu)\sqrt{n}}{S} \qquad \text{.....(4.26)}$$

where μ is the true value.

Problem 5: $\bar{X}$ of 12 determinations = 8.37 and μ = 7.91 say whether this value is significant or not the standard deviation is 0.17. Calculate t value.

Ans:
$$t = \frac{(8.37 - 7.91)\sqrt{12}}{0.17} = 9.4$$

The calculated value for t is 9.4, the result is Lights significant. A static t is the comparison of two experimental means $\bar{X}_A$ and $\bar{X}_B$. Then

$$t = \frac{\bar{X}_A - \bar{X}_B}{S_{pooled}} \times \sqrt{\frac{nm}{n+m}} \qquad \text{.....(4.27)}$$

where S_{pooled} = pooled standard deviation

n and m = no. of values of A and B

$$S_{pooled} = \sqrt{\frac{\left\{\left[(n-1)S_A^2 - (m-1)S_B^2\right]\right\}}{[n+m+2]}} \qquad .....(4.28)$$

The equation 4.27 may not be appropriate in some cases.

Paired t – test: The t is defined by

$$t = \frac{\bar{X}_d}{S_d} \times \sqrt{n} \qquad .....(4.29)$$

where $\bar{X}_d$ = mean difference between paired values

S_d = estimated standard deviation.

Problem 6: Two determinations of poly aromatic hydrocarbon is soil.

No. of deviation in each method = 10

No, of degrees of freedoms = 18

U.V. spectrometry $\bar{X}$ = 28.00 mg. Kg^{-1} S = 0.30 mg. Kg^{-1}

Fluorimetry $\bar{X}$ = 26.25 mg. Kg^{-1} S = 0.30 mg. Kg^{-1}

Find out S_{pooled} and $t_{expt.}$ (as per eq. 4.28)

Ans: ^{S} Spooled = $\sqrt{[9\times(0.3)^2 - 9\times(0.23)^2} / 20–2$

(as per eq. 4.28)

= 0.267 mg. Kg^{-1}

$$t_{expt} = \frac{100}{20}\sqrt{\frac{28.0-26.25}{0.267}\;\frac{100}{20}} \qquad \text{(as per eq. 4.27)}$$

= 14.7

Problem 7: The detn. of Hg by A as gave 400,385,382 ppm for a standard known is 400 ppm. Find t-value.

Ans: $\bar{X}$ = 389 S = 9.64 μ = 400

$$t_{expt} = \frac{389 - 400}{9.64} \times \sqrt{3} \qquad \text{(as per eq. 4.29)}$$

$$= 1.98$$

t_{tab} is 4.3 at 95% probability level.

So $t_{expt} < t_{tab}$, the mean is significant by different from the true value

Analysis of variance (ANOVA): F- tests can be applied to several sets of data to assess and compare different sources of variability associated with a series of results. This enables the effect of each source to be assessed separately and compared with the others using F-tests. The indeterminate or random errors effect all measurements. The additional sources of variability arise are:

(i) random effect factors.

(ii) specific effect from determinate sources.

For one additional effect: One way ANOVA is used

For two additional effects: two way ANOVA is used.

In the use of ANOVA, the examples are

(i) Analysis of heterogeneous material where variation in composition is an additional random factor.

(ii) Analysis of samples by several labs, methods or analysis (additional fixed effect factors)

(iii) Analysis of a material stored under different conditions to investigate stability where the storage conditions provide an additional fixed effect factors.

4.12 Rejecting of Measurements (Reliability of Results)

In a series of similar measurements it sometimes happens that one or more of the numerical values stands out as being considerably different from the others, and the temptation is to reject it in establishing the mean value.

Problem 8: the following values are obtained for detn. of Cd in a sample of dust :4.3, 4.1, 4.0, 3.2 $\mu g.g^{-1}$ .Can the last value 3.2 be rejected?

Ans: Q test may be applied to solve this problem.

$$Q = \frac{|\text{Questinable value - Nearest value}|}{\text{Largest value - Smallest value}}$$

$$Q = \frac{|3.2 - 4.0|}{4.3 - 3.2} = \frac{0.8}{1.1} = 0.727.$$

If $Q_{expt} > Q_{tab}$, then the questionable value can be rejected. In this Q_{expt} 0.727 and Q_{tab} (Critical value) for sample size of four, in 0.831. Hence the result 3.2 $\mu g.g^{-1}$ should be retained.

In the above example if 7 measurements are made with the results: 4.3, 4.1, 4.0, 3.2, 3.9, 4.0 $\mu g.g^{-1}$

$$Q = \frac{|3.2 - 3.9|}{4.3 - 3.2} = \frac{0.7}{1.1} = 0.636$$

The value of Q_{tab} (Critical) for samples size 7 in 0.570 i.e. $Q_{expt} < Q_{tab}$, so 3.2 can be rejected.

The rejection can be considered mathematically justified if the deviation of the suspected value from the mean is at least four times the average deviation of the retained values i.e., if

$$x_i \geq 4d \qquad \text{.....(4.30)}$$

Some writers consider $x_i \geq 2.5d$ .....(4.31)

In either case, the error of the rejection value is called a huge error.

4.13 Regression Analysis

Regression analysis is carried out to find out the relation between two variables measured simultaneously, particularly where the one variable is dependent and another is independent. If we plot a graph between two variables, a straight line comes in case the relationship is linear. By judging the tendency of the points fall in a straight line, we can predict the strength of the association between them which is measured by correlation coefficient. The process of deciding the line of the best fit to summarise a particular set of points on a graph is called regression analysis and equation is regression equation.

Regression equation is $Y = a + bX$ .....(4.32)

Where a and b are constants and fixed for a particular line.

Once the values of a and b are known, Y can be estimated for any corresponding value of X

$$a = \frac{\Sigma y - \Sigma x}{n} \qquad \text{.....(4.33)}$$

The value of

$$b = \frac{b\Sigma xy - \Sigma x \Sigma y}{n\Sigma x^2 - (\Sigma x)^2} \qquad \text{.....(4.34)}$$

The constant 'a' denotes the value of Y when the value of X = 0 and is usually known as intercept. The constant b is called regression coefficient measures the slope of the line. This is shown in Fig. 4.5

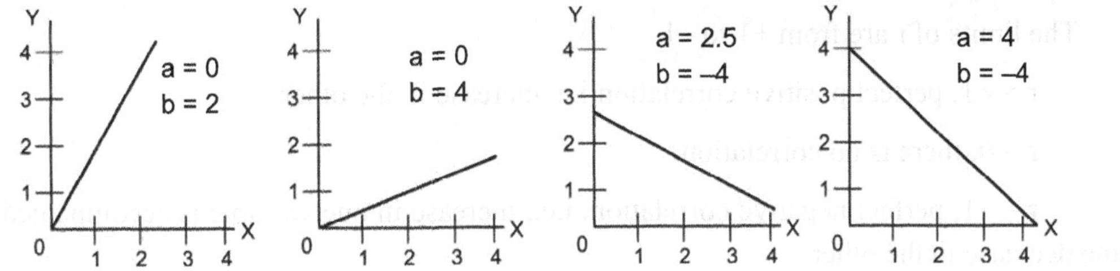

Fig 4.5 Variation in the values of a and b in regression analysis

Correlation: Determination of Correlation is important in finding out the strength of relationship between two independent variables. In other words, it signifies the extent of predictability of one variable from the other. If we plot a graph (Fig 4.6) between two variables, a straight line will indicate a strong relationship while the scattering of the points will show a weak relationship. Statistically this relationship is called correlation coefficient, r

$$r = \frac{n\Sigma XY - \Sigma X\Sigma Y}{\sqrt{[n\Sigma x^2 - (\Sigma X)^2]\times[n\Sigma Y^2 - (\Sigma Y)^2]}} \quad .....(4.35)$$

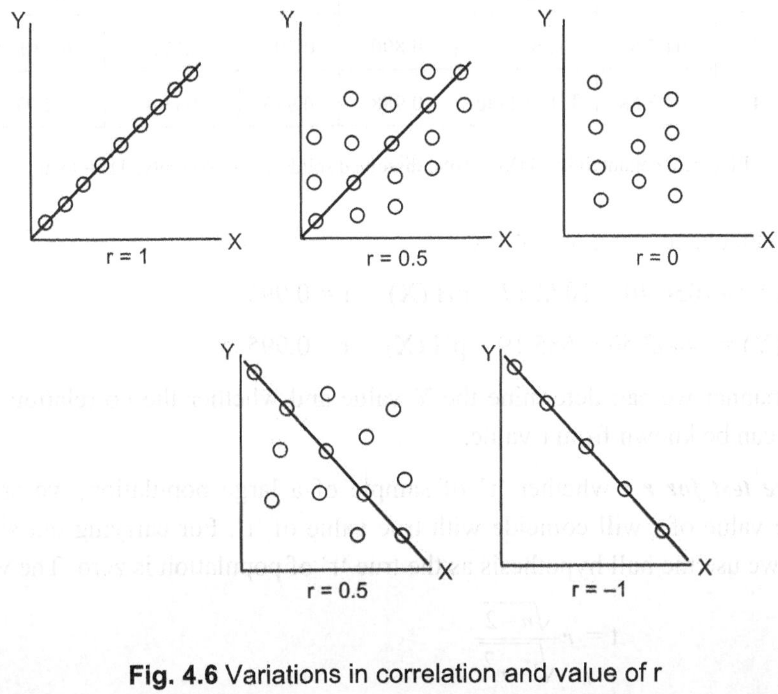

Fig. 4.6 Variations in correlation and value of r

The limits of r are from +1 to –1

r = +1, perfect positive correlation i.e. increase in the other

r = 0, there is no correlation.

r = –1, perfect negative correlation. i.e., increase in one variable is accompanied by the decrease in the other.

Coefficient of Determination, r^2: It gives the measure of the strength of the association and proportion of one variable associated with variation in the other. For r = 0.8 then $r^2 = 0.64$ means 64% variation in the value of Y are associated with the variation in the value of X, the remaining 36% is unknown factor. r^2 ranges from 0 to 1

The calculated values of r and r^2 etc for X and Y are shown in Table 4.3.

Table 4.3

S.No	X	Y	r	r^2	a	b
1	pH 7.68	ECl 216	0.995	0.990	–7034.40	1072.17
2	pH 7.68	TDS 780	0.995	0.990	–4493.50	685.19
3	pH 7.68	AlK 251	0.890	0.792	– 207.23	62.30
4	pH 7.68	TH 382 etc.	0.988	0.976	–933.40	172.97

EC = Electrical conductivity, TDS = Total dissolved solids, Al = alkalinity, TH = total hardness.

r can be calculated using Y = a+bX

EC(Y) = –7034.40 + 1072.17 × pH (X) r = 0.995

TDS(Y) = –4493.50 + 685.19 × pH (X) r = 0.995

In this manner we can determine the Y value and whether the correlation is good or bad with X can be known from r value.

Significance test for r : whether 'r' of sample of a large population, we are not sure whether the value of r will coincide with true value of 'r'. For carrying out significance test for 'r', we use the null hypothesis as the true 'r' of population is zero. The value't' is

$$t = r\frac{\sqrt{n-2}}{\sqrt{1-r^2}} \quad .....(4.36)$$

Non linear Relationships

In many situations, we came across a data when a curvilinear relationship occurs instead of a linear relationship. We find in the graph a clustering of points near either of the ends. In such case it is better to transform curvilinear relationship to linear relationship by plotting a graph between log values of variables. The values of a and b can be calculated from.

$$\log Y = \log a + b \log X \qquad \text{.....(4.37)}$$

log values of Y can be predicted for any value of X and can be converted to original value by taking its antilog.

CHAPTER 5

Principles of Quantitative Analysis (Volumetric and Gravimetric Analysis)

5.1 Introduction

Analytical chemistry is concerned with the determination of the chemical composition of matter. The modern Analytical chemistry covers the identification of a substance, the elucidation of its structure and quantitative analysis of its composition.

The two important steps in analysis are identification and estimation of the constituents of the compound. The identification step is called qualitative analysis. The second step is quantitative analysis. Quantitative analysis can be classified depending upon:

- The method of analysis
- categorized according to the scale of analysis.

The first step is the chemical analysis like volumetry and gravimetry and the others are those involving sophisticated instruments. For getting reproducible results, one cannot always directly depend on these instrumental methods. Before following that methods one must get a true representative sample for analysis which is free from interfering elements. The gravimetric or volumetric methods are used if the samples are present in milligram concentration. If the component to be analysed is present at very low concentration then one can go for optical methods like UV- visible, IR, those involving, luminescence etc. The scale of operations (amount of the sample) in quantitative analytical methods are given in Table 5.1.

Table 5.1

S.No	Class	Volume for Analysis	Quantity of the substance
1.	Macro Analysis	20 ml	100 mg to 500 mg or more
2.	Semi Macro Analysis	1 ml	10 mg to 50 mg
3.	Micro Analysis	0.1 ml	1 mg to 5 mg
4.	Sub-micro or Ultra Micro	0.05 ml	<1 mg
5.	Spot test analysis	Few drops	1μg (0.001 μg = 10^{-6} g)

Various steps in quantitative analysis: The important steps involved in determination are:

- *Sampling*: It should be representative of the mass of the material. The material is reasonably pure and homogeneous.
- *Conversion of desired constituent to measurable form*: This involves methods of separation technique and the selection of this technique decided on the basis of accuracy and the precision required.
- *Measurement of desired constituent.*
- *Calculation and interpretation of analytical data* (Chemo metrics).

5.2 Volumetric Analysis (Titrimetry)

Volumetric methods of analysis in which the final measurement of substance sought is made by direct or indirect measurements of volume. The volumetric methods also known as titrimetric methods are faster, provided good standard indicators are available. In this method a substance to be analysed (Analyte) is allowed to react with another substance (Reagent) of known concentration from the burette in the form of solution. The concentration of the analyte is determined by a careful measurement of the volume of the reagent solution required to react with analyte solution.

The reactions are:

- Neutralization reactions involving the determination of acids and bases.
- oxidation- reductions reaction involving the determination of reducing agents such as Fe^{2+}, $C_2O_4^{2-}$, Sn^{2+}, As^{3+}, and oxidizing agents such as Cu^{2+}, I_2, $Cr_2O_7^{2-}$, MnO_4^- etc.
- Precipitation reactions like AgCl, AgSCN etc.
- Complex formation reactions, which involve the formation of a stable complex ion or a slightly dissociated molecule, e.g., EDTA titrations.

The principles of volumetric method are the use of chemical reaction between analyte and reagent which should be fast and quantitative.

For e.g., the reaction such as m A + n R → Products .....(5.1)

Where m = m, molecules of analyte

n = n, molecules of reagent

5.2.1 Equipment used for Volumetric Analysis

(i) ***Burettes:*** These are used to deliver variable volumes. There are long tubes of clear glass (Borosil) provided with a stopcock (or Poly tetra fluoroethylene / Teflon) at one end and are generally graduated in units of 1/10 millilitre (Fig 5.1a).

The burettes are to be standardized. The burette's capacity varies and is available in 1 ml, 2 ml, 5 ml, 10 ml, 25 ml, 50 ml and 100 ml burettes. Presently advanced burettes are available.

(ii) ***Pipettes:*** Pipettes are designed to deliver a fixed volume (Fig.5.1b). The usual capacities are 5 ml, 10 ml, 15 ml, 20 ml, 50 ml. At tip certain amount of liquid should remain which can never blown and touch the tip of the pipette on top of the solution. The pipettes are also to be calibrated for accurate volumes. Presently advanced automatic pipettes and advance pipettes without sucking are available.

(iii) ***Volumetric flasks:*** There are designed usually to contain a given volume (Fig. 5.1c). The commonly used capacities of the flasks are 100, 250, 500 and 1000 ml. The other capacities are also available.

The capacity of glass vessel varies with temperature and it is therefore necessary to define the temperature, at which its capacity intended to be correct.

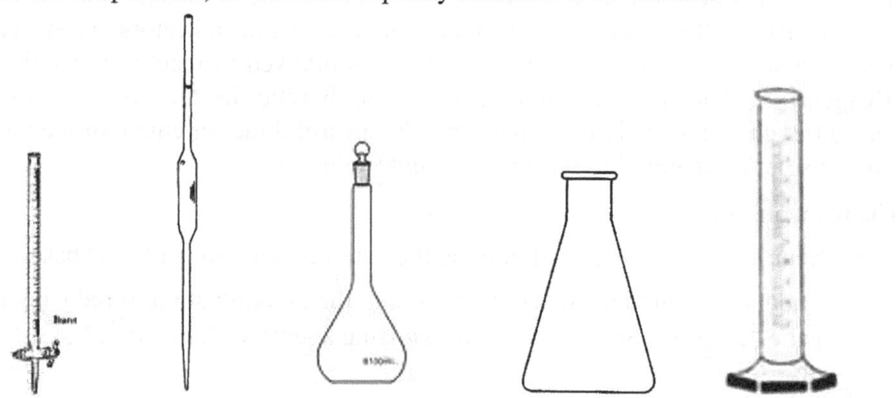

(a) Burette (b) Pipette (c) Volumetric flask or measuring flask (d) Conical flask (e) Measuring jar

Fig. 5.1

(iv) **Conical flasks:** Conical flasks (Fig. 5.1d) are used for taking volumes of the reagents before titration. The various capacities of the conical flasks are 25 ml, 50 ml, 100 ml, 250 ml, 500 ml, 1000 ml.

(v) **Measuring jars:** These are also graduated glass jars (Fig 5.1e) used for rough measuring of volumes in the preparation of secondary standard solutions, percentage solutions etc. The varying capacities from 5 ml to 1000 ml are available.

The other equipments necessary are Beakers of various capacities, weighing bottles for preparation of standard solutions, Balances (Analytical as well as rough).

5.2.2 Conditions for use in Analysis by Titration

A reaction must fulfill the following conditions to be suitable for use in analysis by titration.

- it must go practically to completion when chemically equivalent amounts of the reacting substances are present
- it must be practically instantaneous.
- The end point must be sharply defined by some change in colour or the formation of a precipitate. When the reagents themselves do not make possible, a third substance, known as an indicator is added which reacts only when the main reaction is completed.

5.2.3 Titrations

The process of measuring the volume of a solution that is required (titre) to complete a reaction is known as titration. These titrations are two types: Direct titration methods and indirect titration methods.

***Direct Titration*:** In the direct titration method, at the equivalence point neither the analyte nor the reagent present in excess. So known volume of the analyte directly titrated with reagent taken in the burette.

For e.g., titration of ferrous ammonium sulphate with standard hypo solution in the determination of iron (II)

Indirect Titration: In this method, to a known volume of the analyte, excess reagent (100 to 200%) is added. The reaction mixture is kept for some time is heated so that the reaction is complete. Then the unreacted reagent is determined by titration with another suitable reagent. Knowing the amount of the reagent consumed, the quantity of analyte can be calculated.

For e.g., Determination of formic acid. Excess $KMnO_4$ is added to HCOOH and heated to complete the reaction. After the reaction was over the remaining $KMnO_4$ can be estimated by titrating with standard oxalic acid.

Equivalence point and End point: *In the titration, the point at which the reaction between a analyte (titrand) and the reagent (titrant) just complete is known as equivalence point or stochiometric end point.* The point at which an indicator changes its colour is known as the end point. The visible (experimental end point) coincides with the theoretical end point. The titration is ideal, but in general there is small difference occur which is known as end point error. The end point of the titration is detected by a sudden change in some property of the reaction mixture.

Standard Solution: *A Solution whose concentration is accurately known is called a standard solution.* All titrimetric methods depend on standard solutions which contain an exactly known amount of the reagent in unit volume of solution. The concentration is expressed in terms of normality (gram eq l^{-1}). All reagents are not available in pure form. The substances which are available in pure form with definite composition are called primary standards. The commonly used primary standards are;

- *Acids* C_6H_4 (COOK) (COOO), *Potassium hydrogen phthalate,* C_6H_5COOH: constant boiling HCl, $KHIO_3$,Tl_2CO_3, Succinic acid, Adipic acid H_2 ($C_6H_8O_4$)
- *Bases*: Na_2 CO_3, Mg O and $Na_2B_4O_7$ (borax)
- Oxidising agents: K_2 Cr_2O_7, $(NH_4)_2$ $Ce(NO_3)_6$, $KBrO_3$, I_2
- Realusing agents: $Na_2C_2O_4$, $Fe(N_2H_4N_2H_6)$ $(SO_4)_2$ $4H_2O$, $K_4[Fe(CN)_6]$
- Other O_3, NaCl, KCl, $AgNO_3$, Ag

Directly the standard solution can be prepared by an exactly weighed amount of the pure substance is dissolved and made up to known volume in a volumetric flask. From the known weight and volume, the concentration of the solution is calculated.

The standard solution can be prepared indirectly after standardization. All substances are not available in pure form for *e.g.*, $KMnO_4$ contains MnO_2 impurities, $Na_2S_2O_35H_2O$ (Sodium thiosulphate or Hypo) contains uncertain water content and are known as secondary standard substances. In these cases, a solution of approximate concentration is prepared and then titrated with primary standard solutions i.e. $KMnO_4$ with $Na_2C_2O_4$, $Na_2S_2O_3$ $5H_2O$ with KIO_3 etc. and then calculate the exact concentration for the use in titration.

5.2.4 The Steps Involved in Titrimetric Determination

- A Standard solution of the reagent to be prepared by weighing known amount of the reagent dissolved in fixed volume of water in a volumetric flask.

  For e.g., 100 ml or 250 ml so that the concentration can be calculated.
- The sample solution also called as titrand or analyte or test solution of known volume is pipetted out (with the help of pipette) into a conical flask.

- Add few drops of suitable indicator. The function of the indicator is to show when the reaction between the analyte and the reagent is just complete.
- The titrant solution (standard solution of the reagent) is taken in the burette and gradually added to titrand taken in the conical flask by swirling the contents continuously till the reaction between them is just complete which indicated by the colour change of the indicator (the drop which show the end point to be observed carefully).
- The titre value is recorded (say for e.g., X ml).
- Then calculate the normality and amount of the sample present by using the formula.

$$V_1 N_1 = V_2 N_2 \qquad .....(5.2)$$

Where V_1 = Volume of the burette reading (titrant)

N_1 = Normality of the sample (titrant)

V_2 = Volume of the titrand (Standard solution)

N_2 = Normality of known titrand

V_2, N_2, V_1 can be known, then N_1 we can be calculated

$$N_1 = \frac{V_2 N_2}{V_1} \qquad .....(5.3)$$

The amount of the substance present can be calculated by using the formula

Amount of sample present 1000 ml = normality × equivalent weight

.....(5.4)

$$\text{Amount of sample present in 250 ml} = \frac{\text{normality} \times \text{eq.wt}}{4} \qquad .....(5.5)$$

Like this we can calculate the amount present in the prepared sample solution

5.2.5 Computation of Results

- Burette readings are recorded to a second decimal place.
- All weights are recorded to the fourth decimal place with analytical balance.
- The average two burette readings i.e., 40.25 and 40.28 ml is 40.365 ml but the average is rounded off to 40.27.
- In rounding off the figures, the general rule is that if the last digit is ≥ 5 increase the digit by 1 unit.

For *e.g.,* 26.426 and 26.428 is rounded off to 26.43, while 26.421 and 26.424 is rounded off to 26.42

- In addition and subtraction of numbers, only those digits should be retained to the right of the decimal point that correspond to the least no. that appears in any of the numbers to be added or subtracted.

  For e,g., to add 72.95, 1.052 and 5.436 proceed as

$$72.95 + 1.05 + 5.44 = 79.44$$

- In multiplication also the same procedure

  For e.g., multiply $0.361 \times 0.14 \times 0.228$ as

$$0.36 \times 0.14 \times 0.23 = 0.12$$

In any series of arithmetical operations, one must remember that final result can never show accuracy greater than the least accurate of any number that enters in the computation. Most analytical data are expressed to four significant figures and hence, the result, in general, is at most expressed to four significant figures.

5.2.6 Indicators

The end point in the titration is indicated by indicator due to colour charge of solution. The types of indicators are;

(i) ***Internal indicator:*** The chemical substances which are added to the conical flask in which the titration is carried out.

e.g., phenolphthalein, methyl red, diphenyl benzidine.

(ii) ***External indicator:*** The chemical substances are not added to the conical flask. These are taken on a glacial tile and during titration now and then the drop of the solution from the conical flask is added to the indicator.

e.g., the titration of $K_2Cr_2O_7$ with $FeSO_4$. The external indicator is $K_3[Fe(CN)_6]$. This indicator gives blue colour with Fe(II) and light brown with Fe(III). The end point is indicated by light brown colour.

(iii) ***Self indicator:*** Coloured solution is decolourised at the end point.

e.g., the titration of $KMnO_4$ with oxalic acid. End point is colourless to pink. During titration formed MnO_2 act as self indicator.

(iv) ***Metal ion indicators:*** These contain chelating groups which forms complex with metal ion to produce different colours. At the end point the original colour of the indicator appears.

e.g., Murexide, Xylenenol orange, Eriochrome black – T.

(v) ***Precipitation indicators:*** The chemical substances used in the fractional precipitation of two insoluble salts by the same reagent.

For e.g., K_2CrO_4 is used as indicator in the titration of Ag NO_3 against NaCl in the determination of chloride

(vi) ***Redox indictors:*** The chemical substances used to mark the sudden change in the oxidation potential in the neighbourhood of the equivalence point in the oxidation reduction titration.

(vii) ***Adsorption indicators:*** The chemical substances absorb certain ions at the equivalence point and a typical orientation is developed at the surface which gives a characteristic colour change indicating the end points.

e.g., fluorescein, eosin.

5.2.7 Types of Volumetric Methods

(i) ***Acid–base titrations or Neutralisation titration*:** The titrations involve strong or weak acids or bases.

(ii) ***Redox Titration:*** The titration which involve simultaneous oxidation–reduction reaction.

(iii) ***Precipitation titrations:*** The titrations which involve the formation of precipitate.

(iv) ***Complexometric titration:*** The titration largely involves EDTA titrations, There are specific and also complexation depends on pH.

(a) Acid-base or Neutralisation Titrations (Acidimetry-Alkalimetry in aqueous solutions) :

Principles: Acid-base titrations in aqueous solutions are essentially based on the reaction between H^+ and OH^- ions to form water which is ionized to a slight extent.

$$H^+ + OH^- \rightarrow H_2O \qquad .....(5.6)$$

or hydronium ion H_3O^+ used instead H^+, then

$$H_3O^+ + OH^- \rightarrow 2H_2O \qquad .....(5.7)$$

The hydrogen ion terminology is used for simplicity. Then

$$H_2O \rightarrow H^+ + OH^- \qquad .....(5.8)$$

As per the equilibrium principle

$$\frac{[H^+]+[OH^-]}{H_2O} = K \text{ or } [H^+]\,[OH^-] = Kw$$

Where Kw = ionization constant

$$= 10^{-14} \text{ at } 25\ ^{\circ}C$$

It varies with temperature i.e. from 0.12×10^{-14} at 0 °C to 58.2×10^{-14} at 100 °C

In pure water $[H^+] = [OH^-] = \sqrt{Kw}$ .....(5.9)

Water is said to be neutral then $[H^+] = 10^{-7}$.

If acid is added, the water equilibrium is disturbed

So that $[H^+]$ is not equal to $[OH^-]$ and $[H^+]$ increases and if a base is added to water $[H^+]$ decreases and $[OH^-]$ increases.

For e.g., 0.1 M H^+ ion present in water

$$[0.1]\,[OH^-] = 10^{-14}$$

$$[OH^-] = 10^{-14}/0.1 = 10^{-13}$$

Sorensen suggested the notation pH for H^+ ion concentration.

$$pH = \log 1 / [H^+] = -\log [H^+]$$

$$p\,OH = -\log [OH^-]$$

$$pH + pOH = pKw = 14$$

Problem: What is the pH of a solution that contain $[H^+]$ of $\times 10^{-4}$?

$$pH = \log 1/ 5 \times 10^{-4} = \log 10^4/5 = \log 10^4 - \log 5$$

$$= 4 - 0.699 = 3.33$$

So a solution at room temp is :

Acidic if $pH < 7$ or $pOH > 7$

Basic if $pH > 7$ or $pOH < 7$

Neutral if $pH = pOH = 7$

Acid base titration gives sharp end point with an indicator if the pH at the equivalence point is within 4-10. During titration the pH of the solution changes in a typical manner, showing a sharp change in pH with the volume of the titrant at equivalence point. The acid base reactions

$$\text{Acid} + \text{base} \rightleftharpoons \text{conjugate base} + \text{conjugate acid} \quad .....(5.10)$$

Eg. $HA + H_2O \rightarrow H_3O^+ + A^-$ (base) .....(5.11)

$HA + H_2O \rightarrow BH^+ + OH^-$ (acid) .....(5.12)

where A^- conjugate base

H^+B conjugate acid

Acid dissociation constant

$$K_A = \frac{[H_3O^+]+[A^-]}{[HA]\ [H_2O]} \quad .....(5.13)$$

Base dissociation constant

$$K_B = \frac{[H^+B]+[OH^-]}{[B^-]} \quad .....(5.14)$$

Ionic product of water K_w is given by eq. 5.8

Indicators for Acid and Base Titration

The indicator substances are in general, organic compounds that are weakly acidic or basic in character. For an indicator, the equilibrium is given as;

$$\underset{\text{Acid form}}{H\,In} \rightleftharpoons \underset{\text{basic form}}{H^+ + In^-} \quad .....(5.15)$$

The H^+ ion concentration determines the indicator colour of the form which predominates since if H^+ concentration increases, the equilibrium will be shifted to the left and if it is decreased to right.

Some selected indicators for acid-base titrations are given in Table 5.2

Table 5.2

S.No	Indicator	Colour of acid form	pH Range of titration	Colour of basic form
1	Thymolphthalein	Colourless	9.4-10.6	Blue
2	Thymol Blue	Yellow	8.0-9.6	Blue
3	Phenolphthalein	Colourless	8.3-10.0	Red
4	Bromothymol Blue	Yellow	6.0-7.6	Blue
5	Bromocresol Purple	Yellow	5.2-6.8	Purple
6	Methyl Red	Red	4.2-6.3	Yellow
7	Bromocresol Green	Yellow	3.8-5.4	Blue
8	Methyl Orange	Red	3.2-4.4	Yellow
9	Bromophenol Blue	Yellow	3.0-4.7	Purple
10	Methyl Yellow	Red	2.9-4.0	Yellow

Table 5.2 *Contd...*

S.No	Indicator	Colour of acid form	pH Range of titration	Colour of basic form
	Mixed indicators (produce colour change within a narrow range)			
11	Cyanole (Methyl Orange +Xylene) 2 : 3	Voilet	3.2-4.2	Green
12	Carmine (Methyl Range +Indigo	Voilet	3.8-4.8	Green

The selection of an indicator for particular titration is very important.

Fluorescent indicators: Acid base indicators cannot be used in sparingly coloured or highly turbid Solutions. For this purpose an indicator which shows fluorescence can be used. Important fluorescent indicators are listed in Table 5.3.

Table 5.3 Fluorescent Indicaors.

S.No	Name	Colour of acid form	pH Range	Colour of basic Form
1	Rosin	None	0-3.0	Green
2	Salicylic acid	None	0.2-4.0	Blue
3	α– Naphthylanine	None	3.4-4.8	Blue
4	Dichlorofluorescein	None	4.0-6.6	Green
5	Quinine	Blue violet	9.5-10.0	None
6	Quinoline	Blue	6.2-7.2	None

Titration of Strong Acid-Strong Base

Suppose 25 ml of 0.1N HCl acid titrated with 0.1N NaOH. Initially the pH of HCl solution close to 1. If NaOH is added from burette, neutralization of acid begins. The change in pH is shown in titration curve (Fig. 5.2).

$$HCl + NaOH \rightarrow NaCl + H_2O \qquad .....(5.16)$$

The pH is nearly 3 when 24.5 ml of base added and would increase to 4.4 for the addition of 24.98, when exactly 25 ml of base, pH is 7 and after 25.02, the pH is 9.6. The equivalence point is at 7. *The pH interval between two turning points is called infliction. The inflection range in this is 4.5 to 9.5.*

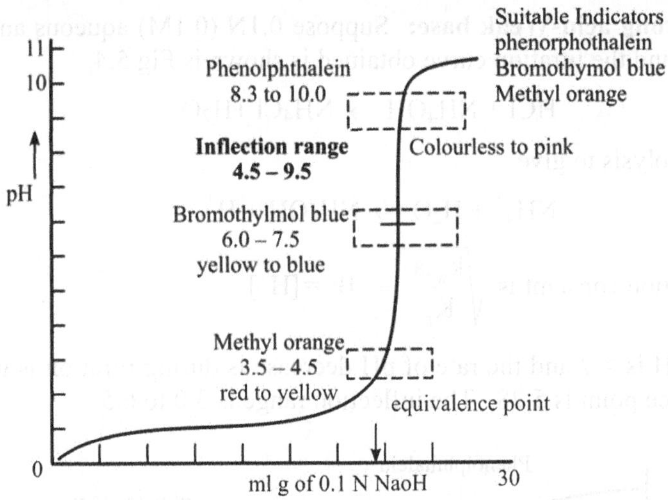

Fig 5.2 Titration curve of 0.1 N HCl with 0.1N NaOH.

Titration of Weak Acid-Strong Base

Suppose 25 ml of 0.1N acetic acid titrated with 0.1N NaOH and the titration curve obtained is shown in Fig. 5.3

$$CH_3COOH + NaOH \rightarrow CH_3COONa + H_2O \quad .....(5.17)$$

$$CH_3COO^- + H_2O \rightarrow CH_3COOH + OH^- \quad .....(5.18)$$

The dissociation constant is $\sqrt{\frac{K_A.K_w}{a}} \simeq [H_A] = [OH^-]$

The initial pH is higher and the rate of pH increase during titration is greater. The pH at equivalence point is 8.7. The inflection range in this titration is 7.8 to 9.7.

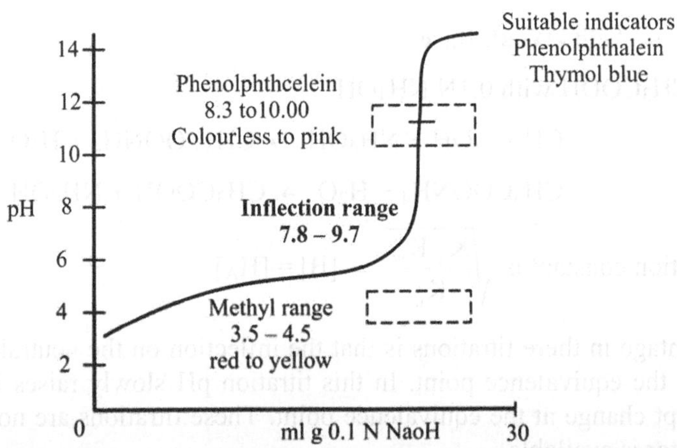

Fig 5.3 Titration curve of 0.1 N CH_3COOH with 0.1 N NaOH.

Titration of strong acid-Weak base: Suppose 0.1N (0.1M) aqueous ammonia solution with 0.1N HCl and the titration curve obtained is shown is Fig 5.4.

$$HCl + NH_4OH \rightarrow NH_4Cl + H_2O \quad .....(5.19)$$

NH_4Cl hydrolysis to give

$$NH_4^+ + H_2O \rightarrow NH_4OH + H^+ \quad .....(5.20)$$

The dissociation constant is $\sqrt{\frac{k_W.a}{K_B}} \simeq B^{\mu} = [H^+]$

The initial pH is > 7 and the rate of pH decrease is during titration is greater. The pH at the equivalence point is 5.28. The inflection range is 3.0 to 6.5

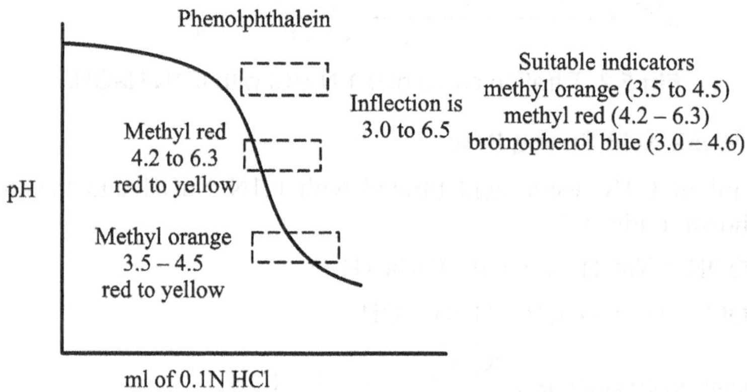

Fig 5.4 Titration curve of 0.1 N HCl with 0.1N NH_4OH.

Titration of Weak Acid -Weak Base

Suppose 0.1N CH_3COOH with 0.1N NH_4OH

$$CH_3COOH + NH_4OH \rightarrow CH_3COONH_4 + H_2O \quad .....(5.21)$$

$$CH_3COONH_4 + H_2O \rightarrow CH_3COOH + NH_4OH \quad .....(5.22)$$

The dissociation constant is $\sqrt{\frac{K_A K_W}{K_B}} \simeq [B] = [H_A]$

The disadvantage in there titrations is that the inflection on the neutralization curve is very small near the equivalence point. In this titration pH slowly raises from 6 to 8 i.e. there is no abrupt change at the equivalence point. These titrations are not practicable as no visual indicator is available.

Titration of Weak Polybasic Acid with Strong Base

A polybasic acid means the basicity of that acid is more than one i.e. which contains more than one ionisable hydrogen. These are also called as polyprotic acids.

For eg., H_2CO_3, H_2SO_3, H_2SO_4 – dibasic

H_3PO_4 – tribasic

Suppose 0.1N phosphoric acid with 0.1N NaOH. Stepwise titration of the acid with alkali is possible. The pH of the first end point will be 4.67 (suitable indicator is Methyl orange or bromocresol green). The pH of the second end point, which is not very sharp, is 9.6 (suitable indicators phenolphthalein or thymolphthalein). The third end point can not be detected because the value of K_3 is very small. The titration curve is shown in Fig 5.5. The dissociation constant is $\sqrt{K_1K_2K_3} \simeq [H_3A] = [A^-]$

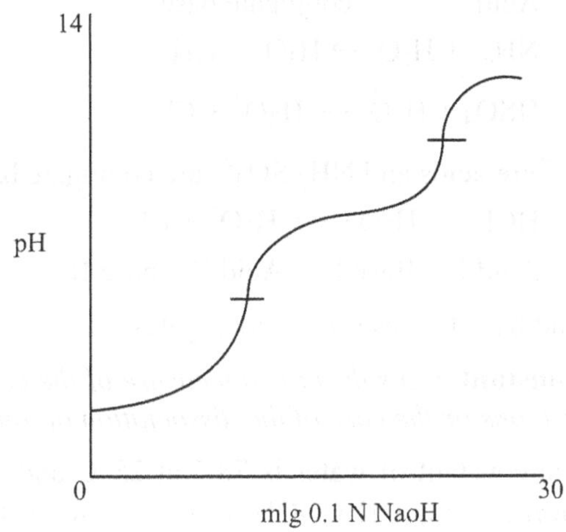

Fig 5.5 Titration 0.1 N H_3PO_4 with 0.1N NaOH.

Acid-base Titrations in Non-aqueous Solutions

The Non-aqueous solutions can be categorized as;

- **Aprotic solvents:** Chemically neutral and virtually unreactive. e.g., CCl_4, C_6H_6.
- **Protophilic solvents:** There chemical substances have high affinity for protons. e.g. Liquid NH_3, amines, ketones.
- **Protogenic solvents:** There chemical substances are acidic in nature and readily donate protons.

  e.g., HF, H_2SO_4

- **Amphiprotic Solvents:** These chemical substances consist of liquids such as water, alcohols and weak organic acids.

$$CH_3OH + CH_3OH \rightleftharpoons CH_3OH_2^+ + CH_3O^- \quad .....(5.23)$$

$$H_2O + H_2O \rightleftharpoons H_3O^+ + OH^- \quad .....(5.24)$$

Autoprotolysis: no other substance is required for the transfer of protons

The conjugate acid and bases:

An acid differs from conjugate base only by proton. Suppose HB acid dissociate to give H^+ and B^-, B^- is the conjugate case of acid HB since B^- is having affinity for H^+.

$$HB \rightleftharpoons H^+ + B^- \quad .....(5.25)$$

Acid conjugate base

For e.g., $NH_4^+ + H_2O \rightleftharpoons H_3O^+ + NH_3$ .....(5.26)

$$HSO_4^- + H_2O \rightleftharpoons H_3O^+ + Cl^- \quad .....(5.27)$$

Here NH_4^+, HSO_4^- are acids and NH_3,SO_4^{2-} are conjugate bases

$$HCl + H_2O \rightleftharpoons H_3O^+ + Cl^- \quad .....(5.28)$$

Acid I Base I Acid II base II

Acid I – base II and base I – base II are conjugates

***The Dielectric Constant** of a solvent is a measure of the ease of separation of two oppositely charged ions or the ease of the dissociation of ion- pair in that solvent.*

For e.g., Dielectric constant of water is 78.5 at 25°C and of ethanol is 24.3. The dissociation of acetic acid, CH_3COOH is greater in water when compared to ethanol. CH_3COOH is very weak acid in ethanol.

The Autoprotolysis Constant:

For Amphiprotic solvent CH_3OH, the autoprotolysis is

$$CH_3OH + CH_3OH \rightarrow CH_3OH_2^+ + CH_3O^- \quad .....(5.29)$$

Then autoprotolysis constant is

$$K_S = [CH_3OH_2^+]\,[CH_3O^-] \quad .....(5.30)$$

This is in methanol is 16.7, in ethanol 19.1. The smaller the value of autoprotolysis constant greater is the inflection on the titration curve.

So, while selecting the non-aqueous solution for Acid-base titration, solvent behaviour, the value of autoprotolysis constant and dielectric constant must be taken into consideration.

Suitable Solvents for Non-aqueous Titrations

Both inorganic and organic solvents solvents are used in non-aqueous determinations. The important solvents are;

- glacial acetic acid (ethanoic acid)
- acetonitrile (methyl cyanide, cyano methane) is frequently used with other solvents such as chloroform and phenol.
- alcohols – used for determination of organic acids.
- dixon.
- dimethyl formamide (DMF)

Indicators for Non-aqueous Titration

Commonly used indicators are;

- *Crystal violet in 0.5% glacial acetic acid:* This is used in the titration of pyridine with perchloric acid. The colour change at the end point is violet to greenish yellow.
- *1-Naphol benzein in 0.2% acetic acid:* This is used in the titration of weak base nitro methane contain acetic anhydride with perchloric acid. The colour change at the end point is yellow to green.
- Oracet blue B in 0.5% acetic acid. This is used in the titration of bases with perchloric acid. The colour change is blue to pink.
- *Quinoldiene red in 0.1% ethanol:* This is used for drug determination in DMF. The colour change is purple to pale green.
- *Thymol blue in 0.2% methanol:* This is used for determination of acid substances in DMF. The colour change is yellow to blue.

The other examples of non-aqueous determination are;

- Determination of a mixture of aniline and ethanolamine with perchloric acid using the indicator crystal violet.
- Determination of saponification value of oils and fats by titration with HCl using phenolphthalein indicator.

Redox Titration (Oxidation-Reduction Titration):

Oxidation: *It is defined as a reaction in which the oxidation state of an element becomes more +ve or less –ve. Also defined as addition of oxygen.*

Reduction: *It is defined as a reaction in which the oxidation state of an element becomes more –ve or less +ve.*

Also defined instead of oxygen, usually through the use of hydrogen.

The change in oxidation state of an element is accomplished by an exchange of an electron between the substance oxidised and the substance reduced.

Oxidation: Reduction processes are mutually dependent. The degree of intensity of oxidising or reducing action is measured by the reduction potential.

Two methods in balancing Ox-red equations are;

- Change in oxidation state, which is measured by oxidation numbers.
- Using of half - reaction.

Oxidant and Reductant

Oxidant (or oxidising agent) is a substance which gains electrons and itself get reduced.

For e.g., $Cu^{2+} + e^- \rightarrow Cu^+$ .....(5.31)

$MnO_4^- + 8H + 5e^- \rightarrow Mn^{2+} + 4H_2O$ .....(5.32)

Reductant (or reducing agent) is a substance which loses electrons and itself get oxidised

For e.g. $Fe^{2+} + (-e^-) \rightarrow Fe^{3+}$ .....(5.33)

$I^- + (-e^-) \rightarrow 1/2\ I_2$ .....(5.34)

Oxidation Numbers Method

The Oxidation numbers

- of all elements in the free state is zero.
- In all compounds i.e., the sum of oxidation numbers of elements composing the compound, is zero.

For e.g. $K\,MnO_4$

$1 + 7 - 8 = 0$

$K_2\,CrO_4$

$2 + 6 - 8 = 0$

- in all ions the sum of oxidation numbers of elements composing the ions is equal to the charge on the ion.

For e.g., $S\,O_4^{-2}$

$+6 - 8 = -2$

$Cr_2\,O_7^{-2}$

$+12 - 14 = -2.$

- Certain elements in combination or ion have fixed oxidation number or valence.

For e.g., H^+, Na^+ have +1

Ca^{2+}, Ba^{2+}, In^{2+}, have +2

Cl^-, OH^- have –1 etc.

- The common exception for this rule is H = –1, hydride.

The ionic equations are, in general preferred in place of molecular equations since they are more characteristic of many reactions.

For e.g., for the equation

$$K_2Cr_2O_7 + 6\,FeSO_4 + 7H_2SO_4 \rightarrow Cr_2(SO_4)_3 + 3\,Fe_2(SO_4)_3 + K_2SO_4 + H_2O \quad .....(5.35)$$

the ionic equation can be written as;

$$Cr_2O_7^{2-} + 6Fe^{2+} + 14H^+ \rightarrow 2Cr^{3+} + 6Fe^{3+} + 7H_2O \quad .....(5.36)$$

Half reaction method

Any oxidation-reduction (ox-red) equation may consist of two separate half reactions.

For e.g., reduction reaction (electrons are gained) of dichromate in acid solution, two electrode reaction is written as;

$$Cr_2O_7^{2-} + 14\,H^+ + 6e^- \rightarrow 2Cr^{3+} + 7H_2O \quad .....(5.37)$$

The electrons flow from other electrode, and must be produced these by a reaction in which electrons are discharged at this electrode.

$$Fe^{2+} \rightarrow Fe^{3+} + e^- \quad .....(5.38)$$

The equation 5.37 and 5.38 combined to give direct reaction between the oxidising and reducing substance

$$Cr_2O_7^{2-} + 14\,H^+ + \cancel{6e^-} \rightarrow 2Cr^{3+} + 7H_2O$$

$$6Fe^{2+} \rightarrow 6Fe^{3+} + \cancel{6e^-}$$

$$Cr_2O_7^{2-} + 14H^+ + 6Fe^{2+} \rightarrow 2Cr^{3+} + 6Fe^{3+} + 7H_2O \quad .....(5.39)$$

Reduction-Oxidation Potentials

Reduction potential is that one ion is more easily reduced (gain electrons) than the other ion.

For eg. $Cu^{2+} + 2e^- \rightarrow Cu$ Electrode potential + 0.34 volts

$Zn^{2+} + 2e^- \rightarrow Zn$ – 0.76 volts.

Oxidation potential is that one ion is more easily oxidised than the other.

$$Zn \rightarrow Zn^{2+} + 2e^{-} \qquad + 0.76 \text{ volts}$$

$$Cu \rightarrow Cu^{2+} + 2e^{-} \qquad - 0.34 \text{ volts}$$

The oxidation potential of any ox- red system is therefore numerically equal to reduction potential but of opposite sign. The standard potentials are listed in table 5.4. The relative oxidising ability decreases by the decreasing +ve potential of the half reaction of oxidising agent.

Table 5.4

S.No	Half Reaction	Standard Reduction Potential.(Volts)
1	$MnO_4^{-} + 4H^{+} + 3e^{-} = MnO_2 + 2H_2O$	+ 1.69
2	$Ce^{4+} + e^{-} = Ce^{3+}$ (Nitrate Medium)	+ 1.61
3	$Cr_2O_7^{2-} + 14H^{+} + 6e^{-} = 2Cr^{3+} + 7\ H_2$	+ 1.33
4	$IO_3^{-} + 6H^{+} + 5e^{-} = ½\ I_2 + 3H_2O$	+ 1.20
5	$Cu^{2+} + I^{-} + e^{-} = Cu\ I$	+ 0.86
6	$Fe^{3+} + e^{-} = Fe^{2+}$	+ 0.77
7	$S_4O_6^{2-} + 2e^{-} = 2S_2O_3^{2-}$	+ 0.08
8	$Cr^{3+} + e^{-} = Cr^{2+}$	– 0.41
9	$Fe^{2+} + 2e^{-} = Fe$	– 0.44

The Nearest equation for an Ox-red electrode reaction at 25°C is

$$E = E_O + \frac{0.059}{n} \log \frac{[OX]}{[red]} \qquad .....(5.40)$$

Where E = observed potential

E_O = characteristic half reaction potential given in table 5.4.

n = number of electrons involved in the reaction.

[ox] = molar concentration. of oxidised form

[red] = molar concentration of reduced form

The quantity $E = E_O$, when the value of [ox] = [red]. Then equation 5.40 becomes

$$E = E_O + 0.059/n \log 1 = E_o - 0 = E_o \qquad .....(5.41)$$

To predict the effectiveness of reaction, it is necessary to know the potentials for the cell systems and by combining any two of the half reactions.

For e.g., $Cr_2O_7^{2-}$ should oxidise every reducing agent which has the electrode potential < 1.36 V. It can oxidise Fe^{2+} = + 0.77, Cu^+ = + 0.15 etc. Some ox-red half reactions involve more than two single ions.

For e.g., MnO_4^- in water is + 0.60 where as in acid + 1.51.

Indicators for Ox-Red Titrations

Some redox indicators are given in table 5.5. The principle of redox indicators different from acid-base indicators. The organic agents used as redox indicators are highly coloured in either their oxidised or reduced form

Table 5.5

No.	Indicator	Colour Change		Transition Potential Volts
		Oxidised form	Reduced form	
1	5- Nitro -1,10-phenolphthalein iron(II) sulphate (nitro ferroin)	Pale Blue	Red	+ 1.25
2	1, 10 phenolphthalein iron (II) sulphate (ferroin)	Pale Blue	Red	+ 1.11
3	Diphenyl amine	Violet	Colour less	+ 0.76
4	p - Ethoxy chrysodine	Yellow	Red	+ 0.76
5	Diphenyl benzidine	Violet	Colour less	+ 076
6	N- phenyl anthranilic acid	Purple Blue	Colour less	+ 1.08
7	Methylene blue	Blue	Colour less	+ 0.53
8	5,6 dimethyl ferroin	Pale Blue	Red	+ 0.97
9	Indigo tetrasulphate	Blue	Colourless	+ 0.36

These indicators are of two types:

(i) Specific indicators are (like starch, KSCN) ones which react with one of the participants in titration. These are not very much preferred.

(ii) True redox indicators respond to the potential of the system. The half reaction is responsible for colour change. The reaction can be shown as;

$$In_{ox} + n\,e^- \rightleftharpoons In_{red} \qquad ....(5.42)$$

$$\therefore \qquad E = E_o - \frac{0.0591}{n} \log \frac{[\text{In red}]}{[\text{In} - \text{ox}]} \qquad .... (5.43)$$

The In_{red} /In_{ox} changes from 0.1 to 10 with the result the indicator will exhibit a detectable colour change. *The potential at which a colour transition takes place depends on the standard potential of a particular indicator system.*

There indicators are:

(i) Self indicator: i.e., reagent serving as its own indicator.

For e.g., the titration of Fe^{2+} with standard $KMnO_4$. The end point is colourless to pink colour

(ii) External indicator: The indicator is not added to the titrand solution but the titration liquid out from time to time perform the spot test. These are being replaced by more satisfactory internal redox indicators. For e.g., titration of Fe^{2+} solution with standard $K_2Cr_2O_7$. The external indicator is $K_3[Fe(CN)_6]$. If the reaction is complete, the solution does not give blue colour.

(iii) Internal indicator: An ideal redox indicator should mark he sudden change in the oxidation potential in the neighborhood of the equivalence point in redox reactions. The best indicators will be those which have an oxidation potential intermediate between that of solution titrated and which exhibit sharp and point.

For e.g., 10-phenanthroline

$$[Fe(phen)_3]^{3+} + e^- \rightleftharpoons [Fe(phen)_3]^{2+} \quad E_o = 1.06V \text{ to } 1.11V \qquad .....(5.44)$$

Pale Blue red

The exchange of electrons proceeds through aromatic rings.

Redox Titration Curves

The titration curves in redox reactions are to plot a graph of e.m.f (electrode potential in volts verses standard cell electrode, SCE) of E cell verses volume of titrant i.e., the change in potential of system is considered during titration (Fig 5.6)

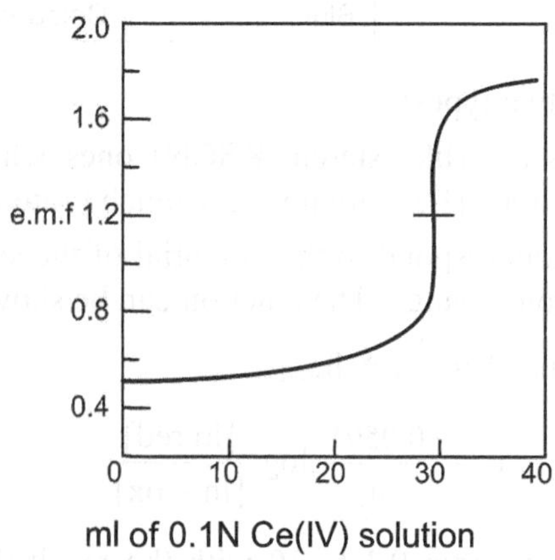

Fig 5.5 Titration curve for Fe(II) with Ce(IV).

For e.g., the titration of Fe^{2+} vs Ce^{4+} is

$$Fe^{2+} + Ce^{4+} \rightleftharpoons Fe^{3+} + Ce^{3+} \quad .....(5.45)$$

The equivalence point potential is 1.06

Examples of oxidising agents are:

- Ceric sulphate, $Ce(SO_4)_2$, ammonium ceric nitrate $(NH_4)_2Ce(NO_3)_6$. Good indicator is 1, 10 phenonthroline
- Potassium permanganate, no indicator is required.

  $$MnO_4^- + 8H^+ + 5e^- \rightleftharpoons Mn^{2+} + 4H_2O \qquad E_o = 1.51 \text{ v} \quad .....(5.46)$$

- Potassium dichromate.

  $$Cr_2O_7^{2-} + 14H + 5e^- \rightleftharpoons Cr^{3+} + 7H_2O \qquad E_o = 1.33 \text{ v} \quad .....(5.47)$$

- Potassium bromate.

  $$BrO_3^- + 6H^+ + 6e^- \rightleftharpoons Br^- + 3H_2O. \qquad E_o = 1.44 \text{ v} \quad .....(5.48)$$

  Good indicators are α - naphthaflavone, quinoline yellow

- Potassium iodate in Andrew's titration.

  $$IO_3^- + Cl^- + 6H^+ + 4e^- \rightarrow ICl + 3H_2O \qquad Eo = +1.20 \text{ v} \quad .....(5.49)$$

  Self indicator i.e., disappearance of violet clouration.

Titration of Ox-red

- Determination of ferrous iron using standard $KMnO_4$ solution
- Determination of total iron in iron ore using standard $KMnO_4$
- Determination of H_2O_2 with standard $KMnO_4$
- Determination of persulphates with standard $KMnO_4$
- Determination of nitrates with standard $KMnO_4$
- Determination of ferrous ion with standard $K_2Cr_2O_7$ indicator is diphenyl benzidine or diphenyl amine in 1% H_2SO_4.
- Determination of chlorate with standard $K_2Cr_2O_7$ using back titration.
- Determination of chemical oxygen demand (COD) with $K_2Cr_2O_7$.

Precipitation Titrations

The precipitation method is based on titration with the reactions accompanied by formation of sparingly soluble compounds. Only few precipitation reactions are used in titration. The oldest methods such those involving Cl^-, Br^-, I^- by Ag(1) (argentometric titrations) are important. These are limited due to the lack of suitable indicators to detect the end point in the reactions and due to unknown composition of the precipitate (ppt).

The conditions necessary for these titrations are:

- The precipitate must be practically insoluble
- The precipitation should be faster
- The titration results should not be distorted appreciably by adsorption (co- precipitation effect)
- The detection of the equivalence point must be possible during titration.

Argentrometry

$$Ag^+ + X^- \rightarrow Ag\,X\downarrow \qquad .....(5.50)$$

Where $X = Cl^-, Br^-, I^-$ or CNS^- etc.

Mercurometry

$$Hg^{2+} + 2X^- \rightarrow Hg\,X_2\downarrow \qquad .....(5.51)$$

Titration Curves

The titration of 50 ml of 0.1N NaCl with 50 ml of 0.1N $AgNO_3$.

(a) at the beginning

$$pCl = -\log [Cl] = -\log 10^{-1} = 1$$

(b) Add 10 ml of NaCl (volume of titrant V_t), in 50 ml of $AgNO_3$ (volume of reagent V_r), then

$$[Cl^-] = \frac{V_r.N_r - V_tN_t}{V_r + V_t} = \frac{[(50\times 0.1)-(10\times 0.1)]}{(50+10)} = 0.067$$

(c) Add 49.9 ml of Ag NO_3

$$[Cl^-] = \frac{(50\times 0.1)-(49.9\times 0.1)}{(50+49.9)} = 1\times 10^{-4}\ pCl = 4.0.$$

(d) At the equivalence point

$$[Ag^+] = [Cl^-] = [Cl^-]\,[Cl^-] = [Cl^-]^2$$

i.e., $[Cl^-]^2 = 1\times 10^{-10}$ so $[Cl^-] = 1\times 10^{-5}$ then $pCl = 5.0$

the solubility product K_{sp} is

$$K_{sp} = [Ag^+] = [Cl^-] \text{ i.e. } [Ag^+] = [Cl^-] = \sqrt{K_{sp}} = \sqrt{1.56\times 10^{-10}}$$

(e) Add 60 ml of $AgNO_3$

$$[Ag^+] = \frac{(60\times0.1)-(50\times0.1)}{(50+60)} = 9.1\times10^{-3} \text{ i.e. } pAg = 2.04$$

Since $\quad pAg^+ + pCl^- = 10$

$pCl^- = 10 - 2.04 = 7.96.$

Titration curve is useful to know the concentration of cation or anion at any point during titration. The titration curve is shown in Fig 5.6

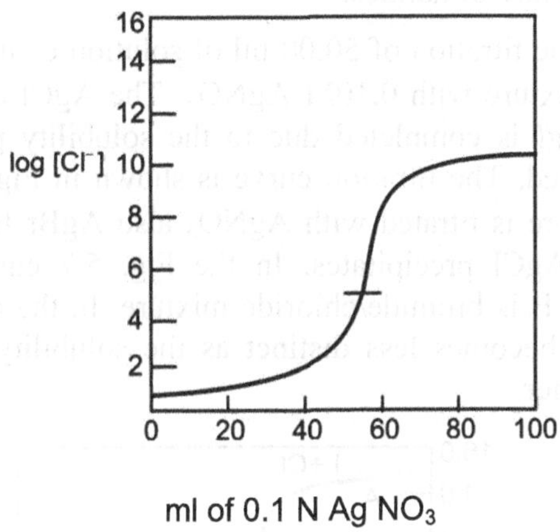

Fig. 5.6 Titration of $AgNO_3$ vs NaCl.

The lower the solubility product of the compound formed in the titration, the wider the range of the break on the titration curve.

For e.g., in the case of AgI, solubility product is 1×10^{-16}

$$K_{sp} = [Ag^+][I^-] = \sqrt{K_{sp}} = \sqrt{1.10^{-16}}$$

i.e., $\quad [I^-] = 10^{-8}$ g .ion. L^{-1}

When compared to AgCl, AgI is having lower solubility product. The change in pI begins at 4 and ends at 12 where as the change in pCL begins at 3 and ends at 7.

The temperature, nature of solvent, common ion effect, effect of diverse ion, effect of pH, effect of hydrolysis and effect of complex will influence the solubility of the precipitate.

Precipitation Titrations and Indicators

(i) Volhard's method: The titration of Ag with NH_4SCN and indictor is ferric alum . One can easily determine SCN^- ion in acidic solution.

$$AgCl + SCN^- \rightarrow AgSCN + Cl^- \quad \ldots(5.52)$$

The end point is white to red

(ii) Mohr's method: The titration of halides with $AgNO_3$: this method is used to determine the halide ions. The indicator is potassium chromate, K_2CrO_4.

The end point is white to brick red.

This method is used in the determination of I^-,Br^-, Cl^- in the mixture of anions. The Ksp values of $I^- = 8.3\times10^{-17}$, $Br^- = 5.2\times10^{-13}$ and $Cl^- = 1.8\times10^{-10}$. On the basis of differences in solubility products of Ag I, Ag Br and AgCl, it is possible to titrate with 0.1M $AgNO_3$ for determination of concentrations of iodide, bromide and chloride in mixture of halides.

For e.g. consider the titration of 50.00 ml of solution containing 0.05M iodide and 0.08M chloride mixture with 0.10M $AgNO_3$. The AgCl cannot precipitate till the precipitation of AgI is completed due to the solubility product of AgCl is high which is not reached. The titration curve is shown in Fig. 5.7. Similarly bromide and chloride mixture is titrated with $AgNO_3$ also AgBr first precipitates and then after completion AgCl precipitates. In the Fig. 5.7 curve A is iodide/chloride mixture and curve B is bromide/chloride mixture. In the case of curve B, the first equivalence point becomes less distinct as the solubility of the two precipitates approach one another.

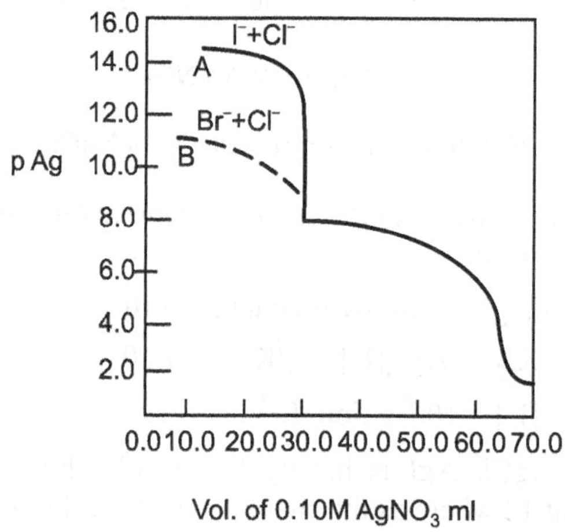

Fig. 5.7 Titration curves for 50.00 ml solution of 0.08M Cl^- and 0.05M I^- or Br^-.

(iii) Fajan's method: The titration of chloride with silver nitrate using adsorption indicators.

$$NaCl + AgNO_3 \rightarrow AgCl + NaNO_3 \quad .....(5.53)$$

The indicator is Fluorescein. The end point is white ppt in greenish yellow medium to red ($AgNO_3$ in neutral or faintly acidic solution). The action of adsorption indicators based upon the properties of colloids.

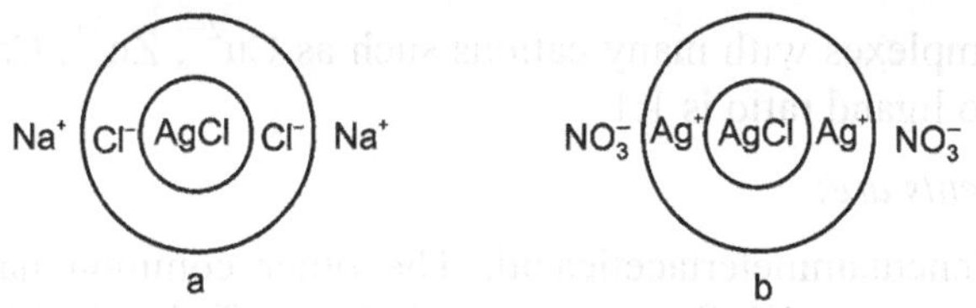

Fig. 5.8 (a) AgCl is precipitated in presence of excess Cl^-.
(b) AgCl is precipitated in the presence of excess Ag^+.

In the case of (b), the Ag^+ ions are present in excess are primarily adsorbed and NO^{3-} ion will be held by secondary adsorption. If fluorescein is also present in solution, the –ve fluorescein ion, which is more strongly adsorbed and will reveal its presence on ppt. not by its own colour, which is that of solution, but by the formation of pink complex of silver and modified fluorosceinate ion on the surface with the first trace of excess silver ions. The different adsorption indicator are listed in Table 5.6

Table 5.6 Different Adsorption indicators.

S.No	Indicator	Use	Colour change at the end point
1	Fluorescein	Cl^-, Br^-, I^-, SCN^- with Ag^+	Yellow green to pink (neutral or weakly basic
2	Dichloro-fluoroscein	Cl^-, Br^-, I^-, SCN^- with Ag^+	Yellowish green to red (pH 4.4-7)
3	Eosin.	Br^-, I^-, SCN^- with Ag^+	Pink to reddish violet (pH 1-2) in acetic acid
4	Rose Bengal	I^-in presence of Cl^- with Ag^+	Red to purple.
5	Diiodo-dimethyl fluorescein	I^- with Ag^+	Orange red to blue red (pH 4-7)
6	Rhodamine 6G	Ag^+ with Br.	Orange pink to reddish violet in 0.3 M H NO_3

Complexometric Titrations

In complexometric titrations, the reactions of complex ions or undissociated neutral molecules in solution are involved. The extreme solubility of the formed complex is the basic requirement.

Examples are: the complexes of metals with EDTA, titrations involving $HgNO_3$, Ag CN etc.

EDTA forms complexes with many cations such as Cu^{2+}, Zn^{2+}, Ca^{2+}, Mg^{2+},Ni^{2+},Co^{2+} etc., and the metal to ligand ratio is 1:1

The Complexing agents are:

(i) **EDTA** Ethylenediaminetertaceticacid: The other common names of EDTA are Versene, Complexone III, Sequesterene, Nullapan, Trilon-B, Idrant III, Chelaton-3. The formula of EDTA is

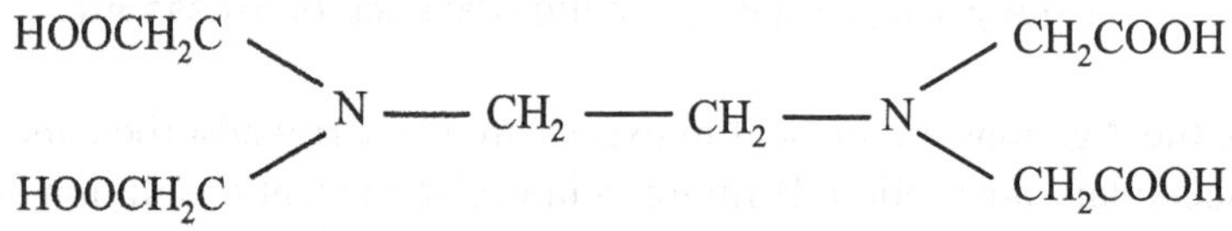

.....(5.54)

It has donar atoms of oxygen as well as nitrogen and it can form six (or five) membered chelate rings simultaneously.

(ii) **Nitrilotriacetic acid III:** $N(CH_2COOH)_3$. The other names are NITA, NTA, complexone 1, used as titrating and masking agent.

(iii) **1,2- di (2-aminoethoxylethane):** Also called as N,N,N[1],N[1](tetra acetic acid), EGTA

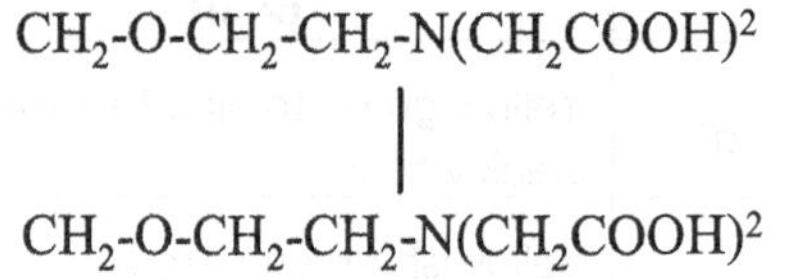

.....(5.55)

(iv) **1,2- di (diaminocyclohexane)** CDTA. The formula is

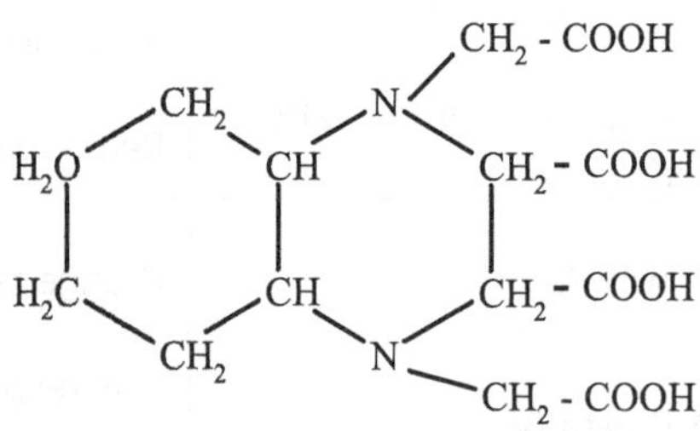

.....(5.56)

EDTA Complexes

EDTA is assigned the formula H_4Y and disodium salt is Na_2H_2Y or H_2Y^{z-} in aqueous solution. The usefulness of EDTA as titrant is because of the presence of 4 (2N and 2 oxygen) or 6(2N and 4 oxygen) atoms available for coordination with metal ions.

$$M^{2+} + H_2Y^{2-} \rightarrow MY^{2-} + 2H^+ \quad .....(5.57)$$

$$M^{3+} + H_2Y^{2-} \rightarrow MY^{-} + 2H^+ \quad .....(5.58)$$

$$M^{4+} + H_2Y^{2-} \rightarrow MY + 2H^+ \quad .....(5.59)$$

The stability constants of some M – EDTA complexes are given in Table 5.7

Table 5.7 Stability constants of M-EDTA complexes.

S.No.	Cation	Stability Constant (K)
1	Mg^{2+}	8.7
2	Ca^{2+}	10.7
3	Sr^{2+}	8.6
4	Ba^{2+}	7.8
5	Mn^{2+}	13.8
6	Fe^{2+}	14.3
7	Co^{2+}	16.3
8	Ni^{2+}	18.6
9	Cu^{2+}	18.8
10	Zn^{2+}	16.7
11	Cd^{2+}	16.6
12	Hg^{2+}	21.9
13	La^{3+}	15.7
14	Lu^{3+}	20.0
15	Sc^{3+}	23.1
16	Ga^{3+}	20.5
17	Th^{4+}	23.2
18	Na^{+}	1.7

Complexometric Titration Curves

EDTA disodium salt acts as quadridentate, and the formula can be written as H_4R. Then the complexation reaction is

$$M^{n+} + R^{4-} \rightleftharpoons [M\,R^{4-n}]^{-} \qquad .....(5.60)$$

Absolute stability constant

$$K_a = \frac{[MR^{4-n}]^{-}}{[M^{n+}][R^{4-}]} \qquad .....(5.61)$$

Step wise dissociation of H_4R can be written as;

$$H_4R + H_2O \rightleftharpoons H_3O+ + H_3R^{-} \qquad K_1 = 1.02 \times 10^{-2}$$

$$H_3R^{-} + H_2O \rightleftharpoons H_3O^{+} + H_2R^{-} \qquad K_2 = 2.1 \times 10^{-3}$$

$$H_3R^{2-} + H_2O \rightleftharpoons H_3O^+ + HR^{3-} \quad K_3 = 6.9 \times 10^{-7} \quad .....(5.62)$$

$$HR^{3-} + H_2O \rightleftharpoons H_3O^+ + R^{4-} \quad K_4 = 5.5 \times 10^{-11}$$

The pH ~ 10.0, H_4R exists as R^{4-}. As the pH lowers, the R^{4-} changes to HR^{3-}, $H_2R_2^{2-}$ and H_3R^-

The fraction of EDTA in R^{4-} C_2 is

$$C = [R^{4-}] + [HR^{3-}] + [H_2R^{2-}] + [H_3R^-] \quad .....(5.63)$$

Where C = analytical concentration.

The fraction of EDTA in R^{4+} forms α_4 is given by

$$\alpha_4 = \frac{R^{4-}}{C_a} = \frac{K_1K_2K_3K_4}{[H_3O^+]^4 + [H_3O^+]K_1 + [H_3O^+]K_1K_2 + [H_3O^+]K_1K_2K_3 + K_1K_2K_3K_4} \quad .....(5.64)$$

Where C_a = total concentration of uncomplexed EDTA

$$R_4^- = \alpha_4 C_a$$

$$K_a = \frac{[MR^{(4-n)-}]}{[M^{n+}][\alpha_4 C_a]}$$

$$K_a\,\alpha_4 = \frac{[MR^{(4-n)-}]}{[M^{n+}]C_a} = K_{eft} \quad .....(5.65)$$

Where K_{eft} = effective conditional stability constant. K_{eft} also varies with pH.

Consider the titration of 50 ml of 0.01M Ca^{2+} buffered at pH 10.0 is titrated with 0.01M EDTA solution. The titration curve is shown in Fig. 5.9

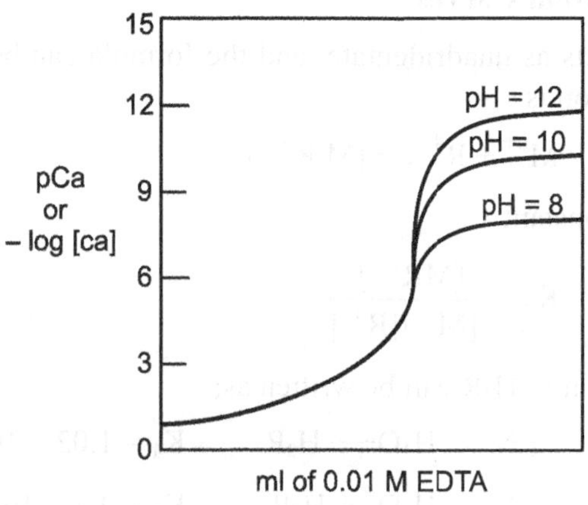

Fig. 5.9 Titration curve for calcium.

There is sharp break in pCa value of the equivalence point. At pH 12 k_{eft} 4.9 × 10^{10} and at pH8 k_{eft} is 2.6 × 10^8. A large break is obtained at pH 12 due to large K_{eft}. The curves are the same only upto equivalence point.

In complexometric titration, the optimum conditions are listed in Table 5.8 for some elements.

Table 5.8 Optiman conditions for some elements.

S. No.	Metal ion	Minimum pH	K_a	α_4	Keft
1	Ca^{2+}	7.3	5.0 × 10^{10}	4.8 × 10^{-4}	1.04 × 10^8
2	Mg^{2+}	10.0	4.9 × 10^8	3.5 × 10^{-1}	1.71 × 10^8
3	Fe^{2+}	5.1	2.1 × 10^4	3.5 × 10^{-7}	0.74 × 10^8
4	Co^{2+}	4.1	2.0 × 10^{16}	3.6 × 10^{-9}	0.72 × 10^8
5	Ni^{2+}	3.2	4.2 × 10^{18}	2.5 × 10^{-11}	1.05 × 10^8
6	Cu^{2+}	3.2	6.3 × 10^{18}	2.5 × 10^{-1}	1.57 × 10^8

Metal ion Indicators

The use of indicator in EDTA titration can be written as;

$$\text{M-In} + \text{EDTA} \rightarrow \text{M} - \text{EDTA} + \text{In} \qquad \text{.....(5.66)}$$

As the reaction proceeds M-In complex is less stable than M-EDTA complex. The indicator colour change is affected by the $[H^+]$ of the solution. For e.g., Erichrome black-T indicator can be written as H_2In^- and its behavior with pH is

$$\underset{\text{Red}}{H_2In^-} \underset{5.3-7.3}{\overset{pH}{\rightarrow}} \underset{\text{Blue}}{H_2In^{2-}} \underset{10.5-12.5}{\overset{pH}{\rightarrow}} \underset{\text{Yellow Orange}}{In^{2-}}$$

In the pH range 7-10 the indicator colour is blue and many metal ions form red complex with indicator.

Eriochrome Black-T or solochrome Black –T or WDFA is sodium 1- (1-hydroxy-2, naphthyl azo) – 6 nitro-2naphthol-4sulphonate (1)

$$M^{2+} + HD^{2-}(\text{blue}) \underset{7-12}{\overset{pH}{\rightarrow}} MD^-(\text{red}) + H^+$$

Table 5.9

S.No	Metal ion indicator	Colour Change		pH Range	Metal ions
		Free indicator colour	Metal Indicator colour		
1	Xylenol Orange. 0.5 g in 100 ml of water	Lemon Yellow	Red	3-5	Bi, Th, Zn, Cd, Pb, Co
2	Pyro catechol Violet 0.1 g in water	Yellow Violet	Blue	2-7	BI, Th, Zn, Cu, Cd, Ni, Co
3	Murexide 0.5 g in water	Blue Violet	Red	11 for Ca only	Cu, Ni, Co, Ca, rare earth elements
4	Erichrome black-T 0.2 g in 15 ml Triethanol amine + 5 ml ethanol or 0.4% in methanol	Blue	Wine Red	7-11	Mg, Mn, Zn, Cd, Al, Fe, Ti, Co, Ni. and Pt
5	Patton and Reeders indicator. 0.1 g mixed with 10 g of Na_2SO_4	Blue	Wine Red	12-14	Ca in presence of Mg
6	Calcon 0.2 g. in methanol	Purple blue	Pink	12.3	Ca in presence of Mg

Types of EDTA Titrations

(i) ***Direct titration:*** Metal ion is directly titrated with standard EDTA solution by using proper indicator and buffer.

(ii) ***Back titration:*** Excess standard EDTA added the metal ion and the excess EDTA is titrated with the standard metal ion.

(iii) ***Replacement or substitution titration***

Titration of mixtures by using masking and damasking agents

EDTA complexes with numerous di-, tri and tetravalent cations so it is a very unselective reagent. The procedures for the increasing selectivity are;

(i) ***Suitable control of pH of the solution***

For e.g., the determinations of Ca in the presence of Mg in strongly alkaline solution using Calcon or Murexide indictor. The determination of Bi and Th in the presence of divalent cations in acidic solution using xylenol orange or Pyrocatechol violet indicator.

(ii) ***Using suitable masking agents***

For e.g., Ca, Mg, Pb, Mn can be determined by EDTA method in the presence of Cd, Zn, Hg(11), Cu, Co, Ni. and platinum metals using cyanide as masking agent since the later metals form stable complexes with cyanide.

Ti(IV), Fe(111), Al can be masked with triethanol amine.

Hg can be masked with iodide ions.

Al, Fe(111), Ti(IV) and Be can be masked with NH_4F

(iii) ***Using suitable demasking agents***

The cyanide complexes Zn and Cd can be demasked with chloral hydrate (formaldehyde –acetic acid solution 3:1)

$$[Zn\,(CN)_4]^{2-} + 4H^+ + 4\,HCHO \rightarrow Zn^{2+} + 4HO.CH_2.CN \quad .....(5.67)$$

For e.g.: A mixture solution of Mg, Zn and Cu can be titrated by the following method.

- Add to the mixture of ions standard EDTA and back titrate with Mg solution using the Erichrome black T indicator. This gives the sum of the cations.
- Treat some portion with excess KCN and titrate. This gives Mg only since cyanide mask Zn and Cu.
- Then demask, zinc cyanide complex with chloral hydrate to release Zn ions and titrate. This gives only Zn. The Cu content then determined from the differences.

Many metal ions (cations) can be determined by using EDTA as titrant and using suitable indicators. Hardness of water can be determined.

Advantages

1. Since EDTA is stable, soluble and has definite composition it is preferable to other titrants.
2. The selectivity of the EDTA titrations can be easily accomplished by carrying at suitable pH.

   At pH 11 – Mg, Cr, Ca, Ba

   pH 4-7- Mn, Fe, Co, Ni, Zn, Cd, Al, Pb, Cu, Ti and V

   pH 1-4- Hg, Bi, Co, Fe, Cr, Ca, In, Se, Ti, V and Th.
3. EDTA disodium salt (Na_2H_2R) is itself a primary standard and does not need further standardization.
4. Complexes are readily soluble in water.
5. Equivalence point is readily reached,
6. The determination of several metals at a semi micro scale of operation is possible.

5.3 Gravimetric Analysis

Gravimetric methods are quantitative methods that are based on determining the mass of a pure compound to which the analyte is chemically related. The determination depends on the transformation of the element or radical to be determined into a pure stable compound, which can be readily converted into a form suitable for weighing. The separation of compound may be effected in a number of ways which are based on mass measurements. They are;

(i) *Precipitation gravimetry:* The analyte is separated from a solution of the sample as a precipitate and is converted to a compound of known composition that can be weighed.

(ii) *Volatalisation gravimetry:* The analyte is separated from other constituents of a sample by conversion to a gas of known chemical composition.

(iii) *Electrogravimetry:* The analyte is separated by deposition on an electrode by an electrical current. The mass of this product then provides a measure of the analyte concentration.

(iv) *Other methods:* Gravimetric titrimetry, Atomic mass spectrometry:

The great advantage of gravimetric method is that the constituent may be seen and examined for the presence of impurities and a correction applied if necessary. The main disadvantage is more time consuming.

5.3.1 Precipitation Methods

The important aspects are;

- the precipitate must be insoluble and separable by filtration
- the physical nature of the precipitate should be such that it can be easily separated from solution by filtration
- the particle size of the precipitate should be of good dimension.
- in the precipitate soluble impurities can be washable.
- the precipitate is a pure substance of definite composition.

For eg., silver is precipitated as AgCl, dried at 130° and weighed as AgCl.

Calcium is precipitated as CaC_2O_4, dried, ignited and weighed as CaO

5.3.2 Properties of Precipitates and Precipitating Reagents

The ideal precipitating reagent would react with the analyte to give a product i.e.,

- easily filtered and washed free of contaminants.

- sufficiently low solubility that no significant loss of analyte occurs during filtration and washing.
- Unreactive with constituents of atmosphere
- known chemical composition after it is dried or if necessary ignited.

Few of the reagents produce precipitates that possess all these desirable properties.

5.3.3 Particle Size and Filterability of Precipitates

Precipitates consisting of large particles are generally desirable for gravimetric work because these particles are easy to filter and wash free of impurities. Precipitates of this type are more purer than the precipitates of fine particles. The particle size of solids formed by precipitation varies enormously. The factors that determine the particle size or the precipitates

- Colloidal suspensions whose tiny particles are invisible to the naked eye (10^{-7} to 10^{-4} cm in diameter). Colloidal particles have no tendency to settle from solution and are not easily filtered.
- Crystalline suspension (temporary dispersion of colloidal particles in liquid phase) whose particles tend to settle spontaneously and are easily filtered.
- The particle size of the precipitate is influenced by the variables such as precipitate solubility, temperature, reactant concentrations and rate at which reagents mixed.
- The particle size is related to a single property of the system called relative super saturation. A supersaturated solution is an unstable solution that contains a higher solute concentration than a saturated solution, with time; super saturation is relived by precipitation of excess solute. The rate of precipitation ® is or relative super saturation is expressed as

$$R = \frac{Q - S}{S} \qquad \text{.....(5.68)}$$

Where Q = concentration of solute at any instant

S = equilibrium solubility

If Q-S/ S is large, the precipitate tend to be colloidal

If Q-S/ S is small, the precipitate is crystalline

For e.g., Separation of Ba as $BaSO_4$ at different reagent concentration are given in Table 5.10

Table 5.10 Separation of Ba as $BaSO_4$

Reagent Concentration (N)	$\frac{Q-S}{S}$	Type of Precipitate
7	175,000	Gelatinous
3	75,000	Gelatinous becomes turbid
1	25,000	Curdy and Colloidal
0.05	1,300	Feathery
0.005	125	Crystalline precipitate
0.001	25	2-3 hours of precipitation
0.0002	5	One month of precipitation

This table indicates that the type of the precipitate and particle size depends on the rate of addition of reagent.

Generally the precipitation is carried out in hot solution as solubility increases with rise in temperature. Precipitation is effected in dilute solutions and the reagent is added slowly with constant stirring, because initially formed particles serve as nuclei for precipitation.

5.3.4 Mechanism of Precipitate Formation

The precipitates form in two ways

(i) ***Nucleation:*** It is a process in which a minimum number of atoms, ion or molecules join together to give a stable solid. Often, these nuclei form on the surface of suspended solid contaminants, such as dust particles. The rate of nucleation is believed to increase enormously with increasing relative super saturation and if it predominates, precipitate contains large no. of small particles.

(ii) ***Particle growth:*** Further precipitation on nuclei results in particle growth. At low relative super saturation, the particle growth predominates. Precipitate contains small number of large particles (Crystalline).

5.3.5 Purity of the Precipitates

If the precipitates are pure, no errors will be there. The precipitate is impure due to the presence of foreign substances and are entered due to:

(i) ***Co-precipitation:*** The contamination of the precipitate by substance that is normally soluble in the mother liquor under the conditions of precipitation is called co-precipitation. It can occur due to:

(a) *Adsorption*: Adsorption at the surface of the precipitate particle exposed to the solution.

(b) *Occlusion*: Occlusion of foreign substance from the primary particles during the process of crystal growth.

Adsorption is maximum on gelatinous precipitates and is least for microcrystalline precipitates.

Eg., In Ag I on acetate of silver and $BaSO_4$ precipitate in alkali nitrates. Mixed crystal formation a type of co-precipitation in which a contaminant ion replaces an ion in the lattice of a crystal.

A drastic but effective way to minimize the effects of adsorption is re-precipitation. In this process, the filtered solid is redissolved and re-precipitated. The first precipitate ordinarily carries down only a fraction of the contaminant present in the original solvent. Thus, the solution containing the redissolved precipitate has a significantly lower contaminant concentration than the original, and even less adsorption occurs during the second precipitation.

For eg., $BaSO_4$ is formed by adding $BaCl_2$ to a solution containing sulphate, lead and acetate ions is found to be severely contaminated by $PbSO_4$ even though acetate ions normally prevent precipitation of $PbSO_4$ by Complexing the lead. Here Pb^{2+} ions replace some of the Ba^{2+} ions in $BaSO_4$ crystals. Other example is $MgKPO_4$ in $MgNH_4PO_4$, Mn S in Cd S etc.

(ii) ***Post-Precipitation*:**This is precipitation which occurs on the surface of the first precipitate after its formation. It occurs with sparingly soluble substances, which form supersaturated solutions, then usually have an ion in common with the primary precipitate.

For eg., the precipitation of CaC_2O_4 in the presence of magnesium. The MgC_2O_4 separates out gradually on CaC_2O_4. Oxalate ion is the common ion.

5.3.6 The Important Steps in Gravimetric Analysis

(i) Weighing the sample.

(ii) Preparation of the sample solution

(iii) Precipitation of the sample by adding the proper reagent for getting pure solid precipitate with definite composition.

(iv) *Digestion (Ageing):* After precipitation process completed, the precipitate is allowed to stand either at room temperature for some time of overnight, or on a water bath or a hot plate for few hours (12-24) for the contents to digest. In this process, the precipitate undergoes considerable degree of self purification. The finer and more soluble particles tend to dissolve and larger ones tend to grow into more compact and perfect crystalline form with resultant elimination of loosely adsorbed impurities. *This process therefore improves both the filtering qualities, the purity of the precipitate and reduce the extent of co-precipitation.*

(v) *Filtration of the Precipitate:* In the process of filtration, the precipitate is isolated from mother liquor (Supernatant liquid above the precipitate) by means of porous material such as Whatman filter paper or sintered glass crucible.

Whatman filter paper is ash *less filter paper* required for all quantitative filtrations where the precipitate is to be determined by ignition. For e.g., CaC_2O_4 precipitate ignited to give CaO, $ZnNH_4PO_46H_2O$ to $Zn_2P_2O_7$, $Fe(OH)_3$ to Fe_2O_3 etc.

For student use Whatman filter paper NO 41or 42 circles of 9 cm diameter is sufficient. Whatman filter paper No 541, 540, 542 mean double acid washed and hardened and ash mean is 0.000041, 0.000044, 0.000050 g respectively. Filter paper is folded and kept in the funnel before filtration. The assembly is shown in Fig. 5.10

Another method of filtration is by suction. This process is used when the precipitates are to be dried at lower temperature in a hot air oven at about 100-180°C. The process is speed up by suction and is faster than filter paper process. For this purpose the sintered glass crucible or Gooch crucible is mostly used. The sintered glass crucibles are made of resistance glass (Pyrex) and have a porous disc of sintered, ground glass fused into the body of the crucible. The fused in fritted filter disc can be obtained in various degrees of porosity.

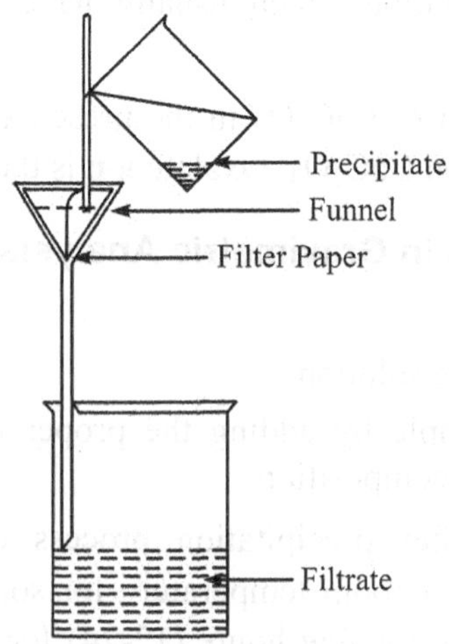

Fig. 5.10 Filtration Assembly using Whatman Filter paper

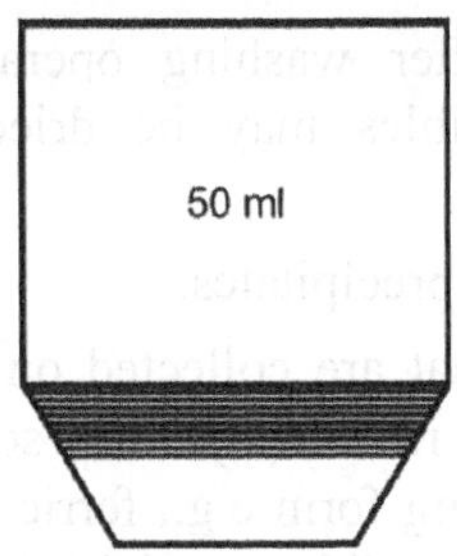

Fig. 5.11 (a) Sintered glass crucible.

The G_4 (25 ml capacity) (5.11 (a) is used mostly which has the porosity 5-10 microns. Some times G_3 is also used which has the porosity 20-30 microns. The filtration assembly using sintered glass crucible is shown in Fig 5.11.

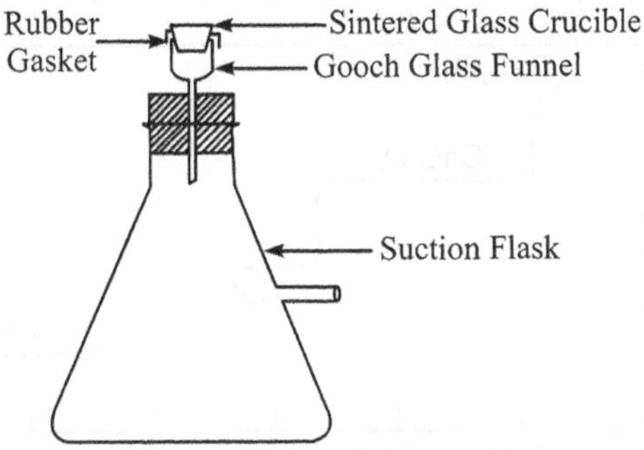

Fig. 5.11 Filter assembly using sintered glass crucible.

(vi) *Washing of the Precipitate:* The objective of washing the precipitate is to remove contamination on the surface of the precipitate. The composition of wash solution will depend on the tendency of the precipitate to undergo precipitation. Very few precipitates can be washed with pure water which may cause solubility loses or may hydrolyse certain precipitates. Wash solutions are divided into three categories.

(i) Solutions which prevent precipitation from becoming colloidal and passing through filter papers. E.g., use NH_4NO_3 for washing precipitation of ferric hydroxide.

(ii) Solutions which would reduce the solubility of precipitates should be avoided.

(iii) Solutions which prevent the hydrolysis of salts of weaker acids and weaker bases.

Washing with many small portions is much more effective than single wash.

(vii) *Drying of precipitates:* After washing operation, many precipitates that are collected on filtering crucibles may be dried directly in a hot air oven at $110^{o} - 180^{o}$.

e.g., AgCl, $ZnNH_4PO_4$ etc., precipitates.

(viii) *Ignition:* The precipitates that are collected on filter paper must be subjected to elevated temperature to get rid of paper. In some cases the precipitate must be converted to a stable weighing form e.g., ferric hydroxide to ferric oxide, calcium oxalate to calcium oxide. Crucibles used in the ignition process are –porcelain, silica or platinum depending on the temperature of ignition and the reactivity of the precipitate. The temperature required to produce a suitable weighing form varies from precipitate to precipitate. This data is obtained by thermo grams for which Thermo Balance is used.

For e.g., Effect of temperature on the weight of CaC_2O_4 is shown in Fig. 5.12

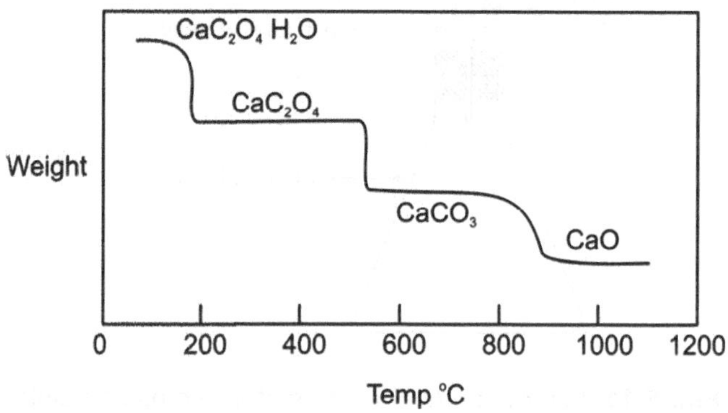

Fig. 5.12

(ix) *Calculations of Gravimetric Analysis:* The final chemical formula of the precipitate is weighed to report the analysis. For conversion of the weight of the precipitate to the corresponding weight of some other chemical substance, calculation is necessary. Gravimetric factor (chemical factor) can be used for calculation. Find out the weight of the precipitate by weighing.

Gravimetric factor = Formula Wt. of substance sought / formula wt of the ppt substance

For.eg. Gravimetric factor of

$$Cl = 35.46(Cl^-)/143.3\ (AgCl)$$

$$= 0.2475$$

$$Ag = 107.88\ (Ag^+)/143.3(AgCl)$$

$$= 0.7528.$$

Any weight of AgCl precipitate is multiplied by 0.2475. We can get the amount of chloride in the sample and if multiplied by 0.7528, we can get the amount of silver. The other examples are SO_4 in $BaSO_4$, the gravimetric factor is 0.4116. i.e. to get the amount of sulphate if we multiply the weight of $BaSO_4$, precipitate with 0.4116. The Gravimetric factor Pb in $PbSO_4$ is 0.6832, Cu in CuSCN is 0.5224.

5.3.7 Inorganic Precipitating Agents

The common inorganic precipitating agents are listed in Table 5.11. These reagents typically form slightly soluble salts or hydrous oxides with analyte. The few reagents are selective.

Table 5.11

S. No.	Precipitating Agent	Element precipitation (Examples)
1	NH_3 (aq)	Be as BeO, Al as Al_2O_3, Fe as Fe_2O_3, In as In_2O_3 etc.
2	H_2S	Zn as ZnO or $ZnSO_4$,Ge as GeO_2,Mo as Mo O_3, Bi as Bi_2S_3 etc.
3	$(NH_4)_2HPO_4$	Mg as $Mg_2P_2O_7$, Zn as $Zn_2P_2O_7$, Zr as $Zr_2P_2O_7$, Bi as $BiPO_4$ etc.
4	H_2SO_4	Pb as $PbSO_4$, Ba as $BaSO_4$, Cd as $CdSO_4$, Sr as $SrSO_4$ etc.
5	$H_2C_2O_4$	Ca as CaO, Th as ThO_2 etc
6	$(NH_4)_2MoO_4$	Pb as $PbMoO_4$, Cd as $CdMoO_4$
7	HCl	Ag as AgCl, Hg as Hg_2Cl_2
8	$AgNO_3$	Cl as AgCl, Br as AgBr, I as AgI
9	$BaCl_2$	SO_4^{2-} as $BaSO_4$
10	$MgCl_2$	PO_4^{3-} as $Mg_2P_2O_7$

5.3.8 Organic Precipitating Agents

Numerous organic reagents have been developed for the gravimetric determination of inorganic species. The organic reagents are more selective when compared to inorganic reagents. These reagents are of two types. One type of these can form coordination compounds by donating more than one lone pair of electrons to the cation. This type is also known as chelating agents, the other type of reagents in which the bonding between the inorganic species and the reagents is largely ionic. The important organic reagents are;

(i) Dimethyl glyoxalin solution for Ni.
(ii) Cupferron selective for Fe (III) and Cu (II).
(iii) 8-hydroxy quinoline (Oxine) for Mg.

(iv) Salicyldoxime for Cu
(v) 1-nitroso- 2naphthol for Co
(vi) Quinoldic acid for Cu
(vii) Mandelic acid for Zr
(viii) Anthranilic acid for Cu
(ix) Pyrogallol for Bi etc.

These are all form chelates with inorganic species or cations.

The examples of other type (ionic) reagents:

(i) Tetraphenylarsonium chloride $[(C_6H_5)_4As]^+Cl^-$. Selective for Ti(III) and form $[(C_6H_5)_4 As] + TiCl_4^-$
(ii) Sodiumtetraphenylborate, $Na^+[B(C_6 H_5)_4]^-$. This is excellent reagent for Potassium.
(iii) Benzidine, $H_2N C_6 H_4$, $C_6 H_4 NH_2$ selective for SO_4^- and form $[C_{12} H_{12} N_2 H_2]^{2+} SO_4^{2-}$ etc.

5.3.9 Precipitation from Homogeneous Solution (PFHS) Method

Precipitation from homogeneous solution is a technique in which a precipitating agent is generated in a solution (insitu) of the analyte by a slow chemical reaction. Excess reagent does not occur locally because the precipitating agent appears gradually and homogeneously throughout the solution and reacts immediately with the analyte. So the relative supersaturation is kept low during the entire precipitation.

The precipitation formed by this PFHS method:

- eliminates the undesirable conc. effects which are inevitably associated with the conventional process.
- is dense and readily filterable.
- the co-precipitation is reduced to minimum.
- it is possible to alter the physical appearance of i.e. formation of larger crystals by varying the rate of chemical reaction producing the precipitate.

The different techniques in this method are;

- Urea hydrolysis method
- Volatalisation of ammonia
- Complexation and replacement
- Hydrolysis of organic reagents to give anions
- Insitu synthesis of organic precipitants
- Diffusion method.

Representative methods for homogeneous generation of precipitating agent are listed in Table 5.12

Table 5.12

S. No.	Precipitating Agent	Reagent	Generation Reaction	Element
1	OH^-	Urea	$(NH_2)_2CO + 3H_2O \rightarrow CO_2 + 2NH_4^+ + 2OH^-$	Al, Ga, Th, Bi, Fe, Sn
2	$PO4^{3-}$	Trimethyl phosphate	$(CH_3O)_3PO + 3H_2O \rightarrow 3CH_3OH + H_3PO_4$	Zr, Hf
3	$C_2O_4^{2-}$	Ethyl oxalate	$(C_2H_5)_2C_2O_4 + 2H_2O \rightarrow 2C_2H_5OH + H_2C_2O_4$	Mg, Zn, Ca
4	SO_4^{2-}	Dimethyl Sulphate	$(CH_3O)_2SO_2 + 4H_2O \rightarrow 2CH_3OH + SO_4^{2-} + 2H_3O^+$	Bs, Ca, Sr
5	H_2S	Thioacetamide	$CH_3CSNH_2 + H_2O \rightarrow CH_3CONH_2 + H_2S$	Mo, Cu, Cd
6	DMG	Biacetyl + hydroxylamine	$CH_3COOCH_3 + 2H_2NOH \rightarrow$ DMG $+ 2H_2O$	Ni

This PFHS methods are more convenient and have many applications in gravimetric analysis than ordinary methods. The gravimetric determinations and descriptions are described in practical part of this book.

The pharmacopoeia usually avoids gravimetric methods of array when the other accurate methods available. Presently included methods are barium sulfate. magnesium trisilicate and sodium aurothiomalate. Earlier pharmacopoeia includes $MgSO_4$, alum, bismuth subcorbonate, lead acetate etc.

CHAPTER 6

Pharmaceutical Aids

6.1 Introduction

These are the chemicals or substances having very little or no therapeutic value but are important in the manufacturing of the various dosage forms such as tablets, ointments, parentral dosage forms etc., These compound are required for such purposes as preservation, stabilization, adsorption, absorption, suspending, colouring, sweeting, prevention of oxidation, acidification, alkalization, complexation etc.,

Various types of pharmaceutical aids include:

1. Buffers
2. Colouring Agents
3. Antioxidants
4. Preservatives
5. Sweetening Agents
6. Pharmaceutical Waters
7. Flavouring Agents
8. Suspending Agents
9. Acidifiers and Alkalisers
10. Absorbents
11. Adsorbents
12. Ionicity adjusters

A number of inorganic chemicals find extensive application in the processing of pharmaceuticals.

Purpose of pharmaceuticals Aids:

(a) Preservation
(b) Stabilization
(c) Acidification / Alkalisation
(d) Suspending
(e) Adsorption
(f) Absorption
(g) Filtration
(h) Prevention of Oxidation
(i) Solubilisation
(j) Complexation

Note: The solubility of drugs may be affected in the presence of phosphate salts.

6.1.1 Acid - Bases

One of the important jobs of the chemists is to devise and evaluate schemes for classifying compounds and reactions. Classification on the basis of acid-base properties has been found to be particularly useful. *Acids and bases possess some opposite characteristics.* Acids and bases are identified using organic dye stuffs. For e.g., acid turns blue litmus red, while the base changes red litmus blue. Schemes that provide structure to the information concerning the type of chemistry towards systematizing its study. One such approach is that of the chemistry of acids and bases. Now briefly examine the development of acid-base concepts.

Arrhenious Concept (Protonic acid-base concept)

S.A. Arrhenious approach was limited to aqueous solutions. On the basis of his theory:

- An acid is a hydrogen containing compounds which yields hydrogen ions in aqueous solution (water) or the substances which give hydrogen ions (protons) when dissolved in water can be called as acids. Eg. HCl, HNO_3, H_2SO_4.
- A base is a hydroxyl compound which yields hydroxide ions in aqueous solutions or the substances which ionize in water to form hydroxyl ions can be called as bases. E.g. NaOH, KOH, Ca $(OH)_2$ NH_3, amines RH_2N, R_2HN, R_3N etc.

Consider the reaction between HCl (gaseous) and water

$$HCl\ (g) + H_2O\ (l) \rightarrow H_3O^+(aq) + Cl^-\ (aq) \qquad .....(6.1)$$

The equation contains H_3O^+, the hydronium ion (generally known as oxonium ion). Similarly

$$HNO_3\ (l) + H_2O\ (l) \rightarrow H_3O^+\ (aq) + NO_3^-\ (aq) \qquad .....(6.2)$$

The solutions of HCl and HNO_3 substances have acidic properties due to the presence of H_3O^+ (solvated hydrogen ion) in the solution.

Finally we can define that an acid is a substance whose water solution contains H_3O^+ ions.

Aqueous solutions of acids are good conductors of electricity.

Consider the reaction between NH_3 (gaseous) and water

$$NH_3\,(g) + H_2O\,(l) \rightarrow NH_4^+\,(aq) + OH^-\,(aq) \qquad .....(6.3)$$

If NaOH dissolves in water, the reaction is not ionization because Na^+ and OH^- already exist in the solid. This process is known as dissolution process.

Finally we can define that a base is a substance whose water solution contains OH^- ions.

The process of neutralization is then the combination of hydrogen ions (H_3O^+) and hydroxyl ions.

Consider the reaction of NaOH solution is added to HCl solution.

$$Na^+\,(aq) + OH^-\,(aq) + H_3O^+(aq) + Cl^-\,(aq) \rightarrow 2H_2O\,(l) + Na^+\,(aq) + Cl^-\,(aq) \qquad .....(6.4)$$

NaCl is soluble in water and there is no change in Na^+ and Cl^- ions.

Net ionic equation is

$$H_3O^+\,(aq) + OH^-\,(aq) \rightarrow 2H_2O\,(l) \qquad .....(6.5)$$

Some acids are not completely dissociated and the ionization is dependent on the concentration of acid. On the basis of dissociation constants we can say the acid is strong or weak. The strong acids can be dissociated completely in water but weak acids cannot For *e.g.*, CH_3COOH have 1.75×10^{-5}, HOCl have 3.5×10^{-8} etc.

Merits

(i) The concept useful in quantitative determination of acid or base strengths

$$HB \rightleftharpoons H^+ + B^-$$

$$K = \frac{a_{H^+} a_{B^-}}{a_{HB}} \qquad .....(6.6)$$

where a_{H^+} = activity of H^+ ions

a_{B^-} = activity of B^- ion in $H^+ B^-$ substance

a_{HB} = activity of acid

(ii) it is valuable in elucidating the behavior of aqueous solutions,

(iii) it correlates the catalytic actions of acids with the concentration of hydrogen ions.

Demerits

(i) It lacks the generality,

(ii) It deals with dissociation and acid base reactions in aqueous medium only,

(iii) It restricts bases merely to hydroxides,

(iv) It cannot explain the acidic behavior of certain salts such as Al Cl_3 in water.

Bronsted-Lowry Concept (The Proton Donar-Acceptor Concept) 1923

J.N. Bronsted and T.M. Lowry felt that a more general acid-base concept was needed which is independent of solvents. In this concept acids and bases are related to by the following equation.

$$\text{acid} = \text{base} + H^+ \quad \text{.....(6.7)}$$

Acid can be defined as it is a hydrogen containing species. It is capable of acting as proton donor. *E.g.*, HCl, H_3PO_4

Base can be defined as it is a species capable of acting as proton acceptor. E.g. PY, PO_4^{3-}.

Thus acids and bases are independent of solvent and bear no relationship to salts. There is no acid without base. A typical acid-base reaction may then be written as

Conjugate pair

$$\text{acid 1} + \text{base 1} \rightarrow \text{acid 2} + \text{base 2}$$

Conjugate pair .....(6.8)

Consider the reaction of acetic acid with ammonia

$$CH_3COOH(aq) + NH_3(aq) \rightarrow CH_3COO^-(aq) + NH_4^+(aq) \quad \text{.....(6.9)}$$

In this reaction acetic acid donates proton to produce its conjugate. The ammonia accepts a proton to produce its conjugate NH_4^+ ion, which can function as proton donor.

The species remaining after a proton is donated is called as a conjugate base. So acetate ion is the conjugate base of acetic acid.

Consider equation

$$H_3O^+ + OH^- \longrightarrow H_2O + H_2O \quad \text{.....(6.10)}$$

Here water may be thought of either the conjugate acid of OH^- ion or conjugate base of oxonium ion.

Conjugate pairs: the pairs of substances which are formed from one another by the gain or loss of a proton are called conjugate acid-base pair as shown in the eq. 6.8 to 6.10

As per this theory

- There is no acid without a base
- The stronger an acid is, the weaker its conjugate will be as a base
- A stronger acid reacts to displace a weaker acid
- The strongest acid in water is H_3O^+ and base is OH^-

Franklin Concept (Solvent System Concept)

Some solvents resemble water in that they undergo self ionization to give a solvated proton and an anion. For *e.g.*,

$$2NH_3\ (l) \rightarrow NH_4^+ + NH_2^- \qquad .....(6.11)$$

Where NH_4^+ ion is acidic species and the amide ion is a basic species.

According to Franklin

Acid is a substance which by dissolution in the solvent forms the same cation as does the solvent itself due to self-ionization. For *e.g.*, in eq. 6.11, an acid is any species which when dissolved in liquid ammonia, increase the concentration of NH_4^+ ion.

Base is a substance which gives on dissociation in the solvent the same anion as does the solvent itself on self-ionization.

For e.g., in eq. 6.11, a base is any species which when dissolved in liquid ammonia, increase the concentration of NH_2^- ion.

NH_3 is a solvent. KNH_2 is base. NH_4Cl protonic acid. The other non-aqueous solvents undergo self ionization (autoprotolysis)

$$2H_2SO_4 \rightarrow H_3SO_4 + HSO_4^- \qquad .....(6.12)$$

Water is fairly strong base in H_2SO_4 since it gives HSO_4^- ion

$$H_2O + H_2SO_4 \rightarrow H_3O^+ + HSO_4^- \qquad .....(6.13)$$

and SO_3 is an acid in H_2SO_4 since it gives protons.

$$SO_3 + H_2SO_4 \rightarrow H^+ + HS_2O_7^- \qquad .....(6.14)$$

HF also undergoes self ionization

$$2HF \rightarrow H^+ + HF_2^- \qquad .....(6.14a)$$

PF_5 is an acid and H_2O, HNO_3 are bases in HF

Aprotonic Acid-Base Concept (Cady and Elrey Solvent Concept)

There are some solvents which contain no hydrogen, aprotonic solvents, undergo self ionization, whose salt solutions are electrically conducting.

Acid is the species which form cation on self ionization

Base is a species which form anion on self ionization.

For e.g., BrF_3 liquid undergo self ionization.

$$2BrF_3 \rightarrow BrF_2^+ + BrF_4^- \quad .....(6.15)$$

Liquid SO_2

$$2SO_2 \rightarrow SO^{2+} + SO_3^{2-} \quad .....(6.16)$$

Lux-Plood Concept

According to this concept

Acid is any material which gains oxide ions or it is an oxide ion acceptor.

Base is any material which gives up oxide ions or it is an oxide ion donor.

$$\text{Base} \rightleftharpoons \text{Acid} + O^{2-} \quad .....(6.17)$$

$$Ca\,O \rightleftharpoons Ca^{2+} + O^{2-} \quad .....(6.18)$$

$$SO_4^{2-} \rightleftharpoons SO_3^{2-} + O^{2-} \quad .....(6.19)$$

Amphoteric substances are those which shows both a tendency to take up or give up oxide ions

$$\underset{\text{acid}}{Na_2O + ZnO} \rightarrow 2Na^+ + Zn\,O_2^{2-} \quad .....(6.20)$$

Usanoich Concept (Positive Negative Concept)

According to this concept:

Acid is any species capable of giving cation, combining with anion or electrons, or neutralizing a base to give salt.

Base is any species capable of giving up anions or electrons, combining with cation or neutralizing an acid to give salt.

This concept includes oxidation-reduction reactions as a special class of acid-base reactions.

$$\underset{\substack{\text{Cl gains electrons}\\ \text{Base}}}{Cl_2} + \underset{\substack{\text{Na loses electrons}\\ \text{acid}}}{2Na} \rightarrow \underset{\text{salt}}{2NaCl} \quad .....(6.21)$$

Lewis Concept (Electron Donor-Acceptor Concept)

G.N. Lewis proposed a more generalized concept of acids-bases.

Acid – As per Lewis concept acid is a substance that can form a covalent bond by accepting an electron pair from base. It is also called as elctrophile.

Base – A base is a substance that has an unshared pair of electrons with which it can form covalent bond with an atom, molecule or ion. It is also called as nucleophile.

$$\underset{\text{Lewis acid}}{BF_3\ (g)} + \underset{\text{Lewis base}}{:NH_3\ (g)} \rightarrow \underset{\text{Adduct complex}}{F_3B \leftarrow NH_3\ (s)} \quad .....(6.22)$$

The bond formed between Lewis acid-base is coordinate covalent bond. This concept has nothing to do with proton transfer. Sometimes, this is known as electronic theory.

Lewis acids are:

- Molecules possessing a central atom with less than octet of electrons (electron deficient molecules).

  B in BF_3, Al in Al Cl_3

- Molecules in which the central atom has available d-orbitals and may acquire more than an octet of valence electrons.

  Si in SiF_4, Sn in $SnCl_4$, P in PF_3, S in SF_4 etc

  $$Si\,F_4 + :\ddot{F}:^- \rightarrow Si\,F_6^{2-} \quad .....(6.23)$$

- The cations of heavy metals with incomplete stable orbitals

  Al^{3+} Cr^{3+} Fe^{3+}, Ag^+ etc

  $$Al^{3+} + 6 : OH_2 \rightarrow Al\,(OH_2)_6 \quad .....(6.24)$$

- Molecules with a multiple bond between atoms of dissimilar electronegativities.

  $$CO_2 \quad :\ddot{O} - \overset{\delta+}{C} = \overset{\delta-}{:O} \quad SO_2 \quad \text{etc}$$

  $$CO_2 + :\ddot{O}H^- \longrightarrow HCO_3 \quad .....(6.25)$$

- Elements with sextet of electrons

  O, S etc

  $$:\ddot{O} + SO_3^{2-} \longrightarrow SO_4 \quad .....(6.26)$$

This concept is very broad and has many of the same aspects as does acid-base theory according to Bronsted-Lowry theory.

They are:

- There is no acid without a base. An electron pair must be donated to one species (acid) by another (base)
- An acid (or base) reacts to displace a weaker acid (or base) from a compound.

  For e.g., $H_3N: BF_3 + BCl_3 \rightarrow H_3N: BCl_3 + BF_3$ .....(6.27)

  BCl_3 is stronger one replaces weaker BF_3
- The interaction of a Lewis acid with Lewis base is a type of neutralization reaction because the acidic and basic characters of the reactants are removed.

Hard and Soft Acid-Base theory (HSAB Theory)

To broaden the scope of Lewis acid–base reaction

$$\underset{\text{Lewis acid}}{A} + \underset{\text{Lewis base}}{:B} \longrightarrow \underset{\text{complex}}{A:B} \quad .....(6.28)$$

and the stability of the complex AB, the Hard and Soft Acid-Base (HSAB) concept was proposed by R.G. Pearson (1963). Hard and soft acids-bases can be defined as:

Soft base (Lewis base) is the one in which the donor atom is of high polarizability and of low electronegativity and is easily oxidized or is associated with empty low-lying orbitals.

E.g., R_2S, RSH, I^-, SCN^-, CN^-, CO, H^-, R^-, R_3P, C_2H_4 etc

Hard base (Lewis base) is the one in which the donor atom is of low polarizability and high electronegativity and is hard to reduce and is associated with empty orbitals of high energy.

E.g., OH^-, F^-, H_2O, SO_4^{2-}, Cl^-, PO_4^{3-}, NO_3^-, ClO_4^-, RO^-, NH_3, RNH_2, N_2H_4 etc

Soft acid (Lewis acid) is the one in which the acceptor atom is of low or zero positive charge, large size and has several easily excited outer electrons.

E.g., Cu^+, Ag^+, Pd^{2+}, Hg^{2+}, $GaCl_3$, RS^+, I^+, O, Cl, Br, I etc

Hard acid (Lewis acid) is the one in which the acceptor atom is of high positive charge, small size and the absence of any outer electrons which are easily excited to higher states.

E.g., H^+, Li^+, Na^+, Ca^{2+}, Mn^{2+}, Al^{3+}, Co^{3+}, Fe^{3+}, Ti^{4+}, BF_3, HCl, $AlCl_3$, SO_3, CO_2, RCO^+ R_2O etc

These are some examples of boarder line Lewis acids and Lewis bases.

The stability correlating principle is that hard acids prefer to coordinate with hard bases and soft acids prefer to coordinate with soft bases. Consider the reaction

$$A_H : B_S + A_S : B_H \rightleftharpoons A_H : B_H + A_S : B_S \qquad .....(6.29)$$

For *e.g.*, $$LiI + C_S F \rightleftharpoons LiF + C_S I \qquad .....(6.30)$$

To explain the correlation principle, the equilibrium constant can be expressed in terms of two strength factors.

$$\log K = S_A S_B + \sigma_A \sigma_B \qquad .....(6.31)$$

where S_A = intrinsic acid strength

S_B = intrinsic base strength

σ_A = softness factor of acid

σ_B = softness factor of base.

This concept is qualitative. The bonding between hard acid and base is primarily ionic.

For *e.g.*, Typical hard acids like Li^+, Na^+, K^+ and hard bases F^-, OH^- form ionic bonds.

The bonding between soft acid and base is primarily covalent. Frequently these interactions involve the bonding between neutral molecules. Orbital overlap that leads to covalent bonding is more favourable when the orbitals of the donor and acceptor atoms are of similar size and energy. C. K. Jorgenson modified the HSAB approach in the case of cobalt complex.

Under normal condition CO^{3+} is hard acid. $[CO(CN)_5]^{2-}$ form stable complex with I^- (sixth ligand), soft base than F^-, hard base. The cyanide complex result the softening of CO^{3+} which occurs when five of the ligands are soft. So $[Co(CN)_5 I]^{3-}$ is stronger than $[Co(CN)_5 F]^{3-}$. In $Co(NH_3)_5$ the reverse is true since Co^{3+} becomes more harder with five NH_3 (hard base) ligands. Though Co^{3+} is normally hard electron pair acceptor, the five CN^- ions attached cause it to behave as soft acid. This effect is known as the symbiotic effect.

This HSAB principle applied by Pearson to a multitude of chemical problems like catalysis, solubility, rates of certain reactions.

6.1.2 Factors Affecting Strength of Acids and Bases

One of the considerations related to acid strength for a polyprotic acid such as H_3PO_4. The first proton is lost from a neutral molecule more easily than the second and third. The stepwise dissociation of H_3PO_4 is

$$H_3PO_4 + H_2O \rightleftharpoons H_3O^+ + H_2PO_4^- \quad K_1 = 7.5 \times 10^{-3} \quad .....(6.32)$$

$$H_2PO_4^- + H_2O \rightleftharpoons H_3O^+ + HPO_4^{2-} \quad K_2 = 6.2 \times 10^{-8} \quad .....(6.33)$$

$$HPO_4^{2-} + H_2O \rightleftharpoons H_3O^+ + PO_4^{3-} \quad K_3 = 1.0 \times 10^{-12} \quad .....(6.34)$$

There is a factor of $\simeq 10^5$ between successive dissociation constants. This relationship is known as pauling's rule. This is obeyed by H_2SO_3 for which $K_1 = 1.2 \times 10^{-2}$ and $K_2 = 1.0 \times 10^{-7}$.

Pauling provided a way of systematising the inductive effect (shifting of electrons) by writing the formulas of oxo-acids of $(OH)_n \times O_m$. The value of m determines the strength of acid (Table 6.1).

Table 6.1

S. No	m value	K_1 value	Example	comment
1	0	10^{-7} or less	B $(OH)_3$ basic acid $K_1 = 5.8 \times 10^{-10}$ HO Cl, hypochlorous acid $K = 2 \times 10^{-9}$	Very weak acid
2	1	10^{-2} or less	H ClO_2 chlorms acid $K = 1.0 \times 10^{-2}$ HNO_2 nitrous acid $K = 4.5 \times 10^{-4}$	Weak acid
3	2	$> 10^3$	H_2SO_4, sulphuric acid. K_1 is large HNO_3 nitric acid K is large.	Strong acid
4	3	$> 10^8$	$HClO_4$ perchloric acid K is very large	Very strong acid

The strength can also be illustrated in chloro-substituted acetic acids

For *e.g.*, 1. CH_3COOH, $Ka = 1.75 \times 10^{-5}$

2. $ClCH_2COOH$, $Ka = 1.4 \times 10^{-3}$

3. $Cl_2CHCOOH$, $Ka = 3.3 \times 10^{-2}$

4. $Cl_3C\ COOH$, $Ka = 2.0 \times 10^{-1}$

The dissociation constants for acids show clearly the effect of replacing hydrogen on methyl group with chlorine atoms. As a result, the chlorine atoms cause a migration of electron density toward the end or molecule where they reside which causes the electron pair in the OH bond to be shifts farther away from the hydrogen atoms. From the dissociation constant values; e. g., 4 is stronger acid than atoms.

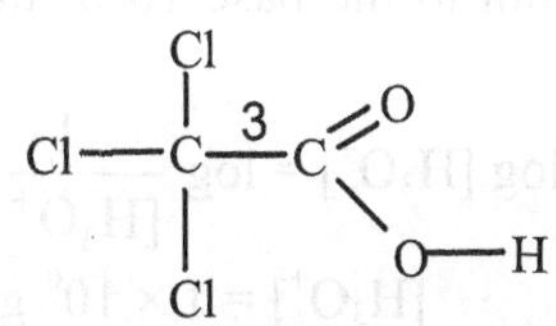

Hydrohalic acids (HF, HCl, HBr, HI) contain large dissociation constants (6.7×10^{-4} to 2.0×10^{9}) than H_2O, H_2S, H_2Se, H_2Te (2×10^{-16} to 2.3×10^{-3}).

Speciation of Acidity and Basicity: The reactivity of acids and bases is a function of their strength. The strength of an acid is a measure of ease of proton donation and of a base is the ease of protonation (Bronsted theory). This discussion on relative strengths is mainly in aqueous solution since water is the common vehicle for liquid pharmaceuticals and biological fluids.

The dissociation constant is depend on the degree of dissociation when the acid dissolved in water. Consider equation

$$HA + H_2O \rightleftharpoons H_3O^+ + A^- \qquad .....(6.35)$$

The equilibrium constant (degree of dissociation or ionisation)

$$K_{eq} = \frac{[H_3O^+][A^-]}{[HA][H_2O]} \qquad .....(6.36)$$

Then acid ionisation constant $K_a = K_{eq}\,[H_2O] = \dfrac{[H_3O^+]\,[A^-]}{[HA]}$ .....(6.37)

The K_a varies directly with strength of acids. So for stronger acids like HCl, HNO_3, H_2SO_4 etc $K_a > 1$ and weaker acids like HCOOH, CH_3COOH has smaller K_a values. Another expression of the strength of acid is the hydrogen ion concentration.

$$2H_2O \rightleftharpoons H_3O^+ + OH^- \qquad .....(6.38)$$

The ionic product of water $K_w = [H_3O^+]\,[OH^-] = 1 \times 10^{-14}$ .....(6.39)

or

$$[H_3O^+] = [OH^-] = 1 \times 10^{-7} \qquad .....(6.40)$$

If the addition of acid or base to water, the concentration of one of the species increases, while other decreases.

For *e.g*, $[H_3O^+]$ or $[H^+] = 1 \times 10^{-3}$ M then $[OH^-] = 1 \times 10^{-11}$M.

This numerical form is not convenient to express acidities. Then Sorenson introduced a pH term for expressing acidities (p from the Germen word for power and H for hydrogen) i.e., a negative logarithm to the base 10 of the hydrogen (or hydronium) ion concentration.

$$pH = -\log [H^+] = -\log [H_3O^+] = \log \frac{1}{[H_3O^+]} \qquad .....(6.41)$$

For *e.g.*, $[H^+] = 1$ mole l^{-1} or $[H_3O^+] = 1 \times 10^{0}$ g equivalent. l^{-1} then

$pH = 0$ or $pOH = 14$

$[H^+] = 0.0001$ mole l^{-1} $H_3O^+] = 1 \times 10^{-4}$ g equivalent l^{-1}

$pH = 4$ and $pOH = 10$

As the concentration of hydrogen ion increases the acidity increases and the P^H decreases.

Relation between K_a and pH

$$K_a = [H_3O^+] \frac{[A^-]}{[HA]} \quad .....(6.42)$$

$$-\log K_a = -\log [H_3O^+] - \log \frac{[A^-]}{[HA]} \quad .....(6.43)$$

Since $-\log K_a = pK_a$

$$pK_a = pH - \log \frac{[A^-]}{[HA]} \quad .....(6.44)$$

$$pK_a = pH + \log \frac{[HA]}{[A^-]} \quad .....(6.45)$$

These equations (6.44 and 6.45) are known as Henderson-Hasselbalch equations. The lower the value of pK_a (-ve) values more stronger the acid. The weaker acids have + ve pK_a values. Hydrogen ion concentration may be considered as a measure of acidity.

Relation between k_b (dissociation constant of base) and pOH

The stronger base has larger K_b. The Hydroxide ion concentration is a measure of basicity of compound of water.

$$K_b = K_{eq} [H_2O] = \frac{[BH^+][OH^-]}{[B:]}$$

$$Pk_a + pk_b = pk_w = 14 = pH + pOH \quad .....(6.46)$$

The conclusions from the above points are:

(a) If an acid has $K_a = 1 \times 10^{-5}$ (strong acid), its conjugate base in water have $K_b = 1 \times 10^{-9}$ (weak base)

(b) If $pk_b = 4.5$ then the conjugate acid $pK_a = 9.5$

(c) The strong basic solution $pOH = 2$ then $pH = 12$.

6.1.3 Percent of Ionization of Acids-Bases

This concept of percent of ionization of a barometer acid or base is important to pharmacists when considering the biopharmaceutical parameters of absorption, distribution and excretion of drug molecule. More of the drugs fall into the weak acid or base category, the pH adjustments affect the percent ionization.

Consider equation for calculation of percent ionization of acids and bases.

$$pK_a - pH = \log \frac{[HA]}{[A^-]} \qquad .....(6.47)$$

$$pk_b - pH = \log \frac{[BH^+]}{[B:]} \qquad .....(6.48)$$

If pk_a and pH is same, then

$$pk_a - pH = 0 = \log \frac{1}{1} = \log 1$$

$$\% \text{ ionization} = \frac{1}{1+1} \times 100 = 50\%$$

If pka = 4 and pH = 3, then

$$Pk_a - pH = 4 - 3 = 1$$

$$\therefore \qquad \log \frac{[HA]}{[A^-]} = 1$$

The log ratio of acid to conjugate base is

$$\text{Log} \frac{[HA]}{[A^-]} = \log \frac{10}{1} = 1$$

$$\% \text{ ionization} = \frac{1}{(10+1)} \times 100 = 9.09\%$$

6.1.4 Mechanism of Acid-Base Reaction

These are of two types – S_N1 and S_N2. Lewis bases are sometimes called nucleophilic reagents. Hence the reactions are often called nucleophilic substitution reactions.

$$Y: + A - X \rightarrow A - Y + X:$$

For the above reaction the possible mechanism is S_N1 (dissociation) mechanism. The coordination number of A decreases first

$$A - X \rightarrow A + X: \text{ unimolecular reaction followed by coordination} \qquad .....(6.49)$$

$$A + Y \rightarrow A - Y \qquad .....(6.50)$$

S_N2 (displacement) mechanism. The coordination number of A increases first.

$$Y: + A - X \rightarrow [Y... A...X] \rightarrow Y - A + X: \quad \text{bimolecular reaction} \quad .....(6.51)$$

In S_N1, the rate of substitution should be independent of the nucleophile reagent since the latter is not involved in the initial dissociation.

In S_N2, the rate is dependent on the nucleophilic character of Y.

For *e.g.*, $Y + [Co\,(en)_2\,Cl_2]^+ \rightarrow [Co\,(en)_2\,Cl\,Y]^+ + Cl^-$ .....(6.52)

Weak nucleophiles like SCN^-, Br^-, Cl^-, NO_3^- reacts at the same rate S_N1

Strong nucleophiles like CH_3O^-, N_3^-, NO_2^- reacts at S_N2 type.

e.g., NO_2^- + HO Cl →

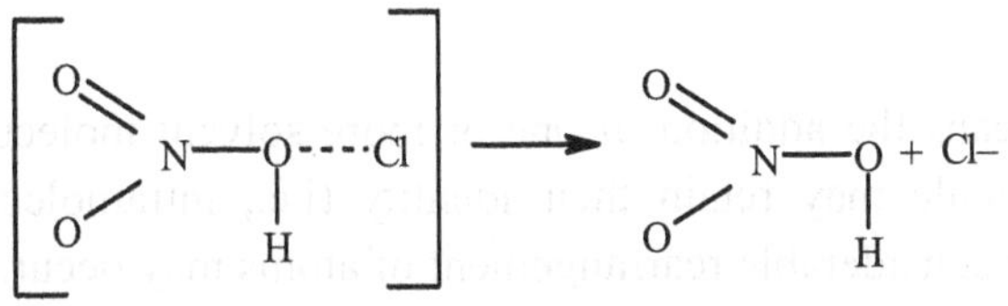

.....(6.53)

6.1.5 Salts of Acids and Bases

A weak acid have a strong conjugate base and a weak base have strong conjugate acid, and vice-versa. This information is useful in considering the properties of salts formed between acid and base.

For *e.g.*, HCl (Strong acid) + NaOH (strong base) → H_2O + NaCl (salt) .....(6.54)

NaCl is a neutral salt which is unable to impose either acidic or basic properties on an aqueous solution. Other examples are $Na_2\,SO_4$, $Na\,NO_3$ etc.

Consider the reaction between weak acetic acid ($pK_a = 4.7$) and strong base NaOH and the salt of sodium acetate is formed.

CH_3COOH (Weak acid) + NaOH (strong base) → CH_3COONa (salt) + H_2O .....(6.55)

The CH_3COONa salt ionises in water to provide acetate ion which is a conjugate base of acetic acid and the solution is alkaline and the pH is increased.

$$CH_3COO^- + Na^+ + H_2O \rightleftharpoons CH_3COOH + Na^+ + OH^- \quad .....(6.56)$$

Consider the reaction of HCl, strong acid and NH_4OH weak base, the salt formed is NH_4Cl.

$$HCl + NH_4OH \rightarrow NH_4Cl + H_2O \quad .....(6.57)$$

NH_4Cl salt on hydrolysis gives NH_4^+ ion which is a conjugate acid of NH_4OH and the solution is acidic and the pH is decreased.

Consider the reaction of CH_3COOH weak acid and NH_4OH weak base, the salt formed ammonium acetate

$$CH_3COOH + NH_4OH \rightarrow \underset{\text{salt}}{CH_3COONH_4} + H_2O \qquad .....(6.58)$$

The salt on hydrolysis form slightly ionised compounds, but the acidic or basic character of resulting solutions is dependent upon the relative strength of acidic or basic species formed.

6.1.6 Solvation and Solvolysis

Solvation

Solvation means the addition of one or more solvent molecules to another species. The solvent molecule may retain their identity (i.e., intramolecular solvent bonds are not broken), or a considerable rearrangement of atoms may occur.

$$CaCl_2 + 6NH_3 \rightarrow Ca\,(NH_3)_6\,Cl_2 \qquad .....(6.59)$$

$$BF_3 + 2H_2O \rightarrow H_3O\,(F_3\,BOH) \qquad .....(6.60)$$

$$B_2H_6 + 2NH_3 \rightarrow [BH_2\,(NH_3)_2]\,BH_4 \qquad .....(6.61)$$

Solvolysis

A solvolysis may be considered a solvation in which the solvent molecule is split into two. Either one or both solvent ions may add to other species.

There are three types of solvolysis

(i) Acid solution is formed

$$SO_3 + H_2O \rightarrow HSO_4^- + H^+ \qquad .....(6.62)$$

$$NH_4^+ + H_2O \rightarrow NH_4\,OH + H^+ \qquad .....(6.63)$$

$$Fe^{3+} + H_2O \rightarrow Fe\,OH^{2+} + H^+ \qquad .....(6.64)$$

This is not only applicable to water or protic acid but also to aprotic solvents

$$SbF_5 + BrF_3 \rightarrow BrF_2^{\,+} + SbF_6^- \qquad .....(6.65)$$

$$AlCl_3 + COCl_2 \rightarrow COCl^+ + AlCl_4 \qquad .....(6.66)$$

(ii) Base solution is formed

$$CaC_2 + 2H_2O \rightarrow Ca^{2+} + 2OH^- + C_2H_2 \qquad .....(6.67)$$

$$\text{KH phthalate} + HOAC \rightarrow H_2\,(\text{phthalate}) + K^+ + OAC^- \qquad .....(6.68)$$

$$KNO_3 + 4HF \rightarrow K^+ + H_2\,NO_3^{\,+} + 2HF_2^- \qquad .....(6.69)$$

(iii) In some reactions both ions of the solvent becomes attached to another species with the breaking of the bond.

$$O=C=O+H_2O \longrightarrow O=C(OH)_2 \quad .....(6.70)$$

$$O_3P-O-PO_3+H_2O \rightarrow 2HPO_4^{2-} \quad .....(6.71)$$

$$Cl_2+H_2O \rightarrow HOCl+HCl \quad .....(6.72)$$

6.1.7 Role of Acids and Bases in Pharmacy

Acids bases and their reactions play an important role in pharmacy practice. Therapeutically they can be used in the control and adjustment of the pH of the GI tract, body fluids and urine e.g.,

- Dilute HCl is used in the treatment of achlorhydria (a condition in which stomach is not able to secrete gastric acid) so as to achieve the acidic pH in stomach.
- Sodium bicarbonate is used as antacid to reduce the acidity because of the increased secretion of gastric acid further it can also be used in the treatment of metabolic acidosis.

The conjugate pairs of acids and bases are used as buffers e.g. buffer acid as a proton donor and buffer base as a proton acceptor.

They are used in various analytical procedures, which involve acid-base titrations (acidimetry and alkalimetry titration) *e.g.* NaOH can be assayed by its titration with HCl or H_2SO_4. They can also be used in various pharmaceutical preparations and analytical procedures as acidifier or alkalizer to achieve the required pH e.g.,

- In the limit test for iron, ammonia solution is used to obtain an alkaline pH.
- H_2SO_4 is used to provide acidic pH, in the assays involving use of $KMnO_4$. The acid solution of $KMnO_4$ reacts to reduce the permanganate ion (Mn^{7+}) to the manganous ion (Mn^{2+}) with the evolution of oxygen.

$$2KMnO_4+3H_2SO_4 \rightarrow K_2SO_4+2MnSO_4+3H_2O+5[O] \quad .....(6.73)$$

Acid base neutralization reactions are used in the conversion of drugs to chemical forms (*e.g.* Hydrochloride salts, sulphate salts and sodium salts etc.), which are suitable for product formulation.

6.1.8 Uses and Formulation of Acids

1. Boric Acid:

Chemical Formula: H_3B0_3 *Mol. Wt.* 61.83

It contains not less than 99.5% and not more than 100.5% of H_3BO_3, calculated with reference to the dried substances.

Preparation: It can be prepared by decomposing boiling solution of native borates *e.g.* borax, colemanite, resonite etc.

a. *From Borax*: A hot conc. solution of borax is treated with sulphuric acid or HCl.

$$Na_2B_4O_7 + H_2SO_4 + 5H_2O \rightarrow Na_2\,SO_4 + 4H_3BO_3 \quad .....(6.74)$$

After decomposition the hot liquid is filtered and is kept aside so as to crystallize the boric acid. Crystals of boric acid are collected by filtration and are washed so as to make it free from sulphate, then it is allowed to dry at ordinary temperature.

b. *From Colemanite:* Colemanite (calcium borate, $Ca_2B_6O_{11}.5H_2O$) is powdered and suspended in boiling water. SO_2 gas is then passed through the suspension when boric acid is formed. On cooling boric acid crystallizes out.

$$Ca_2B_6O_{11} + 2SO_2 + 9H_2O \rightarrow 2CaSO_3 + 6H_3BO_3 \quad .....(6.75)$$

Identification Tests: It occurs in the form of pearly, lamellar, triclinic crystals, which are soluble in 25 parts of cold water and in 4 parts of glycerol.

a. On igniting, solution of boric acids in methanol containing few drops of sulphuric acid, a flame having a green border is produced. This is due to the formation of volatile methyl ortho-borate.

$$H_3BO_3 + 3CH_3OH \rightarrow B\,(OCH_3)_3 \uparrow + 3H_2O \quad .....(6.76)$$

b. The dilute solution of boric acid in boiling distilled water (30 g in 90 ml) is when cooled a faintly acidic solution is produced. This solution is found to have pH between 3.8 and 4.8. Further free boric acid changes the colour of litmus to red but it does not produce any effect on methyl orange.

Test for purity: It should be tested for clarity and colour of 3.5% w/v solution of boric acid in water, arsenic, heavy metals, sulphate, and loss on drying and for solubility in ethanol.

As per IP (1996), the 1 g of boric acid shall dissolve almost completely in 10 ml of boiling ethanol (95%). This test is done to check the absence of metallic borates and insoluble impurities.

Assay: Boric acid is a much weaker acid than carbonic acid or even hydrogen sulphide. The value of pK_a for the ionization of the first proton at 25 °C is 9.19. This shows that it is a very weak acid and because of this one does not get accurate results on titrating it with a standard base in aqueous solution. With strong alkalis; it forms salts known as metaborates.

$$H_3BO_3 + NaOH \rightarrow B(OH)_4^- + Na^+ \rightarrow NaBO_2 + 2H_2O \quad .....(6.77)$$

(Sodium metaborate)

When boric acid is titrated with sodium hydroxide the end point is not sharp due to the excessive hydrolysis of sodium meta-borate formed during titration.

$$NaBO_2 + H_2O \rightarrow H_2BO_3 + NaOH \quad .....(6.78)$$

However, the hydrolysis can be checked by adding polyhydroxy compounds such as glycerol, mannitol and catechol etc, to the titration mixture. These compounds react with meta-borate ion to give complex compounds resulting in the free ionization of boric acid during titration with strong alkali. Thus, the addition of such compounds makes boric acid behave as a strong monobasic acid and the end point can be easily detected.

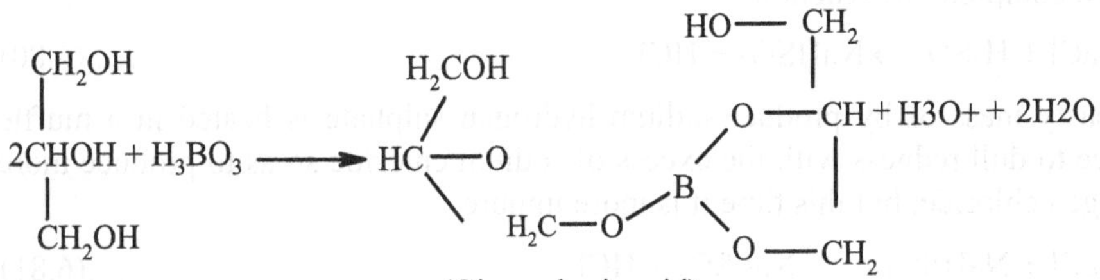

(Glycerin) (Glyceroboric acid) .....(6.79)

Method: About 2 g of the sample is weighed accurately and dissolved in a mixture of 50 ml of water and 100 ml of glycerin, previously neutralized to phenolphthalein solution. Contents are then titrated with 1M sodium hydroxide using phenolphthalein solution as indicator.

Uses:

- It is a weak bacteriostatic agent, mainly used as local anti-infective.
- It is used as an eyewash in the form of solutions in concentrations from 2.5 to 4.5% as it is non-irritating when applied to the intact skin and mucous antiseptic ointment for treating diaper rash.
- It is also added to various dusting powders for its local anti-infective properties.
- It is used to provide acidic media and buffered media for other drugs.
- It is used in different topical medications to maintain an acidic pH in the medium.
- It is used as a buffer to maintain the pH around 6 in various ophthalmic preparations. In these solutions it also helps in the maintenance of isotonicity.

- It is used to prepare Boroglycerin Glycerite ($C_3H_5BO_3$), which is used as a suppository base.

Note: Due to its toxic effects it is not used in preparations meant for internal use.

2. Hydrochloric acid:

Chemical Formula: HCl *Mol. Wt.* 36.46

It is also known as chlorhydric acid and muriatic acid and spirit of salt. It contains not less than 35.0% w/w and not more than 38.0% w/w of HCl.

Preparation

(a) *From Sodium Chloride*: It can be prepared by reacting concentrated sulphuric acid with sodium chloride. The weighed amount of salt is taken in the cast-iron pan of a salt cake furnace, now an equal amount of concentrated sulphuric acid is allowed to run over it. Hydrogen chloride gas is evolved. Gently heat the furnace so as to complete the reaction.

$$NaCl + H_2SO_4 \rightarrow NaHSO_4 + HCl \qquad .....(6.80)$$

The pasty mass of by product sodium hydrogen sulphate is heated in a muffle furnace to dull redness with the excess of sodium chloride so as to produce more hydrogen chloride, but this time it is more impure.

$$NaCl + NaHSO_4 \rightleftarrows Na_2\,SO_4 + HCl \qquad .....(6.81)$$

The hydrogen chloride so obtained is passed by a pipeline into the tower containing lumps of coke, down which water falls. The crude concentrated acid is collected as it contaminated with iron, sulphuric acid, sulphur dioxide, arsenic and other impurities.

(b) *From hydrogen and chlorine gases:* Pure hydrochloric acid can be prepared using hydrogen and chlorine gases produced as by products during electrolysis of brine solution (sodium chloride solution) in the preparation of caustic soda.

In this process the chlorine is burned in large diameter silica tubes in the presence of hydrogen, to give hydrogen chloride, which is then absorbed in water to give hydrochloric acid.

$$H_2 + Cl_2 \rightarrow 2HCl \qquad .....(6.82)$$

Identification Tests

(a) Addition of hydrochloric acid to potassium permanganate results in the evolution of chlorine gas.

$$2KMnO_4 + 16HCl \rightarrow 2MnCl_2 + 2KCl + 5Cl_2 \uparrow + 8H_2O \qquad .....(6.83)$$

In this test the hydrochloric acid is oxidized by potassium permanganate, a strong oxidizing agent, resulting in the evolution of chlorine gas. In the above test the potassium permanganate can be replaced by manganese dioxide.

(b) It gives the reactions of chlorides

(i) Due to the chloride ion it precipitates the metal ions such as Ag, Pb and Hg (I) in the form of insoluble chlorides of metals.

$$AgNO_3 + HCl \rightarrow AgCl \downarrow + HNO_3 \quad .....(6.84)$$

(ii) It reacts with a mixture of potassium dichromate and sulphuric acid and results in the evolution of chromyl chloride, which turns the colour of the strip moistened with solution of diphenyl carbazide in alcohol to violet red.

$$4Cl^- + Cr_2O_7^{2-} + 6H^+ \rightarrow 2CrO_2Cl_2 \uparrow + 3H_2O \quad .....(6.85)$$

Test for purity: It has to be tested for arsenic, heavy metals, bromide and iodide, free chlorine, sulphate, sulphite and residue on evaporation (non-volatile matter).

(a) *Test for bromide and iodide*: 5 ml of the acid is diluted with 10ml of water and to this 1ml of chloroform is added. Now with constant shaking drop wise chlorinated lime solution is added, the chloroform layer does not become brown or violet.

In this test chlorinated lime act as oxidizing agent and liberates bromine or iodine from bromides or iodides. If any bromine or iodine is formed, it is extracted with chloroform. This changes the colour of chloroform layer.

(b) *Test for free chlorine*: Take 60 ml acid and add 50 ml of carbon dioxide free water, then add 1 ml of a 10% w/v solution of potassium iodide and 0.5 ml of starch solution. The mixture is allowed to stand in the dark for 2 minutes. Any blue colour produced disappears on the addition of 0.2 ml of 0.01M sodium thiosulphate.

In the test the free chlorine is detected by the liberation of iodine from potassium iodide. The amount of liberated iodine is detected by sodium thio-sulphate.

$$2KI + Cl_2 \rightarrow 2KCl + I_2 \quad .....(6.86)$$

Starch is added to make adsorption complex with iodine, which is of blue colour.

$$I_2 + 2Na_2S_2O_3 \rightarrow 2NaI + Na_2S_4O_6 \quad .....(6.87)$$

Sodium thiosulphate Sodium tetrathionate

Assay: The assay is based upon acid-base titration. It involves titration of strong acid (HCl) with alkali hydroxide (NaOH) using methyl red as indicator.

$$HCl + NaOH \rightarrow NaCl + H_2O \quad .....(6.88)$$

Method: Take about 2 g accurately weighed acid is a conical flask. To this add 30 ml of water mix the contents and titrate with 1M sodium hydroxide using methyl red solution as indicator.

Each ml of 1M sodium hydroxide is equivalent to 0.03646 g of HCl.

Uses

- It is mainly used as a pharmaceutical aid more specifically as an acidifying agent.
- Because of its strong acid character it can react with weakly basic organic molecules to form usually water soluble hydrochloride salts for extraction or other separation purposes, e.g. alkaloids, which are usually sparingly soluble in water, can be treated with HCl to form salts which are freely soluble in water.
- Now-a-days in various acid-base titrations hydrochloric acid is generally preferred to sulphuric acid because with certain indicators it gives a sharper end point.
- In the determination of calcium carbonate and calcium hydroxide, sulphuric acid cannot be used because that gives precipitates of sparingly soluble calcium sulphate, while hydrochloric acid gives completely insoluble calcium chloride. For the same reason hydrochloric acid is used in the processes involving the titration of barium hydroxide. Very dilute hydrochloric acid can be used as gastric acidifier in achlorhydria.

3. Nitric Acid:

Chemical Formula: HNO_3 *Mol. Wt.* 63.01

It contains not less than 69.0 percent and not more than 71.0 percent, by weight of HNO_3.

Preparation

(a) *From sodium nitrate:* In the laboratory, it can be prepared by heating a mixture of sodium nitrate and sulphuric acid.

$$NaNO_3 + H_2SO_4 \rightarrow NaHSO_4 + HNO_3 \quad .....(6.89)$$

(b) *By Ostwald process:* The method involves the catalytic oxidation of ammonia to NO, followed by oxidation of NO to NO_2 and conversion of NO_2 with water to HNO_3.

$$4NH_3\,(g) + 5O_2 \xrightarrow[\text{5 atmospheres, 850°C}]{\text{Platinum / Rhodium catalyst}} 4NO\,(g) + 6H_2O\,(g) \quad .....(6.90)$$

The NO and air (O_2) are cooled and the mixture of gases is absorbed in a counter current of water.

$$2NO(g) + O_2\,(g) \rightleftarrows 2\,NO_2\,(g)$$

$$2NO_2\,(g) + H_2O(1) \rightarrow HNO_3 + HNO_2 \quad .....(6.91)$$

$$2HNO_2 \rightarrow H_2O + NO_2 + NO$$

$$3NO_2 + H_2O \rightarrow 2HNO_3 + NO$$

Overall Reaction:

$$NH_3 + 2O_2 \rightarrow HNO_3 + H_2O \quad .....(6.92)$$

This process gives dilute nitric acid (50% -60% by weight), which is concentrated by distillation till it forms a constant boiling mixture (B.P. 121 °C). This is the ordinary concentrated HNO_3 and is 68% in strength. Fuming HNO_3 (98% HNO_3) can be obtained by distilling this acid with concentrated H_2SO_4. Fuming nitric acid may also be obtained by dissolving excess of NO_2 in concentrated HNO_3. Crystals of pure HNO_3 may be obtained by cooling fuming HNO_3 in a freezing mixture.

Identification Test

(a) It gives positive tests for nitrate ions. Mix equal volumes of nitric and sulphuric acid. Cool the mixture and slowly add a solution of ferrous sulphate. A brown colour is produced at the junction of two liquids.

$$2HNO_3 + 3H_2SO_4 + 6FeSO_4 \rightarrow 3Fe_2(SO_4)_3 + 4H_2O + 2\,NO\uparrow \quad .....(6.93)$$

$$NO + FeSO_4 \rightarrow \underset{\text{(Brown ring)}}{[Fe(NO)]SO_4} \quad .....(6.94)$$

(b) When nitric acid is heated with sulphuric acid and metallic copper, brownish red fumes are evolved. In this test nitric acid is decomposed by concentrated sulphuric acid to give brownish red fumes of NO_2, which intensify on addition of copper turnings in the mixture.

$$4HNO_3 \xrightarrow{\text{Conc. } H_2SO_4} 4NO_2\uparrow + O_2\uparrow + 2H_2O \quad .....(6.95)$$

(c) It gives xanthoproteic test. It produces a yellow stain on animal tissues. This yellow stain is because of nitration of aromatic acids, phenylalanine, tyrosine and tryptophan present in the protein of the skin.

Test for purity: It has to be tested for residue on ignition, chloride, sulphate, iron and heavy metals.

Assay: Its assay is based upon simple acid base titration.

A weighed sample is titrated against 1N sodium hydroxide using methyl red as an indicator.

$$HNO_3 + NaOH \rightarrow NaNO_3 + H_2O \quad .....(6.96)$$

Each ml of 1N NaOH is equivalent to 0.06301 g of HNO_3.

Uses: It is the most important oxoacid of nitrogen. Besides as a useful acidifying agent, it is also used for its oxidizing and nitrating properties. When mixed with conc. H_2SO_4, the nitronium ion NO_2^+ is formed, which is the active species in the nitration of organic compounds. This is an important step in making important nitro compounds. It is an excellent oxidising agent particularly when hot and concentrated.

H^+ ions are oxidising, but the NO_3^- ion is an even stronger oxidising agent in acid solution. Thus metals such as copper and silver, which are insoluble in HCl, dissolve in HNO_3. Non-metals can also be oxidized with nitric acid. This reaction is used to produce sulphuric acid and phosphoric acid by treating elemental sulphur and phosphorous respectively with acid.

It can be used as a source of nitrate ion.

It can be used externally to destroy warts and chancres.

4. **Phosphoric acid (Orthophosphoric; Concentrated phosphoric acid):**

Chemical Formula: H_3PO_4 *Mol. Wt.* 98.00

It contains not less than 84.0% w/w and not more than 90.0% w/w of H_3PO_4

Preparation

(a) *By Furnace method:* Pure H_3PO_4 can be prepared by this method. Molten phosphorous is burnt in a furnace with air and steam, it results in the formation of P_2O_5 (phosphorous pentoxide) by reaction between P and O_2. This P_2O_5 is immediately hydrolysed by hot water to give liquid phosphoric acid

$$4P + 5O_2 \rightarrow 2P_2O_5 \qquad .....(6.97)$$

$$P_2O_5 + 3H_2O \rightarrow 2H_3PO_4 \qquad .....(6.98)$$

The liquid is transferred to a dish and evapourated to syrup. Concentrated acid contains about 85% by weight of H_3PO_4. 100% pure (anhydrous) H_3PO_4 is seldom used, but it can be prepared by evaporation at low pressure.

(b) *In laboratory* it can be prepared by the action of concentrated HNO_3 on phosphorous

$$P + 5HNO_3 \rightarrow H_3PO_4 + 5NO_2 + H_2O \qquad .....(6.99)$$

(c) *On large scale* H_3PO_4 is prepared by digesting crushed mineral phosphate with dilute sulphuric acid.

$$Ca_3(PO_4)_2 + 3H_2SO_4 \rightleftharpoons 3CaSO_4 + 2H_3PO_4 \qquad .....(6.100)$$

The $CaSO_4$ is hydrated to gypsum ($CaSO_4.2H_2O$), which is separated from the solution along with other undissolved matter by filtration. The H_3PO_4 is concentrated by evaporation.

Test for identification

(a) Its aqueous solution is strongly acidic.

(b) Its neutralized (neutralization is done with NaOH) aqueous solution gives following reactions of phosphates

$$2NaOH + H_3PO_4 \rightarrow Na_3PO_4 + 2H_2O \qquad .....(6.101)$$

(c) *On adding silver nitrate solution*: a light yellow precipitate of silver phosphate (Ag_3PO_4) is formed. (Metaphosphates and pyrophosphates give white precipitates with silver nitrate). The colour of the precipitate is not changed by boiling and they are readily soluble in ammonia solution and in dilute nitric acid.

$$Na_3PO_4 + 3AgNO_3 \rightarrow Ag_3PO_4 \downarrow + 3NaNO_3 \quad \text{.....(6.102)}$$

(Yellow ppt)

$$Ag_3PO_4 \rightleftharpoons 3Ag^+ + PO_4^{3-} \quad \text{.....(6.103)}$$

$$Ag^+ + 2NH_3 \rightleftharpoons [Ag(NH_3)_2]^+ \quad \text{.....(6.104)}$$

(Soluble diammine silver complex)

(d) *On adding ammoniacal magnesium sulphate:* solution a white crystalline precipitate of magnesium ammonium phosphate is formed.

$$Mg^{2+} + NH_4^+ + PO_4^{3-} \rightarrow MgNH_4\,PO_4 \downarrow \quad \text{.....(6.105)}$$

Magnesium Ammonium Phosphate

(e) On adding dilute nitric acid and ammonium molybdate solution and warming the solution a can they yellow precipitate of a complex ammonium 12-molybdophosphate is slowly formed.

$$Na_3PO_4 + 3HNO_3 \rightarrow 3NaNO_3 + H_3PO_4 \quad \text{.....(6.106)}$$

$$H_3PO_4 + 12(NH_4)_2\,MoO_4 + 21\,HNO_3 \rightarrow$$

$$(NH_4)_3\,[PO_4.Mo_{12}O_{36}] + 12H_2O + 21NH_4NO_3 \quad \text{.....(6.107)}$$

Canary yellow ppt

Test for purity: It has to be tested for clarity and colour of the solution, arsenic, heavy metals, iron, chloride, sulphate, alkali phosphates, aluminium and calcium, hypophosphorous acid and phosphorous acid.

Impurities of calcium, iron and sulphate are usually present in phosphoric acid prepared directly from natural phosphates. As a result of incomplete oxidation, the acid prepared from phosphorous may contain phosphorous acid impurity. Impurities of phosphorous and hypophosphorous acids can be detected with silver nitrate which is reduced to silver by these impurities.

Assay: It contains three replaceable H atoms, and is tribasic. It undergoes stepwise dissociation

$$H_3PO_4 \rightleftharpoons H^+ + H_2PO_4^- \quad K_{a1} = 7.5 \times 10^{3-}\ (pK_{a1} = 2.12) \quad \text{.....(6.108)}$$

$$H_2PO_4^- \rightleftharpoons H^+ + HPO_4^{2-} \quad K_{a2} = 6.2 \times 10^{8-}\ (pK_{a2} = 7.21) \quad \text{.....(6.109)}$$

$$HPO_4^{2-} \rightleftharpoons H^+ + PO_4^{3-} \quad K_{a3} = 1 \times 10^{12-} \quad (pK_{a3} = 12.32) \quad \text{.....(6.110)}$$

The K_a values of the above reactions show the decrease in acid strength with each successive ionization. In the simple ionization process very little of the phosphate trianion, PO_4^{3-} is produced.

The official assay method (as given in IP 1996) includes titration of only the first two protons. The titration is done to the Na_2HPO_4 end point with 1M sodium hydroxide using dilute phenolphthalein solution as indicator.

$$H_3PO_4 + 2NaOH \rightarrow Na_2\,HPO_4 + 2H_2O \qquad .....(6.111)$$

Further in order to get more accurate end point the assay is carried out in the presence of sodium chloride.

Method: Weigh accurately about 1 g, add a solution of 10 g of sodium chloride in 30 ml of water and titrate with 1M sodium hydroxide using dilute phenolphthalein solution as indicator.

1 ml of 1M sodium hydroxide is equivalent to 0.04900 g of H_3PO_4.

Uses:

- It is used as pharmaceutical aid (acidifying agent) but the solubilities of the various phosphate salts produced limits its use as an acidifying agent.
- It is nonvolatile and does not possess oxidizing properties, thus it can be used wherever a non-oxidizing acid is required *e.g.* in the synthesis of HBr from NaBr and in the synthesis of HI from NaI, it will substitute sulphuric acid which cannot be used because of its oxidizing action.
- On treatment with NaOH it gives mixtures of NaH_2PO_4 and Na_2HPO_4, which are used in the preparation of phosphate buffer system.

5. Sulphuric Acid: (Oil of Vitriol)

Chemical Formula: $H_2\,SO_4$ *Mol. Wt.* 98.07

It contains not less than 95% and not more than 98%, by weight, of H_2SO_4.

Preparation

(a) *From* SO_2

Sulphuric acid is commercially the most important acid and is manufactured by two processes, namely – chamber process and contact process. Each of the two processes is suited to certain concentration and purity of the product. By chamber process sulphuric acid of about 78% strength can be prepared and by contact process about 100% pure sulphuric acid can be produced.

(i) *Lead chamber process*: It involves oxidation of sulphur dioxide by atmospheric oxygen in the presence of oxides of nitrogen (NO_2) as catalyst. This process is now obsolete.

(ii) *Contact process*: It involves the oxidation of sulphur dioxide by air at low temperature (optimum temp is 400-450 °C), high pressure (about 1.6 to 1.7 atm), excess of oxygen and presence of catalyst (platinum, ferric oxide or vanadium pentoxide) to give sulphur trioxide, which when dissolved in 98% sulphuric acid gives oleum ($H_2S_2O_7$) which on dilution with water gives sulphuric acid.

$$2SO_2 + O_2 \xrightleftharpoons{\text{Catalyst}} 2SO_3 \qquad .....(6.112)$$

$$SO_3 + H_2SO_4\,(98\%) \rightarrow H_2S_2O_7 \qquad .....(6.113)$$

(Oleum or pyrosulphuric acid)

$$H_2S_2O_7 + H_2O \rightarrow 2H_2SO_4 \qquad .....(6.114)$$

In the above process, the SO_3 can be mixed with water to give H_2SO_4, but the reaction is violent and produces a dense chemical mist which is difficult to condense.

(b) *From Sulphur*: On laboratory scale it can be prepared by boiling sulphur with nitric acid.

$$S + 6HNO_3 \xrightarrow{\text{boil}} H_2SO_4 + 6NO_2 + 2H_2O \qquad .....(6.115)$$

Identification Test

1. It gives tests for sulphate ion

(a) On treatment with $BaCl_2$ solution, white precipitate is produced which is insoluble in hydrochloric acid and in nitric acid.

$$SO_4^{2-} + BaCl_2 \rightarrow BaSO_4 + 2Cl^- \qquad .....(6.116)$$

(b) With lead acetate solution, it gives a white precipitate which is soluble in ammonium acetate solution.

$$SO_4^{2-} + (CH_3COO)_2Pb \rightarrow PbSO_4 \downarrow + 2CH_3COO^- \qquad .....(6.117)$$

***Test for purity*:** It has to be tested for residue on ignition, chloride, arsenic, heavy metals and reducing substance.

***Assay*:** A weighed sample of acid is titrated against 1N sodium hydroxide using methyl orange as an indicator.

The assay is based upon simple acid-base titration method in which strong acid (sulphuric acid) is titrated with strong base (sodium hydroxide) in the presence of methyl orange as an indicator.

Each ml of 1N sodium hydroxide is equivalent to 0.04904 g of H_2SO_4.

Uses:

It is used as a pharmaceutical aid (acidifying agent).

It is used to form water soluble salts of basic organic drug molecules.

It can also be used as a dehydrating agent in many reactions such as esterification and nitration where water has to be removed. Sulphuric acid is applied in drying gases which do not react with the acid (e.g. SO_2, Cl_2, HCl etc.)

In inorganic qualitative analysis, it is used to detect the certain basic radicals such as barium, strontium and lead as it forms insoluble sulphates with these basic radicals (sulphate salts of most metals are soluble in water).

$$Ba^{2+} + H_2SO_4 \rightarrow BaSO_4 \downarrow + 2H^+ \quad .....(6.118)$$

It is used to sulphonate fatty acids to make detergents.

6.1.9 Uses and Formulation of Bases

1. Strong Ammonia Solution (Ammonium hydroxide, stronger ammonia water):

Chemical Formula: NH_3 *Mol. Wt.* 17.03

Strong ammonia solution is a solution of NH_3, containing not less than 27.0% and not more than 31.0% (w/w) of NH_3. On exposure to air it loses ammonia rapidly.

Preparation:

1. *From binary metal compounds:* It can be prepared from binary metal compounds such as Mg_3N_2, Ca_3P_2 by the action of water or dilute acids.

$$Mg_3N_2 + 6H_2O \rightarrow 3Mg(OH)_2 + 2NH_3 \quad .....(6.119)$$

2. *From Nitrogen (Haber Process):* On commercial scale it can be prepared by the Haber process, which involves the direct combination of N_2 and H_2.

$$N_2 + 3H_2 \rightarrow 2NH_3 \quad .....(6.120)$$

The reaction is allowed to take place at 450-550 °C under a pressure of 100-1000 atmosphere. Catalyst is platinum rhodium gauze.

At 20°C and one atmosphere pressure 53.1g NH_3 dissolves in 100g of water. Therefore, its 27.0% solution can be prepared by dilution of the above mentioned solution.

In solution ammonia forms ammonium hydroxide (NH_4 OH) and behaves as a weak base.

$$NH_3 + H_2O \rightleftharpoons NH_4^+ + OH^- \quad .....(6.121)$$

The chemical reactions and various other properties of ammonia solution is due to the presence of NH_3.

***Identification Test*:**

(a) Dip a glass rod in hydrochloric acid and bring it near the surface of the NH_3 solution. Dense white fumes are produced. The test is based upon reaction of NH_3 with HC1.

$$NH_3 + HCl \rightarrow NH_4Cl \quad .....(6.122)$$

(b) Because of its ability to give hydroxyl ions, aqueous ammonia had basic character and thus turns red litmus blue and phenolphthalein pink.

$$NH_3\ (aq.) + H_2O \rightleftharpoons NH_4^+ + OH^-$$

(c) With Nessler's reagent it gives reddish brown precipitate.

$$HgCl_2 + 2KI \rightarrow 2KCl + HgI_2 \downarrow \text{(Red ppt.)} \quad .....(6.123)$$

$$HgI_2 + 2KI \rightarrow \underset{\text{(Nesseler's reagent)}}{K_2HgI_4}$$

$$K_2HgI_4 + NH_3 + 3KOH \rightarrow H_2N.\ HgO.\ HgI \downarrow + 7KI + 2H_2O \quad .....(6.124)$$

***Test for purity*:** It has to be tested for heavy metals, limit of nonvolatile residue and readily oxidisable substances.

***Assay*:** Weigh accurately about 3 g in a flask containing 50 ml of 1N H_2SO_4 and titrate the excess of acid with 1N NaOH using solution of methyl red as indicator.

Each ml of 1N H_2SO_4 is equivalent to 0.01703 g of NH_3

Uses

- It is used as a pharmaceutical aid. Mainly, it is used in making ammonia water by dilution and as a chemical reagent.
- It is a good laboratory reagent. It is used as a base (Bronsted base) to form ammonium salts of acids, and to make certain solutions alkaline e.g. in the limit test for iron.
- It combines with certain cations like Ag^+, Cu^{2+}, Hg^{2+} etc. and forms very stable soluble complex cations.

$$\underset{}{AgCl} + \underset{\text{(Excess)}}{2NH_4OH} \rightarrow \underset{\substack{\text{(Silverammine Chloride)} \\ \text{Soluble}}}{[Ag(NH_3)_2]Cl} + H_2O \quad .....(6.125)$$

- It is also used in preparing ammonical silver nitrate solution (Tollen's reagent).

2. Calcium Hydroxide:

Chemical Formula: $Ca(OH)_2$ *Mol. Wt.* 74.09

It contains not less than 95.0% and not more than 100.5% of $Ca(OH)_2$.

Preparation

From calcium oxide: It is manufactured by the process of slaking from quick lime or calcium oxides (CaO), through the addition of water in limited amounts. The calcium oxide absorbs water and there is evolution of large amount of heat. The CaO lumps swell and disintegrate into a fine powder of $Ca(OH)_2$. The calcium hydroxide so obtained is mixed with excess of water and allowed to settle. The calcium hydroxide so obtained is mixed with excess of water and allowed to settle. The supernatant liquid is decanted and the residue is air dried.

$$CaO + H_2O \rightarrow Ca(OH)_2 \qquad .....(6.126)$$

Test for Identification

(a) When mixed with from 3 to 4 times its weight of water it forms a smooth magma. The clear supernatant liquid from the magma is alkaline to litmus.

(b) Mix 1 g with 20 ml of water, and add sufficient 6N acetic acid to make solution. The resulting solution gives following tests for calcium.

(i). To 5 ml of above solution add 1 ml of glacial acetic acid and 0.5 ml of potassium ferrocyanide solution. The resulting solution remains clear. On adding about 30 mg of NH_4Cl, a white, crystalline precipitate is formed.

$$2Ca^{++} + K_4Fe(CN)_6 \rightarrow Ca_2Fe(CN)_6 + 4K^+ \qquad .....(6.127)$$

$$Ca_2\,Fe(CN)_6 + 2NH_4Cl \rightarrow \underset{\text{White ppt.}}{(NH_4)_2Ca[Fe(CN)_6]\downarrow} + CaCl_2 \qquad .....(6.128)$$

(ii) Take 5 ml of the prepared solution and add 0.2 ml of a 2% w/v solution of ammonium oxalate, a white precipitate is obtained that is only sparingly soluble in dilute acetic acid but is soluble in hydrochloric acid.

$$(NH+_4)_2\,(COO)_2 \rightleftharpoons 2NH_4^+ + \begin{matrix} COO^- \\ | \\ COO^- \end{matrix} \qquad .....(6.129)$$

(Calcium Oxalate)

.....(6.130)

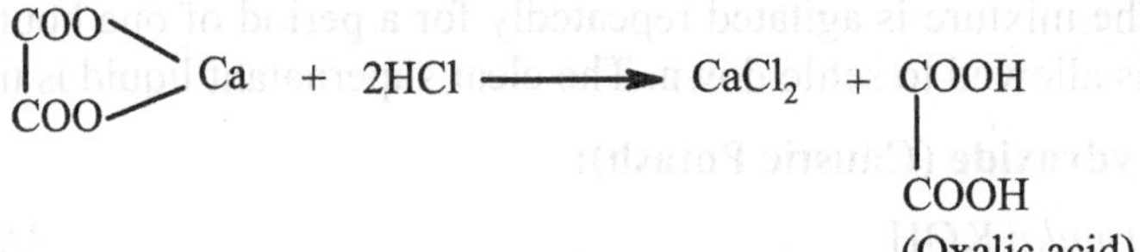

(Oxalic acid) .....(6.131)

3. Take 5 ml of the prepared solution, add 5 ml of ammonium carbonate solution; a white precipitate is formed which after boiling and cooling the mixture, is only sparingly soluble in ammonium chloride solution.

$$Ca^{2+} + (NH_4)_2CO_3 \rightarrow \underset{\text{(White ppt.)}}{CaCO_3 \downarrow} + 2NH_4^+ \quad .....(6.132)$$

Test for Purity: It is tested for limit of acid – insoluble substances, carbonate, heavy metals, chloride, sulphate, phosphate and limit of magnesium and alkali salts.

Assay: Its assay is based upon the principle of complexometric titrations in which calcium ions form a complex with disodium EDTA. The pH of the solution is adjusted to 12 by using sodium hydroxide.

Method: weigh accurately about 1.5 g of $Ca(OH)_2$ and transfer it to a beaker and gradually add 30 ml of 3N hydrochloric acid. When solution is complete, transfer the solution to a 500 ml volumetric flask, rinse the beaker thoroughly, adding the rinsing to the flask, dilute with water to volume, and mix. Pipette 50 ml of the solution into a suitable container, add 100 ml of water, 15 ml of 1N sodium hydroxide and 300 mg of hydroxyl naphthol blue. Triturate and titrate the contents with 0.05M disodium ethylenediaminetetraacetate to a blue end point.

Each ml of 0.05M disodium ethylenediaminetetraacetate is equivalent to 3.705 mg of $Ca(OH)_2$.

Uses:

- It is used in various pharmaceutical preparations because of its high hydroxide ion concentration. It is used to prepare calcium soaps of fatty acids, which have emulsifying properties and can be used in the preparation of suspensions or mixing of other ingredients.
- It can also be used as a topical astringent and as a fluid electrolyte (as a source of calcium).
- *It can absorb* CO_2; for this property it is used in soda lime (A mixture of calcium hydroxide and sodium hydroxide) to absorb CO_2 from expired air in metabolic function test and in closed circuit anesthetic machine. It is used to prepare bleaching powder and for softening of water.

Available as calcium hydroxide solution (A solution containing, not less than 140 mg of $Ca(OH)_2$ in each 100 ml). It is prepared by adding 3 g of calcium hydroxide to 1000 ml of purified water. The mixture is agitated repeatedly for a period of one hour. The excess calcium hydroxide is allowed to settle down. The clear supernatant liquid is used.

3. Potassium Hydroxide (Caustic Potash):

Chemical Formula: KOH *Mol. Wt.* 56.11

It contains not less than 85.0% of total alkali, calculated as KOH, including not more than 3.5% of K_2CO_3 as per USP and not more than 4.0% of K_2CO_3 as per B.P.

Preparation

1. *By electrolysis of KCl solution*: It can be prepared by the electrolysis of KCl solution in a diaphragm cell. The caustic liquor obtained is evaporated and the resulting fused potassium hydroxide is either allowed to solidify and then broken up, or cast into sticks, or made into the more convenient pellets.

$$KCl_{(aq)} \xrightleftharpoons{\text{Electrolysis}} K^+ + Cl^- \quad .....(6.133)$$

$$H_2O \xrightleftharpoons{\text{Electrolysis}} H^+ + OH^- \quad .....(6.134)$$

At cathode $2H^+ + 2e^- \rightarrow H_2$ .....(6.135)

At Anode $2Cl^- \rightarrow Cl_2 + 2e^-$ .....(6.136)

In solution $K^+ + OH^- \rightarrow KOH$ .....(6.137)

2. By the action of lime on potassium carbonate

$$Ca(OH)_2 + K_2CO_3 \rightarrow CaCO_3 + 2KOH \quad .....(6.138)$$

3. It may also be prepared by the action of $Ba(OH)_2$ on potassium sulphate.

$$K_2SO_4 + Ba(OH)_2 \rightarrow 2KOH + BaSO_4 \downarrow \quad .....(6.139)$$

Identification Tests: A Solution (1 in 25) responds to the following tests for potassium.

(a) It gives a violet colour to a nonluminous flame, (the presence of small quantities of sodium masks the colour).

(b) On adding sodium hydrogen tartrate to the neutral aqueous solution of the compound, a white crystalline precipitate of potassium hydrogen tartrate is produced. The precipitate is soluble in 6N ammonium hydroxide and in solutions of alkali hydroxides and carbonates. Stirring or rubbing the inside of the test tube with a glass rod can accelerate the formation of the precipitate. The addition of a small amount of glacial acetic acid or alcohol also helps precipitation.

$$\underset{\text{(Sodium hydrogen tartrate)}}{NaHC_4H_4O_6} + KOH \rightarrow \underset{\text{(Potassium hydrogen tartrate)}}{KHC_4H_4O_6} + NaOH \quad .....(6.140)$$

(c) A fairly dilute aqueous solution of potassium hydroxide gives a yellow precipitate of potassium cobaltinitrite, with sodium cobaltinitrite.

$$3KOH + Na_3[CO(NO_2)_6] \rightarrow K_3[CO(NO_2)_6] + 3NaOH \quad .....(6.141)$$

Potassium cobaltinitrite

(Yellow ppt.)

(d) To the aqueous solution of sample add platinic chloride (Chloroplatinic acid) solution in the presence of hydrochloric acid; a yellow crystalline precipitate of potassium chloroplatinate is formed.

$$H_2PtCl_6 + 2K^+ \rightarrow K_2[PtCl_6] + 2H^+ \quad .....(6.142)$$

Note: Ammonium salts also give positive responses to all the above precipitation tests for potassium. Therefore, ammonium salts should be removed by ignition, before testing for potassium.

Test for purity: It has to be tested for insoluble substances, heavy metals. Potassium hydroxide can be adulterated with sodium (in the form of sodium hydroxide). Sodium because of its lower equivalent actually increases the apparent percentage of total alkali. Antimonate test can be done for the detection of significant amounts.

Antimonate Test for the detection of sodium as impurity in potassium compounds:The test is done with aqueous-alcoholic solution of the KOH sample. An alkaline solution of the potassium antimonate gives with sodium salts a white, amorphous precipitate of the sodium antimonite, which rapidly changes to the crystalline pyroantimonate.

$$KSbO_3 . 2H_2O + Na^+ \rightarrow NaSbO_3 . 3H_2O + K^+ \quad .....(6.143)$$

Potassium Antimonate — Sodium Antimonate

$$2NaSbO_3.3H_2O \xrightarrow{H_2O} Na_2H_2Sb_2O_7 . 6H_2O \quad .....(6.144)$$

Crystalline pyroantimonate

The test is of some sensitivity if done in aqueous-alcoholic solution, because, solubility of the pyroantimonate is about 1 in 350 in cold water and it is almost insoluble in alcohol.

Assay: The assay is done by the modified Winkler method, Winkler method is the most satisfactory process for the determination of alkali hydroxide and carbonate when present together.

Method: Dissolve about 1.5 g of potassium hydroxide, accurately weighed in 40 ml of carbon dioxide free water. Cool the content to 15 °C, add phenolphthalein and titrate with 1N sulphuric acid. At the discharge of the pink colour of the indicator, record the volume of acid solution required, then add methyl orange and continue the titration to a persistent pink colour.

Each ml of 1N sulphuric acid is equivalent to 56.11 mg of total alkali, calculated as KOH, and each ml of acid consumed in the titration with methyl orange is equivalent to 138.2 mg of K_2CO_3.

In the assay, firstly the hydroxide in the solution is determined by titration with N/10(0.1N) sulphuric acid, using phenolphthalein as indicator. In the titration to phenolphthalein end point two titrations occur.

$$2KOH + H_2SO_4 \rightarrow K_2SO_4 + 2H_2O \quad .....(6.145)$$

$$2K_2CO_3 + H_2SO_4 \rightarrow K_2SO_4 + 2KHCO_3 \quad .....(6.146)$$

The titration is done slowly and with constant shaking, the indicator changes colour before any of the $KHCO_3$ is acted upon by the acid. Methyl orange is now added, and the titration is continued to the pink end point.

$$2KHCO_3 + H_2SO_4 \rightarrow K_2SO_4 + CO_2 + H_2O \quad .....(6.147)$$

Uses

- It is used as a pharmaceutical aid in several pharmaceutical preparations.
- Potassium hydroxide is a very strong base. It is mainly used as a base or alkaline reagent.
- It possess caustic or corrosive effect on tissues, therefore, great care shall be taken while handling potassium hydroxide, as it rapidly destroys tissues.
- It is used to manufacture the soft soap.
- In I.P. 1996 the acid value and saponification value of fatty substances are defined in terms of potassium hydroxide.
- It can be used in place of sodium hydroxide in soda lime as it can absorb CO_2.
- Its aqueous and alcoholic solutions can be used for titrating acids. In the form of alcoholic solution (alcoholic caustic potash) it can be used in organic chemistry, especially in the elimination of hydrogen halide.

$$C_2H_5Br + KOH\ (Alc) \rightarrow C_2H_4 + KBr + H_2O \quad .....(6.148)$$

It is available as potassium hydroxide and potassium hydroxide solution.

4. Sodium Carbonate (Carbonic acid disodium salt, Disodium carbonate):

Chemical Formula: Na_2CO_3

Sodium carbonate is anhydrous or contains one molecule of water of hydration. It contains not less than 99.5% and not more than 100.5% of Na_2CO_3 calculated on the anhydrous basis.

Preparation: It exists in various forms, namely anhydrous sodium carbonate Na_2CO_3 (Soda ash); monohydrate $Na_2CO_3 . H_2O$ (crystal carbonate); heptahydrate $Na_2CO_3.7H_2O$ and decahydrate $Na_2CO_3.10H_2O$ (washing soda or sal soda).

(a) *From Solvay (Ammonia Soda) Process*: Most of the Na_2CO_3 is produced synthetically by the solvay (ammonia-soda) process.

In this process, brine (NaCl), ammonia and carbon dioxide are used as raw materials. The chemical reactions involved are as below:

$$NH_3 + H_2O + CO_2 \rightarrow NH_4HCO_3 \quad .....(6.149)$$

$$NaCl + NH_4HCO_3 \rightarrow \underset{\text{(Sodium bicarbonate)}}{NaHCO_3} \downarrow + NH_4Cl \quad .....(6.150)$$

$$2NaHCO_3 \xrightarrow{150^o} Na_2CO_3 + CO_2 + H_2O \quad .....(6.151)$$

$$CaCO_3 \xrightarrow[\text{Lime Kiln}]{1100°C \text{ in}} CaO + CO_2 \quad .....(6.152)$$

$$CaO + H_2O \rightarrow Ca(OH)_2 \quad .....(6.153)$$

$$2NH_4Cl + Ca(OH)_2 \rightarrow 2NH_3 + CaCl_2 + 2H_2O \quad .....(6.154)$$

The first stage in the process is to purify saturated brine, and then react it with gaseous ammonia. Now the ammoniated brine is carbonated with CO_2 to form $NaHCO_3$. Because of the common ion effect this $NaHCO_3$ is insoluble in the brine solution and therefore can be filtered off. On heating to 150 °C $NaHCO_3$ is decomposed to anhydrous Na_2CO_3 (called light soda ash because it is fluffy solid with a low packing density of about 0.5g cm^{-3}). CO_2 is removed by heating the solution, and the CO_2 is reused. Limestone ($CaCO_3$) is heated to provide Lime (CaO) and CO_2, which is required in the reaction. On mixing with water lime gives $Ca(OH)_2$ which drives off NH_3 from NH_4Cl.

Thus the materials consumed are NaCl and $CaCO_3$, and there is formation of one useful product, Na_2CO_3, and one by-product, $CaCl_2$. As very small amount of $CaCl_2$ is required so only a small amount is recovered from solution, and the rest is wasted.

(b) *From natural deposit*: It can also be obtained from a natural deposit called Trona, ($Na_2CO_3 . NaHCO_3 . 2H_2O$), obtained from dried up lake beds in Egypt. Trona is sometimes called sodium sesquicarbonate and this is converted to sodium carbonate by heating.

$$2(Na_2CO_3 . NaHCO_3 . 2H_2O) \xrightarrow{\text{heat}} 3Na_2CO_3 + CO_2 + 5H_2O \quad .....(6.155)$$

(c) *By electrolysis of sodium chloride*: It can be prepared by the electrolysis of sodium chloride. By this sodium is produced at cathode and chlorine is formed at anode. Sodium reacts with water and gives sodium hydroxide. This solution on treatment with carbon dioxide gives sodium carbonate.

(d) *The monohydrate form of Na_2CO_3* can be prepared by crystallizing a concentrated solution of this salt at a temperature above 35 °C. Stirring is done so as to make small crystals. Crystals contain about 15% water of crystallization.

The largest use of Na_2CO_3 is for glass making and for this heavy ash is required which is chemically $Na_2CO_3 . H_2O$. It can be prepared by recrystallisation of 'light ash' produced in the Solvay process, from hot water.

Identification Tests

(a) A 1 in 10 solution is strongly alkaline to phenolphthalein.

(b) It gives tests for sodium and for carbonate.

(i) *Tests for Sodium:*

(a) To the aqueous solution of the compound add potassium carbonate and heat to boiling; no precipitate is formed. Now add freshly prepared potassium antimonate solution and heat to boiling. Allow to cool in ice and if necessary scratch the inside of the test tube with a glass rod, a dense, white precipitate is formed.

Sodium salts give a white amorphous precipitate of sodium antimonate on reaction with alkaline solution of potassium antimonate. The amorphous precipitate immediately changes to crystalline form due to the formation of sodium pyroantimonate.

$$KSbO_3 . 3H_2O + Na^+ \rightarrow \underset{\text{(Sodium Antimonate)}}{NaSbO_3 .} + 3H_2O + K^+ \quad .....(6.156)$$

$$2NaSbO_3 . 3H_2O \xrightarrow{H_2O} \underset{\text{(Sodium pyroantimonate)}}{Na_2H_2Sb_2O_7 . 6H_2O} \quad .....(6.157)$$

(b) An acidified solution (acidification is done with acetic acid) of the substance on reaction with a large excess of magnesium uranyl acetate solution gives a yellow crystalline precipitate.

$$Na^+ + 3(CH_3COO)_2\,UO_2 + (CH_3COO)_2Mg + CH_3COOH \rightarrow$$

$$\underset{\text{(Sodium magnesium uranyl-acetate)}}{NaMg(UO_2)_3\,(CH_3COO)_9} + H^+ \quad .....(6.158)$$

(ii) *Tests for carbonate:*

(a) Carbonates and bicarbonates effervesce with acids, evolving a colourless gas, which when passed into calcium hydroxide produces a white precipitate immediately.

$$Na_2CO_3 + 2HCl \rightarrow H_2\,CO_3 + 2NaCl \quad .....(6.159)$$

$$H_2CO_3 \rightarrow CO_2 + H_2O \quad .....(6.160)$$

$$Ca(OH)_2 + CO_2 \rightarrow CaCO_3 \downarrow + H_2O \quad .....(6.161)$$

(b) A cold solution of the compound is coloured by phenolphthalein, while a similar solution of bicarbonate remains unchanged or is only slightly coloured.

***Test for purity*:** It has to be tested for water, heavy metals and organic volatile impurities.

***Assay*:** Sodium carbonate is a strong base (its 1M solution has a pH of 11.6). This high alkalinity in solution is because of the hydrolysis of the carbonate anion, which is a strong base, and because of the hydrolysis of bicarbonate ion.

$$CO_3^{2-} + H_2O \rightleftharpoons HCO_3^- + OH^- \quad .....(6.162)$$

$$HCO_3^- + H_2O \rightleftharpoons H_2CO_3 + OH^- \quad .....(6.163)$$

Thus it can be assayed by titrating its solution with a strong acid such as H_2SO_4, using methyl orange as indicator.

$$Na_2CO_3 + H_2SO_4 \rightarrow Na_2SO_4 + H_2O + CO_2 \uparrow \quad .....(6.164)$$

***Method*:** Transfer the anhydrous sodium carbonate obtained in the test for water to a flask with the aid of 50 ml of water, add methyl orange and titrate with 1N sulphuric acid.

Each ml of 1N sulphuric acid is equivalent to 52.99 mg of Na_2CO_3.

Uses

- It is mainly used as a pharmaceutical aid for its basicity, in various pharmaceutical preparations.
- It is used to form sodium salts of acidic drugs. It is also used in the softening of water.
- It may also be used in the preparation of carbonates of various metals.

5. Sodium Hydroxide (Caustic Soda):

Chemical Formula: NaOH *Mol. Wt.* 40.00

It contains not less than 97.0% and not more than 100.5% of total alkali calculated as NaOH.

Preparation

(a) *From electrolysis of* NaCl: On large scale it is produced by the electrolysis of a concentrated aqueous solution of NaCl (brine) using either a diaphragm cell or a mercury cathode cell. Since chlorine is one of the by products, it may react with NaOH forming NaCl and sodium hypochlorite, therefore to avoid reaction between NaOH and Cl_2 specially designed electrolytic cells e.g. Nelson cell (or diaphragm cell) and Castner-Kellner cell (or mercury cathode cell) are used.

$$NaCl\ (aq) \rightleftharpoons Na^+ + Cl^- \quad .....(6.165)$$

$$H_2O \rightleftharpoons H^+ + OH^- \quad .....(6.166)$$

(i) *Reaction in Nelson Cell*:

At Cathode : $2H^+ + 2e^- \rightarrow H_2$ .....(6.167)

At Anode : $2Cl^- \rightarrow Cl_2 + 2e^-$ .....(6.168)

The solution remained behind contains Na^+ and OH^- and thus about 10-15% NaOH is obtained. It is concentrated under vacuum to 50% when NaCl (being less soluble) is almost separated. The filtrate is evapourated and dry mass is fused and cast into sticks.

(ii) *Mercury Cathode Process*:

$NaCl(aq) \rightleftharpoons Na^+ + Cl^-$ .....(6.169)

At Cathode (mercury): $2Na^+ + 2e^- \rightarrow 2Na$ .....(6.170)

$Na + Hg \rightarrow NaHg$ .....(6.171)
(Sodium Amalgam)

At Anode (graphite): $2Cl^- \rightarrow Cl_2 + 2e^-$ .....(6.172)

Sodium amalgam is removed from cell. It is then decomposed in a separate cell by water giving NaOH, hydrogen and mercury.

$$Hg_x Na_2 + 2H_2O \rightarrow 2NaOH + xHg + H_2\uparrow \quad .....(6.173)$$

(b) *From Sodium Carbonate (Causticizing Process):* It is an old process. A 10% solution of Na_2CO_3 is heated with a little excess of milk of lime ($Ca(OH)_2$).

$$Na_2CO_3 + Ca(OH)_2 \rightarrow CaCO_3\downarrow + 2NaOH \quad .....(6.174)$$

The sodium hydroxide can be separated by filtration. The filtrate is collected and evapourated to get molten NaOH. This is then converted into sticks, scales, pellets or masses.

Note: Commercial sodium hydroxide, besides other impurities, contains some amount of sodium carbonate due to the absorption of atmospheric CO_2. However it may be purified by dissolving it in alcohol in which impurities including Na_2CO_3 are insoluble. The filtrate may be used as pure NaOH solution or may be evapourated in a silver basin to get solid NaOH.

Test for identification

(a) The pH of its 0.01% w/v solution is not less than 11.0.

(b) Its neutralized (neutralization is done with HCl) aqueous solution gives reactions of sodium.

(i) To the above solution add 15% w/v solution of potassium carbonate and heat to boiling, no precipitate is produced. Now add few ml of freshly prepared potassium antimonate solution and heat to boiling. Allow to cool in ice and if necessary scratch the inside of the test-tube with a glass rod, a dense white precipitate is formed. The reactions are

$$Na^+ + KSbO_3.3H_2O \xrightarrow{H_2O} NaSbO_3.3H_2O + K^+ \quad .....(6.175)$$

$$\underset{\text{Sodium antimonite}}{2NaSbO_3 .3H_2O} \xrightarrow{H_2O} \underset{\text{Sodium pyroantimonate}}{Na_2H_2Sb_2O_7.6H_2O\downarrow} \quad .....(6.176)$$

(ii) Acidify the above prepared solution with acetic acid and add a large excess of magnesium uranyl acetate solution, a yellow crystalline precipitate will be produced.

$$Na^+ + 3(CH_3COO)_2UO_2 + (CH_3COO)_2Mg + CH_3COOH \rightarrow$$
$$\underset{\text{Sodium Magnesium Uranyl Acetate}}{NaMg(UO_2)_3 (CH_3COO)_9 + H^+} \quad .....(6.177)$$

Test for purity: It is tested for clarity and colour of solution, arsenic, heavy metals, iron, carbonate, chloride, sulphate and potassium.

Assay: Commercial sodium hydroxide contains soluble carbonate, which is calculated as Na_2CO_3 and are determined in the assay of NaOH. Therefore the assay of NaOH is based upon modified Winkler method which is considered as the most satisfactory method for the determination of alkali hydroxide and carbonate when present together.

The mixture is titrated with a standard solution of HCl or H_2SO_4 by the selective use of indicators. With HCl, the reaction proceeds in the following way:

$$NaOH + HCl \rightarrow NaCl + H_2O \quad .....(6.178)$$

$$Na_2CO_3 + HCl \rightarrow NaCl + NaHCO_3 \quad .....(6.179)$$

This $NaHCO_3$ further reacts with HCl.

$$NaHCO_3 + HCl \rightarrow NaCl + H_2O + CO_2 \quad .....(6.180)$$

The phenolphthalein is added as first indicator, which loses its colour of alkaline medium when NaOH and half Na_2CO_3 are neutralized. In fact, it does not give colour in $NaHCO_3$ solution.

Now use of methyl orange as second indicator is made which shows the complete neutralization, that is a red colour is obtained at end point.

***Method*:** Weigh accurately about 2 g, dissolve it in about 80 ml of carbon dioxide free water, add 0.3 ml of phenolphthalein solution and titrate with 1M hydrochloric acid. Add 0.3 ml of methyl orange solution and continue the titration with 1M hydrochloric acid.

Each ml of 1M HCl used in the 2nd part of the titration is equivalent to 0.1060 g of Na_2CO_3.

Each ml of 1M HCl used in the combined titration is equivalent to 0.0400 g of total alkali, calculated as NaOH.

Uses

- It is the most important alkali used in pharmaceutical industry. It is used for a wide variety of purposes including making many inorganic and organic compounds, neutralizations and making soaps. It is used to make water soluble sodium salts of various organic drugs.
- It is also used in making soda lime. It is used as a saponifying agent.
- It is a powerful cautery and breaks down proteins of the skin thereby damage tissues, therefore, it has been used to remove warts.

6.2 Buffers

A buffer is a mixture of a weak acid and its conjugate base or a weak base and its conjugate acid that resists changes in pH of a solution.

A very important aspect of pharmaceutical chemistry is the control of pH in solutions. Though the pH of a solution is adjusted by acid or base, the pH is not constant indefinitely. Some factors like gases in the air such as CO_2, NH_3 or the alkalinity in the glass of cheap variety.

The resistance of a solution to changes in pH upon the addition of small amounts of acid or alkali is termed as buffer action. A solution which possesses such properties is known as buffer solution. The buffer solution can also be defined as that can resists changes in pH with dilution or with addition of acids or bases.

Generally, buffer solution can be prepared from a conjugate acid / base pair. Buffer solutions usually consists of solution containing a mixture of a weak acid HA (acetic acid) and its sodium / potassium salt, A^- (sodium acetate), or of a weak base (ammonia) and its salt (ammonium chloride). The buffer pair complement each other.

For *e.g.*, when small amounts of hydrogen ion are introduced into the medium, they will react with conjugate base or basic member of the buffer pair to form the weak acid which will ionize slightly. Similarly when small amounts of hydroxide ion (base) are introduced into the medium, they will react with the weak acid to form the conjugate base and water. In either case the buffer solution decrease the large changes in pH.

6.2.1 Calculation of the pH of Buffer Solution

Consider the buffer solution containing weak acid HA and its conjugate base A^-. It may be acidic, neutral or basic; depending the position of two comparative equilibria

$$HA + H_2O \rightleftharpoons H_3O^+ + A^- \quad K_a = \frac{[H_3O^+][A^-]}{[HA]} \qquad .....(6181)$$

$$A^- + H_2O \rightleftharpoons OH^- + HA \qquad K_b = \frac{[OH^-][HA]}{[A^-]}$$

$$= \frac{k_w}{K_a} \qquad .....(6.182)$$

In equation 6.181, the equilibrium lies farther to the right than 6.182, the solution is acidic.

To find the pH of a solution containing HA ad Na A, we need to express in analytical concentrations, C_{HA} ad C_{Na} A.

$$[HA] = C_{HA} - [H_3O^+] + [OH^-] \qquad .....(6.183)$$

$$[A^-] = C_{Na}\,A + [H_3O^+] - [OH^-] \qquad .....(6.184)$$

Since there is inverse relationship between $[H_3O^+]$ and $[OH^-]$

$$[HA] \simeq C_HA \qquad .....(6.185)$$

$$[A^-] \simeq C_{Na}\,A \qquad .....(6.186)$$

$$[H_3O^+]\,[H^+] = K_a \frac{C_HA}{C_{Na}A} \qquad .....(6.187)$$

Henderson–Hasselbalch equation

$$-\log [H^+] = -\log K_a + \log \frac{C_{Na}A}{C_HA} \qquad .....(6.188)$$

$$pH = pK_a + \log \frac{C_{Na}A}{C_HA} \qquad .....(6.189)$$

This is used to calculate the pH of buffer solutions.

Problem 1: Calculate the pH of a buffer solution containing 0.400M formic acid and 1.00M is sodium formate.

$$HCOOH + H_2O \rightleftharpoons H_3O^+ + HCOO^- \qquad K_a = 1.80 \times 10^{-4}$$

$$HCOO^- + H_2O \rightleftharpoons HCOOH + OH^- \qquad K_b = \frac{K_w}{K_a} = \frac{1.0\times10^{-14}}{1.80\times10^{-4}}$$

$$= 5.56 \times 10^{-11}$$

$$[HCOO^-] = C_{HCOO-} = 1.00M$$

$$[HCOOH] = C_{HcooH} = 0.04M$$

As per eq. 6.187

$$[H^+] = 1.8 \times 10^{-4} \times \frac{0.40}{1.00} = 7.20 \times 10^{-5}\ M$$

$[H^+] << C_{HCOOH}$ and $[H^+] << C_{HCOO^-}$ is valid.

Ans: $pH = -\log[7.20 \times 10^{-5}] = 4.14$

Problem 2: Calculate the pH of solution i.e. 0.20MNH_3 and 0.30M NH_4Cl, K_a for $NH_4^+ = 5.70 \times 10^{-10}$.

$$NH_4^+ + H_2O \rightleftharpoons NH_3 + H_3O^+ \qquad K_a = 5.70 \times 10^{-10}.$$

$$NH_3 + H_2O \rightleftharpoons NH_4^+ + OH^- \qquad K_b = \frac{K_w}{K_a} = \frac{1.0\times10^{-14}}{5.7\times10^{-10}} = 1.75 \times 10^{-5}$$

$$[NH_4Cl] \simeq C_{NH_4}Cl = 0.30\ M$$

$$[NH_3] \simeq C_{NH_3} = 0.20\ M$$

$$[H^+] = \frac{K_a \times C_{NH_4}Cl}{C_{NH_3}} = \frac{5.70\times10^{-10}\times0.30}{0.20} = 8.55 \times 10^{-10}$$

Ans: $pH = -\log(8.55 \times 10^{-10}) = 9.07$

6.2.2 Buffer Capacity

The capacity of a buffer is a measure of its effectiveness in resisting changes in pH upon the addition of acid or base. The greater the concentration of an acid and conjugate base, the grater the capacity of the buffer. The buffering capacity is maintained for mixtures within the range 1 acid: 10 salt and 0 acid: 1 salt and the $\simeq$pH range of weak acid buffer is

$$pH = pK_a \pm 1 \qquad .....(6.190)$$

Buffer capacity can be defined quantitatively as the number of moles of strong base required to change the pH of 1 litre of solution by one pH unit. The preparation of buffer solution of definite pH is a simple process once the acid (or base) of appropriate dissociation constant is found. Small variation in pH is obtained by variation in the ratio of acid to the salt concentration.

The buffer capacity, β, of a solution is also defined as the number of moles of a strong acid or a strong base that causes one litre of the buffer solution to undergo a one unit pH change.

$$\beta = \frac{d_{cb}}{d_{pH}} = -\frac{d_{ca}}{d_{pH}} \qquad \text{.....(6.191)}$$

d_{cb} = number of mole L^{-1} of strong base

d_{ca} = number of mole L^{-1} of strong acid.

Adding acid always decrease pH, so $\frac{d_{ca}}{d_{pH}}$ is negative and buffer capacity is always positive.

Problem

It is desired to prepare 100 ml of buffer of pH 5.00 acetic acid and their salts. What acid-salts ratio should be used?

Pk_a of acetic acid = 4.74.

$$pH = p\,k_a + \log\frac{[OAC^-]}{[HOAC]}$$

$$5.0 = 4.74 + \log\frac{[OAC^-]}{[HOAC]}$$

Taking anti logs, $[OAC^-]/[HOAC] = 1.8/1$, i.e. 1.8 times as much salt as acid.

Acetic acid-sodium acetate buffer mixtures are given in Table 6.2

Table 6.2

S.No.	Acetic acid X ml	Sodium acetate Y ml	pH
1.	9.5	0.5	3.48
2.	9.0	1.0	3.80
3.	8.0	2.0	4.16
4.	7.0	3.0	4.39
5.	6.0	4.0	4.58
6.	5.0	5.0	4.76
7.	4.0	6.0	4.93
8.	3.0	7.0	5.13
9.	2.0	8.0	5.36
10.	1.0	9.0	5.74
11.	0.5	9.5	6.04

6.2.3 Properties of Buffer Solutions

The factors which affect the pH of a buffer solution:

(i) *The effect of dilution:* The pH of buffer solution remains essentially independent of dilution, until the concentrations of the species C_H A and C_{Na} A it contains, is decreased to the point where the approximations used to develop [HA] $\simeq$ C_{HA} and $[A^-]$ $\simeq$ C_{Na} A become invalid. Fig 6.1 shows the effect of dilution on the pH of buffered and unbuffered solution.

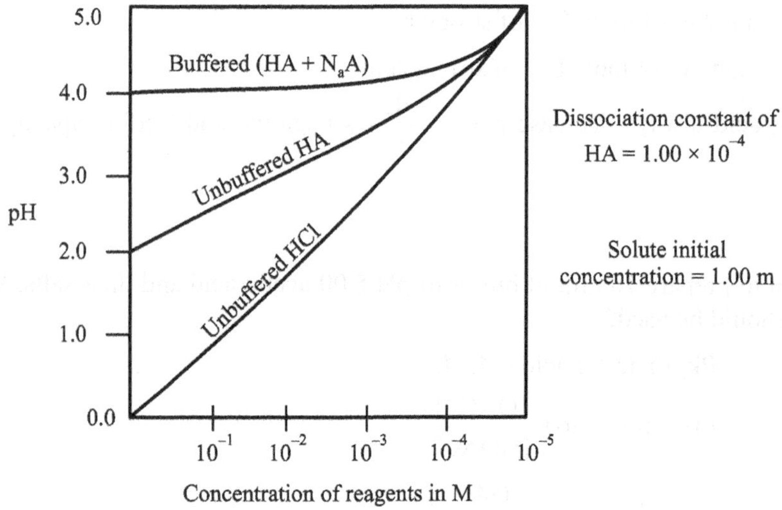

Fig. 6.1

For each initial solute concentration is 1.00 M. The resistance of buffered solution to changes in pH during dilution is clear.

(ii) *The effect of added acids and bases*: Buffers do not maintain pH at an absolutely constant value but changes in pH are relatively small when small amounts of acid or base are added.

For *e.g.,* Addition of 100 ml of 0.0500M NaOH to 400 ml of NH_3 (0. 200M) NH_4 Cl (0.300M) buffer converts part of the NH_4^+ in the buffer to NH_3.

$$NH_4^+ + OH^- \rightleftharpoons NH_3 + H_2O \qquad .....(6.192)$$

then

$$C_{NH_3} = \frac{(400 \times 0.200) + (100 \times 0.05)}{500} = \frac{85.0}{500} = 0.170M \qquad .....(6.193)$$

$$C_{NH_4}Cl = \frac{400 \times 0.300) + (100 \times 0.05)}{500} = \frac{115}{500} = 0.230M \qquad .....(6.194)$$

Dissociation constant = 5.70×10^{-10}

$$\text{So } [H_3O^+] = k_a \times \frac{C_{NH_4}Cl}{C_{NH_3}} \qquad .....(6.195)$$

$$= 5.70 \times 10^{-10} \times \frac{0.230}{0.170} = 7.71 \times 10^{-10} \text{ M}$$

$$pH = -\log 7.71 \times 10^{-10} = 9.11$$

pH of the buffer solution = 9.07.

then change in pH (ΔpH) = 9.11 – 9.07 = 0.04

addition of HCl converts part of NH_3 to NH_4^+

$$NH_3 + H_3O^+ \rightleftharpoons NH_4^+ + H_2O \qquad .....(6.196)$$

$$C_{NH_3} = \frac{400 \times 0.200 - 100 \times 0.0500}{500} = \frac{75}{500} = 0.150 \text{ M} \qquad .....(6.197)$$

$$C_{NH_4^+} = \frac{400 \times 0.300 + 100 \times 0.0500}{500} = \frac{125}{500} = 0.250 \text{ M} \qquad .....(6.198)$$

$$\text{So} \quad [H_3O^+] = 5.70 \times 10^{-10} \times \frac{0.250}{0.150} = 9.5 \times 10^{-10}$$

$$pH = -\log 9.50 \times 10^{-10} = 9.02$$

$$\Delta pH = 9.02 - 9.07 = -0.05$$

The change in pH is small, i.e., for acid is 0.05 and for base 0.04.

6.2.4 Physiological Buffers

The great physiologist C. Bernard was first emphasized that the fluids of the body provide an 'internal environment' in which body cells live and perform their many functions; protected from the inconsistency of the external environment. The maintenance of a nearly constant pH in the blood and other fluids of the body is important. The principal buffers in the blood are proteins, bicarbonates, phosphates, heamoglobin and oxyheamoglobin.

6.2.5 Universal Buffer Mixtures

A mixture of 6.008 g of citric acid, 3.893 g potassium dihydrogen phosphate, 1.769 g of boric acid and 5.266 g of pure diethyl barbituric acid is dissolved in water and made to 1 liter. The pH values (2 to 12) of mixtures of 100 ml of this solution with various volumes of 0.2N NaOH (X) solution are shown in Table 6.3.

Table 6.3

pH	X ml	pH	X ml	pH	X ml
2.6	2.0	6.2	41.2	9.8	79.3
3.0	6.4	6.6	46.0	10.2	82.0
3.4	10.1	7.0	50.6	10.6	83.9
3.8	13.7	7.4	55.8	11.0	86.0
4.2	17.6	7.8	61.7	11.4	89.7
4.6	22.4	8.2	65.6	11.8	95.0
5.0	27.1	8.6	69.3	12.0	99.6
5.4	31.8	9.0	72.7		
5.8	36.5	9.4	75.8		

6.2.6 Buffer Solutions

According to I.P., buffer solutions have been classified into two categories: Standard buffer solutions other buffer solutions.

1. ***Standard Buffer Solutions:*** They are the solutions of standard pH. They are used for reference purposes in pH measurement and for carrying out many pharmacopoeial tests which require adjustments to or maintenance of a specific pH.

   Standard buffer solutions having pH ranges between 1.2 and 10 can be prepared by appropriate combinations of the solutions described below:

   (i) *Boric acid and potassium chloride, 0.2M:* Dissolve 12.366 g of boric acid and 14.911 g of potassium chloride in water and make up the volume to 1000 ml with water.

   (ii) *Disodium hydrogen phosphate, 0.2M:* Dissolve 71.630 g of disodium hydrogen phosphate in water and make up the volume to 1000 ml with water.

   (iii) *Hydrochloric acid, 0.2M:* Dilute hydrochloric acid with water to contain 7.292 g of HCl in 1000 ml and standardize it.

   (iv) *Potassium chloride, 0.2M:* Dissolve 14.911 g of potassium chloride in water and make up the volume to 1000 ml with water.

   (v) *Potassium dihydrogen phosphate, 0.2M*: Dissolve 27.218 g of potassium dihydrogen phosphate in water and make up the volume to 1000 ml with water.

(vi) *Potassium hydrogen phthalate, 0.2M:* Dissolve 40.846 g of potassium hydrogen phthalate in water and make up the volume to 1000 ml with water.

(vii) *Sodium hydroxide, 0.2M:* Dissolve sodium hydroxide in water to produce a 40-60% w/v solution and allow to stand. Siphon off the clear supernatant liquid avoiding absorption of CO_2, dilute with water to contain 8.0 g NaOH in 1000 ml and standardize it.

All the crystalline reagents except boric acid should be dried at 110°-120° for one hour before use. The water used for the preparation of buffer solution should be carbon dioxide free.

Composition of standard buffer solutions

(i) *Hydrochloric acid buffer*: Transfer 50 ml of the 0.2M potassium chloride to a 200 ml volumetric flask. Add the specified volume of 0.2M hydrochloric acid as given in Table 6.4 and then make up the volume with water.

Table 6.4

pH	0.2M HCl, ml	pH	0.2M HCl, ml	pH	0.2M HCl, ml
1.2	85.0	1.6	32.4	2.0	13.0
1.3	67.2	1.7	26.0	2.1	10.2
1.4	53.2	1.8	20.4	2.2	7.8
1.5	41.4	1.9	16.2		

(ii) *Acid phthalate buffer*: Transfer 50 ml of the 0.2M potassium hydrogen phthalate to a 200 ml volumetric flask. Add the specified volume of 0.2M hydrochloric acid as given in Table 6.5 and then make up the volume with water.

Table 6.5

pH	0.2M HCl, ml	pH	0.2M HCl, ml	pH	0.2M HCl, ml
2.2	49.5	3.0	22.3	3.8	2.9
2.4	42.2	3.2	15.7	4.0	0.1
2.6	35.4	3.4	10.4		
2.8	28.9	3.6	6.3		

(iii) *Neutralized phthalate buffer or phthalate buffer*: Transfer 50 ml of the 0.2M potassium hydrogen phthalate to a 200 ml volumetric flask. Add the specified volume of the 0.2M sodium hydroxide as given in the Table 6.6 and then make up the volume with water.

Table 6.6

pH	0.2M NaOH, ml	pH	0.2M NaOH, ml	pH	0.2M NaOH, ml
4.2	3.0	4.8	16.5	5.4	34.1
4.4	6.6	5.0	22.6	5.6	38.8
4.6	11.1	5.2	28.8	5.8	42.3

(iv) *Phosphate buffer*: Transfer 50 ml of the 0.2M potassium disodium phosphate to a 200 ml volumetric flask. Add the specified volume of 0.2M sodium hydroxide as given in the Table 6.7 and then make up the volume with water.

Table 6.7

pH	0.2M NaOH, ml	pH	0.2M NaOH, ml	pH	0.2M NaOH, ml
5.8	3.6	6.6	16.4	7.4	39.1
6.0	5.6	6.8	22.4	7.6	42.5
6.2	8.1	7.0	29.1	7.8	44.5
6.4	11.6	7.2	34.7	8.0	46.1

(v) *Alkaline borate buffer*: Transfer 50 ml of the 0.2M boric acid and potassium chloride to a 200 ml volumetric flask. Add the specified volume of 0.2M sodium hydroxide as given in the Table 6.8 and then make up the volume with water.

Table 6.8

pH	0.2M NaOH, ml	pH	0.2M NaOH, ml	pH	0.2M NaOH, ml
8.0	3.9	8.8	15.8	9.6	36.9
8.2	6.0	9.0	20.8	9.8	40.6
8.4	8.6	9.2	26.4	10.0	43.7
8.6	11.8	9.4	32.1		

The standard pH values given in the tables are reproducible within ± 0.02 unit at 25°C. The prepared solutions should be stored in glass-stoppered bottles made up of chemically resistant, alkali-free glass and should be used within three months of preparation. If the solution becomes cloudy or shows any other sign of deterioration it should be discarded.

2. ***Other buffer solutions***: These are actual pharmaceutical buffers, which are used to maintain the pH within the specified limits in pharmaceutical preparations. The I.P.1996 enlists a number of such buffer solutions with their pH value and specified method of preparation. They are:

Acetate Buffers

1. Acetate Buffer (pH 2.8)
2. Acetate Buffer (pH 3.4)
3. Acetate Buffer (pH 3.5)
4. Acetate Buffer (pH 3.7)
5. Acetate Buffer (pH 4.0)
6. Acetate Buffer (pH 4.4)
7. Acetate Buffer (pH 4.6)
8. Acetate Buffer (pH 4.7)
9. Acetate Buffer (pH 5.0)
10. Acetate Buffer (pH 5.5)
11. Acetate Buffer (pH 6.0)
12. Acetate Buffer solution

Acetic acid-Ammonium Acetate Buffer

13. Acetic-Ammonia Buffer, Ethanolic (pH 3.7)
14. Albumin Phosphate Buffer (pH 7.2)

Ammonia-Ammonium Chloride Buffer

15. Ammonia Buffer (pH 9.5)
16. Ammonia Buffer (pH 10.0)
17. Ammonia Buffer (pH 10.9)
18. Barbitone Buffer (pH 7.4)
19. Barbitone Buffer (pH 8.6)
20. Boric Buffer (pH 9.0)

21. Buffer Solution (pH 2.5)
22. Carbonate Buffer (pH 9.7)
23. Chloride Buffer (pH 2.0)
24. Citro-Phosphate Buffer (pH 5.0)
25. Citro-Phosphate Buffer (pH 6.0)
26. Citro-Phosphate Buffer (pH 7.0)
27. Citro-Phosphate Buffer (pH 7.2)
28. Citro-Phosphate Buffer (pH 7.6)
29. Cupric Sulphate Solution, Buffered (pH 4.0)
30. Diethanolamine Buffer (pH 10.0)
31. Glycine Buffer (pH 11.3)
32. Glycine Buffer Solution
33. Imidazole Buffer (pH 6.5)
34. Imidazole Buffer (pH 7.4)
35. Palladium Chloride Solution, Buffered
36. Phosphate Buffer (pH 2.0)
37. Phosphate Buffer (pH 2.5)
38. Phosphate Buffer (pH 3.6)
39. Phosphate Buffer, Mixed (pH 4.0)
40. Phosphate Buffer (pH 4.9)
41. Phosphate Buffer (pH 5.0)
42. Phosphate Buffer, Mixed (pH 5.5)
43. Phosphate Buffer (pH 6.5)
44. Phosphate Buffer, Mixed (pH 6.8)
45. Phosphate Buffer, 0.2M Mixed (pH 6.8)
46. Phosphate Buffer, Mixed (pH 7.0)
47. Phosphate Buffer with Azide, Mixed (pH 7.0)
48. Phosphate Buffer, 0.067M Mixed (pH 7.0)
49. Phosphate Buffer, 0.033M Mixed (pH 7.5)
50. Phosphate Buffer, 0.02M (pH 8.0)

51. Phosphate Buffer, 0.025M Standard
52. Saline, Phosphate-Buffered (pH 6.4)
53. Saline, Phosphate-Buffered (pH 7.4)

6.2.7 Selection of Pharmaceutical Buffer

The selection of a pharmaceutical buffer depends on chemical and pharmacological factors.

(i) ***Chemical factors are***

- The buffer system should not react with other chemical ingredients of a pharmaceutical preparation i.e., it should not change the solubility of other ingredients, should not indulge in undesirable acid-base or oxidation-reduction reactions and should not undergo complexation with active components of the pharmaceutical preparation.
- The buffer system must have reasonable chemical stability. The volatile species such as CO_2 and NH_3 should not be used for the preparation of buffers since their loss will change the pH and will affect the buffer capacity of the system. The basic buffer i.e. a mixture of weak base and its salt can absorb CO_2 from the atmosphere resulting in the lowering of the pH.
- For an acid buffer, the buffer capacity is maximum when the ratio of concentration of salt/acid is one. In other words when pH = pK_a. So, a pharmaceutical buffer should be selected in which buffer acid is having pK_a near the middle of the desired pH range.

(ii) ***Pharmacological factors are***

- The buffer system should not be toxic especially when used in preparations intended for internal administration. Toxicity is an important factor and restricts the use of some of the buffer systems e.g., borate buffer systems because of their toxicity are not employed in internal preparations and are mainly used in ophthalmic preparations, and preparations intended for topical application.
- The buffer system should not affect the therapeutic ability of the active ingredients of the preparation.
- The buffer system should not support the growth of microorganisms. Many buffer systems are susceptible to microbial growth as they can provide the nutrients for the same. Such systems can be preserved by adding low concentrations of suitable antimicrobial agents.

6.3 Antioxidants

Antioxidants are the compounds, which are added to pharmaceutical preparations containing easily oxidizable substances. They prevent oxidation and subsequent deterioration of the formulation. Chemically they are reducing agents which prevent oxidation either by getting oxidise themselve in place of active component as they under go oxidation more readily than the active component or by reducing the already oxidized active component back to the normal oxidation state.

In brief the antioxidant action is based upon oxidation-reduction reaction in which the antioxidant itself gets oxidized. Oxidation-reduction reactions can be considered similar to Bronsted acid-base reactions, as the "conjugate pairs" of oxidised and reduced forms of a chemical compound can be separated from the chemical equation. In redox reaction there is transfer of electrons from one compound to the other, therefore the loss or gain of electrons is used to balance the oxidation states on both sides of the half-reaction e.g. if Ox_A is the oxidised form of compound A and the Red_B is the reduced form of compound B then the half reaction can be written as,

$$Ox_A + e^- \rightleftarrows Red_A \qquad .....(6.199)$$

$$Red_B \rightleftarrows Ox_B + e^- \qquad .....(6.200)$$

The overall redox reaction can be given by

$$Ox_A + Red_B \rightleftarrows Red_A + Ox_B$$

It is possible to determine the tendency of a chemical substance to under go oxidation-reduction reaction, this can be done by using electro chemical cells in which the electron transfer takes place through a system of electrodes. The electrical potential developed in the electrochemical cell can be measured by the voltmeter and the electrode potential can be determined by using Nernst equation.

$$E_{cell} = E^0_{cell} - \frac{RT}{nF} \ln\left[\frac{Ox}{Red}\right] \qquad .....(6.201)$$

where, E_{cell} is the potential of the cell in volts,

E°_{cell} is the standard potential.

R is the gas constant.

T is the absolute temperature.

F is the Faraday constant.

n is the no. of electrons transferred and

[Ox]/[Red] is the ratio of concentration of oxidized and reduced forms respectively.

For a cell at 298.73 °K and converting to the logarithms to the base 10, the above equation can be rewritten as,

$$E_{cell} = E^{\circ}_{cell} - (0.0591/n) \log \{[Ox]/[Red]\} \quad .....(6.202)$$

The value of E°_{cell} is determined from a table of standard electrode potentials enlisting the potentials for various half reactions.

6.3.1 Selection Criteria of Antioxidants

Following criteria should be considered while selecting an antioxidant.

- It should be able to produce desire redox reaction. This can be assessed by using standard electrode potentials and the Nernst equation.
- It should be physiologically and chemically compatible.
- It should be physiologically inert.
- It should be non-toxic both in the reduced and oxidized forms.
- It should not create any solubility problem for various components of the formulation.
- It should be effective in low concentration and should provide prolonged stability to the formulation.
- Utmost care must be taken in selecting antioxidants for formulations containing strong oxidizing agents as very strong reducing agents can form explosive mixtures with very strong oxidizing agents.

Active Component + Antioxidant + Oxidizing Agent
(Reduced form) (Reduced form) (Oxidized form)

↓

Active Component + Antioxidant + Oxidizing Agent .....(6.203)
(Reduced Form) (Oxidized form) (Reduced form)

6.3.2 Hypophosphorous Acid

Chemical Formula: H_3PO_2 *Mol. Wt.* 66.00

Hypophosphorous acid contains not less than 30.0 percent and not more than 32.0 percent of H_3PO_2.

Preparation

From hypophosphite of calcium, barium or potassium: It can be prepared by the reaction of hypophosphite of calcium, barium or potassium with excess of sulphuric acid, oxalic acid and tartaric acid respectively.

$$Ca(H_2PO_2)_2 + H_2SO_4 \rightarrow 2H_3PO_2 + CaSO_4\downarrow \quad .....(6.204)$$

$$Ba(H_2PO_2)_2 + H_2C_2O_4 \rightarrow 2H_3PO_2 + BaC_2O_4\downarrow \quad .....(6.205)$$

(oxalic acid) (Barium oxalate)

$$KH_2PO_2 + C_4H_6O_6 \rightarrow H_3PO_2 + KHC_4H_4O_6\downarrow \quad .....(6.206)$$

(Tartaric acid) (Pot. Bitartrate)

Identification Tests

(i) When strongly heated it evolves spontaneously flammable phosphine.

$$3H_3PO_2 \rightarrow 2PH_3\uparrow + 2H_3PO_3 \quad .....(6.207)$$

In this test the acid on heating is decomposed giving phosphine and phosphorous acid.

(ii) Its solution gives a white precipitate with mercuric chloride solution. This precipitate becomes grey when an excess of hypophosphite is present.

$$H_2PO_2^- + 4HgCl_2 + 2H_2O \rightarrow 2Hg_2Cl_2\downarrow + H_3PO_4 + 3H^+ + 4Cl^- \quad .....(6.208)$$

$$H_2PO_2^- + 2Hg_2Cl_2 + 2H_2O \rightarrow 4Hg\downarrow + H_3PO_4 + 3H^+ + 4Cl^- \quad .....(6.209)$$

(Excess) (Metallic Mercury grey in colour)

In this test the hypo phosphorous acid is acting as strong reducing agent.

Thus mercuric chloride is reduced to mercurous chloride and mercury.

The acid itself is oxidized to orthophosphoric acid.

(iii) Its solutions when acidified with sulphuric acid and warmed with cupric sulphate gives a red precipitate.

$$3H_2PO_2^- + 4Cu^{2+} + 6H_2O \rightarrow 4CuH\downarrow + 3H_3PO_4 + 5H^+ \quad .....(6.210)$$

Red ppt.

Test for purity: It is tested for barium, oxalate and heavy metals.

Assay: Hypophosphorous acid is monobasic as it contains only one ionisable hydrogen (one which is bonded to oxygen atom). Its assay is based upon acid base titration, and is done by titrating it with sodium hydroxide using phenolphthalein indicator.

$$H_3PO_2 + NaOH \rightarrow NaH_2PO_2 + H_2O \quad .....(6.211)$$

Method: Take about 7 ml of hypophosphorous acid into a tared, glass-stoppered flask and weigh accurately. Dilute with about 25 ml of water, add phenolphthalein and titrate with sodium hydroxide.

Each ml of IN sodium hydroxide is equivalent to 66.00 mg of H_3PO_2.

Uses

- Hypophosphorous acid and its salts do not show any important pharmacological action therefore they don't have any medicinal use.
- It is a very strong reducing agent as the P in it is in the oxidation state (+1). So it is used as antioxidant in Ferrous Iodide Syrup, Hydroiodic Acid Syrup and in Diluted Hydroiodic Acid to prevent the formation of free iodine. In Ferrous Iodide Syrup it also prevents the formation of ferric ions.
- Its salts (hypophosphites) can be used as preservatives *e.g.* ammonium hypophosphite can be used as preservative in many preparations.

6.3.3 Sulphur Dioxide

Chemical Formula: SO_2 *Mol. Wt.* 64.06

Sulphur dioxide contains not less than 97.0% by volume of SO_2.

Preparation

(i) *From Sulphur*: It can be manufactured by burning S in air or oxygen.

$$S + O_2 \rightarrow SO_2 \quad .....(6.212)$$

(ii) *From Metal sulphide*: It can be manufactured by roasting various metal sulphide ores (particularly FeS_2 and to a smaller extent CuS and ZnS) with air.

$$4FeS_2 + 11O_2 \rightarrow 2Fe_2O_3 + 8SO_2 \uparrow \quad .....(6.213)$$

$$2ZnS + 3O_2 \rightarrow 2ZnO + 2SO_2 \uparrow \quad .....(6.214)$$

(iii) By heating copper turnings with conc. H_2SO_4

$$Cu + H_2SO_4\,(conc.) \xrightarrow{\Delta} CuSO_4 + SO_2 + 2H_2O \quad .....(6.215)$$

(iv) By the reduction of H_2SO_4 with carbon or sulphur

$$C + 2H_2SO_4 \rightarrow CO_2 + 2SO_2 + 2H_2O \quad .....(6.216)$$

$$S + 2H_2SO_4 \rightarrow 3SO_2 + 2H_2O \quad .....(6.217)$$

(v) On large scale by the action of mineral acid on sulphites

$$Na_2SO_3 + 2HCl \rightarrow 2NaCl + H_2O + SO_2 \quad .....(6.218)$$

Test for purity: Sulphur dioxide is used mostly in the form of gas in pharmaceutical applications. However it is usually packaged under pressure. Hence the following purity tests shall be done in liquid form.

(i) *Water*: It can be determined by the Karl Fischer Method.

(ii) *Non-volatile residue*: For the determination of the non-volatile residue the weighed sample is allowed to evaporate and any residue obtained is weighed. The weight of the residue should not exceed 7.5 mg (0.0025%).

(iii) *Sulphuric acid*: To determine this, the residue obtained in the test for non-volatile residue is mixed with water, and the resulting mixture is titrated with 0.10N sodium hrydroxide (not more than 1.3 ml is required).

Identification tests: On bringing a filter paper previously moistened with acidified potassium dichromate solution, near SO_2, the colour of the paper turns green. In this test SO_2 reduces acidic potassium dichromate to green chromium sulphate $Cr_2(SO_4)_3$.

$$K_2Cr_2O_7 + H_2SO_4 + 3SO_2 \rightarrow K_2SO_4 + Cr_2(SO_4)_3 + H_2O \quad .....(6.219)$$

Assay: The assay of SO_2 include collection of the measured volume of gaseous SO_2 over mercury. The temperature and pressure is noted. Now excess of 0.1N sodium hydroxide solution is introduced into the air space over mercury. After shaking the contents, the solution is titrated with 0.1N iodine using starch as indicator, appearance of pale blue colour is the end point.

In the assay of SO_2, the sodium hydroxide reacts with SO_2 and forms sodium bisulphite; which is then titrated with I_2.

$$SO_2 + NaOH \rightarrow NaHSO_3 \quad .....(6.220)$$

$$HSO_3^- + I_2 + H_2O \rightarrow SO_4^{2-} + 2I^- + 3H^+ \quad .....(6.221)$$

Each ml of 0.1N iodine is equivalent to 1.094 ml of SO_2 at a temp. of 0° and a pressure of 760 mm of mercury.

6.3.4 Sodium Bisulphite

Chemical Formula: $NaHSO_3$ *Mol. Wt.* 104.06

Synonym: Sodium acid sulphite, Sodium Hydrogen Sulphite, Sodium pyrosulphate

Sodium bisulphite is unstable in the solid state, so the official preparation is a mixture of sodium bisulphite ($NaHSO_3$) and sodium metabisulphite ($Na_2S_2O_5$) in varying proportions. It gives not less than 61 .6% and not more than 67.4% SO_2.

Preparation: It is manufactured by passing SO_2 into an aqueous solution of Na_2CO_3 until the solution is saturated.

$$Na_2CO_3 + 2SO_2 + H_2O \rightarrow 2NaHSO_3 + CO_2 \uparrow \quad .....(6.222)$$

Sodium bisulphite can be crystallized from the aqueous solution or precipitated with ethanol.

Identification test: They liberate SO_2 on treatment with dilute acids, which smells like burning sulphur.

$$NaHSO_3 + HCl \rightarrow NaCl + SO_2 + H_2O \quad .....(6.223)$$

This SO_2 reduces acidic $K_2Cr_2O_7$ to green $Cr_2(SO_4)_3$ and acidic $KMnO_4$ to colourless $MnSO_4$.

$$K_2Cr_2O_7 + H_2SO_4 + 3SO_2 \rightarrow K_2SO_4 + Cr_2(SO_4)_3 + H_2O \quad .....(6.224)$$

$$2KMnO_4 + 2H_2O + 5SO_2 \rightarrow K_2SO_4 + 2MnSO_4 + 2H_2SO_4 \quad .....(6.225)$$

Test for purity: It has to be tested for arsenic, heavy metals and iron.

Assay: Sodium bisulphite contains S in the (4+) oxidation state and is a moderately strong reducing agent. Its assay is based upon its reaction with I_2, and determination of the remaining I_2 with sodium thiosulphate.

$$NaHSO_3 + I_2 + H_2O \rightarrow NaHSO_4 + 2HI \quad .....(6.226)$$

$$2Na_2S_2O_3 + I_2 \rightarrow Na_2S_4O_6 + 2Na^+ + 2I \quad .....(6.227)$$

Uses

- It is mainly used as an antioxidant. It is used to prevent oxidation of drugs which have phenol or catechol ring to quinones and similar substances.
- It may also be used to prepare water soluble derivatives of normally insoluble drugs.
- Reduction of its solution containing SO_2, with Zn dust, or electrolytically give Sodium dithionites. This contain S in the oxidation state (+III).

$$2HSO_3^- + SO_2 \xrightarrow{Zn} \underset{\text{(Dithionite)}}{S_2O_4^{2-}} + SO_3^{2-} + H_2O \quad .....(6.228)$$

Structure of sodium dithionite

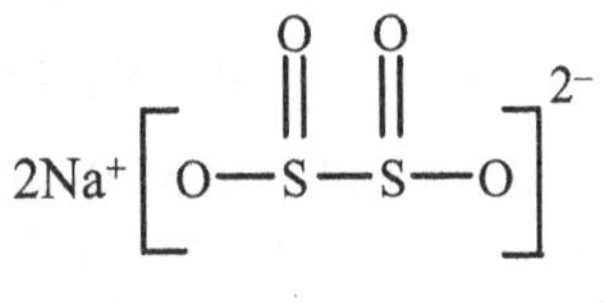

6.3.5 Sodium Metabisulphite

Chemical Formula: $Na_2S_2O_5$ *Mol. Wt.* 190.10

Synonym: Sodium Pyrosulphite, Sodium Disulphide

Preparation

It can be prepared by saturating a hot concentrated solution of sodium hydroxide with sulphur dioxide and allowing crystallization to occur. Presumably sodium bisulphite ($NaHSO_3$) is first formed, being unstable in the solid state, loses water and the solid sodium metabisulphite separates out.

$$NaOH + SO_2 \rightarrow NaHSO_3 \quad .....(6.229)$$

$$2NaHSO_3 \rightarrow Na_2S_2O_5 + H_2O \quad .....(6.230)$$

Identification

(i) Gives reactions of sodium salts.

(ii) A solution decolourizes iodinated potassium iodide solution and the resulting solution gives the reactions of sulphates.

In this test the sodium metabisulphite reduces molecular I_2 to iodide (I^-), which is colourless, and itself converts to bisulphate.

$$2I_2 + Na_2S_2O_5 + 3H_2O \rightarrow 4HI + 2NaHSO_4 \quad .....(6.231)$$

Test for purity: It shall be tested for arsenic, heavy metals, iron and thiosulphate.

Thiosulphate is detected by heating a mixture of sample and hydrochloric acid, appearance of an opalescence or turbidity due to S indicates the presence of thiosulphate.

$$Na_2S_2O_3 + 2HCl \rightarrow 2NaCl + SO_2 + S + H_2O \quad .....(6.232)$$

While in case of sodium metabisulphite

$$Na_2S_2O_5 + 2HCl \rightarrow 2NaCl + 2SO_2 + H_2O \quad .....(6.233)$$

Assay: It contains not less than 95.0% and not more than 100.5% of $Na_2S_2O_5$.

It can be assayed by reducing 0.1M iodine solution (added in excess) in slightly acid solution. The excess of iodine is then determined by titration with sodium thiosulphate.

$$Na_2S_2O_5 + H_2O \rightarrow 2Na\,H\,SO_3 \quad .....(6.234)$$

$$2Na\,H\,SO_3 + 2I_2 + 2H_2O \rightarrow 2Na\,H\,SO_4 + 4HI \quad .....(6.235)$$

$$2I_2 + Na_2S_2O_5 + 3H_2O \rightarrow 4HI + 2NaHSO_4 \quad .....(6.236)$$

$$2Na_2S_2O_3 + I_2 \rightarrow 2NaI + Na_2S_4O_6 \quad .....(6.237)$$

Method: Weigh accurately about 0.1 g and dissolve in 50.0 ml of 0.1M iodine, add 1 ml of hydrochloric acid and titrate the excess of iodine with 0.1M sodium thiosulphate using starch solution added towards the end of the titration as indicator.

Each ml of 0.1M iodine is equivalent to 0.004753 g of $Na_2S_2O_3$.

Uses

- It is widely used as an antioxidant and reducing agent in various formulations. It is added to the solutions of drugs that contain the phenol or catechol nucleus to prevent oxidation of these nucleus to quinones and like substances e.g. it is used to stabilize injections containing salts of adrenaline or morphine.
- When dissolved in water it immediately converts to bisulphite, so, this salt can be used when bisulphite is specified.

$$Na_2S_2O_5 + H_2O \rightarrow 2NaHSO_3 \quad .....(6.238)$$

6.3.6 Sodium Thiosulphate

Chemical Formula: $Na_2S_2O_3.5H_2O$ *Mol. Wt.* 248.17

Synonym: Sodium Hyposulphite.

It is disodium salt of thiosulphuric acid, $H_2S_2O_3$ (a highly unstable oxyacid of sulphur).

In the structure of sodium thiosulphate the sulphur is in two oxidation state. One of the sulphur atom is oxidized and is in a +VI state (due to this it resists further oxidation). The other sulphur atom is in zero oxidation state, therefore it can act as a reducing agent or as an antioxidant.

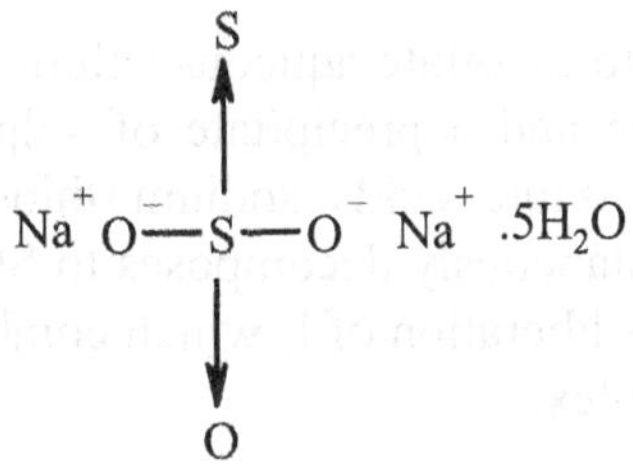

Preparation

1. It can be made by boiling sodium sulphite solution with S in absence of air. The unreacted S is removed, filtrate is evapourated to give crystals of sodium thiosulphate.

$$\underset{\text{(Sodium Sulphite)}}{Na_2SO_3} + S \rightarrow Na_2S_2O_3 \quad .....(6.239)$$

2. By oxidizing polysulphides with air

$$2Na_2S_3 + 3O_2 \xrightarrow{\text{Heated in air}} 2Na_2S_2O_3 + 2S \quad .....(6.240)$$

3. By passing SO_2 gas into Na_2S solution

$$2Na_2S + 3SO_2 \rightarrow 2Na_2S_2O_3 + S$$

4. By passing SO_2 gas through the concentrate of waste liquor of sodium sulphide which contains Na_2S, Na_2SO_3 and sodium sulphate.

$$Na_2SO_3 + 3SO_2 + 2Na_2S \rightarrow 3Na_2S_2O_3 \quad .....(6.241)$$

$$Na_2CO_3 + 4SO_2 + 2Na_2S \rightarrow 3Na_2S_2O_3 + CO_2 \quad .....(6.242)$$

Identification

(i) With silver nitrate its aqueous dilute solution gives a white precipitate which quickly becomes yellowish and finally black. Mixing of sodium thiosulphate with solutions containing other metal cation (eg. Ag^+) results in the precipitation of the metal thiosulphate. In acid solution these precipitates may darken due to the formation of the corresponding sulphide (eg. Ag_2S).

$$2AgNO_3 + Na_2S_2O_3 \rightarrow Ag_2S_2O_3\downarrow + 2NaNO_3 \quad .....(6.243)$$

$$Ag_2S_2O_3 + H_2O \rightarrow Ag_2S\downarrow + H_2SO_4 \qquad .....(6.244)$$

(ii) On adding few drops of iodine to its aqueous solution, the colour of iodine solution is discharged.

The presence of excess of sulphur makes sodium thiosulphate a useful reducing agent. Iodine very rapidly oxidizes thiosulphate ions ($S_2O_3^{2-}$) to tetra thionate ions ($S_4O_3^{2-}$) and the I_2 is reduced to I^- ions (hence colour of iodine is discharged).

$$2Na_2S_2O_3 + I_2 \rightarrow \underset{\text{(sodium tetra thionate)}}{Na_2S_4O_6} + 2NaI \qquad .....(6.245)$$

(iii) On adding hydrochloric acid to its dilute aqueous solution, a gas is evolved which turns starch iodate paper blue and a precipitate of sulphur is produced. This is because, when treated with dilute acids sodium thiosulphate is converted to thiosulphuric acid which spontaneously decomposes to SO_2 and S. The iodate ions (IO_3^-) oxidize SO_2 and there is liberation of I_2 which combines with starch to give a blue coloured adsorption complex.

(a) $$Na_2S_2O_3 + 2HCl \rightarrow 2NaCl + H_2S_2O_3 \qquad .....(6.246)$$

$$H_2S_2O_3 \rightarrow H_2O + SO_2 + S \qquad .....(6.247)$$

$$Na_2S_2O_3 + 2HCl \rightarrow 2NaCl + H_2O + SO_2 + S\downarrow \qquad .....(6.248)$$

(b) $$5SO_2 + 2KIO_3 + 4H_2O \rightarrow 2KHSO_4 + 3H_2SO_4 + I_2 \qquad .....(6.249)$$

(c) $$I_2 + \text{Starch} \rightarrow \text{Blue colour complex} \qquad .....(6.250)$$

(d) Its aqueous solution gives reaction of sodium salts.

Test for purity: It shall be tested for arsenic, chloride, heavy metals, sulphide, sulphate and sulphite and clarity and colour of solution.

Sulphide can be tested by adding sodium nitroprusside solution to the aqueous solution of the sodium thiosulphate. The solution does not become violet.

Assay: It contains not less than 99.0% and not more than 101 .0% of $Na_2S_2O_3.5H_2O$. In solution, sodium thiosulphate reacts quantitatively with iodine to form sodium iodide and sodium tetrathionate. So it can be assayed by titrating its solution with iodine using starch mucilage as indicator. Appearance of blue colour is the end point as iodine gives a deep blue coloured adsorption complex with starch.

$$Na_2S_2O_3 + I_2 \rightarrow Na_2S_4O_6 + 2NaI \qquad .....(6.251)$$

Method: Weigh accurately about 0.5 gm, dissolve in 20 ml of water and titrate with 0.05ml iodine using starch solution added towards the end of the titration as indicator.

Each ml of 0.05M iodine is equivalent to 0.02482 gm of $Na_2S_2O_3.5H_2O$

Uses: The presence of excess of sulphur makes it a useful reducing agent. Thus it reduces halogen.

- It is mainly used for iodine titrations in volumetric analysis.
- In conjunction with sodium nitrite it is used as an antidote for cyanide poisoning. It is also used as an effective antidote in mercury, iodine, lead and bismuth poisoning.
- It is used as an antioxidant but its use as an antioxidant is usually limited to solutions containing iodides. Sometimes in combination with acids (acid causes the precipitation of S and evolution of SO_2) it is used to treat various dermatological problems.
- It is used as an antichlor e.g., in the bleaching operations to destroy any excess Cl_2. Similarly it is sometimes used to remove the taste from heavily chlorinated drinking water.

6.3.7 Sodium Nitrite

Chemical Formula: $NaNO_2$ *Mol. Wt.* 69.00

Synonym: Nitrous acid sodium salt.

It contains not less than 97.0% and not more than 101.0% of $NaNO_2$, calculated on the dried basis.

Preparation

(i) It can be prepared by strong heating of sodium nitrate either on its own or with Pb.

$$2NaNO_3 \xrightarrow{Heat\ (500\ ^oC)} 2NaNO_2 + O_2 \quad .....(6.252)$$

$$NaNO_3 + Pb \xrightarrow[Heat\ (500\ ^oC)]{} NaNO_2 + PbO \quad .....(6.253)$$

(ii) It can be manufactured by absorbing oxides of nitrogen in Na_2CO_3 solution.

$$Na_2CO_3 + NO_2 + NO \rightarrow 2NaNO_2 + CO_2 \quad .....(6.254)$$

(iii) It can also be made by chemical reduction of $NaNO_3$.

$$2NaNO_3 + C \rightarrow 2NaNO_2 + CO_2 \quad .....(6.255)$$

(iv) More commonly it is made by reduction of sodium nitrate with metallic lead at a lower temperature.

$$NaNO_3 + Pb \rightarrow NaNO_2 + PbO\downarrow \quad .....(6.256)$$

Identification Tests

(i) To the aqueous solution of the sample add 2 ml of 15% w/v solution of potassium carbonate. Heat the contents to boiling, no precipitate is produced. Add 4 ml of a freshly prepared potassium antimonate solution and heat to boiling. Allow cooling in ice and if necessary scratch the inside of the test tube with a glass rod; a dense, white precipitate is formed.

$$\underset{\text{(Pot. antimonate)}}{Na^+ + KH_2SbO_4} \rightarrow \underset{\text{(Sod. pyroantimonate)}}{NaH_2SbO_4 \downarrow + K^+} \quad .....(6.257)$$

(ii) Add a large excess of magnesium uranyl acetate solution to the acidified (with acetic acid) solution of the sample, a yellow, crystalline precipitate is formed.

$$Na^+ + 3UO_2 (CH_3COO)_2 + Mg(CH_3COO)_2 + CH_3COOH$$

$$\rightarrow \underset{\text{(yellow ppt.)}}{[NaMg(UO_2)_3] (CH_3COO)_9 \downarrow} + HCl \quad .....(6.258)$$

(iii) Sodium nitrite gives brownish red fumes when treated with diluted mineral acids or acetic acid

Test for purity: It has to be tested for loss on drying, pH, arsenic, lead and heavy metals.

Assay: Take sodium nitrite, accurately weigh and dissolve it in water. Transfer solution into a mixture of potassium permanganate, water and sulphuric acid (Take care that while adding the sodium nitrite solution, the tip of the pipette is beneath the surface of the permanganate mixture). Warm the liquid to 40 °C and allow it to stand for few minutes. Now add oxalic acid. Heat the mixture to about 80 °C and titrate with potassium permanganate.

Each ml of 0.IN potassium permanganate is ≡ 3.450 mg of $NaNO_2$

Note: The nitrite solution must be run into the potassium permanganate solution, and not vice versa, because the acidification of a solution of sodium nitrite results in the production of nitrous acid, which decomposes immediately with loss of oxides of nitrogen. The potassium permanganate already present in the solution prevents this loss by immediately oxidizing nitrous acid.

$$NaNO_2 + H_2SO_4 \rightarrow HNO_2 + NaHSO_4 \quad .....(6.259)$$

$$HNO_2 + O \rightarrow HNO_3 .....(6.250) \quad .....(6.260)$$

Uses:

- Chemically, nitrites can act as both oxidizing and reducing agents. When nitrites acts as a reducing agent then nitrite is oxidized into nitrate.

$$\underset{\text{(N oxidation state +III)}}{3HNO_2} + KClO_3 \rightarrow \underset{\text{(N oxidation state +V)}}{HNO_3} + KCl \quad .....(6.261)$$

- When nitrites react as oxidizing agent then nitrite is reduced into nitric oxide (NO) in acidic solution or molecular nitrogen in neutral to alkaline solution.

$$2HNO_2 + 2KI + H_2SO_4 \rightarrow I_2 + K_2SO_4 + 2NO \quad .....(6.262)$$

$$NaNO_2 + NH_4Cl \rightarrow NaCl + 2H_2O + N_2\uparrow \quad .....(6.263)$$

- It is not used specifically as an antioxidant in pharmaceutical preparations. However it is used as a reducing agent when combined with sodium carbonate. This combination is available in anti-rust tablets which are used to prevent rusting of various surgical instruments.
- Nitrite ions relax the smooth muscles of the blood vessels thus produce vasodilator effect. Instead of sodium nitrite, organic derivatives containing nitrite and nitrate groups eg. Isosorbide dinitrite, are used as coronary vasodilator.
- Sodium nitrite is used as a food additive in cured meats and fish. Here it serves three functions: colour development (reductive decomposition of NO_2^- gives NO, which forms a red complex with hemoglobin and improves the look), flavour production and preservation (NO_2^- ions inhibit the growth of bacteria, particularly Clostridium botulinum). However, the nitrites may react with amines and be converted into potentially carcinogenic nitrosamines ($R_2N\text{-}N\text{=}0$), it has resulted in a great deal of concern about the safety of using nitrites, and the levels that should be allowed to remain in meat products. Some countries have banned the use of nitrite as a flavouring and colouring agent but they only permit the level necessary as an anti botulinogenic agent.
- It is official in USP as an antidote in the cyanide poisoning where it causes the oxidation of the ferrous (Fe^{2+}) ion of hemoglobin to the ferric ion of methemoglobin, which then binds with the serum cyanide.
- In laboratory it is used to prepare diazo dye or in the detection of primary aromatic amines.
- The NO_2^- ion may act as a chelating ligand.

6.3.8 Nitrogen

Chemical Formula: N_2 *Mol. Wt.* 28.01

It contains not less than 99.0% by volume of N_2.

Preparation

(i) Can be prepared on lab scale by heating a mixture of NH_4Cl and $NaNO_2$.

$$NH_4C1 + NaNO_2 \rightarrow \underset{\text{(Unstable)}}{[NH_4NO_2]} + NaCl \quad .....(6.264)$$

$$NH_4NO_2 \xrightarrow{\text{Heat}} N_2 + 2H_2O \quad .....(6.265)$$

(ii) By oxidation of ammonia

$$8NH_3 + 3Cl_2 \rightarrow 6NH_4Cl + N_2 \quad .....(6.266)$$

$$2NH_3 + 3CuO \rightarrow 3Cu + N_2 + 3H_2O \quad .....(6.267)$$

(iii) By passing vapours of HNO_3 on strongly heated copper.

$$5Cu + 2HNO_3 \rightarrow 5CuO + N_2 + H_2O \quad .....(6.268)$$

(iv) By heating ammonium dichromate.

$$(NH_4)_2 Cr_2O_7 \rightarrow Cr_2O_3 \downarrow + N_2 + 4H_2O \quad .....(6.269)$$

(v) Commercially it can be obtained by liquefying air followed by fractional evaporation. By doing this the Nitrogen, having a lower b.p. (77.3° K), boils off more readily than the less volatile oxygen (b.p. 90.04° K).

Identification Tests: Insert a burning wood splinter into a test tube filled with nitrogen. The flame will extinguish, because it is an inert gas and neither burns nor supports the combustion.

Test for purity: It has to be tested for odour and carbon monoxide.

Uses

- Because of the inert nature it can be used to prevent the air oxidation of the pharmaceuticals and chemicals, by displacing the air in the containers and in the reacting vessels, *e.g*,.
- It can used to replace air during manufacturing and packaging of parenterals, multivitamin preparation and various oil preparations.
- It is used as a diluent to dilute the action of oxygen before administering later to patient.
- It can also be used in the preparation of ammonia, nitric acid, nitrides, cyanamides and other nitrogen compounds.
- Nitrogen is an essential constituent of all proteins found in animals and plants.
- It is used to provide an inert atmosphere.

6.4 Water (Solvent)

It is one of the most widely and abundantly used substances in pharmaceutical manufacturing. It is required for a variety of purposes ranging from manufacturing process to the preparation of the final dosage forms. It is a highly associated liquid having high boiling point (100°C), a high dielectric constant (at 25°C dielectric constant 78.5) and a high specific heat (at 14.5 ^{0}C specific heat 1 calorie). Its good solubilising capacity for salts is due to high dielectric constant. It is a chemically stable compound. Even at 1700°C, less than 1% is dissociated into its elements. The k_w for H_2O is 10^{-14}. Despite this relative non reactivity it acts as a solvent.

These days in pharmaceutical systems pharmacists are concerned with dissolution of relatively non polar drugs in aqueous or mixed polar aqueous solvents. Organic compounds having low molecular weight with polar groups capable of making hydrogen bonds with water are soluble in water. Polar groups -OH, -CHO, -COH, -CHOH, $-CH_2OH$, -COOH, $-NO_2$, -CO, $-NH_2$ and - SO_3H increase the solubility of organic compounds in water.

Indian pharmacopoeia prescribes standards for the various types of water to be used in the manufacture of pharmaceutical preparations. The starting material for most forms of water is drinking water. Drinking water itself may be used in the manufacture of drug substances but not in the preparation of dosage forms, or in the preparation of reagents and test solutions.

The pharmaceutical manufacturer must continuously review the needs and upgrade their water system. The water used in the manufacture should be pure and of drinkable quality (potable water), free from pathogenic microorganisms.

Quality water is needed in the manufacture of the products-

- Parenteral products (LVPs & SVPs)
- Liquid orals
- Biological products
- Over the counter products
- Diagnostic preparations
- Radiopharmaceuticals
- Veterinary products
- Medical devices.

6.4.1 Classes / Grades of Water

Generally two grades of water are used, namely purified water IP (PW) and water for injection (WFI). According to USP XIX, purified water means water obtained from any one of the methods such as:

- ion – exchange treatment
- reverse osmosis (RO)

- distillation
- electrodialysis
- ultrafiltration (U/F).

The selection of a system dependent on factors such as raw quality, extent of use, flow rate, cost, etc. Water is one of the most difficult products to maintain the required standard, especially as the quality increases.

The classes of water normally encountered in the pharmaceutical industry are given in Table 6.9.

Table 6.9

Class of Water	Quality of Water	Applications / Uses
Well water	Not chemically pure, not micro- biologically pure	Non-manufacturing operations, such as lawn sprinklers, fire protection systems, other utilities.
Potable water (or) city water	treated water (chlorination), microbiologically pure	Used for drinking, processing operations (cleaning and sanitization purpose).
Purified water USP	treated potable water, specific levels of purity, no chlorides, no microbial contamination.	Processing operations, non – sterile water operations, solid dosage forms, oral liquids, creams.
Specially purified grade (WFI)	No chlorides, no added substances, no microbial contamination	Parenteral use.
FDA or water grade		For cleaning and initial rinse of parenteral area.

6.4.2 Treatment of Water

Deionization

Deionization does not improve the microbial quality. The bacterial population may increase and also produce pyrogenic problems. Some preventive actions include strict scheduling of regeneration process (i.e after production of for injection, a double pass RO system must be used to ensure high degree of consistent water quality.

Storage of Water

Not only producing the water of pharmaceutical grade, but also storage of such water is also critical. A few methods are as follows:

- Effective microbial control can be maintained by storing water at 80°C. However, this approach is expensive and produces some hazards to personnel & material.
- Ultra violet radiation may be used, but it has limited applications because of the many factors, which can reduce its effectiveness.
- Written instructions should be developed for sanitization procedure for water treatment, equipment on a regular and prescribed basis.

Analysis of Water

Samples of water for manufacturing should be taken on a daily basis and test results must be transmitted to production department prior to start of manufacture. Samples are taken at designated use outlets. Quality control testing should be carried out using test methods and meeting specifications outlined in the USP.

Notification of reports should be sent to manufacturing areas and permanent records are kept indicating assay results and approvals.

Certain quantities or specified time of operation and the installation of bacteria retentive filters. Procedures should be written to ensure that water treatment equipment is properly operated, monitered, maintained and sanitized at regular intervals.

Distillation

It is probably safest but also most expensive method of treating water in terms of high capital and operating expenses.

Hence, distillation may be adopted and this water is used in most critical applications.

Reverse Osmosis

Reverse osmosis treatment removes a large portion of dissolved salts, particulate matter, bacteria and pyrogens. Conflicting reports are available regarding the removal of bacteria and pyrogens by RO process. It is essential to make sure and verify before it is used in most critical applications.

Ultrafiltration

Water filtration generally is adopted on two major considerations, prefiltering to remove large particles and microfiltering to remove bacteria. Ultrafiltration utilizes membranes with porosities ranging from 0.0001 to 0.01 μm, which will remove both particulate matter and bacteria. The problem of filters is that the bacteria retained on the filter may multiply. Therefore, filters and the down stream piping must be effectively sanitized on a regular basis. There should be written procedures for filter inspection, replacement, integrity testing and monitoring on a scheduled basis.

Purified Water

Purified water IP (PW) must not contain any added substance and therefore, microbial control of this water is difficult, unless it is handled in a manner similar to that of water for injection (WFI). The use of this class of water should be restricted to those operations that mandate the use of purified water. It should be used for critical bulk batch applications, non-sterile formulations etc.

If deionization equipment (DI) is already available, purified water can be produced by DI followed by distillation or reverse osmosis (RO) or ultrafiltration (U/F). The still RO or U/F unit, storage-tank and piping, distribution system should be periodically treated by an appropriate means to control microbial growth.

In solid dosage forms, oral liquids and creams, purified water is used for process and find rinse of cleaned equipment.

6.4.3 Water for Injection (WFI)

Water for injection is the most difficult quality to achieve. It must satisfy the specifications, which are stringent chemical, microbial and pyrogenic requirements. The water must be chloride free and should not contain any added substance, which precludes the use of residual chlorine as an anti – microbial agent.

There are two acceptable methods of production in USP XXI, distillation and RO to produce USP water

***Official Water as given in I.P.* 1996:** It has been categorized into

- Purified water
- Water for injection.
- Water for injection in bulk.
- Sterile water for injection.

Purified water: It can be prepared by distillation, by means of ion exchange or by any other suitable method. It is a clear colourless, odorless and tasteless liquid. It contains no added substance.

Test for purity: It has to be tested for acidity or alkalinity, ammonium, calcium and magnesium, heavy metals, chloride, nitrate, sulphate, oxidisable substances and residue on evaporation.

The test for sulphate and chloride is done to ensure almost complete absence of these impurities. Oxidisable matter is detected by boiling the water with very dilute acidified permanganate.

Uses: It is used as pharmaceuticals aid (solvent). It is used in preparation of dosage forms for internal administration. It is not intended for parenteral administration.

Water for injection: It is apyrogenic but not sterile distilled water. It is obtained by distilling potable water or purified water from a neutral glass, quartz or suitable metal still fitted with effective device for preventing the entrainment of droplets. The first portion of the distillate is discarded and remainder is collected. It contains no added substance.

Uses: It is used as a solvent for dissolving or diluting substances. It is also used in the preparation of injectables (injectable preparations in the final container should be made sterile and thereafter protected from microbial contamination).

Water for injection in bulk: It is a clear colourless, odorless, and tasteless liquid.

Test for purity: It complies with the tests given under purified water with an additional test for bacterial endotoxins.

Sterile water for injection: It is water for injection which is sterilized by heat under conditions that ensure that the water remains pyrogenic and is packed in suitable containers so as to permit the withdrawal of a nominal volume (The water for injection shall be sterilised within 12 hours of collection and distributed in sterile containers). It is a clear, colourless and odorless liquid which is stored in single dose container of not larger than one litre capacity.

Test for purity: It has to be tested for clarity and colour, acidity or alkalinity, oxidisable substances, ammonium, calcium and magnesium, heavy metals, chloride, nitrate, sulphate, bacterial endotoxins, sterility, and residue on evaporation.

Uses: It is mainly used as a solvent for injectable preparations such as powders for injection that are distributed dry because of limited stability of their solutions.

Pharmaceutical Water

Water is one of the most widely and abundantly used substances in pharmaceutical manufacturing

Its use rage from manufacturing process to the preparation of final dosage forms

The quantity of water, chemically and microbiologically needs to be controlled carefully for pharmaceutical use.

Based on its source, preparation and application, water can be classified as

(a) Natural water (I.P),

(b) Potable water (I.P),

(c) Purified water (I.P),

(d) Water for injection (I.P),

(e) Water for injection (I.P),

(f) Sterile water for injection in bulk (I.P),

(g) Bacteriostatic water for injection (U.S.P).

6.5 Pharmaceutically Acceptable Glass

Greatest attention has to be paid to the containers in which pharmaceuticals are stored or maintained even for short periods of time. The containers are in intimate contact with the pharmaceutical. None of the containers presently available is totally non reactive. Both the chemical and physical characteristic affect the stability of the pharmaceutical, but the physical characteristics are given primary consideration in the selection of a protective container. The most commonly used materials for pharmaceutical containers are glass and plastics. Although challenged by plastics, glass is still the most important material for the containers.

The chemistry of the glass is extremely complicated. It is an amorphous, hard, brittle, transparent, supercooled liquid of infinite viscosity. It is not a true solid. Glass is a generic term used for vitreous silicate materials which are prepaid by fusing a base e.g., Na_2CO_3 and $CaCO_3$, with pure silica. Upon cooling, a clear vitreous mass is formed. This vitreous material does not have a clearly defined melting point and it softens gradually over a temperature range as a result of somewhat haphazard arrangement of the silicon-oxygen bonds. Certain other compounds may be added to vitreous silicates to impart special properties e.g., manganese dioxide is added to hide the blue-green colour of the iron usually present in silica, borates are added to reduce the coefficient of expansion, compounds of alkaline earth metals (e.g., $CaCO_3$, CaO, $BaCO_3$) are added for getting glass a high refractive index, calcium phosphate is added for getting opalescent glass and colouring materials e.g., Fe_2O_3 to get yellow colour, chromic oxide to get green colour, manganese oxide to get purple colour, and cobalt oxide to get blue colour. Potassium is added to get a brown light-resistant glass.

Glass containers must be strong enough so that they can withstand the physical shocks of handling and shipping and the high pressure during the process of autoclaving and sterilisation.

They must also be able to withstand the thermal shock resulting from large temperature changes during processing (i.e. it shall have a low coefficient of thermal expansion). To inspect the contents, the container must be transparent. Keeping them in amber glass containers protects light sensitive preparations.

Advantages of glass containers: Offer transparency, sparkle, easy cleaning, effective closure and reclosure where applicable, high speed handling, good rigidity and stack ability and also the correct type of glass is usually inert.

Disadvantages of glass containers: The two main disadvantages are fragility and heavy weight. These can be partially reduced by surface coatings to increase the surface lubricity and careful design, which includes the avoidance of sharp angles and the use of adequate radii.

As the basic structural network of glass is of silicon oxide tetrahedron, boric oxide will enter into this structure, but most of the other oxides do not. They are only loosely bound and are relatively free to migrate. These loosely bound oxides may be leached into the solution in contact with the glass. The leaching process is accelerated by heat. The leaching may result in rise in pH of the solution because of the hydrolysis of the dissolved oxides. Some of the glass compounds are attacked by solutions and result in removal of glass flakes into the solution. Such problems can be minimized by the proper selection of the glass composition.

The pharmacopoeia specifies the type of glass container to be used for certain materials and include tests for these types of glasses.

Types of glass: On the basis of results from the official tests, U.S.P. classifies glass compounds into four types.

Type I- A borosilicate glass

Type II- A soda lime treated glass

Type-III- A soda lime glass

NP- General purpose soda lime glass not suitable for containers for parentrals

The U.S.P. provides the Powdered Glass test and Water Attack tests for evaluating chemical resistance of glass. These tests measure the amount of alkaline constituents leached from the glass by purified water under controlled elevated temperature conditions. Powdered Glass test is done on ground, sized glass particles, and the Water Attack test is done on whole container.

Type I, III and NP glasses are tested for Powdered Glass test whereas type II glass is tested for Water Attack test.

Type I glass offer maximum chemical resistance while resistances offered by NP glass is least. The selection of the type of glass to be used for various products depends upon the chemical resistance offered by glass.

Type I glass is suitable for all products (buffered and unbuffered aqueous solutions) although SO_2 treatment sometimes is used for even greater resistance to glass leachables.

Type II glass is suitable for buffered aqueous solution with pH below 7 or for a solution which does not react with the glass.

Type III glass is suitable for anhydrous liquids or dry substances.

NP glass is not suitable for parentrals. They are used for tablets, oral solution and suspensions and liquids for external use.

6.6 Pharmaceutical Excipients

Definition

The substances other than active constituents which are used in the preparation of a product or dosage form is called as additives (or) excipients. Also called as adjuncts or formulation aids (or) pharmaceutical aids.

The term excipient comes from the Latin word excipients (excipere) which means to receive, to gather and to take out.

Additives or excipients are of various origins.

- **Animal:** Eg, lactose, gelatin, stearic acid etc
- **Plant:** Eg, starches, sugars, cellulose, alginates
- **Minerals:** Eg, calcium, phosphate, silica etc
- **Synthetic:** Eg, PEG's, polysorbates, povidone etc

Additives often lack a trade name.

In order to carry out the numerous functions required, new classes of excipients have now become available, derived from old and new materials, alone or in combination, adopted to manufacture of high – performance product.

Looking at the matter from this angle, excipients can no longer be considered mere inert supports for the active principles, but essential functional components of a modern formulation.

The requirement of excipients in the society is in developing stages as the growth of industrialization in the community.

6.7 Colouring Agents

Definition

Colouring agent is defined as substance that imparts colour to the formulation.

They not only make the product attractive but are also vital to carry product that is intended to alter the colour of any part of the body in case of cosmetics.

6.7.1 Need of Colour in Pharmaceuticals

There is evidence that the colouring of pharmaceuticals has been practiced since antiquity, and although the use of colourants in medicinals produces no direct therapeutic benefit, the physiological effects of colour have long been recognized.

The colouring of pharmaceutical dosage form is extremely useful for the identification during manufacturing and distribution. Many patients rely on colour to recognize the prescribed drug and proper dosage. Unattractive medication can be made more acceptable to the patient by the proper dosage, careful selection of colour. Colour, in combination with flavouring agents, can be used to provide taste masking of disagreeable components of a pharmaceutical preparation.

These attributes assist in improving patient compliance, which is known to be significant problem when proving drug therapy for use in the home.

Although most of the medical error problems are related to practices by doctors and hospitals, some medical errors are related to mistakes made in identification of drug products by pharmacists and patients. Many drugs on market look similar (i.e., small, round, white, uncoated tablets) identified only by a small embossed or printed logo and identifier. This makes it very easy for a pharmacist or patient to confuse for one drug for another. Hence this may prove to be serious, depending on the drugs. The use of different colours for different drugs and strengths can help to eliminate this problem.

6.7.2 Classification

Colouring agents are classified as:

1. Soluble dyes
2. Insoluble pigments and lacquers
3. Colour lakes

1. *Soluble dyes*

(i) *Natural dyes* Eg, Alkannin, carmine, chlorophyll, carotene

(ii) *Synthetic Soluble dyes*

They include- Alanine derivatives

Coal tar isolates

Eg., Naphthol, Orange-1, Tratrazin, Indigotin, Amaranth, Quinazarin green.

2. *Insoluble Pigments*

They are:

- Natural Pigments: Examples are
  Yellow ochre, terradisiena red bolous, umber.
- Synthetic Pigments: Examples are
  Synthetic iron oxide
- White synthetic pigments like bismuth carbon, titanium dioxide, zinc oxide, bismuth oxychloride.
- Blue synthetic pigments- Cobalt compounds

- Coal tar dyes as pigments
- Other synthetic pigments – Cadmium sulfide and Prussian blue.

3. ***Colour Lakes: Examples are***
   - Florentine lake – obtained from precipitation of carmanine and brazilin on Aluminum hydroxide
   - Lake from Alizarin red
   - Lakes prepared from coal tar dyes

6.8 Preservatives

Definition

Preservatives are the substances that prevent the microbial growth in preparations thereby prevent spoilage and deteriation of preparation.

Preservatives are included in the pharmaceutical preparations to control the microbial growth of the formulation.

6.8.1 Ideal Characteristics

These

- should possess a broad spectrum of antimicrobial activity encompassing gram + ve and gram – ve bacteria and fungi,
- should be chemically and physically stable over the shelf life of the product.
- should have low toxicity,
- should be readily soluble at its effective concentration
- should be without odour or colour,
- should be non-volatile,
- should be able to retain its activity in the presence of metallic salts of aluminium, zinc and iron.

6.8.2 Classification

Preservatives are classified based on their activity in pH.

1. **Acidic Preservatives**

   They are less dissociated in acidic pH and are more effective in acidic formulation

   E.g., (i) Benzoic acid in 0.1-0.5% concentration, optimum pH < 4.5

   (ii) Sorbic acid in 0.05-2% concentration, optimum pH – 4.5.

2. **Alkaline Preservatives**

   They are less dissociated in alkaline pH and more effective in alkaline formulations

   Eg. (i) Benzalkonium chloride in 0.002-0.02% concentration optimum pH 4-10

   (ii) Cetrimide in 0.005% concentration

3. **Neutral Preservatives**

   They are effective in both acidic and alkaline pH.

   Eg. (i) Sodium benzoate in 0.1-0.2% concentration,

   (ii) Methyl paraben in 0.1-0.25% conc^n,

   (iii) Propyl paraben in 0.1-0.25% conc^n

6.8.3 Methyl Paraben

Chemical Formula: $C_8H_8O_3$ *Mol. Wt.* 152.15

Synonym – methyl chemosept, Nipagin H etc

Characteristics

- exhibits antimicrobial activity between pH 4-8,
- more active against yeast, moulds than against bacteria,
- more active against gram + ve than gram – ve bacteria,
- activity of parabens increases with increasing chain length of the alkyl moiety,
- activity is improved when used in combination

eg., methyl paraben (0.18%) + propyl paraben (0.02%) are used.

Advantages: This is

- less toxic,
- non irritant to skin,
- odourless and donot cause discolouration of preparation during storage.

Disadvantage

- less soluble in water and their solid form cannot be easily incorporaled into some formulations,
- this activity against bacteria is less when compared to moulds and yeasts,
- due to poor solubility of parabens, paraben salts, particularly sodium salts are frequently used in formulations. This may cause pH of poorly buffered formulation to become more alkaline.
- incompatibility with bentonite, talc, tragacanth, sodium alginate, essential oils reported.

6.9 Sweetening Agents

Sweetening agents are employed in formulation designed for oral administration specifically to increase the palatability of the therapeutic agent.

The main sweetening agents employed in oral preparations are sucrose, liquid glucose, sorbital, saccharin and aspartame.

6.10 Flavouring Agent

A flavouring agent or a flavour is a mixture of aromatic compounds, which are blended to form a body which tends to improve the palatability of dosage forms.

6.11 Suspending Agents

A suspension may be defined as a dispersion having finely divided insoluble material suspended in liquid media. Generally, a substance has to be added for overcoming agglomeration of the dispersed particles and for increasing the viscosity of the medium so that the particles settle down very slowly. Such a substance is called a suspending agent.

- Also known as stabilizing agents for suspensions,
- Also include thickening agents or emulsifying agents,
- They act mainly by increasing the viscosity of external phase and thus reduce the rate of sedimentation of the dispersed particles,
- Some also form protective coat around the individual particles, making them less sensitive to electrolyte concentration.

Commonly used suspending agents are:

1. Bentonite, I.P
2. Colloidal silica
3. Aluminium stearate

Classification

They are classified into three main types

1. Polysaccharides Natural: e.g., Acacia, tragacanth, starch and sodium alginate

   Semisynthetic : e.g., methyl cellulose
sodium CMC
hydroxy propyl methyl cellulose
hydroxy ethyl cellulose
hydroxy propyl cellulose
microcrystalline cellulose.

2. Inorganic agents eg., clay bentonite, hectorite
3. Synthetic agent eg., carbomer (carbopol), colloidal silica dioxide.

1. Natural

(i) Acacia

It is generally used as mucilage which is 35% of dispersion in water. Commonly used with tragacanth.

- Acts by increasing the viscosity, forms protective coat around the dispersed particles and is less sensitive to electrolyte concentration,
- Viscosity is maximum at a pH of 3-9,

Disadvantages

- more prone to microbial growth hence preservative is necessary,
- It has peroxidase enzyme which causes oxidation of oxygen sensitive drugs. Hence heating to 100°C destroys the peroxidase enzyme.

(ii) Tragacanth

Action is similar to acacia.

- Tragacanth 6% concentrated mucilage is employed at pH 4-6

Advantage

Effective in lesser amounts

Peroxidase enzyme absent.

2. Semi-synthetic

(i) Sodium carboxymethylcellulose

Also called as carmellose sodium.

- It contains sodium salt of carboxy methyl ether derivative of cellulose,
- It is available in various viscosity grades,
- Generally used in 0.25-1% concentration as suspending agent,
- Being anionic, it is incompatible with cationic compounds,
- It can tolerate alcohol concentration upto 50%.

(ii) Methyl Cellulose

- High viscosity grades methylcellulose 2500 and 4500 are used as suspending agents,
- It is soluble in cold water but insoluble in hot water,
- Stable over a wide pH of 3-11,
- It is used in mixtures containing 70% alcohol,
- Methyl cellulose is resistant to bacterial attack.

3. **Inorganic Agents**

Clays

Clays are inorganic materials, mainly hydrated silicates derived from natural sources. These form highly thixotropicgels,

These gels must be preserved with suitable antimicrobial agents

(i) Bentonite

Formula: $Al_2O_3\ 4SO_2.\ H_2O$ and little amount of Mg, Fe and $CaCO_3$

- Fine, pale buff (or) cream colour, hygroscopic powder with a faint earthy taste,
- Although insoluble in water it swells upto 12 times its original volume,
- Used at a concentration of 0.5-5% for suspending powders in aqueous solution

  e.g,, Calamine lotion

(ii) Hectoxite

It is a white powder,

- It contains lithium (or) fluorine,
- It is also known as aluminium hydroxide,
- It has similar properties to bentonite but swells up to 36 times its original volume,
- Used at concentration of 1-2% and in suspensions to external used.

Synthetic Agnets

1. Carbomers

Also known as carboxyvinyl polymers.

These are high molecular weight polymers of acrylic acid with cross linkage of allyl sucrose.

- Generally it is low viscous but neutralization with NaOH produce high viscous gel,
- Most viscous gels are produced at a pH of 6-11,
- It is sensitive to oxidation when exposed to light. It can act as suspending even in low concentration,
- Mostly used for external lotion and gel preparations,
- They function mainly as protective colloids and alter the viscosity of medium.

2. Colloidal Silicon dioxide
    - It acts as suspending agents by performing aggregates which associate to form three dimensional network thus preventing sedimentation.
    - 1.5- 4% suitable.

Bentonite

Chemical Formula: $Al_2O_3SiO_2. H_2O$

It is a colloidal hydrate aluminium silicate which occurs naturally. It is obtained from naturally occurring sources.

Properties: It occurs as a very fine, pale buff or cream coloured powder. It is odourless, free from grit and has slightly earthy taste. It is almost insoluble in water but swells to about 12 times its volume when added to water. It neither dissolves nor swells in organic solvents.

Identification: When a sample of bentonite is fused with anhydrous sodium carbonate and extracted with water followed by repeated extractions with dil. HCl, it yields a residue of silica and the acid solution after neutralization which gives reaction characteristic of aluminium.

Tests for Purity: It is tested for pH, gel formation, swelling factor, fineness of powder, loss on drying.

Uses: Good pharmaceutical aid, protective colloid to stabilize the emulsions, suspending agent. It is also used as a base for many pharmaceutical preparations including plasters and ointments.

Colloidal Silica

It is also known as colloidal silicon dioxide.

Chemical Formula: SiO_2 *Mol. Wt.* 60.08

It is prepared by vapor phase hydrolysis of some silicon compounds. It occurs as sub-microscopic fine powder, of particle size about 15 nm. It is light, white and noon-gritty, insoluble in water and most acids. It may be dissolved by hot alkalis forming silicates. It is used as diluent / lubricant / granulating agent is some tablet formulations and as suspending / thickening agent for suspensions, stabilizer for some emulsion. It also fined use as anti caking agent for hygroscopic powders and granules.

Aluminium Stearate

Chemically it is an aluminium monostearate, but is usually a mixture of aluminium stearate and aluminium palmitate, with an aluminium content of 14.5 to 16.5% calculated as Al_2O_3.

It occurs as a fine, white bulky powder having a faint characteristic odour, insoluble in water, alcohol and ether. It forms gel with fixed oils and minerals oils, when heated. It finds use for preparing such oily gels containing medicaments. It also finds use in the preparation of oily suspension of sodium benzyl penicillin, which has been a slow release depot preparation.

CHAPTER 7

Impurities in Pharmaceutical and their Limit Tests

7.1 Introduction

The control of pharmaceutical impurities is currently a critical issue to the pharmaceutical industry. Impurities in pharmaceuticals are the unwanted chemicals that remain with the active pharmaceutical ingredients (APIs), or develop during formulation, or upon aging of both API and formulated APIs to medicines. The presence of these unwanted chemicals even in small amounts may influence the efficacy and safety of the pharmaceutical products. Impurity profiling (i.e., the identity as well as the quantity of impurity in the pharmaceuticals), is now getting receiving important critical attention from regulatory authorities. The different pharmacopoeias, such as the British Pharmacopoeia (BP) and the United States Pharmacopoeia (USP), are slowly incorporating limits to allowable levels of impurities present in the APIs or formulations.

Chemically a compound is impure if it contains undesirable foreign matter i.e. impurities. Thus chemical purity is freedom from foreign matter. It is virtually impossible to have absolutely pure chemical compounds and even analytically pure chemical compounds contain minute trace of impurities. The chemical purity may be achieved as closely as desired provided sufficient care is observed at different levels in the manufacturing of a pharmaceutical. The level of purity of the pharmaceutical substance depends partly on the cost-effectiveness of the process employed, methods of purification, and stability of the final product. Setting higher standards of purity for pharmaceutical substances than that of desirable and pharmacologically safe level will unduly result in wastage of money, material, labour and time. Purification of chemical compounds is a very expensive process hence one has to strike a balance in order to obtain a pharmaceutical substance at reasonable cost yet sufficiently pure for all pharmaceutical purposes.

Hence, different pharmacopoeias prescribe Test for purity for different substances. 'Purified Water' of the Indian Pharmacopoeia may be prepared by distillation or by means of ion- exchange and it may have pyrogens. 'Sterile Water for Injection' on the other hand must be free from pyrogens and must be sterile. It may have slightly acidic pH or may be very slightly alkaline. Hence, different limits are prescribed for impurities in water prepared for different purposes. Investigation of impurity profile of an active pharmaceutical ingredient is of crucial importance for medical safety reasons. The level of impurities must be reduced typically to less than 1%, and each individual impurity of 0.1% and above must be identified. Therefore a lot of efforts during a development of synthetic technologies are focused to minimization of the impurities. To achieve this, identity and origin of each impurity is essential.

7.2 Sources of Impurities

The impurities in chemical substances can be traced to many different sources like raw materials employed, the manufacturing process and stability of the final product. Impurities may also arise from physical contamination and improper storage conditions. The various sources of impurities in pharmaceutical substances are as follows:

(i) Raw materials

Pharmaceutical substances are either isolated from natural sources or synthesized from chemical starting materials. The natural sources include mineral sources, plants, animals and microbes. It is essential to verify the identity of the source material and to establish its quality otherwise impurities associated with the raw materials may be carried through the manufacturing process to contaminate the final product. In nature minerals rarely occurs in a reasonably pure from. Almost always mixtures of closely related substances occur together e.g., aluminum ores are usually accompanied by alkali and alkaline earth compounds, barium and magnesium impurities are found in calcium minerals, zinc accompanies magnesium or iron compounds, lead and heavy metals are found as impurities in many sulphide ores, among the acid radicals or anions, bromides and iodides are often found as impurities in chlorides, bismuth salts contains silver copper and lead as impurities Rock salt used for the preparation of sodium chloride is contaminated with small amounts of calcium and magnesium chlorides, so that sodium chloride prepared from rock salt will definitely contain traces of calcium and magnesium compounds impurities.

(ii) Method of Manufacture

The Process or method of manufacture may introduce new impurities into the final product arising due to contamination by reagents, catalysts and solvents reaction vessels and reaction employed at various stages of the manufacturing process. Furthermore, for certain drugs a multiple step synthesis procedure is involved, produces intermediate compounds. The new impurities may also arise from the side reactions and the byproducts formed.

(a) Reagents Employed in the Manufacturing Process

Calcium carbonate contains 'soluble alkali' as impurity which arises from the sodium carbonate (Na_2CO_3) employed in the process. Calcium carbonate is prepared by the interaction of a soluble calcium salt with a soluble carbonate. Therefore, the final product ($CaCO_3$) is liable to contain small amount of 'soluble alkali' as impurities which were not removed by the washing process.

$$\underset{\text{Soluble}}{CaCl_2 + Na_2CO_3} \rightarrow \underset{\text{Precipitate}}{CaCO_3 \downarrow} + \underset{\text{Soluble}}{2\,NaCl} \quad .....(7.1)$$

Anions like Cl^- and SO_4^{-2} are common impurities in many substances because of the use of hydrochloric acid and sulphuric acid respectively in processing. Barium ion may be an impurity in hydrogen peroxide so, hydrogen peroxide employed as reagent in the manufacturing process can contaminate the final product.

(b) Regents used to eliminate other Impurities

Barium is used in the preparation of potassium bromide to remove sulphate which in turn arises from the bromine used in the process. It is likely that potassium bromide will now be contaminated by traces of barium.

(c) Solvents

Most of the pharmaceutical substances are prepared in solvated crystalline form. Small amounts of solvents employed in preparation, and purification of reaction intermediates or the final product may also result in the Contamination of the pharmaceutical substances. Water is the cheapest solvent available and is used quite frequently in the preparation of inorganic pharmaceuticals. Water can be the major source of impurities as different types of water containing different types and amount of impurities are available. Various types of water which are available are

Tap water: Containing impurities of Ca^{2+}, Mg^{2+}, Na^+, Cl^-, CO_3^{-2} and SO_4^{-2} in trace amounts. The use of tap water on large scale will lead to the contamination of the final product with these impurities because the impurities will remain in the product even after washings.

Softened water: It is almost free from divalent cations (Ca^{2+} Mg^{2+}) but contains more of Na^+ and Cl^- ions as impurities because of the usual chemical water softening process: Therefore, the final products obtained using softened water as solvent will not have Ca^{2+} and Mg^{2+} impurities but still contain Na^+ and Cl^- impurities.

Demineralized water: It is prepared by means of ion-exchange and is free from Na^+, Ca^{2+}, Mg^{2+}, Cl^-, SO_4^{-2} and CO^{-2} etc. It may have pyrogens, bacteria and organic impurities. So, it is a better solvent than tap water or softened water but the economic factors discourage its use on large scale.

Distilled water: It is free from all organic and inorganic impurities and is therefore the best as a solvent but it is quite expensive. As it is free from all impurities, it does not pass on any impurities to the final products.

(d) Reaction Vessels

The reaction vessels employed in the manufacturing process may be metallic such as copper, iron, cast iron, galvanized iron, silver, aluminum, nickel, zinc and lead. Glass and silica are also used in the construction of the chemical plants but these days many of these are replaced by stainless steel and variety of other alloys. Some solvents and reagents employed in the process may react with the metals of reaction vessels, leading to their corrosion and passing traces of metal impurities into the solution, contaminating the final product. Similarly, glass vessels may give traces of alkali to the solvent. Lead (Pb) may be found as impurity in commercial sulphuric acid which has been manufactured by lead chamber process. Also, substances prepared by same electrolytic process, may contain electrode material as an undesirable impurity e.g. antimony, bismuth etc.

(e) Intermediates

Sometimes, an intermediate substance produced during the manufacturing process may contaminate the final product e.g. Sodium bromide is prepared by reaction of sodium hydroxide and bromine in slight excess.

$$6\ NaOH + 3\ Br_2 \rightarrow NaBrO_3 + 5\ NaBr + 3\ H_2O \qquad .....(7.2)$$

The sodium bromate an intermediate product is reduced to sodium bromide by heating the residue (obtained by evaporating the solution to dryness) with charcoal.

$$\underset{\text{Sodium bromate}}{NaBrO_3} + 3\ C \rightarrow \underset{\text{Sodium bromide}}{NaBr} + 3\ CO \qquad .....(7.3)$$

If sodium bromate is not completely converted to the sodium bromide then it is likely to be present as an impurity.

(f) Atmospheric Contamination

Atmosphere may contain dust (aluminum oxide, sulphur, silica, soot etc.) and some gases like carbon dioxide, sulphur dioxide, arsine and hydrogen sulphide. These may contaminate the final product during the manufacturing process. Some substances which are susceptible to action by atmospheric carbon dioxide and water may get contaminated with them during their preparation e.g. sodium hydroxide readily absorbs atmospheric carbon dioxide when exposed to atmosphere.

$$2NaOH + CO_2 \rightarrow Na_2CO_3 + H_2O \quad .....(7.4)$$

Calcium hydroxide solutions can absorb carbon dioxide from the atmosphere to form calcium carbonate.

$$Ca(OH)_2 + CO_2 \rightarrow CaCO_3 + H_2O \quad .....(7.5)$$

(g) Manufacturing Hazards

If the manufacturer is able to control and check impurities from the all above mentioned sources there exists certain manufacturing hazards which can lead to product contamination. The various manufacturing hazards can lead to:

Contamination from the particulate matter: The unwanted particulate matter can arise by a number of ways, such as accidental inclusion of dirt or glass, porcelain, plastic or metallic fragments from sieves, granulating, tabletting and filling machines and the product container. The particulate contamination mainly arises from the wear and tear of the equipments. It may also arise from the bulk materials used in the formulation or from dirty or improperly maintained equipments.

Cross-contamination of the product: This manufacturing hazard has to be considered in the preparation of solid dosage forms. Cross-contamination of product can occur by air-born dust arising out of handling of powders, granules and tablets in bulk. Cross-contamination is dangerous particularly in case of steroidal and other synthetic hormones and therefore, it should be carefully controlled. Precautions, such as use of face mask and special extraction equipment can minimize these undesirable contaminations.

Contamination by Microbes: Many products, like liquid preparations and creams intended for topical applications are liable to contamination by microbes from the atmosphere during manufacturing. For all products intended for parenteral administration and ophthalmic preparations, sterility testing is done and it provides an adequate control for microbial contaminations in such preparations. Microbial contamination can be controlled by adding suitable antimicrobial and antifungal agents.

Errors in the manufacturing process: Sometimes in a liquid preparation, there is incomplete solution of the solute. This ought to be detected by the normal analytical methods as it can lead to major error. A proper check on the efficiency of mixing, filling, tabletting, sterilization etc., should be exercised in order to obtain a product of maximum purity and desired quality. Special precautions are required to be observed to avoid mixing and filling errors in the preparation of low dosage forms ($\geq$ 5mg) such as tablets and capsules containing highly potent medicaments.

Errors in the packaging: Similar looking products, such as tablets of the same size, shape and colour, packed in similar containers can result in mislabeling of either or both of the products. Adequate care should be taken to avoid the handling of such products in the close proximity.

(h) Instability of the Product

(a) Chemical Instability

Impurities can also arise during storage because of chemical instability of the pharmaceutical substance. Many pharmaceutically important substances undergo chemical decomposition when storage conditions are inadequate. This chemical decomposition is often catalyzed by light, traces of acid or alkali, traces of metallic impurities, air oxidation, carbon dioxide and water vapour. The nature of the decomposition can easily be predicted from the knowledge of chemical properties of the substance. All such decompositions can be minimized or avoided by using proper storage procedures and conditions. The photosensitive substances should be protected from light by storing them in darkened glass or metal containers thereby inhibiting photochemical decomposition. Materials susceptible to oxidation by air or attack by moisture should be stored in sealed containers and if necessary the air from the containers can be displaced by an inert gas such as nitrogen.

(b) Changes in Physical Properties

Pharmaceuticals may undergo changes in physical properties during storage. There can be changes in crystal size and shape, sedimentation, agglomeration and caking of the suspended particles. These physical changes are not always avoidable and may result in significant changes in the physical appearance, pharmaceutical and therapeutic effects of the product. Particle size and consequently surface area is a critical factor in determining the bioavailability of the low solubility drug such as griseofulvin. Physical changes such as sedimentation and claying in case of multidose suspension may constitute a safety hazard leading to the possibility of under dosage and later to over dosage of the drugs. Similarly increase in the globule size of the injectable emulsions on storage may lead to fat embolism.

(c) Reaction with Container Material

The possibility of reaction between the container material and the contents cannot be ruled out as it constituents a safety hazard. Preparations susceptible to reaction with metal surfaces e.g. salicylic acid ointment must not be packed in metal tubes. Solutions of substances which are alkali-sensitive e.g., atropine sulphate injection must be packed in glass ampoules which comply with the test of hydrolytic resistance therefore such preparations must not be packed in containers made from soda glass. Plastic containers and closures must be carefully evaluated because of their tendency to give undesirable additives, such as plasticizers, particularly in the presence of non-aqueous solvents. Plastic containers intended for injectables should be sufficiently translucent to allow visual inspection of the contents and if they are having higher than 500 ml capacity, they must also comply with the test limiting animal toxicity in the cat, ether-soluble extractive and metal additives with special reference to barium and heavy metals like lead, tin and cadmium. Rubber closures are more susceptible to absorb medicaments, antioxidants and bactericides from solution, unless they are appropriately pretreated by immersion in solutions of the concerned compounds.

7.3 Effect of Impurities

It can be seen that almost pure substances are difficult to get and some amount of impurity is always present in the material. The impurities present in the substances may have the following effects.

- Impurities which have toxic effect can be injurious when present above certain limits.
- Impurities, even when present in traces, may show a cumulative toxic effect after a certain period.
- Impurities are sometimes harmless, but are present in such large proportions, that the active strength of the substance is lowered. The therapeutic effect of the drug is decreased.
- Impurities may bring about a change in the physical and chemical properties of the substance, thus making it medicinally useless.
- Impurities may cause technical difficulties in the formulation and use of the substances.
- They bring about an incompatibility, with other substances.
- They may lower the shelf life of the substances.
- Impurities, though harmless in nature, may bring about changes in odor, colour, taste etc., thus making the use of the substance unethical, as well as unhygienic.

Table 7.1 Describes the descriptive term and its meaning.

Descriptive Term	Approximate Volume of the Solvent for 1 Part of the Solute
Very soluble	less than 1 part
Freely soluble	from 1 to 10 parts
Soluble	from 10 to 30 parts
Sparingly soluble	from 30 to 100 parts
Slightly soluble	from 100 to 1000 parts
Very slightly soluble	from 1000 to 10,000 parts
Insoluble or practically insoluble	more than 10,000 parts

The following terms are used in the IP for defining the conditions of temperature.

- ***Cold:*** Any temperature not exceeding 8°C and usually between 2°C and 8°C. A refrigerator is a cold place in which the temperature is maintained thermostatically between 2°C and 8°C.
- ***Cool:*** Any temperature between 8°C and 25°C. An article directed to be stored in cool place, may alternatively be stored in a refrigerator, unless otherwise specified in the monograph.
- ***Room temperature:*** The temperature prevailing in the working area.
- ***Warm:*** Any temperature between 30°C and 40°C.
- ***Excessive heat:*** Any temperature above 40°C.
- ***Protection from freezing:*** The label of container bears this instruction where, in addition to the risk of breaking of the container, freezing results in a loss of strength or potency or in destructive alteration of the characteristics of an article.
- ***Storage under non-specific conditions:*** When no specific storage conditions are indicated in the monograph, the storage conditions include protection from moisture, freezing and excessive heat.

7.4 Identification Test

The purpose of identification test is to ensure the correct labeling of the substances, identification tests are specific but they are not necessary sufficient in establishing the absolute proof of identity of the substances. If an articles taken from a labeled container do not meet to the requirements of a prescribed identification test indicates that the articles is either mislabeled or substituted. In same monographs, more than one identification tests are given. In such cases, if the articles complies with the either one or the other identification test, in sufficient to verify the identity of the article.

Identification tests are generally based upon the combination of simple chemical test and measurement of the appropriate physical constants. There is considerable overlap between identification tests and the limit tests. Limit tests are designed to ensure that the undesirable impurities are within the prescribed limits. Identification tests whether physical or chemical, provided they are sufficiently specific, can be used as the basis of a quantitative estimation or in the design of specific limit tests. Practically, a single identification test may contribute to identification as well as standardization of the substance.

Chemical tests, used for identification are basically qualitative confirming to the presence of the substance under investigation. They may be far too general or lack specificity but can be considered sufficiently specific when used in conjugation with the other requirement of the monograph.

Physical constants such as melting point, boiling point, solubility, weight per ml, refractive index, optical rotation, viscosity etc have characteristic values for a given substance. They can be used in identification, checking quality and maintaining standard of purity.

7.5 Test for Purity

'Test for purity' for substances have been prescribed by the pharmacopoeias of the various countries in order to ensure reasonable freedom from the undesirable impurities. The so called 'Test for purity' are infact the tests for the presence of impurities in the substance and fix the limits of tolerance for these undesirable impurities. Test for purity are not framed to guard against all the possible impurity rather they provide appropriate limitation of the potential impurities only.

The guiding factor for fixing a limit of tolerance for the various impurities is the amount of impurity that is likely to be harm. Arsenic and lead are quite dangerous even in trace amounts therefore very small limits of tolerance have been fixed for their presence in all pharmaceutical substances. Another factor is the practicability of the commercial method of production of the substance meeting the requirements of a particular standard of purity. It would be useless to fix the limits of tolerance which can only be attained at a very high cost. There are cases in which the limits fixed in the pharmacopoeia were later relaxed because they were found to be too difficult to attain by the available methods of manufacturing.

The ultimate objective is that the pharmaceutical substances if not completely free from the undesirable or toxic impurities should be of reasonable good purity ensuring therapeutic safety. The presence of sodium bromide (NaBr) in the more expensive potassium bromide (KBr) is not likely to cause any harm to the patient but at the same time the KBr should be of sufficiently good pharmaceutical quality and purity not containing excess amount of sodium bromide. Some of the tests which may be undertaken to ascertain the purity of a substance are:

(a) ***Clarity of solution:*** The degree of clarity or opalescence of solution is measured by direct comparison with a reference solution having standard opalescence. The comparison is against a black background by viewing vertically downward under diffused light. A solution is considered clear if its clarity is the same as that of water or of the solvent employed in the preparation of the solution being examined.

(b) ***Colour of Solution:*** In Indian Pharmacopoeia, the color standards are based on three primary colorimetric solutions: yellow, red and blue prepared from ferric chloride, cobaltous chloride and cupric sulphate respectively. These primary solutions are mixed in various proportions with or without 1% w/v hydrochloric acid to give five reference color solutions which are yellow, greenish yellow, brownish yellow, brown, and red color of the solution is compared with the reference color solution by viewing vertically downwards through the columns of liquids in diffused light. A solution is said to be colorless when it has the same appearance as water or as solvent employed in the preparation of the solution to be examined.

(c) ***Acidity or Alkalinity:*** Pharmaceutical substances prepared using chemical reactions involving acids and alkalis may possess some degree of acidity or alkalinity resulting from improper purification by inadequate washings after their separation. The limits for acid or alkali impurities are fixed for various pharmaceutical substances and the test for acidity and alkalinity is of great help in determining the extent of such impurities.

(d) ***Loss on Ignition:*** It is the loss of weight in % w/w resulting from a volatile part of any test material that is driven off under specified conditions. It is applied to thermo stable substances which contain thermo labile impurities that decompose and lose a volatile product e.g., zinc carbonate decomposes losing carbon dioxide. The substance is heated, cooled and weighed repetitively until a constant weight is attained. The loss on ignition in this case should nit be more than 2% w/w.

(e) ***Loss on Drying:*** It is the loss of weight in % w/w resulting from water and volatile matter that is lost under specified conditions. The temperature to which the substance is subjected varies considerably according to the nature of the substance. The temperature applied should not be so high as to cause decomposition of the substance but at the same time it should be sufficiently high to produce the desired results within a reasonable time. It is usually applied by drying the substance to constant weight at 105°C.

(f) ***Moisture Content:*** Sometimes the determination of the moisture content of the substance is a good measure of the purity of the substance especially in case of crude drugs.

(g) ***Ash Values:*** The determination of ash values in crude drugs, organic compounds and certain inorganic compounds provides valuable information regarding the extent of heavy metals and minerals impurities.

7.6 Assay

An assay method should be specific for the substance or chemical species being examined. Nevertheless non-specific assay methods are quite commonly employed particularly in acid-base titrations. Many inorganic salts are assayed by simply determining the content of one of the ions present e.g., sodium sulphate to assayed by determining its sulphate content by precipitating the sulphate as barium sulphate. Although non-specific an assay method can be considered as sufficiently specific when used in conjugation with other requirement of the monograph.

Assay Tolerances

Assay tolerances play an importance role in fixing standards for pharmaceutical substances. They include the limits of error of the actual assay process for the active pharmaceutical ingredients, the limits of tolerance of manufacturing process for the particular dosage form and the sampling errors. The limits of error of the assay process depend upon the method employed.

7.7 Limit Tests

Limit tests are quantitative or semi-quantitative tests designed to identify and control small amount of impurities, which are likely to be present in the substance. They involve simple comparisons of opalescence, turbidity or color produced in test with that of fixed standards.

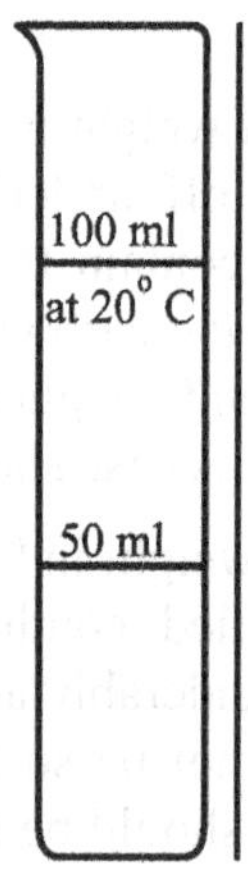

Fig.7.1 Nesslers cylinder.

Some of the limit tests are performed in a special apparatus known as Nessler cylinder. Nessler cylinders are matched tubes of clear, colorless glass with a uniform internal diameter and a flat, transparent base. They have a nominal capacity of 50 ml. The overall height of Nessler cylinder is about 150 mm and the external height to the 50 ml mark is about 110 to 124 mm. In general, limit tests are performed in an aqueous solution or sometimes the solution is prepared as specified in the monograph of the Pharmacopoeia. An identical pair of Nessler cylinder should be used i.e. a pair made of the same glass, having same diameter and same height of the graduation mark from the base. Comparison is made by placing the two Nessler cylinders side by side and viewing transversely against proper background in case of limit test chlorides, sulphates and iron. Whereas for heavy metals comparison is done by placing the two Nessler cylinders close together and viewing down through the solution against a light background.

7.7.1 Limit Test for Chloride

Principle: This test, which is mainly used to control chloride impurity in inorganic substance, depends upon the simple reaction between silver nitrate and soluble chlorides to give insoluble silver chloride in the presence of dilute nitric acid. The insoluble silver chloride makes the solution opalescent and the extent of opalescence depending upon the amount of chloride present in the substance, is compared with a standard opalescence produced in a standard solution having a known amount of chloride by adding silver nitrate and same volume of dilute nitric acid as used in the test solution.

$$\underset{\text{(soluble)}}{Cl^-} + AgNO_3 \xrightarrow{\text{dil.HNO}_3} \underset{\text{(Insoluble)}}{AgCl\downarrow} + NO_3^- \quad .....(7.6)$$

If the opalescence produced in the test is less intense than that of standard opalescence, the sample passes the limit test for chloride and vice versa.

Dilute nitric acid in used to prevent the precipitation of other acid radicals with silver nitrate solution. It acts by providing common ion i.e. nitrate.

Method (I.P. 1996): Specified quantity of the substance to be examined is dissolved in water or a solution is prepared as directed in the individual monograph and transferred to a Nessler cylinder. To this solution 10 ml of dilute nitric acid is added, except when nitric acid is used in the preparation of the solution and the volume is made up to 50 ml with water. Then, 1 ml of 0.1 M silver nitrate is added and the solution is stirred immediately and allowed to stand for 5 minutes protected from light. Any opalescence produced when viewed transversely against a black background is not more intense than that obtained by treating a mixture of 10.0 ml of chloride standard solution and 5 ml of water in the same manner.

Preparation of standard chloride solution: Dilute 5 volumes of a 0.0824% w/v Sodium chloride to 100 volumes with water.

7.7.2 Limit Test for Sulphate

Principle: This test is designed to control sulphate impurity primarily in inorganic substance and is based up on the simple reaction between barium chloride and soluble sulphate in the presence of acetic acid to give insoluble barium sulphate.

$$\underset{\text{(soluble)}}{SO_4^{2-}} + BaCl_2 \text{ ..... } CH_3COOH \rightarrow \underset{\text{(Insoluble)}}{BaSO_4} + 2Cl_4^- \quad \text{.....(7.7)}$$

The opalescence produced in the test is compared with that of standard opalescence obtained from a standard sulphate solution containing known amount of sulphate produced in the same manner. If the test opalescence is less intense than that of standard opalescence, the sample passes the limit test for sulphate and vice versa.

Note: The solubilities of barium sulphate precipitates are very much affected by the concentration of the acid. The acidity of the solution is controlled by using acetic acid. Ethanolic sulphate standard solution (10 ppm SO_4^{2-}) is added to increase the sensitivity of the test. The ionic concentration is adjusted in such a manner that the solubility product of barium sulphate gets exceeded and traces of barium sulphate present assist rapid and complete precipitation by seeding. Ethanol prevents super saturation and helps in producing a more uniform opalescence.

Method (I.P. 1996): Take 1.0 ml of a 25.0% w/v solution of barium chloride in a Nessler cylinder. To this add 1.5 ml of ethanolic sulphate standard solution (10 ppm SO_4^{2-}), mix and allow to stand for 1 minute. Add a solution of the specified quantity of the substance being examined in 15 ml of water or 15 ml of the solution prepared as directed in the individual monograph and 0.15 ml of 5M acetic acid. Make up the volume to 50 ml with water, stir immediately with a glass rod and allow to stand for 5 minutes. Any opalescence produced when viewed transversely against a black background is not more intense than that obtained by treating 15 ml of sulphate standard solution (10 ppm SO_4^{2-}) in the same manner.

Preparation of 25% w/v barium chloride solution: Dissolve 25g of barium chloride in sufficient distilled water to produce 100 ml.

7.7.3 Limit Test for Iron

Principle: The limit test for iron is based on the formation of pale pink to deep reddish purple colour by reaction of iron with thioglycollic acid in the presence of citric acid in a solution made alkaline with ammonia solution. The colour is due to the formation of the co-ordination compound, ferrous thioglycolate $Fe(HSCH_2COO)_2$ The original state of oxidation of the iron is immaterial since thioglycollic acid also reduces ferric iron to the ferrous state.

$$2Fe^{3+} + 2HSCH_2\ COOH \rightarrow 2Fe^{2+} + 2H^+ + HOOCCH_2SSCH_2COOH \quad \text{.....(7.8)}$$

The colour produced is compared with the standard colour containing a definite quantity of iron by viewing transversely through the Nessler cylinders.

Citric acid is added in the official process to prevent the precipitation of iron by ammonia. Iron gets precipitated in the form of Fe $(OH)_2$ and $Fe(OH)_3$ depending upon the oxidation state of the iron in the presence of ammonia. Citric acid makes soluble complex with iron and ammonia will not be able precipitate iron and it will just provide alkaline medium.

Method (I.P. 1996): The specified, quantity of the substance being examined is dissolved in water or a solution is prepared as directed in the individual monograph and transferred to a Nessler cylinder. 2 ml of a 20% w/v solution of iron-free citric acid and 0.1 ml of thioglycollic acid are added and mixed. The solution is made alkaline with iron-free ammonia solution and diluted to 50 ml with water. The solution is stirred with a glass rod and allowed to stand for 5 minutes. Any colour produced is not more intense than that obtained by treating 2.0 ml of iron standard solution (20 ppm Fe) in the same manner in place of the solution being examined.

Preparation of iron standard solution (20 ppm Fe): Dilute 1 volume of a 0.1726% w/v solution of ferric ammonium sulphate in 0.05M sulphuric acid to 10 volumes with distilled water. It contains iron in the ferric state.

7.7.4 Limit Test for Lead

Principle: The limit test for lead is based on the reaction between lead and diphenylthiocarbazone (dithizone) in an alkaline medium to form lead-dithizonate complex. The lead present as an impurity in the substance is separated by extracting an alkaline solution with dithizone extraction solution. The interference by other metal ions is eliminated by adjusting the optimum pH for the extraction by using reagents like ammonium citrate, potassium cyanide and hydroxylamine hydrochloride.

The original colour of dithizone in chloroform is green while the lead-dithizonàte complex is violet in colour. The intensity of the violet colour of the complex depending upon the quantity of lead present in the solution is compared with that of standard colour produced by treating standard solution containing definite amount of lead in the similar manner.

$$\text{Lead} + \text{Dithiazone} \rightarrow \text{Lead Dithizonate Complex} \qquad .....(7.9)$$

Method (I.P. 1996): Prepare the sample as directed in the monograph and transfer to a separator. Unless otherwise directed in the monograph, add 6 ml of ammonium nitrate solution and 2 ml hydroxylamine hydrochloride solution. To this, add two drops of phenol red solution and the make the solution alkaline (red in colour) by adding strong ammonia solution. If necessary, cool the solution and add 2 ml of potassium cyanide solution. Immediately extract the solution with 5 ml portions of dithizone extraction solution, until it retains green colour. Combine the dithizone extracts and transfer to a second separator. Shake the combined dithizone extracts for 30 seconds with 30 ml of 1% v/v nitric acid solution and discard the chloroform layer. To the nitric acid solution, add exactly 5 ml of dithizone standard solution and shake for 30 seconds. The violet colour of the chloroform layer should not be more intense than that obtained by treating a volume of standard lead solution (1 ppm Pb) equivalent to the amount of lead permitted in the substance being examined in the same manner as that of solution being examined.

Preparation of standard lead solution (1 ppm Pb): Dissolve 0.400 g of lead nitrate in water containing 2 ml of dilute nitric acid and add sufficient water to produce 250.0 ml. This gives **standard lead solution (1% Pb).** Standard lead solution (1 ppm Pb) is prepared by diluting 1 volume of standard lead solution (1% Pb) to 1000 volumes with water.

Preparation of dithizone extraction solution: Dissolve 30 mg of dithizone in 1000 ml of chloroform and add 5 ml of ethanol (95%). The solution is stored in refrigerator. Before use, the solution is shaken with about half of its volume of 1% v/v nitric acid solution and acid is discarded.

Preparation of Dithizone standard solution: Dissolve 10 mg of dithizone in 1000 ml of chloroform.

7.7.5 Limit Test for Heavy Metals

The limit test for heavy metals is designed to determine the content of metallic impurities that are coloured by hydrogen sulphide or sodium sulphide under the condition of the test. The heavy metals (metallic impurities) may be iron, copper, lead, nickel, cobalt, bismuth, antimony etc. The limit for heavy metals is indicated in the individual monograph in term of ppm of lead i.e. the parts of lead per million parts of the substance being examined. The Indian pharmacopoeia had adopted four methods (methods A, B, C, and D) for the limit test for heavy metals.

Principle: Method A, B and C are based upon the reaction of the heavy metal ion with hydrogen sulphide (in method A and B) or sodium sulphide (in method C) leading to the formation of heavy metal sulphides. The metal sulphides remain distributed in a colloidal state and give rise to a brownish colouration. The colour produced in the test solution is compared with that of standard solution containing a definite amount of the lead.

$$\text{Heavymetals} + H_2S/Na_2S \rightarrow \underset{\text{(Brownish colouration)}}{\text{Heavy metal sulphides}} \quad \text{.....(7.10)}$$

Method D is based upon the precipitation of relatively insoluble and characteristically coloured sulphides of heavy metals when aqueous solutions are treated with alkali metal sulphides (NaSH). NaSH is generated immediately before use by heating thioacetamide with sodium hydroxide solution. In this test there is a formation of brown colour because of the precipitation of metal sulphides at about pH 3.5 in colloidal form which is stabilized by the glycerin.

The colour is compared by keeping the two Nessler cylinders side by side and viewing vertically downwards against a white background. The usual limit for heavy metals as per I.P. is 20 ppm.

Method (I.P. 1996)

Method A

It is applicable for the substance, which give clear colorless solution under specified conditions of test.

Standard solution: Pipette 1.0 ml of standard lead solution (20 ppm Pb) into a Nessler cylinder labeled as "Standard" and dilute to 25 ml with water. Adjust the P^H between 3.0 and 4.0 with dilute acetic acid or dilute ammonia solutions, dilute to 35 ml with water and mix.

Test solution: Take 25 ml of the solution prepared as directed in the individual monograph into a Nessler cylinder labeled as "Test" or dissolve the specified quantity of the substance in water to produce 25 ml. Adjust the pH between 3.0 and 4.0 with dilute acetic acid or dilute ammonia solution, dilute to 35 ml with water and mix.

Procedure: Add 10 ml of freshly prepared hydrogen sulphide to each of the Nessler cylinder containing test solution and standard solution respectively. Mix, dilute to 50 ml with water and allow to stand for 5 minutes. Compare the colour by viewing vertically downwards over a white surface. The colour produced with the test solution in not more intense than that produced with the standard solution.

Preparation of standard lead solution (20 ppm Pb): Dilute 1 volume of standard lead solution (0.1% Pb) to 50 volumes with water.

Method B

It is applicable for those substances, which do not give a clear colorless solution under the specified conditions.

Standard solution: Same as for Method A

Test solution: Weigh the quantity of the substance as specified in the individual monograph in a crucible, wet the sample by adding sufficient sulphuric acid, and ignite at low temperature to thorough charring. To the charred mass add 2 ml of nitric acid and 5 drops of sulphuric acid and heat until white fumes are no longer evolved. Ignite at 5 00°C to 600°C until the carbon is completely burnt off. Cool, add 4 ml of dilute hydrochloric acid, digest on a water-bath for 15 minutes and evaporate to dryness on a water bath. Moisten the residue with 1 drop of hydrochloric acid, add 10 ml of hot water and digest for 2 minutes. Make the solution just alkaline to the litmus paper by drop wise addition of ammonia solution. Dilute to 25 ml with water and adjust the pH between 3.0 and 4.0 with dilute acetic acid. Filter, rinse the crucible and filter with 10 ml of water. Combine the filtrate and washing to Nessler cylinder labeled as "Test", dilute to 35 ml with water and mix.

Procedure: Same as for Method A.

Method C

It is applicable to those substances that give clear colorless solution in sodium hydroxide.

Standard solution: Pipette 1.0 ml of standard lead solution (20 ppm Pb) into a Nessler cylinder labeled as "Standard", add 5 ml of dilute sodium hydroxide solution, dilute to 50 ml with water and mix.

Test solution: Place 25 ml of the solution prepared as directed in the individual monograph into a Nessler cylinder labeled as 'Test' or dissolve the specified quantity of the substance in 20 ml of water and 5 ml of dilute sodium hydroxide solution. Dilute to 50 ml with water and mix.

Procedure: Add 5 drops of sodium sulphide solution to each the Nessler cylinder containing the standard solution and the test solution respectively. Mix and allows to stands for 5 minutes and compare the colour by viewing vertically down words over a white surface. The colour produced with the test solution is not more intense than that produced with the standard solution.

Method D

Standard Solution: Pipette 10.0 ml of either standard lead solution (1 ppm Pb) or standard lead solution (2 ppm Pb) into a small Nessler cylinder labeled as "Standard". Add 2.0 ml of the test solution and mix.

Test Solution: Prepare as directed in the individual monograph and pipette 12 ml into a small Nessler cylinder labeled as "Test".

Procedure: Add 2 ml of acetate buffer pH 3.5 to each of the above Nessler cylinders, mix, add 1.2 ml of thioacetamide reagent and allow to stand for 2 minutes. Compare the colour by viewing vertically downwards over a white surface. The colour produced with the test solution is not more intense that than produced with the standard solution.

Preparation of standard lead solution (2 ppm Pb): Dilute 1 volume of standard lead solution (20 ppm Pb) to 10 volumes with water.

Preparation of thioacetamide reagent: To 0.2 ml of 4% w/v thioacetamide solution, add 1 ml of a mixture of 15 ml of 1M sodium hydroxide, 5 ml of water and 20 ml of glycerin (8 5%). Heat on a water bath for 20 seconds, cool and use immediately.

Preparation of acetate buffer pH 3.5: Dissolve 25 g of ammonium acetate in 25 ml of water and add 38 ml of 7M hydrochloric acid. Adjust the P^H to 3.5 by using 3M hydrochloric acid or 6M ammonia and dilute to 100 ml with water.

7.7.6 Limit Test for Arsenic

Principle: The limit test for arsenic is based on the reduction of the arsenic in the arsenious state to the arsine gas (AsH_3) with zinc and hydrochloric acid. The arsine gas stains the mercuric chloride paper yellow. The sample is dissolved in acid whereby the arsenic present as impurity in the sample gets converted to arsenic acid. The arsenic acid is reduced to arsenious acid by reducing agents like stannous acid, potassium iodine etc. The nascent hydrogen formed during the reaction further reduced arsenious acid to the arsine gas. The arsine gas reacts with mercuric chloride paper to produce a yellow stain.

$$As^{3+} \rightarrow \underset{\text{(Arsenic acid)}}{H_3AsO_4} \qquad \text{.....(7.11)}$$

$$\underset{\text{(Arsenic acid)}}{H_3AsO_4} \rightarrow \underset{\text{(Arsenious acid)}}{H_3AsO_3} \qquad \text{.....(7.12)}$$

$$\underset{\text{(Arsenious acid)}}{H_3AsO_3 + 3H_2} \rightarrow \underset{\text{(Arsine)}}{AsH_3(g)} + 3H_2O \qquad \text{.....(7.13)}$$

$$\underset{\text{(Arsine)}}{2AsH_3} + HgCl_2 \rightarrow \underset{\text{(Yellow stain)}}{Hg(AsH_2)_2} + 2HCl \qquad \text{.....(7.14)}$$

The depth of the yellow stain depending upon the amount of arsenic present in the sample, is compared with that of standard stain produced from a known amount of arsenic.

The Apparatus (I.P. 1996)

Apparatus for limit test for arsenic (Gut-zeit apparatus)

The apparatus consists of a 100-ml bottle or conical flask closed with a rubber or ground- glass stopper through which passes a glass tube (about 20 cm × 05 mm). The lower part of the tube is drawn to an internal diameter of 1.0 mm, and 15 mm from its tip is a lateral orifice 2 to 3 mm in diameter. When the tube is in position in the stopper the lateral orifice should be at least 3 mm below the lower surface of the stopper. The upper end of the tube has a perfectly flat surface at right angles to the axis of the tube. A second glass of the same internal diameter and 30 mm long, with a similar flat surface, is placed in contact with the first and is held in position by two spiral springs or clips. Into the lower tube insert 50 to 60 mg of lead acetate cotton, loosely packed, or a small plug of cotton and a rolled piece of lead acetate paper weighing 50 to 60 mg. Between the flat surfaces of the tubes place a disc or a small square of mercuric chloride paper large enough to cover the orifice of the tube (15 mm × 15 mm).

Preparation: The solution of water soluble substances is prepared with water and stannated hydrochloric acid AsT. The solution of substances such as metallic carbonates, which effervesce with acids, is obtained with brominated hydrochloric acid AsT. The substances which are insoluble e.g.., $BaSO_4$, bentonite or kaolins are diffused in water.

Method: The solution of the substances to be examined is prepared as specified in the individual monograph and transferred to the wide mouthed bottle. To this add 1 g of potassium iodide (5 ml of 1 M KI), 5 ml of stannated hydrochloric acid solution and 10 g of zinc AsT. Immediately place the glass tube in position and immerse the bottle in a water-bath at a temperature such that a uniform evolution of gas is maintained. The most suitable temperature for the test is about 40°C. The reaction is allowed to continue for 40 minutes. After 40 minutes, the yellow stain produced on the $HgCl_2$ paper is compared with the standard stain produced by treating 1.0 ml of the **arsenic standard solution (10 ppm As)** diluted to 50 ml with water in the same manner. If the intensity of the yellow stain produced in the test solution is less than that of standard stain, the sample passes the limit test for arsenic and vice-versa. The stain produced on paper fades on keeping and therefore the stains should be compared immediately.

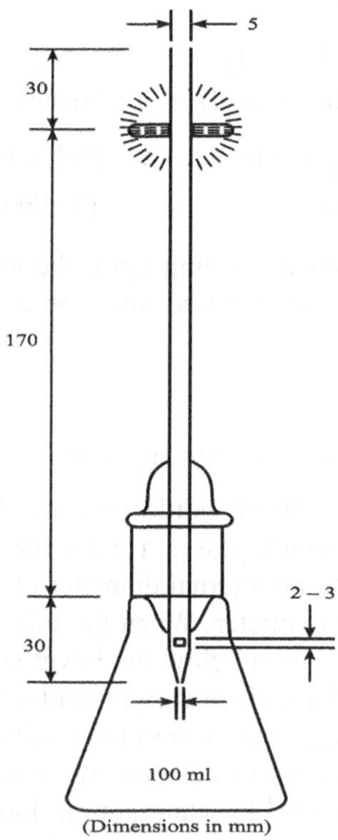

Fig 7.2 Apparatus for limit test of arsenic

Hydrochloric acid AsT: Hydrochloric acid, low in arsenic, commercially available.

Bromine solution: Dissolve 9.6 ml of bromine and 30g of potassium bromide in sufficient amount of water to make 100 ml.

Brominated hydrochloric acid AsT: It is prepared by adding 1 ml of bromine solution to 100 ml of hydrochloric acid.

Stannous chloride solution: It can be prepared by either of the following two methods.

1. Dissolve 330 g of stannous chloride in 100 ml of hydrochloric acid and add sufficient water to make 1000 ml.
2. Dissolve 60 ml of hydrochloric acid with 20 ml of water, add 20 g of tin and heat gently until evolution of gas ceases. Add sufficient water to make 100 ml. Store over a little of the undissolved tin remaining in the solution and protected from air.

Stannous chloride solution AsT: It is prepared by adding stannous chloride solution to an equal volume of hydrochloric acid AsT, reducing to the original volume by boiling and filtering through a fine-grain filter paper.

Stannated hydrochloric acid AsT: It is prepared by adding 1 ml of stannous chloride solution AsT to 100 ml of hydrochloric acid AsT.

Preparation of standard arsenic solution (10 ppm As): Dissolve 0.330 g of arsenic trioxide in 5 ml of 2M sodium hydroxide and dilute to 250.0 ml with water. Dilute 1 volume of this solution to 100 volumes with water.

Zinc AsT: It is the granulated zinc which complies with the following additional test:

To 10 g of the granulated zinc add 15 ml of the stannous chloride solution AsT and 5 ml of 0.1 M potassium iodide. Apply the general test but allow the reaction to continue for one hour. No visible stain should be produced on the mercuric chloride paper. Repeat the test by adding 0.1 ml of standard arsenic solution (10 ppm As); a faint but distinct yellow stain is produced.

CHAPTER 8

Major Intracellular and Extracellular Electrolytes

8.1 Introduction

About 56% of the adult human body is fluid. Although most of this fluid is inside the cells and is called intracellular fluid, about one third is in the space outside the cells and is called extracellular fluid. The extracellular fluid is in constant motion throughout the body. In the extracellular fluid are the ions and nutrients needed by the cells for the maintenance of cellular life. Therefore, all the cells live in essentially the same environment, the extracellular fluid, for which reason the extracellular fluid is called internal environment of the body.

The body fluids are solutions of inorganic and organic solutes. The concentration balance of the various components is maintained in order for the cell and tissue to have a constant environment. In order for the body to maintain this internal homeostasis, (homeostasis means maintenance of static or constant conditions in the internal environment) there are regulatory mechanisms which control pH, ionic balance, osmotic pressure etc. The volume and composition of the body fluids vary tremendously from one compartment to another, and are maintained remarkably constant despite the vicissitude of daily life and the stress imposed by disease. Disturbance of fluid and electrolyte metabolism involve four properties of the body fluid-volume, hydrogen ion concentration (pH) and the concentration of other specific ions. The total body water is divided into three compartments

(a) the intercellular compartment

(b) the extracellular compartment, which consists of the plasma and the interstitial fluid

(c) the transcellular compartment, which includes the fluid within the gastrointestinal tract, humor of the eye and the excretory system of the kidneys and glands, pericardial, peritoneal, synovial, cerebrospinal fluid.

All the body fluids intracellular, extracellular (interstitial, plasma or vascular) contains electrolytes. The electrolyte concentration varies in these fluids, it is 45-50% of body weight in intercellular fluid, interstitial fluid makes 12-15 % and plasma makes 4-5% of body weight. About 40% of intracellular fluid (4l) is dense connective tissue i.e. bone and cartilage and does not take part in quick exchange of electrolytes with the remaining body. The rest of the interstitial fluid IF (6.6l) and plasma (3.5l) comprise the active part of the extracellular fluids. These fluid compartments are separated from each other by membranes which are permeable to water and many organic and inorganic solutes. They are nearly impermeable to macromolecules e.g. proteins and selectively permeable to certain ions e.g. Na^+, K^+ and Mg^+ as a result, each of these fluid compartments has distinct solute pattern and the solution in each compartment is ionically balanced. For electro neutrality to exist in extracellular fluid, the sum of the concentration of cations must be equal to the sum of the concentration of anions.

The extracellular fluid contains large amounts of sodium, chloride and bicarbonate ions, plus nutrients for the cell such as oxygen, glucose, fatty acids and amino acids. The intracellular fluid contains large amounts of potassium, magnesium and phosphate ions. Measurement of electrolyte concentrations (plasma) is usually limited to Na^+, K^+, Cl^-, and HCO_3^-. The sum of the concentration of sodium and unmeasured cations (Ca^{2+}, Mg^{2+}, K^+) equals the sum of the concentration of Cl^- and HCO_3^- and unmeasured anions (phosphates, proteins, sulphates, derivatives of organic acids). The difference between the concentration of unmeasured cations and anion is known as anionic gap. Variation in this gap is a useful diagnostic indication to disorders of acid base balance.

The electrolyte balance of the body is maintained by a regulation between the intake and output of water. The intake of water includes the fluid taken orally and the release of water during the oxidation and other metabolic process in body.

Water is eliminated from body by urine, expiration (lungs), perspiration and feces. Excessive loss of water results in concentration of body fluids which causes rise in osmotic pressure, as a result water moves out from intracellular compartment to maintain the osmotic pressure in extracellular fluid. This results in dehydration of cells. Loss of water above 20% may prove to be fatal.

8.2 Calcium

About 99% of body calcium is found in bones and the remaining is present in extracellular fluid compartment. Only 10% of the ingested calcium is absorbed from the intestinal tract and the remainder is excreted with feces. The concentration of calcium in plasma averages about 9.4mg/dl, (9-l0mg/dl). The calcium level in plasma is regulated within narrow limits by parathyroid hormone. The calcium in plasma is present in three forms

(a) About 40% is combined with plasma proteins and is non diffusible through the capillary membrane.

(b) About 10% is combined with other substances of plasma and interstitial fluid (citrate, phosphate for instance) and is diffusible through the capillary membrane in such a manner that it is not ionized.

(c) The remaining 50% calcium present in plasma is diffusible through the capillary membrane and ionized. The plasma and interstitial fluid have a normal calcium ion concentration of about 1.2 mmole/l^{-1} (or 2.4 meql^{-1} because it is a divalent ion), a level only half of the total plasma calcium concentration.

Calcium is important for blood clotting and contraction of various smooth muscles. In cardiovascular system (CVS) calcium is essential for contraction coupling in cardiac muscles as well as for the conduction of electric impulse in certain regions of heart. Calcium also plays role in maintaining the integrity of mucosal membrane, cell adhesion and function of the individual cell membrane as well.

Hypercalcemia: When the level of calcium rises above normal, the nervous system is depressed, and the reflux action of CNS can become sluggish. It also decreases the QT interval of the heart which can lead to cardiac arrhythmia. It causes constipation and lack of appetite and depresses contractility of the muscle walls of the GIT. The depressive effect begins to appear when blood calcium level rises above 12mg/dl and beyond 17 mg/dl calcium phosphate crystals are likely to precipitate throughout the body. This situation occurs due to hypoparathyroidism, vitamin D deficiency, Osteoblastic metastasis, steatorrhea (fatty stools), Cushing syndrome (hyper active adrenal cortex), acute pancreatitis and acute hypophosphatemia.

Hypocalcemia: Change in blood pH can influence the degree of calcium biding to plasma proteins. With acidosis less calcium is bound to plasma proteins. When calcium ion concentration falls below normal, the excitability of the nerve and muscle cells increases markedly.

8.3 Sodium

The sodium and its associated anions, mainly chloride, account for more than 90% of the solute in extracellular fluid compartment. The concentration of sodium is 142 meql^{-1} in extracellular fluid, and 10 meql^{-1} in intracellular fluid. Plasma sodium is a reasonable indictor of plasma osmolarity under many conditions. When plasma sodium is reduced below normal level a person is said to have hyponatremia. When plasma sodium is elevated above normal level a person is said to have hypernatremia.

Hyponatremia: Decreased plasma sodium concentration can result from loss of sodium chloride from the extracellular fluid. Conditions that cause hyponatremia owing to loss of sodium chloride include excessive sweating, diarrhea and vomiting and over use of diuretics that inhibit kidney to conserve sodium. Addison's disease, which results from decreased secretion of hormone aldosterone (impairs the ability of kidneys to reabsorb sodium) can be one of the causes of hyponatremia

Hypernatremia: Hypernatremia is increased plasma sodium level which also increases osmolarity, can be due to excessive water loss from extracellular fluid, secretion of sodium- retaining hormone aldosterone (cushings syndrome) excessive treatment with sodium salts.

8.4 Potassium

Potassium is major intracellular cation present in a concentration approximately 23 times higher than the concentration of potassium present in Extracellular fluid compartment. Extracellular fluid potassium concentration is normally precisely regulated at 4.2meq/l. This is because many of the cell functions are sensitive to change in the extracellular fluid potassium concentration. Increase in potassium concentration can cause cardiac arrhythmias and higher concentrations can lead to cardiac arrest by fibrillation. About 95% of body potassium is contained in the cells and only 2% in extracellular fluid. Maintenance of potassium balance depends primarily on its excretion by kidney because only 5-10 percent is excreted in feces. Both, elevated and low levels of potassium, can be fatal,

Hypokalemia occurs due to high intake of potassium or in kidney damage while **Hyperkalemia** due to vomiting, diarrhea, burns, diabetic coma, over use of thiazide diuretics, alkalosis etc.

8.5 Chloride

Chloride, major extracellular anion is principally responsible for maintaining proper hydration, osmotic pressure, and normal cation anion balance in vascular and interstitial compartment. The concentration of chloride is 103 $meql^{-1}$ in extracellular fluid, and 4 $meql^{-1}$ in intracellular fluid.

Decreased chloride concentration can be the result of salt losing nephritis, leading to lack of tubular reabsorption of chloride, metabolic acidosis such as found in diabetes mellitus, in renal failure and prolonged vomiting. Increased concentration of chloride may be due to dehydration, decreased renal blood flow found with congestive heart failure (CHF) or excessive chloride uptake.

8.6 Phosphate

Phosphate is the principal anion of intracellular fluid compartment. Inorganic phosphate in the plasma is mainly in two forms HPO_4^- and $H_2PO_4^-$. The concentration of HPO_4^- is

1.05 mmolel^{-1} and the concentration of $H_2PO_4^-$ 0.26 mmolel^{-1}. When the total quantity of the phosphate in extracellular fluid raises so does the concentration of each of these ions. When pH of the extracellular fluid becomes more acidic there is relative increase in $H_2PO_4^-$ and decrease in HPO_4- and vice versa. Phosphorous is essential for proper metabolism of calcium, normal bone and tooth development. HPO_4^- and $H_2PO_4^-$ makes an important buffer system of body.

8.7 Bicarbonate

Bicarbonate is the second most prevalent anion in extracellular fluid compartment. Along with carbonic acid it acts as body's most important buffer system. Each day kidney filters about 4320 mill equivalents of bicarbonate and under normal conditions all of this is reabsorbed from the tubules, thereby conserving the primary buffer system of the extracellular fluid. When there is reduction in the extracellular fluid hydrogen ion concentration (alkalosis) the kidneys fail to reabsorb all the filtered bicarbonate thereby increasing the excretion of bicarbonate. Because bicarbonate ions normally buffer hydrogen in the extracellular fluid, this loss of bicarbonate is as good as adding a hydrogen ion to the extracellular fluid. Therefore, in alkalosis, the removal of bicarbonate ions raises the extracellular fluid hydrogen ion concentration back towards normal. In acidosis the kidneys do not excrete the bicarbonate in the urine but reabsorb all the filtered bicarbonate and produces new bicarbonate which is added back to the extracellular fluid. This reduces the extracellular fluid hydrogen ion concentration back towards normal.

8.8 Replacement Therapy

The basic objective of replacement therapy is to restore the volume and composition of the body fluids to normal one. Volume contraction is a life threatening condition because it impairs the circulation. Blood volume decreases, cardiac output falls and the integrity of microcirculation is compromised. In volume depletion of sufficient magnitude to threaten life, a prompt infusion of isotonic sodium chloride solution is indicated. In an extreme case, intravenous therapy at the rate of 100 ml per minute for the first 1000ml has been considered necessary for the successful treatment of cholera. A general rule is to replace one half of the estimated volume loss in the first 12-24 hours of treatment.

8.8(a) Sodium Replacement

Sodium Chloride:

Chemical Formula: NaCl *Mol. Wt.* 58.44

I.P. Limit: Sodium chloride contains not less than 99.5 % and not more than 100.5 % calculated with reference to dried substance. It contains no added substances. It occurs as colorless cubic crystals or as white crystalline powder having saline taste. It is freely soluble in water, and slightly more soluble in boiling water, soluble in glycerin and slightly soluble in alcohol.

Test for identification

For Sodium: To sample solution add 15 % w/v potassium carbonate heat, no precipitate. Add potassium antimonite solution, heat to boiling, cool and if necessary scratch the inside of test tube with a glass rod, a dense white precipitate is produced.

For Chloride: Dissolve sample in water, acidify with dilute nitric acid and add silver nitrate solution shake, and allow to stand, a curdy white precipitate is formed which is insoluble in nitric acid but, soluble after being well washed with water, in dilute ammonium hydroxide solution from which it is reprecipitated by the addition of dilute nitric acid.

Preparation: On commercial scale it is prepared by evaporation of sea water in shallow pans. It contains impurities of sodium carbonate, sodium sulphate, magnesium chloride, magnesium sulphate, calcium chloride etc. these impurities are removed by dissolving the salt in water in a cemented tank; some alum and lime are added. The suspended impurities are allowed to settle down. The clear solution is decanted into iron pans and concentrated. The crystals of sodium chloride settle down which are then collected and dried.

Assay: The assay of sodium chloride is dependent on the modified Volhard's method in which indirect volumetric precipitation titration is involved. An acidified solution of sodium chloride with nitric acid is treated with a measured excess amount of standard solution of silver nitrate in the presence of nitrobenzene. Some of the silver nitrate is consumed in the reaction with sodium chloride. The remaining unreacted $AgNO_3$ is determined by titration with standard solution of ammonium thiocyanate using ferric alum (ferric ammonium sulphate) as indicator. The end point is obtained as a permanent brick red color due to formation of ferric thiocyanate.

Procedure: Accurately weigh the substance (0.1 gm) and dissolve in 50 ml water. Add 50 ml of 0.1N $AgNO_3$, 3 ml HNO_3, 5 ml nitro benzene, 2 ml ferric ammonium sulphate and mix thoroughly. The solution is titrated with ammonium thiocyanate until the color becomes brick red.

$$1\text{ml of } 0.1\text{N } AgNO_3 \equiv 0.005844\text{gm of NaCl}$$

Uses:

Used as fluid and electrolyte replenisher, manufacture of isotonic solution, flavor enhancer.

- Isotonic solutions are used in wet dressings, for irrigating body cavities or tissues
- Hypotonic solutions are administered for maintenance therapy when patients are unable to take fluids and nutrients orally for one to three days.
- Hypertonic solution/injection is used when there is loss of sodium in excess.
- Official preparations of Sodium chloride

Sodium Chloride Injection I.P.

Sodium chloride injection is a sterile isotonic solution of sodium chloride in water for injection. It contains not less than 0.85 % and not more than 0.95 % w/v of sodium chloride. It contains no antimicrobial agents. It is a clear, colorless solution with pH between 4.5-7.0.

Sodium Chloride Hypertonic Injection I.P. *(Hypertonic saline)*

It is a sterile solution of sodium chloride in water for injection. It contains not less than 1.52% and not more than 1.68 % w/v of sodium chloride. It contains no antimicrobial agents. It is a clear, colorless solution with pH between 5-7.5.

It complies with the test for pyrogens.

Compound Sodium Chloride Injection I.P. *(Ringer injection)*

It contains not less than 0.82 % and not more than 0.9 % w/v of sodium chloride, not less than 0.0285 %, not more than 0.0315 % w/v of potassium chloride and not less than 0.03 % and not more than 0.036% w/v of calcium chloride in water for injection. It contains no antimicrobial agents. It is a clear, colorless solution with pH between 5-7.5.

Sodium Chloride and Dextrose Injection

It is a sterile solution of sodium chloride and dextrose in water for injection. It contains not less than 95% and not more than 105 % w/v of the stated amount of sodium chloride and dextrose as given in table 8.1.

Table 8.1 Combinations of **Sodium Chloride and Dextrose.**

% of Sodium Chloride	% of Dextrose	% of Sodium Chloride	% of Dextrose
0.11	5	0.45	5
0.18	5	0.45	10
0.20	5	0.90	2.5
0.225	5	0.90	5
0.3	5	0.90	10
0.33	5	0.90	25
0.45	2.5		

It is clear colorless or faintly straw colored solution with pH between 3.5-6.5.

8.8(b) Potassium Replacement

Potassium Chloride:

Chemical Formula: KCl *Mol. Wt.* 74.56

I.P. Limit: Potassium chloride contains not less than 99 % calculated with reference to dried substance. It occurs as Carnallite ($KCl.MgCl_2$) $6H_2O$ contaminated with magnesium sulphate and chlorides. It occurs as white crystalline solid, cubic crystals. It is less soluble in water than sodium chloride, and slightly more soluble in boiling water, soluble in glycerin and insoluble in alcohol.

Test for Identification

For potassium: To 1ml of solution add 1ml dilute acetic acid and 1ml of 10% w/v sodium cobalt nitrite, a yellow color is produced.

For Chloride: Substance in water is added with dilute solution of silver nitrate, shake the solution and allow to stand , on standing white precipitate is obtained which is insoluble in nitric acid but soluble after being washed with water; in dilute ammonium hydroxide, from which it is reprecipitated by the addition of dilute nitric acid.

Preparation

1. It is prepared by fusing carnallite whereby liquefied magnesium chloride hexahydrate is separated from the solid potassium chloride.
2. The crushed carnallite is dissolved by boiling with liquor leaving other impurities undissolved. These are filtered off and the filtrate is crystallizes to get cubic crystals of potassium chloride.
3. It is also prepared in laboratory by reacting HCI with potassium carbonate or bicarbonate

$$K_2CO_3 + 2HCl \rightarrow KCl + H_2O + CO_2 \quad .....(8.1)$$

$$KHCO_3 + HCl \rightarrow KCl + H_2O + CO_2 \quad .....(8.2)$$

Assay: The assay is based on Mohr's method of direct volumetric precipitation titration. An aqueous solution of the substance is titrated against a standard solution of silver nitrate using solution of potassium chromate as indicator.

$$KCl + AgNO_3 \rightarrow AgCl + KNO_3 \quad .....(8.3)$$

When whole of potassium chloride has been precipitated as AgCl, further addition of silver nitrate solution gives brick red color with the indicator. The end point is change of color from yellow to red.

Procedure: Accurately weigh the specified (0.25g) amount of potassium chloride and dissolve in 50 ml of water. Titrate the solution with 0. 1N silver nitrate solution using potassium chromate solution as indicator.

$$2AgNO_3 + K_2CrO_4 \rightarrow Ag_2CrO_4 + 2KNO_3 \quad .....(8.4)$$

1ml of 0.1N silver nitrate ≡ 0.007455g of KCl

Use: Electrolyte replenisher in potassium deficiency, familial periodic paralysis, Meniere's syndrome (disease of inner ear), antidote in digitalis intoxication, myasthenia gravis.

Contraindication: renal impairment with oligouria, acute dehydration.

Potassium Chloride injection: Ringer injection

8.8(c) Calcium Replacement

Calcium Lactate:

Chemical Formula: $C_6H_{10}CaO_6.\ 5H_2O$ *Mol. Wt.* 308.30 (Pentahydrate)

I.P. Limit: Potassium chloride contains not less than 97% and not less than 103% of calcium chloride dihydrate. It occurs as white odourless powder. The pentahydrate effloresces and becomes anhydrous at 120°. Aqueous solutions are prone to become moldy. It is soluble in water, practically insoluble in alcohol.

Test for Identification

For Calcium: Dissolve substance in 5 M acetic acid and add 0.5 ml of potassium Ferro cyanide solution. The solution remains clear. Add ammonium chloride white crystalline precipitate is formed.

For Lactate: To sample solution add bromine water, 1 M H_2SO_4 and heat on water bath stirring occasionally until the color is discharged. Add ammonium sulphate mixture of 10% solution of sodium nitro prusside in ammonia solution. Allow to stand for 10 mins, a dark ring appears at the interface of two liquids.

Preparation

1. It is obtained by neutralizing a hot solution of lactic acid with calcium carbonate in slight excess. The hot liquid is filtered and filtrate is evaporated to crystalline product.

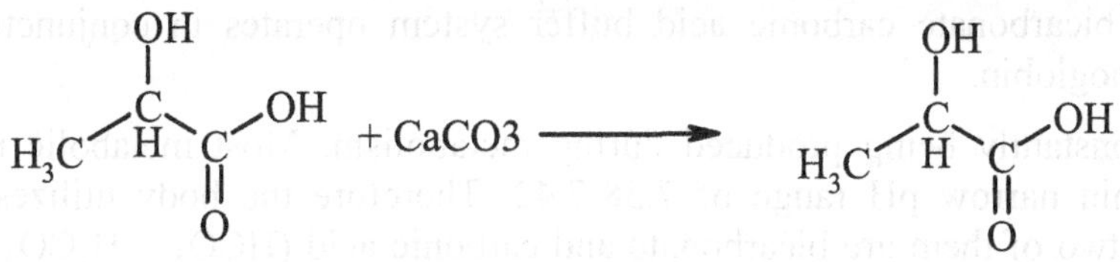

.....(8.5)

2. It is also obtained by fermenting hydrolyzed starch with a suitable mold in the presence of calcium carbonate (Or)
3. By fermentation of mother liquor resulting from the production of milk sugar and chalk. The mixture is digested for a week at about 30°. The product is purified by crystallization.

Assay: The assay is based on complexometric method of titration wherein disodium EDTA as titrant and calcon mixture as indication. The end point is change of color from pink to blue.

Procedure: Accurately weigh specified amount of sample and dissolve in water (50 ml), titrate the solution with 0.05 M disodium EDTA to within few ml of the expected end point. Add sodium hydroxide solution and calcon mixture and continue titration till end point is observed. The color of solution changes from pink to blue.

1 ml of 0.05 M disodium EDTA ≡ 0.005004gm of calcium

Use: An excellent source of calcium in oral treatment of calcium deficiency.

8.9 Physiological Acid Base Balance

Abnormalities of the pH of body are frequently encounter and are of major clinical importance. Acedemia and alkalemia refer respectively to an abnormal decrease or increase in the pH of the blood. Acidosis and alkalosis refer respectively to clinical state that can lead to either acedemia or alkalemia. However in each condition the extent to which there is an actual change in pH depends in part on the degree of compensation which varies in most clinical disturbances. It is most convenient to evaluate clinical disturbances of pH by reference to $HCO_3^- - H_2CO_3$ system

Because it is in buffer system of extracellular fluid, this results from a number of factors:

- There is considerably more bicarbonate present in extracellular fluid than any other buffer component.
- There is a limitless supply of carbon dioxide.
- Physiological mechanisms operate to maintain the extracellular pH function by controlling fluid.

- The bicarbonate–carbonic acid buffer system operates in conjunction with haemoglobin.

Acids are constantly being produced during metabolism. Most metabolic reactions occur only within narrow pH range of 7.38-7.42. Therefore the body utilizes several buffer systems, two of them are bicarbonate and carbonic acid (HCO_3^- : H_2CO_3) present in plasma and kidney and monohydrogen phosphate/dihydrogen phosphate (HPO_4^{2-} : $H_2PO_4^-$) found in cells and kidney.

RBC's have hemoglobin buffer system which is most effective single buffer system for buffering the carbonic acid produced during metabolic process. For each millimole of oxygen that dissociates from hemoglobin (Hb) 0.7 millimole of H^+ are removed.

Carbon dioxide, the acid anhydride of carbonic acid is continuously produced in the cells. It diffuses into the plasma and reacts with water to form carbonic acid. The increased carbonic acid is buffered by plasma proteins. Most CO_2 enters the erythrocytes where it either rapidly forms $H_2CO_3$by the action of carbonic anhydrase or combines with Hb.

The tendency to lower the pH of the erythrocytes due to increased concentration of H_2CO_3 is compensated by Hb.

$$CO_2 + H_2O \xrightarrow{\textit{Carbonic anhydrase}} H_2CO_3 \qquad .....(8.6)$$

The bicarbonate anion then diffuses out of erythrocytes and chloride anion diffuses in. This has been named as chloride shift. Te bicarbonate in plasma, along with the plasma carbonic acid now acts as efficient buffer system

$$H_2CO_3 + K^+ + HbO_2^- \rightarrow K^+ + HCO_3^- + HHb + O_2 \qquad .....(8.7)$$

The normal HCO_3^-/ H_2CO_3 ratio is 27/1.35 meq/l (20:1) corresponding to pH 7.4. In lungs there is reversal of the above process due to the large amount of O_2 present. Oxygen combines with the protonated deoxyhemoglobin releasing proton. These combine with HCO_3^- forming H_2CO_3 which then dissociates to CO_2 and water. The carbon dioxide is exhaled by the lungs. Thus by regulating breathing it is possible for the body to exert a partial control on the HCO_3^-/H_2CO_3 ratio.

$$O_2 + H^+Hb + K^+ + HCO_3^- \rightarrow K^+HbO_2^- + H_2CO_3 \qquad .....(8.8)$$

$$\downarrow \textit{Carbonic anhydras}$$

$$CO_2 + H_2O$$

The phosphate buffer system is also effective in maintaining physiological pH. At pH 7.4 the HPO_4^{-2}/$H_2PO_4^-$ ratio is approximately 4:1. In kidney, the pH of urine can drop to 4.5-4.8 corresponding to $HPO4^{-2}$/$H_2PO_4^-$ ratio of 1:99- 1:100. The acid is excreted from kidney as follows:

- Sodium salt of mineral or organic acids are removed from the plasma by glomerular filtration.
- Sodium is preferentially removed from the renal filtrate or tubular fluid in the tubular cells. The process known as sodium hydrogen exchange.
- The sodium bicarbonate returns to plasma (eventually being removed in the lungs as CO_2) and protons enter tubular fluid, forming acids of the anions that originally were sodium salts.

8.10 Factors Altering the pH of Extra Cellular Fluid

A. ***Acidosis:*** Acidosis is defined as increase in either potential and/or nonvolatile hydrogen ion (H^+) content of body. Increase in the H^+ concentration of plasma is known as acedemia and is manifested by fall in the pH of blood. In case there is no rise in H^+ concentration of plasma, such state of acidosis (without academia) is known as **compensated acidosis.**

Types and Causes of Acidosis

Metabolic acidosis: it occurs due to excess production of proton in the body which may be because of

- Acceleration of normal metabolic process i.e. excessive catabolism e.g. in fever
- Administration of drugs which are proton donors e.g. salicylates, chlorides
- Excessive loss of alkaline fluid from the intestine, as in diarrhea
- Administration of large quantity of saline.

Metabolic acidosis is treated with sodium salts of bicarbonate, lactate, acetate and occasionally citrate. When there is bicarbonate deficient, administration of bicarbonate increases the HCO_3^-/H_2CO_3 ratio. Lactate, acetate and citrate ions are normal components of metabolism and are degraded to carbon dioxide and water by TCA cycle (Citric acid cycle or Krebs cycle).

Renal Acidosis: where increase in H^+ is due to defective renal excretion of H^+. Seen in Tubular disorders, Addison's disease, drugs which interfere with tubular secretion of H^+ e.g. carbonic anhydrase inhibitors

Respiratory Acidosis: Is due to increase in retention of carbon dioxide leading to rise in plasma carbonic acid content. It occurs due to chronic lung disease, respiratory muscle paralysis, by drugs that depress respiratory center.

B. ***Alkalosis:*** Alkalosis is reduction in the total hydrogen ion content of the body. Alkalemia is reduction of hydrogen ion content in plasma and is manifested by increase in the pH of blood. In case there is no decrease in H^+ concentration of plasma, such state of alkalosis (without alkalemia) is known as **compensated alkalosis.**

Metabolic alkalosis: Due to renal damage that cannot excrete an appreciable amount of alkali. Occurs due to alkali ingestion in presence of renal damage, excessive vomiting which causes loss of H^+ and Cl^- ions. Metabolic alkalosis has been treated with ammonium salts. Its action is in kidney where it retards the Na^+- H^+ exchange.

Contraction alkalosis: seen following administration of mercurial diuretics which cause excessive loss of Cl^- and sodium.

Respiratory alkalosis: Respiratory alkalosis is caused by hyperventilation which washes away large amount of carbon dioxide formed in metabolism causes lowering of arterial CO_2 and reduction in ratio of bicarbonate ion and carbonic acid with fall in hydrogen ion concentration. It occurs due to high altitude, fever, encephalitis, hypothalamic tumor, drugs like salicylate.

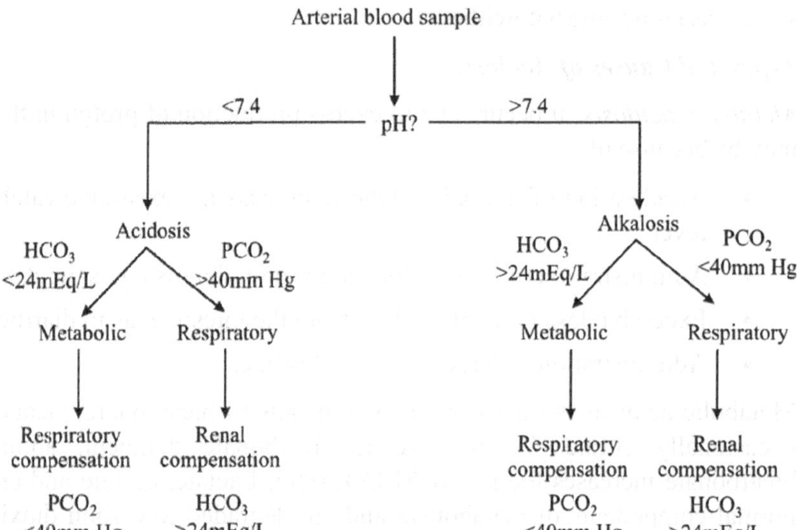

Fig. 8.1 Analysis of acid base disorder.

If the compensatory responses are markedly different than those shown at the bottom of the figure, one should expect a mixed acid base disorder

1. Sodium bicarbonate:

Chemical Formula: $NaHCO_3$ *Mol.Wt.* 84.01

Synonym: Sodabicarb.

I.P. limit: Sodium bicarbonate contains not less than 99.0 % and not more than 101 % of sodium bicarbonate.

Sodium bicarbonate occurs as a white odourless crystalline powder or granules. It begins to lose carbon dioxide at 50° and at 100° it is converted into sodium carbonate. It is soluble in water (1 in 12); partially soluble in alcohol. The aqueous solution is alkaline to litmus; alkalinity increases on standing, agitation or heating. It is stored in well closed containers.

Sodium bicarbonate when mixed with calcium or magnesium salts, cisplatin, dobutamine hydrochloride or oxytetracyclin forms insoluble precipitates. The following drugs are susceptible to inactivation on mixing with sodium bicarbonate; adrenaline hydrochloride, benzyl penicillin potassium, carmustine, glycopyrronium bromide; isoprenaline hydrochloride and suxamethonium chloride.

It is stable in dry air, but slowly decomposes in moist air.

Preparation

1. It is prepared by passing strong brine containing high concentration of ammonia through a carbonating tower where it is saturated with carbon dioxide under pressure. The ammonia and CO_2 reacts to form ammonium bicarbonate which is allowed to react with sodium chloride to precipitate sodium bicarbonate. It is then separated by filtration
2. By passing carbon dioxide through a saturated solution of sodium carbonate.

$$Na_2CO_3 + H_2O + CO_2 \rightarrow 2NaHCO_3 \qquad .....(8.9)$$

Chemical Properties

1. When sodium bicarbonate is heated, it is decomposed into the normal carbonate, carbon dioxide and water.

$$2NaHCO_3 \rightarrow Na_2CO_3 + H_2O + CO_2 \qquad .....(8.10)$$

2. A solution of sodium bicarbonate is alkaline due to hydrolysis (pH 8.2)

$$2NaHCO_3 \rightarrow Na^+ + H_2CO_3 + OH^- \qquad .....(8.11)$$

Sodium bicarbonate is slightly alkaline and fails to turn phenolphthalein red. On the other hand, in sodium carbonate the carbonate ion is so extensively hydrolyzed that the solution is quite alkaline (pH is 11.6)

$$CO_3^{2-} + H_2O \rightarrow HCO_3^- + OH^- \qquad .....(8.12)$$

3. When a mercuric chloride solution is added to a solution of sodium bicarbonate there is no immediate formation of precipitate. After some time a reddish precipitate of HgO is formed.

$$2NaHCO_3 + HgCl_2 \rightarrow 2NaCl + Hg(HCO_3)_2 \qquad .....(8.13)$$

$$Hg(HCO_3)_2 \rightarrow HgO + H_2O + 2CO_2 \qquad ......(8.14)$$

4. When the bicarbonate is treated with an acid, carbon dioxide is liberated;

$$NaHCO_3 + HCl \rightarrow NaCl + H_2O + CO_2 \quad .....(8.15)$$

Test for purity: Tests for alkalinity; aluminum; calcium; insoluble matter; arsenic; iron; lead; chloride; sulphate; ammonium compounds.

- For detecting the presence of aluminum, calcium and insoluble matter an aqueous solution is boiled with ammonia solution and filtered. The residue is ignited and weighed.
- An aqueous solution after addition of nitric acid complies wit the limit test for chloride.
- An aqueous solution after addition of hydrochloric acid complies with the limit test for sulphates.
- Evolution of ammonia on heating the substance with sodium hydroxide indicates the presence of ammonium compound.
- An aqueous solution after addition of hydrochloric acid complies with the limit test for iron.
- Heavy metals are determined by comparing the colour produced with the substance and with standard lead solution after treatment with hydrogen sulphide solution.
- Simultaneous administration of sodium bicarbonate with other drugs inhibits the activity of the drug. Such a therapeutic incompatibility is found when sodium bicarbonate and sodium salicylate are used in equivalent amounts.

Test for identification: It gives the reactions of sodium, and of bicarbonates.

For Sodium: To sample solution add 15 % w/v potassium carbonate, heat, no precipitate, add potassium antimonite solution, heat to boiling, cool and if necessary scratch the inside of test tube with a glass rod, a dense white precipitate is produced.

For bicarbonate: To sample add magnesium sulphate no precipitate is produced. On boiling a white colored precipitate is formed.

Assay: A solution of weighed amount of sample dissolved in water is titrated with 0.5 N hydrochloric or sulphuric acid, using methyl orange as indicator.

Each ml of 0.5 N hydrochloric acid is equivalent to 0.042 g of $NaHCO_3$.

It is a direct titration method; the end point is yellow to pink. The equivalence point of this titration is at about pH 3.6 which corresponds to the colour change of methyl orange (pH 2.8-4.0, red-yellow). The reaction at the equivalence point is acidic because of the presence of carbonic acid.

Uses: Sodium bicarbonate is an electrolyte replenisher, and systemic alkalinizing agent used in the treatment of metabolic acidosis (increase in acidity), diarrohoea, acute poisoning from acidic drugs (phenobarbitone and salicylates), and as an antacid to relieve dyspepsia. Solutions of sodium bicarbonate are used as eye lotions, to aid the removal of crusts in blepharitis, as eardrops, to soften and remove ear wax, and as lubricating fluid for contact lenses.

Administration of sodium bicarbonate by mouth can cause stomach cramps and flatulence. Its large quantities may cause systemic alkalosis, vertigo (loss of power of balancing) and jerky muscular movement.

Sodium bicarbonate is available as mint-flavored soda-mint tablets. Ti is self-medicated, inexpensive and easily available drug. It is absorbed from the intestine which produces effects all over the body. It produces carbon dioxide gas in stomach and may cause perforation of a deep ulcer. Its onset of action is quick but the duration of action is short. It may cause rebound acidity due to short duration of action and systemic effects. When taken with milk it may cause milk alkali syndrome, characterized by deposition of calcium of milk on the kidney and increased blood urea.

2. Sodium acetate:

Chemical Formula: $CH_3COONa.\ 3H_2O$

I.P. Limit: Sodium acetate contains not less than 99.0 per cent of $CH_3COONa.\ 3H_2O$

Preparation: It is prepared by neutralization of acetic acid with sodium carbonate or sodium hydroxide, and then crystallizing the product.

Characters: It occurs as colourless, transparent crystals or a white granular powder or white flakes: odourless or with a slight odour of acetic acid; m.p. 58°; becomes anhydrous at 120, decomposes at higher temperature. It effloresces in warm dry air. It is soluble in water (1 in 0.8), and alcohol (1 in 19). A 5 % solution in water has a pH of 7.5 to 9.2. It is kept in airtight containers.

Tests for purity: Tests for arsenic; calcium and magnesium; heavy metals; iron; chloride; sulphate; reducing substances; pH; clarity and colour of solution ; loss on drying.

For determining calcium, and magnesium, a mixture of dilute ammonia buffer solution, mordant black II is titrated with 0.05 M EDTA. When reducing substances are absent, the pink colour is not entirely discharged on treatment of potassium permanganate with an acidified solution of the substance with dilute sulphuric acid.

Incompatibility: Aqueous solutions of substance react with oxygen to produce slight pink colour. It can be prevented by addition of a solution of sodium metabisulphite.

Test for identification: It gives the reactions of sodium and of acetates.

For Sodium: To sample solution add 15 % w/v potassium carbonate heat, no precipitate, add potassium antimonite solution, heat to boiling, cool and if necessary scratch the inside of test tube with a glass rod, a dense white precipitate is produced.

For acetate

1. Heat the sample with equal quantity of oxalic acid, acidic vapours with the characteristic odour of acetic acid are liberated.
2. To 1 gm of sample add sulphuric acid, ethanol and warm. Ethyl acetate is evolved which is recognizable by its odour.

Assay: Accurately weighed amount of sample (0.25 g) is dissolved in glacial acetic acid (50 ml.), acetic anhydride (5 ml) and kept for 30 minutes and titrated with 0.1N perchloric acid using 1-naphtholbenzein solution as indicator. The end point is change of colour from yellowish green to dark green. A blank determination is also performed and necessary correction made.

Each ml of 0.1 N perchloric acid is equivalent to 0.0 1361 of $CH_3COONa.\ 3H_2O$

Uses: An effective buffer in metabolic acidosis, it is used as pharmaceutical aid (for peritoneal dialysis fluid); acidulate in food; and as an effective buffer in metabolic acidosis.

3. Potassium acetate:

Chemical Formula: CH_3COOK *Mol. Wt.* 98.15

I.P. Limit: Potassium acetate contains from 99 to 101.0% of CH_3COOK. It occurs as colourless crystals or a white crystalline powder; odourless with a faint acetic acid like odour. It is deliquescent in moist air. It is soluble in water and alcohol. A 5 % solution in water has a pH of 7.5 to 9.5.

Potassium acetate should be kept in a well-closed container.

Tests for purity: Tests for aluminium: arsenic; calcium; heavy metals; magnesium; sodium; chloride; nitrate; sulphate; readily oxidizable substances; loss on drying; and alkalinity

For determining nitrate, an aqueous solution is treated with sodium chloride, indigo carmine solution and nitrogen-free sulphuric acid. A blue colour is produced which persists for at least 10 minutes.

Test for identification: It gives reactions characteristic of potassium salts and of acetates.

For potassium: To 1ml of solution add 1ml dilute acetic acid and 1ml of 10% w/v sodium Cobalt nitrite, a yellow color produced.

For acetate

1. Heat the sample with equal quantity of oxalic acid, acidic vapours with the characteristic odour of acetic acid are liberated.
2. To 1 gm of sample add sulphuric acid, ethanol and warm. Ethyl acetate is evolved which is recognizable by its odour.

   The presence of readily oxidizable substances is found out by treating an aqueous solution with sulphuric acid and potassium permanganate. The pink colour is not completely discharged. The salt complies with the limit test for arsenic, calcium, heavy metals, magnesium and chloride.

Assay: Non-aqueous titration is carried out using perchloric acid and crystal violet solution as indicator. Accurately weighed amount of sample is dissolved in glacial acetic acid (50 ml), acetic anhydride (5 ml) and kept for 30 minutes and titrated with 0.1N perchloric acid using 1-naphtholbenzein solution as indicator. The end point is change of colour from yellowish green to dark green. A blank determination is also performed and necessary correction made.

Each ml of 0.1 N perchloric acid is equivalent to 9.814mg of CH_3COOK

Uses: It is used in solutions for haemodialysis and peritoneal dialysis as an alkalizer. It is also used as a food preservative.

4. Sodium citrate:

Chemical Formula: $C_6H_5O_7Na_3$ *Mol. Wt.* 258.06

I.P. Limit: Sodium citrate is trisodium 2-hydroxypropane- 1,2,3,-tricarboxylate dihydrate. It contains about 99% of $C_6H_5Na_3O_7$.

Preparation: It is prepared by mixing of calculated amounts of hot solution of citric acid and sodium carbonate and crystallizing the product.

$$3Na_2CO_3 + 2\,H_3C_6\,H_5O_7 \rightarrow 2Na_3C_6\,H_5O_7 + 3CO_2 + 3H_2O \quad .....(8.16)$$

Characters: it occurs as white, granular crystals or a white crystalline powder; slightly deliquescent in moist air. It is freely soluble in water; practically insoluble in ethanol. It is stored in air-tight containers. Sterilized solutions of sodium citrate on keeping cause separation of small solid particles from a glass container. A solution containing such particles must not be used.

Tests for purity: Tests for heavy metals; oxalate; sulphate; readily carbonizable substances; water; acidity or alkalinity; clarity and colour of solution.

- Acidity or alkalinity is determined by neutralizing an aqueous solution with hydrochloric acid or sodium hydroxide using phenolphthalein as indicator.
- For determining oxalate, zinc and phenyl hydrazine are treated with acidified solution. Hydrochloric acid and potassium hexa cyano ferrate solution are added. Any pink color produced is not more intense than that obtained by treating at the same time and in same manner a reference solution of oxalic acid.
- Readily carbonisable substances are detected by heating the salt with sulphuric acid at about 90° for one hour. On cooling, the solution is not more intensely colored than a reference solution.
- Sodium citrate complies with the limit test for chloride and sulphate

Test for identification: Aqueous solution gives reactions characteristic of sodium salts and citrates.

For Sodium: To sample solution add 15 % w/v potassium carbonate, heat, no precipitate, add potassium antimonite, solution heat to boiling, cool and if necessary scratch the inside of test tube with a glass rod, a dense white precipitate is produced.

For Citrate: To a neutralized solution of sample add calcium chloride solution no precipitate is produced, boil the solution white precipitate, soluble in 6M acetic acid is produced.

Assay: A solution of the accurately weighed substance in anhydrous acetic acid is heated and after cooling to room temperature titrated with 0. 1N perchloric acid using 1-naphtholbenzein solution as indicator, until a green color is produced. A blank determination is also performed.

Each ml of 0.1 N perchloric acid is equivalent to 8.602mg of $C_6H_5Na_3O_7$

Uses: It is used as systemic alkalizing substance. Sodium citrate has anticlotting properties and is employed in mixtures as the acid citrate in the anticoagulation and preservation of blood for transfusion purposes. It is also used for dentifrices as desensitizing agent and added to milk for infant feeding to prevent the formation in the stomach of large curds. It also has a diuretic effect due to increased body salt concentration.

5. Potassium Citrate:

Chemical Formula: $C_6H_5O_7K_3$. H_2O *Mol. Wt.* 324.42

I.P. Limit: Potassium citrate is the monohydrate of tripotassium 2-hydroxy-propane-1,2,3,- tricarboxylate. It contains about 99% of $K_3C_6H_5O_7$.

Preparation: It is prepared by mixing of calculated amounts of hot solution of citric acid and potassium carbonate and crystallizing the product.

$$3K_2CO_3 + 2H_3C_6H_5O_7 \rightarrow 2\,K_3C_6H_5O_7 + 3\,H_2O \quad .....(8.17)$$

Potassium citrate occurs as transparent, odorless, hygroscopic crystals or a white granular powder, taste is saline. It is soluble in water (1 in 1) and glycerol (1 in 25), practically insoluble in alcohol. Aqueous solutions are slightly alkaline and may be incompatible with acidifying agents. It is stored in airtight containers.

Test for purity: Tests for acidity or alkalinity; arsenic; heavy metals; lead, sodium chloride, sulphates, oxalates, readily carbonizable substances and water

- Acidity or alkalinity is determined by neutralizing aqueous solution with either 0.1N sulphuric acid or 0.1N sodium hydroxide to thymol blue solution. A clear solution obtained after addition of potassium antimonite solution to the aqueous solution indicated the absence of sodium.

- For determining oxalate an aqueous mixture of the substance, hydrochloride acid, alcohol and calcium chloride remains clear. Readily carbonisable substance are detected by heating the substance with sulphuric acid for a few min at 80-90°; the solution is not intensely colored, than a mixture of ferric chloride, copper sulphate, cobalt chloride and hydrochloric acid.

Test for identification: A solution (1 in 20) gives the reactions of potassium and of citrate

For potassium: To 1ml of solution add 1ml dilute acetic acid and 1ml of 10 % w/v sodium cobalt nitrite a yellow color produced.

For Citrate: To a neutralized solution of sample add calcium chloride solution, no precipitate is produced, boil the solution, white precipitate soluble in 6M acetic acid is produced.

Assay: A weighed amount (0.15gm) dissolved in glacial acetic acid is heated to 50° and cooled. To this is added 1-naphthaolbenzein solution and titrated with 0.1N perchloric acid until a green color is obtained. A blank determination is performed.

Each ml of 0.1 N perchloric acid is equivalent to 0.01021g of $K_3C_6H_5O_7$.

Uses: it is used as systemic alkalizer and gastric antacid. It is used to relieve painful irritation caused by cystitis (inflammation of gall bladder)

6. Ammonium Chloride

Ammonium chloride is a sterile solution of ammonium chloride in water for injection

I.P. Limit: It contains not less than 99.5 % and not more than 105 % with reference to dried substance. HCl may be added to adjust pH. The NH_4^+ cation possess certain pharmacological activities

1. Acid base equilibrium of the body
2. Diuretic effect
3. Expectorant effect

1. Ammonium ion plays important role in maintenance of the acid base equilibrium of the body particularly in combating acidosis. By excreting ammonium ions the kidney saves base i.e. sodium for the body
2. The diuretic effect of ammonium chloride is produced by conversion of ammonium cation to urea. Ammonium chloride is contraindicated in patients with impaired renal and hepatic functions.

Preparation: It is prepared by neutralizing hydrochloric acid with ammonia and evaporating the solution to dryness, followed by crystallization.

$$NH_3 + HCl \rightarrow NH_4Cl \qquad \text{.....(8.18)}$$

Test for Identification

1. Heat a few mg of sample being examined with sodium hydroxide solution. Ammonia is evolved which is recognized by its order and by its action on moist litmus paper which turns blue
2. Substance in water, is added dilute ammonia solution and silver nitrate, shake the solution and allow to stand, on standing white precipitate is obtained which is insoluble in nitric acid but soluble after being washed with water, in dilute ammonium hydroxide from which it is reprecipitated by the addition of dilute nitric acid.

Test for purity: Test for chloride content, pyrogen is performed and pH between 4-6 is checked.

Assay: The distilled solution, obtained on addition of ammonium chloride and sodium hydroxide solution and heating, is titrated with 0.lN sulphuric acid using methyl red as indicator. A blank determination is also performed.

Each ml of 0.1 N sulphuric acid is equivalent to 0.00549 g of NH_4Cl

OR

Weigh accurately 1gm substance and add water and 5 ml formaldehyde solution previously neutralized to dilute phenolphthalein solution, add 20 ml of water again. After 2 minutes, titrate slowly with 0.1N NaOH using phenolphthalein as indicator.

Each ml of 0.1 N NaOH is equivalent to 0.00549 g of NH_4Cl.

Uses: It is used in acid base therapy, as a diuretic. It is also used to correct hypochloremic alkalosis due to prolonged use of mercurial diuretics.

8.11 Electrolyte Combination Therapy

Combinations of glucose and saline solutions are usually sufficient in short term therapy for restoring electrolyte loss. But in severe deficit of electrolytes due to heavy blood loss or chronic diarrhea, solutions containing additional electrolytes are usually required. The combination products are of two types

1. Fluid maintenance therapy
2. Electrolyte replacement therapy

Maintenance therapy with intravenous fluids is required to supply normal necessity of water and electrolyte to patient who cannot take them orally. All maintenance therapies should contain at least 5% dextrose. General electrolyte composition of maintenance therapy includes:

Table 8.2

Electrolyte	Concentrations (meq/l⁻)
Sodium	25-30
Potassium	15-20
Chloride	22
Bicarbonate	20-23
Magnesium	3
Phosphorous	3

Replacement therapy is required when there is excess loss of water and electrolytes caused by fever, severe vomiting and diarrhea.

Two types of solutions are used in replacement therapy:

(i) Solution for rapid initial replacement

(ii) Solution for subsequent replacement

The electrolyte concentrations in solutions for rapid initial replacement are almost similar to the electrolyte concentrations found in extracellular fluids. The electrolyte concentrations of these solutions are given as

Table 8.3

Electrolyte	Concentrations (meql^{-1}) for rapid initial replacement	Concentrations (meql^{-1}) for subsequent replacement
Sodium	130-150	40-121
Potassium	4-12	16-35
Chloride	98-109	30-103
Bicarbonate	28-55	16-53
Calcium	3-5	0-5
Magnesium	3-5	0-13
Phosphorous	-	0-13

CHAPTER 9

Gastrointestinal Agents

9.1 Introduction

Agents used to treat gastrointestinal disturbance are known as gastrointestinal agents. Various inorganic agents used to treat GIT disorders include:

- Products for altering gastric pH i.e. acidifying agents and antacids
- Protective's and adsorbents
- Saline cathartics or laxatives

9.1.1 Acidifying Agents

The pH of stomach is 1.5-2 when empty and rises to pH 5-6 when food is ingested. The pH of stomach is so low because of the secretion of HCl. Gastric HCl act by destroying the bacteria in the ingested food and drinks. It softens the fibrous food and promotes the formation of the proteolytic enzyme pepsin. This enzyme is formed from pepsinogen at acidic pH (>6). Pepsin helps in the metabolism of proteins in the ingested food. Therefore lack of HCl in the stomach can cause achlorhydria. Two types of achlorhydria are known:

- where the gastric secretion is devoid of HCl, even after stimulation with histamine phosphate
- Where gastric secretion is devoid of HCl, but secreted upon stimulation with histamine phosphate.

The cause of achlorhydria in first case may be subtotal gastrectomy, atrophic gastritis and carcinoma and etc., while in later case it may be chronic nephritis, tuberculosis, hyperthyroidism, chronic alcoholism, pellagra etc. The symptoms vary with associated disease but they generally include mild diarrhoea or frequent bowl movement, epigastric pain and sensitivity to spicy food.

Achlorhydria can be treated by various acidifying agents like ammonium chloride, dilute HCl, calcium chloride etc.

A. Dilute Hydrochloric Acid

(Refer to page 176)

9.1.2 Antacids

Antacids are the substances which reduce gastric acidity resulting in an increase in the pH of stomach and duodenum. Gastric acidity occurs due to excessive secretion of HCl in stomach due to various reasons.

The pH of the stomach is 1.5-2.5 when empty and raises to 5-6 when food is ingested. Low pH is due to the presence of endogenous HCl, which is always present under physiological conditions. When hyperacidity occurs the result can range from:

- gastritis (a general inflammation of gastric mucosa)
- peptic ulcer or esophageal ulcer (lower end of esophagus)
- gastric ulcer (stomach)
- duodenum ulcers

Peptic ulcers occur due to defective oesophageal sphinter as in hiatal hernia. Gastric ulcers occur in lesser curvature and are found in first portion of duodenum.

Symptoms include uncomfortable feeling from over eating, heart burn, and growing hungry between meals. Complications involved are hemorrhage (being more common with duodenal ulcers), perforation. Depending upon the severity and location of an ulcer treatment will range from diet and antacids and /or anticholinergic therapy to complete bed rest to surgery. Small meals after short interval help in reducing acidity, stimulants of gastric acid must be avoided like coffee, alcohol, spicy food, oil or fried food.

Antacid Therapy

Antacids are alkaline bases used to neutralize the excess gastric HCl associated with gastritis or peptic ulcer. Since gastric HCl secretion is continuous, so is the administration of antacids.

Role of Antacids

- Primarily in pain relief
- Higher doses given continuously can promote ulcer healing
- Superior to H_2 blockers in bleeding peptic ulcers

Criteria for Antacids

- The antacid should not be absorber/or cause systemic alkalosis
- It should not be constipative or laxative
- It should exert effect rapidly and over a long period of time

- The antacid should buffer in the range of pH 4-6
- Reaction of antacid with HCl should not cause large evolution of gas.

Side Effects of Long Term Antacid Therapy

- If pH raises too high rebound acidity to neutralize the alkali occurs.
- Antacids which absorbed systemically exert alkaline effect on body's buffer system.
- Some antacids cause constipation while others have laxative effect.
- Sodium containing antacids are problem for patients on sodium restricted diet.

Systemic Antacids

Systemic antacids are antacids which get systemically absorbed e.g. sodium carbonate is water soluble and potent neutralizer, but it is not suitable for the treatment of peptic ulcer because of risk of ulcer perforation due to production of carbon dioxide in the stomach.

Systemic absorption leads to alkalosis, may worsen edema and congestive heart failure because of sodium ion load.

Non Systemic Antacids

They are insoluble and poorly absorbed systemically. In Magnesium salt, Magnesium carbonate is most water soluble and reacts with HCl at a slow rate, while Magnesium hydroxide has low solubility and has the power to absorb and inactivate pepsin and to protect the ulcer base. Aluminium hydroxide is a weak and slow reacting antacid. The aluminium ions relax smooth muscles and cause constipation. It absorbs pepsin at pH >3 and releases it at lower pH. It also prevents phosphate absorption. Calcium carbonate is a potent antacid with rapid acid neutralizing capacity, but on long term use, it can cause hypercalcemia, hypercalciuria and formation of calcium stone in kidney.

Every single compound among antacid have some side effect especially when used for longer period or used in elderly patients. To avoid certain side effects associated with antacids, **combinations of antacids** are used such as:

(i) Magnesium and aluminium containing preparation e.g. magnesium hydroxide a fast acting antacid with aluminium hydroxide which is a slow acting antacid.

(ii) Magnesium and calcium containing preparation where one is laxative and the later one is constipative in nature.

Compounds used as Antacids

(i) Sodium Bicarbonate (Baking soda)

Chemical Formula: $NaHCO_3$ *Mol. Wt.* 84.01

I.P. limit: It contains not less than 99% and not more than 101% of $NaHCO_3$

Properties: White crystalline powder, odorless, with saline and slight alkaline taste, Stable in dry air, sparingly soluble in water, insoluble in alcohol.

Preparation

(a) By passing strong brine containing high concentrations of ammonia through a carbonating tower where it is saturated with carbon dioxide under pressure. The ammonia and carbon dioxide reacts to form ammonia bicarbonate which is allowed to react with NaCl to precipitate $NaHCO_3$ which is separated by filtration.

$$NH_3 + H_2O + CO_2 \rightarrow NH_4HCO_3 \quad .....(9.1)$$

$$NH_4HCO_3 + NaCl \rightarrow NaHCO_3 \quad .....(9.2)$$

(b) It can also be prepared by covering sodium carbonate crystals with water and passing carbon dioxide to saturation.

$$Na_2CO_3 + H_2O + CO_2 \rightarrow NaHCO_3 \quad .....(9.3)$$

Test for identification: To 5ml of 5%w/v solution in carbon dioxide free Water add 0.1 ml phenolphthalein solution a pale pink color is obtained. On heating a gas is evolved and the solution turns red.

For Sodium

1. To sample solution add 15% w/v potassium carbonate, heat, no precipitate is obtained add potassium antimonite solution heat to boiling, cool and if necessary scratch the inside of test tube with a glass rod, a dense white precipitate is produced.
2. Acidified the sample solution with 1M acetic acid and add excess of magnesium uranyl acetate solution yellow crystalline precipitate is obtained.

For bicarbonate: to sample add magnesium sulphate no precipitate is produced. On boiling a white colored precipitate is formed.

Assay: Weigh accurately 1gm and dissolve in 20ml of water, titrate the solution with 0.5N sulphuric acid using methyl orange as indicator. Each ml of 0.5N sulphuric acid $\equiv$ 0.0425gm of $NaHCO_3$

Use: It is used as antacid, and in electrolyte replacement.

(ii) Aluminium Hydroxide:

Chemical Formula: Al $(OH)_3$ *Mol. Wt.* 78.0

Aluminium hydroxide gel is an aqueous suspension of hydrated aluminium oxide with different amounts of basic aluminium carbonate and bicarbonate.

I.P. limit: It contains not less than 3.5% and not more than 4.4% of Al_2O_3

Properties: Aluminium hydroxide is a white, light odorless, tasteless amorphous powder. It is soluble in dilute mineral acids and in solution of alkali hydroxides but practically insoluble in water. It forms gel on prolonged contact with water at pH 5.5-8.0. It absorbs acids and carbon dioxide. The aluminium hydroxide gels are ideal buffers in the pH 3-5 range due to its amphoteric nature.

Preparation: It is prepared by dissolving sodium carbonate in hot water and the solution is filtered. To the filtrate add clear solution of alum (aluminium salt, chloride or sulphate) in water with constant stirring. Add more of water and remove all gas. The Aluminium Hydroxide precipitate out, collect the precipitate, wash and suspend in sufficient purified water flavored with 0.01% peppermint oil and preserve with 0.1% sodium benzoate.

$$Al_2\left(SO_4\right)_3 + 3Na_2CO_3 + 3H_2O \rightarrow 2Na_2SO_4 + Al\left(OH\right)_3 + 3CO_2 \quad .....(9.4)$$

Test for identification: A solution in 2N HCl gives the characteristic reactions of aluminium salts. To sample add 5drops of freshly prepared 0.05%w/v solution of quinalizarin in 1% w/v solution of NaOH heated to boiling, cool, acidify with excess of acetic acid a reddish violet color is produced.

Assay: Accurately weigh 5gm and dissolve in 3ml HCl by warming on water bath, cool to below 20 °C and dilute to 100ml with water. To 20ml of this solution add 40ml of 0.05M disodium EDTA, 80ml water, 0.1 5ml methyl orange/red and neutralize by the drop wise addition of 1M sodium hydroxide. Again warm on water bath for 30 min, add 3gm hexamine and titrate with 0.05M lead nitrate using 0.5ml xylenol orange as indicator.

Each ml of 0.05M disodium EDTA $\equiv$ 0.002549 gm of Al_2O_3

Uses: Aluminium hydroxide is used as antacid in the management of peptic ulcer, gastritis, gastric hyperacidity. It is also used as skin protectant and mild astringent.

(iii) Aluminium Phosphate:

Chemical Formula: $AlPO_4$ *Mol. Wt.* 122.0

Aluminium phosphate consists of hydrated aluminium orthophosphate.

I.P. limit: It contains not less than 80.0% of Al PO_4

Properties: It occurs in nature as the minerals angelite, evansite, lucinite, sterretite etc. It is white infusible powder with some aggregates. It is insoluble in solutions of alkali hydroxide or acetic acid, water, alcohol etc; very slightly soluble in concentrated HCl and nitric acid. A 4% suspension in water has a pH of 5.5- 6.5. Aluminium phosphate gel is a white, viscous suspension from which small amount of water may separate on standing. The gel has a pH in the range of 6.0-7.2.

Preparation

1. It may be prepared by treating aluminium sodium oxide ($NaAlO_2$) with phosphoric acid (H_3PO_4)

$$NaAlO_2 + 2H_3PO_4 \rightarrow 2AlPO_4 + 2H_2O + 2NaH \quad .....(9.5)$$

2. It can also be prepared by drying under suitable conditions the product of interaction in aqueous of an aluminium salt with an alkali phosphate such as sodium phosphate.

Test for Identification

For aluminum: A solution in 2N HCl gives the reactions characteristic of aluminium salts. To sample add 5drops of freshly prepared 0.05%w/v solution of quinalizarin in 1% w/v solution of NaOH heated to boiling, cool, acidify with excess of acetic acid a reddish violet color is produced.

For phosphate: To neutral sample solution add silver nitrate solution, a light yellow precipitate forms, the color of which is not changed by boiling and is readily soluble in 10M ammonia and dilute HNO_3.

Assay: To acidify solution of weighed amount of the aluminium phosphate (0.8gm), 7.7gm disodium acetate is added, pH is adjusted to 4.5 with glacial acetic acid and dithizone in ethanol is added and sufficient quantity of ethanol is added. Titrate the solution with 0.05M zinc chloride until color turns red. A blank titration is also performed and the volume consumed by sample is calculated. Each ml of 0.05M disodium EDTA ≡ 6.098 mg of $AlPO_4$

Use: It is used as antacid.

(iv) Dihydroxy Aluminium Ammonium Acetate:

Chemical Formula: $C_2H_6Al\ NO_4\ xH_2O$

I.P limit: It contains not less than 35.5% and not more than 38.5% of aluminum trioxide calculated on dried bases, contains small amounts of aluminum oxide and amino acetic acid.

Properties: It exists as very fine powder, bland taste, insoluble in water but forms suspension with water.

Preparation: It is prepared by adding solution of aluminum isopropoxide in isopropanolol to an aqueous solution of glycine.

Test for Identification

For aluminium: A solution of sample in 2N HCl gives the characteristic reactions of aluminium salts. To the sample, add 5drops of freshly prepared 0.05%w/v solution of quinalizarin in 1% w/v solution of NaOH heated to boiling, cool, acidify with excess of acetic acid, a reddish violet color is produced.

For acetate: Heat the sample with an equal amount of oxalic acid, acidic vapours with characteristic smell of acetic acid are liberated

Assay: Transfer about 2.5gm of accurately weighed dihydroxy aluminium ammonium acetate and add 15rnl of HCl and warm if necessary, to dissolve the sample completely. Transfer the solution with the aid of water to 500ml volumetric flask and dilute with water to volume and mix. Transfer 20ml and add 25ml 0.05M disodium EDTA and 20ml of acetic acid ammonium acetate buffer. Heat near boiling for 5mim, cool and add 50ml of alcohol and 2ml of dithizone. Titrate with 0.05M zinc sulphate until green violet color turns rose pink. Perform blank and calculate the volume consumed by sample. Each ml of 0.05M disodium EDTA $\equiv$ 2.549 mg of $Al_2 O_3$

Use: It is used as antacid.

(v) Dihydroxy Aluminium Sodium Carbonate:

Chemical Formula: $CH_2Al NaO_5$ *Mol. Wt.* 143.99

I.P limit: It contains not less than 90% and not more than 110.0% of $Al_2 O_3$

Preparation: It is prepared by reaction between an alkoxide and sodium bicarbonate in water.

Test for Identification

For aluminum: A solution in 2N HCl gives the reactions characteristic of aluminium salts. To sample add 5drops of freshly prepared 0.05%w/v solution of quinalizarin in 1% w/v solution of NaOH heated to boiling, cool, acidify with excess of acetic acid, a reddish violet color is produced.

For Carbonate: Suspend sample in 2ml of water in a test tube, add 2M acetic acid close the tube immediately with a stopper fitted with a glass tube bent at two right angles, heat gently and collect the gas in 5ml of 0.1 M barium hydroxide a white precipitate is formed which is dissolves on addition of excess of dilute HCl.

For Sodium: To sample solution add 15% w/v potassium carbonate, heat, no precipitate is formed add potassium antimonite solution heat to boiling, cool and if necessary scratch the inside of test tube with a glass rod, a dense white precipitate is produced.

Assay: Weigh accurately powder not less than 200mg add 10ml of 2N HNO_3 cover and boil for 1 min, add 25ml of 0.1M disodium EDTA again boil for 1min cool, then add 10ml acetic acid ammonium acetate buffer and 50ml of acetone and 2ml of dithizone, adjust pH to 4.5 by the addition of ammonium hydroxide and titrate with 0.05M zinc sulphate maintaining pH at 4.5. Perform blank and calculate the volume consumed by sample.

Each ml of 0.05M disodium EDTA $\equiv$ 5.098 mg of Al_2O_3

Use: It is used as antacid.

(vi) Calcium Carbonate:

Chemical Formula: $CaCO_3$ *Mol. Wt.* 100

Synonym: precipitated chalk.

Calcium carbonate is found in nature as limestone, marble, calcite, vaterite, aragonite and shell of sea animals.

I.P. limit: It contains not less than 98% and not more than 100.5% with reference to dried substance.

Properties: It occurs as a white, odorless tasteless microcrystalline powder which is stable in air. It exists in two crystal form and both are of commercial importance, one Aragonite and other is Calcite.

Precipitated chalk is prepared as a fine precipitate by adding a solution of ammonium carbonate and ammonia or sodium carbonate to a solution of calcium nitrate.

Preparation

1. It can be prepared by mixing and boiling calcium and sodium carbonate solution and allowing the resulting precipitate to settle. The precipitate is collected, washed with boiling water until free from chloride and dried.

$$CaCl_2 + Na_2CO_3 \rightarrow CaCO_3 + 2Na\,Cl \qquad \text{.....(9.6)}$$

2. By passing carbon dioxide through lime water

$$CaO + CO_2 + H_2O \rightarrow CaCO_3 \qquad \text{.....(9.7)}$$

Test for Identification

For Calcium: Dissolve substance in 5M acetic acid and add 0.5ml of potassium ferrocyanide solution. The solution remains clear. Add ammonium chloride white crystalline precipitate is formed,

For Carbonate: Suspend sample in 2ml water in a test tube, add 2M acetic acid close the tube immediately with a stopper fitted with a glass tube bent at two right angles, heat gently and collect the gas in 5ml of 0.1 M barium hydroxide a white precipitate is formed which is dissolves on addition of excess of dilute HCl.

Assay: Accurately weigh 0.1gm, dissolve in 3ml of dilute HCl, and add 10 ml of water. Boil the solution for 10mim, cool, dilute with 50 ml with water. Titrate the solution with 0.05M disodium EDTA to with a few ml of the expected end point and add 8ml of NaOH solution and 0.1 g of calcon mixture. Continue the titration until the color changes from pink to blue. Each ml of 0.05M disodium EDTA $\equiv$ 0.005004g of $CaCO_3$

Uses: It is used as fast acting antacid, in calcium deficiency, denitrifies and in combination with magnesium containing antacids due to its constipative properties.

(vii) Tribasic Calcium Phosphate:

Chemical Formula: $CaO_3P_2O_5.\ H_2O$ *Mol. Wt.* 328.2

I.P limit: Tribasic calcium phosphate consists of variable mixture of calcium phosphate having approximate composition of $CaO_3P_2O_5.\ H_2O$ not less than 34% and not more than 40% of calcium and an amount of phosphate equivalent to not less than calcium phosphate calculated with reference to ignited substance.

Properties: It is a white odourless, tasteless, amorphous powder practically insoluble in water, alcohol or acetic acid but readily soluble in dilute HCl and HNO_3

Preparation

1. It is manufactured from bones which are calcined until white, powdered and digested with sulphuric acid. The insoluble tribasic calcium phosphate is converted to soluble phosphoric acid and insoluble calcium sulphate. The solution is filtered and the filtrate treated with calcium hydroxide to precipitate calcium phosphate.
2. Decomposition of calcium chloride and sodium phosphate in presence of aqueous ammonia at high temperature yield calcium phosphate. The white precipitate is filtered washed and freed from chloride and dried.

Test for Identification

For Calcium: Dissolve substance in 5M acetic acid and add 0.5ml of potassium Ferro cyanide solution. The solution remains clear. Add ammonium chloride white crystalline precipitate is formed.

For phosphate: To neutral sample solution add silver nitrate solution, a light yellow precipitate forms, the color of which is not changed by boiling and is readily soluble in 10M ammonia and dilute HNO_3.

Assay:

For calcium: Weigh accurately 0.2gm and dissolve in HCl, triethanol amine and hydroxyl naphthol blue indicator are added and titrate with 0.05M disodium EDTA until blue colour is obtained. Each ml of 0.05M disodium EDTA ≡ 0.002004 g of calcium.

For phosphate: Acidify the aqueous solution of substance 0.2gm with dilute nitric acid, filter and add strong ammonium solution to produce slight precipitate. The precipitate is dissolved in dilute nitric acid, ammonium molybdate is added and precipitate is filtered, washed with potassium nitrate solution and redissolved in 1N sodium hydroxide add phenolphthalein and titrate the excess alkali with 1N sulphuric acid. Each ml of 1N NaOH ≡ 0.006743 gm of calcium phosphate.

Use: It is used as antacid, as non hygroscopic diluent, as an abrasive in tooth pastes.

(viii) Magnesium Carbonate:

Chemical Formula: $MgCO_3$ *Mol. Wt.* 508

Magnesium carbonate is a hydrated basic magnesium carbonate containing 40-45% of magnesium oxide. It occurs in nature as the meniral magnate and lansfordite.

Heavy Magnesium Carbonate: 15 g occupy a volume of about 30ml

Light Magnesium Carbonate: 15 g occupy a volume of about 150ml

I.P. limit: It contains not less than 40% and not more than 45% of magnesium oxide

Properties: Both heavy and light magnesium carbonates are hydrated. Both are white, odorless powder practically insoluble in water and alcohol but solubilizes in dilute acids with strong effervescence.

Preparation: It is prepared by mixing hot solution of magnesium sulphate and sodium carbonate. The mixture is evaporated to dryness and the residue consisting of magnesium carbonate and sodium sulphate is digested for half an hour with boiling water. The precipitate of magnesium carbonate is collected on filter paper, washed with water until free from sulphate and then dry.

$$5MgSO_4\,7H_2O + 5Na_2CO_3\,10H_2O \rightarrow (MgCO_3)_4\,Mg\,(OH)_2\,5H_2O + 5Na_2SO_4 + 5O_2 + 79\,H_2P \quad .....(9.8)$$

Test for Identification

For Carbonate: Suspend sample in 2m*l* water in a test tube, add 2M acetic acid close the tube immediately with a stopper fitted with a glass tube bent at two right angles, heat gently and collect the gas in 5ml of 0.1 M barium hydroxide a white precipitate is formed which is dissolves on addition of excess of dilute HCl.

For Magnesium: to solution of sample add dilute nitric acid solution a white precipitate is produced that is redissolved by adding 1ml of 2M ammonium chloride, add 0.25M disodium hydrogen phosphate a white crystalline precipitate is produced.

Assay: Accurately weigh 15g of magnesium carbonate and dissolve in a mixture of 20ml of water and 2ml 2M HCl. To this solution add 50ml of water and 10ml strong ammonia ammonium chloride solution titrate this with 0.05M disodium EDTA using mordant black II mixture as indicator until blue color is obtained. Each ml of 0.05M disodium EDTA ≡ 0.002015g of MgO

Uses: It is used as antacid and mild laxative. It is used as pharmaceutical aid (dispensing volatile oil for use in inhalants).

(ix) Magnesium Oxide:

Chemical Formula: MgO *Mol. Wt.* 40.3

Synonym: Magnesia.

I.P. limit: It contains not more than 98% of magnesium oxide

It occurs in nature as mineral periclase. It occurs in two varieties heavy magnesium oxide which is relatively dense white powder with 15 g occupying volume of about 30 ml while light magnesium oxide is very bulky with 15 g occupying volume of about 150 ml.

Properties: Both heavy and light magnesium oxides are odorless taste slightly alkaline, practically insoluble in water yield a solution which is alkaline. It readily dissolves in dilute acids with slight effervescence. In presence of acid, the oxide forms the magnesium hydroxide; therefore the chemistry and pharmacology are same as those of magnesium hydroxide.

Preparation

1. It can be prepared by heating gently magnesium carbonate to redness.

$$(MgCO_3)_4 Mg(OH)_2\ 5H_2O \rightarrow MgO + CO_2 + H_2O \quad .....(9.9)$$

2. It is also prepared by heating light magnesium carbonate to redness.

$$(MgCO_3)_4\ Mg(OH)_2\ 3H_2O \rightarrow 4MgO + 3CO_2 + 3H_2O \quad .....(9.10)$$

Test for identification

For Magnesium: to solution of sample add dilute nitric acid solution a white precipitate is produced that is redissolved by adding 1ml of 2M ammonium chloride, add 0.25M disodium hydrogen phosphate a white crystalline precipitate is produced.

Assay: The assay of magnesium oxide is performed by complexometry. Accurately weigh 15g of magnesium oxide and dissolve in a mixture of 20ml of water and 2ml 2M HCl. To this solution add 50ml of water and 10ml strong ammonia ammonium chloride solution titrate this with 0.05M disodium EDTA using mordant black II mixture as indicator until blue color is obtained. Each ml of 0.05M disodium EDTA $\equiv 0.002015 g\ MgO$

Uses: It is used as antacid and laxative. It is ingredient of universal antidote along with tannic acid and charcoal. It is used for compounding and preserving fluid extract because of its absorptive power.

(x) Magnesium Phosphate:

Chemical Formula: $Mg_3(PO_4)_2.\ 5H_2O$ *Mol. Wt.* 352.93

I.P. limit: It contains not less than 98% and not more than 101.5% substance ignited at 425°C to constant weight.

Properties: It is white, odourless, tasteless powder readily soluble in dilute mineral acids but practically insoluble in water. The chemistry is same as tribasic calcium phosphate.

Test for Identification

For Magnesium: to solution of sample add dilute nitric acid solution a white precipitate is produced that is redissolved by adding 1 ml of 2M ammonium chloride, add 0.25M disodium hydrogen phosphate a white crystalline precipitate is produced.

For phosphate: To neutral sample solution add silver nitrate solution, a light yellow precipitate forms, the color of which is not changed by boiling and is readily soluble in 10M ammonia and dilute HNO_3.

Assay: Weigh accurately 200mg previously ignited to 425°C to constant weight and dissolve in mixture of 25ml of water and 10ml of 2N HNO_3. Filter if necessary, wash any precipitate add sufficient 6Nammonium hydroxide to the filtrate to produce a slight precipitate and then add 1ml of 2N HNO_3. Adjust temperature to 50° add 7ml ammonium molybdate and maintain temperature for 30min with occasional stirring. Wash the precipitate once or twice with water by decantation. Transfer precipitate to the filtrate, wash with KNO_2 solution until the last washing is not acid to litmus. Transfer precipitate and filter to precipitate vessel add 50ml of water and 40ml of 1N NaOH, agitate until the precipitate is dissolved, add phenolphthalein and titrate excess of alkali with 1N H_2SO_4.

Each ml of 1N NaOH $\equiv$ 5.716mg of magnesium phosphate.

Uses: It is used as antacid and laxative.

(xi) Magnesium Trisilicate:

Chemical Formula: $2MgO3SiO_2 . xH_2O$ *Mol. Wt.* 260.86(Anhydrous)

I.P. limit: It contains not more than 29% of magnesium oxide and not more than 65% of silicon dioxide both calculated with reference to the ignited substance.

Properties: Fine, white, odorless, tasteless powder, free from grittiness. Magnesium silicate hydrate is a compound of magnesium oxide and silicon dioxide with varying amount of water.

Preparation: It is prepared by precipitation from solution of magnesium sulphate and sodium silicate.

$$2MgSO_4 + 2Na_2O3SiO_2 + XH_2O \rightarrow 2MgO\ 3SiO_2XH_2O + Na_2SO_4 + H_2O \quad .....(9.11)$$

Test for identification

For Magnesium: to solution of sample add dilute nitric acid solution a white precipitate is produced that is redissolved by adding 1ml of 2M ammonium chloride, add 0.25M disodium hydrogen phosphate a white crystalline precipitate is produced.

For silicate: in a lead or platinum crucible mix by means of a copper wire to obtain a thin slurry the prescribed amount (0.25gm) of substance with 10mg of sodium fluoride and a few drops of sulphuric acid cover the crucible with thin transparent plate of plastic under which a drop of water is suspended and warm gently within a short time a white ring is formed around the drop of water

Assay

For Magnesium Oxide: Weigh accurately 1gm substance and dissolve in 35ml of water allow to stand for 15min on water bath. Cool the contents to room temperature filter and wash the residue with water and dilute the combined filtrate and washing to 250ml with water. Neutralize 50ml of this solution with about 8ml of 10M NaOH and then add 10ml ammonia buffer pH 10, 50mg mordant black II mixture. Heat the contents to 40 °C and titrate with 0.05M disodium EDTA until color changes to deep blue. Each ml of 0.05M disodium EDTA $\equiv$ 0.002015g of MgO

For silicon dioxide: weigh accurately 0.7g of substance add 10 ml of 1M sulphuric acid, 10ml water and heat on water bath for 1.5hrs. Shaking frequently and replacing the evaporated water. Allow to cool decant on to an ashless filter paper

(7cm in diameter). Wash the precipitate by decantation with three quantities each of 5ml of hot water, transfer it to the filter paper and wash it with hot water, until 1ml of the filtrate remains clear on addition of 2ml of barium chloride solution and 0.5ml of 2N HCl. Ignite the filter paper and its contents in a tarred platinum crucible at 900 °C to constant weight. The residue is silicon dioxide.

Uses: It is used as non systemic antacid and adsorbent.

9.1.3 Protectives and Adsorbents

Adsorption is the adhesion of atoms, ions, biomolecules or molecules of gas, liquid, or dissolved solids to a surface. This process creates a film of the adsorbate (the molecules or atoms being accumulated) on the surface of the adsorbent. It differs from absorption, in which a fluid permeates or is dissolved by a liquid or solid. The term sorption encompasses both processes, while desorption is the reverse of adsorption.

Compounds used as Protectives and Adsorbents

(i) Bismuth Subcarbonate:

Chemical Formula: $(BiO)_2CO_3.\ 1/2H_2O$ *Mol. Wt.* 518.98

I.P. limit: It contains not less than 80% and not more than 82.5% of bismuth calculated with reference to dried substance.

Preparation: To a solution of bismuth nitrate an excess of 20% sodium carbonate solution is added with constant stirring and the solution is allowed to stand for sometime. Filter the solution and wash the residue till washings are neutral. The solid is collected and dried at about 50°C

$$2Bi + 8HNO_3 \rightarrow 2Bi\ (NO_3)_2 + 2NO_2 + 4H_2O \quad .....(9.12)$$

$$2Bi\left(NO_3\right)_2 \xrightarrow{Na_2CO_3} \left(BiO\right)_2 CO_3 \quad .....(9.13)$$

Test for identification

For Bismuth: To 0.5g sample add 10ml 2N HCl and heat to boiling for 1min, cool and filter, if necessary. To 1ml add 20ml of water a white or slightly yellow precipitate is formed which on addition of 0.05-0.1ml of sodium sulphate solution turns brown

For Carbonate: Suspend sample in 2ml water in a test tube, add 2M acetic acid close the tube immediately with a stopper fitted with a glass tube bent at two right angles, heat gently and collect the gas in 5ml of 0.1 M barium hydroxide a white precipitate is formed, which is dissolves on addition of excess of dilute HCl.

Assay: Weigh accurately 0.5 g of sample and dissolve in 3ml HNO_3 and dilute with 250ml of H_2O. Add strong ammonia solution until cloudiness is first observed; add

0.5ml of HNO_3 and heat to 70 °C maintain the solution at this temperature till it becomes clear. Add about 50mg of xylenol orange mixture and titrate with 0.1M disodium EDTA until color changes from pinkish violet to lemon yellow. Each ml of 0.lM disodium EDTA ≡ 0.02090gm of bismuth.

Uses: Topically as protective in lotions and ointments, Internally as astringent and absorbent. It is also used as adsorbent in enteritis, diarrhoea, dysentery, ulcerative colitis, in wound dressing.

(ii) Bismuth Subgallate:

Chemical Formula: $C_7H_5BiO_6$ *Mol. Wt.* 394.01

I.P limit: It contains not less than 52% and not more than 57% of bismuth trioxide calculated with reference to the substance dried to constant weight at 105°C.

Bismuth subgallate is a basic salt of bismuth. It is amorphous, bright yellow powder, odorless, tasteless. Practically insoluble in water, ether, ethanol but readily dissolves in hot mineral acids with decomposition and in solution of alkali hydroxides.

Test for Identification: To 0.5g sample add l0ml 2N HCl and heat to boiling for 1min cool and filter, if necessary. To 1ml add 20ml of water a white or slightly yellow precipitate is formed which on addition of 0.05.0.1ml of sodium sulphate solution turns brown

Preparation: It is prepared from bismuth nitrate and gallic acid in acetic acid medium. The acetic acid can be replaced by mannitol or glycol.

Assay: Dry about 1gm at 105°C to constant weight, add nitric acid drop wise with stirring and warm until solution is complete. Evaporate the solution to dryness and carefully ignite the residue to constant weight. The residue represents the quantity if Bi_2O_3 in the weight of substance taken.

Use: It is used as astringent, antacid and protective.

(iii) Kaolin:

Chemical Formula: Al_2O_3 $2SiO_2$ $2H_2O$

Heavy kaolin is purified natural hydrated aluminium silicate of variable composition.

Light kaolin is native hydrated aluminum silicate freed from most of its impurities by elutriation and dried. It may contain a suitable dispersing agent.

Preparation: Kaolin is widely distributed in nature contaminated with ferric oxides. It is prepared when the rock is mined, evacuated and the impurities are washed with water and then powdered. The rock is elutriated with water and large sized particles are separated. The turbid liquid is allowed to settle; heavy kaolin containing large particles and colloidal kaolin containing particles of small size are separated and dried. For pharmaceutical use it is purified by treatment with HCl and H_2SO_4 or both and then washed with water.

Test for Identification: Fuse 2g of substance with 4gm anhydrous sodium carbonate. Warm residue with water and filter, acidify the filtrate with HCl evaporate to dryness and warm the residue with dilute HCl, residue of silica is obtained and the acid solution after neutralization gives reaction for aluminum.

For aluminium: To 0.5 g in a metal crucible add 1gm HNO_3 and 3g anhydrous sodium carbonate, heat to melt and allow cool, adding 20ml of boiling water to this residue and filtering. To filtrate add 1ml of 10M NaOH and filter. To filtrate add 3ml of ammonium chloride solution a gelatinous white precipitate is obtained.

For silicate: Fuse 1g of substance with 2g anhydrous sodium carbonate and warm the residue with l0ml of water, filter, wash with water and reserve the residue. To combined filtrate and washings add 3ml of HCl a gelatinous precipitate is obtained.

Use: It is used as adsorbent in diarrhoea caused by agents capable of being absorbed e.g., due to food poisoning. Also used in chronic ulcerative colitis. As poultice, dusting powder, clarifying and decolorizing medium, as filtering medium, as tablet diluent.

(iv) Activated Charcoal:

Absorbing power should not be less than 40% of weight of phenazone calculated with reference to the dried substance.

Preparation: Decolorizing charcoal is obtained from magnesium matter by suitable carbonization process intended to confer a high absorbing power.

Test for identification: When heated to redness burns slowly without flame.

Assay: To 0.3g of substance add 25ml of freshly prepared 1% w/v solution of phenazone shake thoroughly for 15min. filter the solution and discard the first 5ml of the filtrate. To l0ml of the filtrate add 1gm potassium bromide, 20ml of 2M HCl and titrate with 0.0167M potassium borate using 0.1ml of ethoxydrysoidine HCl solution as indicator. The endpoint is change of color from reddish pink to yellow pink. Repeat the titration with phenazone only using l0ml of phenazone. Calculate the % of phenazone absorbed using expression 2.353 (a-b) /w

Where **w** is weight in gms of sample, **a** is volume consumed during first titration, **b** is volume consumed by phenazone only.

9.1.4 Saline Cathartics

Saline cathartics or purgatives are agents that quicken and increase evacuation from the bowl. Laxatives are mild cathartics. Cathartics are used:

- To ease defecation in patients with painful hemorrhoids or other rectal disorders and to avoid excessive straining and concurrent increase in abdominal pressure in patients with hernias (or)
- To avoid potentially hazardous rise in B.P. during defecation in patients with hypertension, cerebral coronary or other arterial disease (or)
- To relieve acute constipation (or)
- To remove solid material from intestinal tract prior to certain roentgenographic studies.

Laxative should only be used for short term therapy as prolonged use may lead to loss of spontaneous bowl rhythm upon which normal evacuation depends, causing patient to become dependent on laxatives, the so called laxative effect.

Constipation is the infrequent or difficult evacuation of the feces. It may be due to a person resisting the natural urge to defecate, causing the fecal material which remains in the colon to lose fluid and to become relatively dry and hard. Constipation can also be due to intestinal atony, intestinal spasm, emotions, drugs and diet. Many a time constipation can be helped by eating food such as natural laxatives or food with large roughages. Four types of laxatives are known:

1. Stimulants
2. Bulk forming
3. Emollient
4. Saline cathartics

Stimulants act by local irritation on the intestinal tract which increases peristaltic activity. They include phenolphthalein, cascara extract, rhubarb extract, senna extract, podophyllin, castor oil, bisacodyl, calomel etc.

Bulk forming laxatives are made from cellulose, sodium carboxyl methyl cellulose and karaya gum.

The **emollient laxatives** act either as lubricants facilitating the passage of compacted fecal material or as stool softeners. E.g., mineral oil, d-octyl sodium sulfosuccinate, an anionic surface active agent.

Saline cathartics act by increasing the osmotic load of the GIT. They are salts of poorly absorbable anions – $H_2PO_4^-$ (biphosphate), – HPO_4^{2-} (phosphate), sulphates, tartarates, and soluble magnesium salt.

Saline cathartics are water soluble and are taken with large quantities of water. This prevents excessive loss of water from body fluids and reduces nausea vomiting if a too hypertonic solution should reach the stomach. They act in the intestine and a full cathartic dose produces a water evacuation within 3-6 hrs. Because of their quick onset of action they are given early in the morning before breakfast.

They are used for bowel evacuation before radiological, endoscopic and surgical procedures and also to expel parasite and toxic materials.

Small amounts of these drugs may be absorbed in the blood causing occasional toxicity. The absorption of magnesium may cause marked CNS depression while that of sodium worsens the existing congestive cardiac failure (CCF).

Compounds used as Saline cathartics

(i) Sodium Acid Phosphate:

Chemical Formula: $NaH_2PO_4.2H_2O$ *Mol. Wt.* 156.01

Synonym: sodium biphosphate.

I.P limit: It contains not less than 98.0% and not more than 100.5% of NaH_2PO_4 calculated with reference to the dried substance.

Properties: Colorless, odorless, crystalline powder with saline acidic taste. Freely soluble in water and practically in soluble in alcohol. Slightly deliquescent.

Preparation:

1. It is prepared by adding phosphoric acid to hot concentrated solution of disodium phosphate until liquid ceases to give precipitate with barium chloride. The solution is then concentrated to the crystallization point.

$$NaHPO_4 + H_3PO_4 \rightarrow 2NaH_2PO_4 \quad .....(9.14)$$

2. By reaction with phosphoric acid with calculated quantity of sodium hydroxide.

$$H_3PO_4 + NaOH \rightarrow 2NaH_2PO_4 \quad .....(9.15)$$

Test for Identification

For phosphate: To neutral sample solution add silver nitrate solution, a light yellow precipitate forms, the color of which is not changed by boiling and is readily soluble in 10M ammonia and dilute HNO_3.

For sodium: To 2ml of solution add 2ml of 15% w/v of K_2CO_3 heat to boil, no precipitate is produced. Add 3ml of potassium antimonite solution and heat to boil. Allow to cool in ice and if necessary scratch the inside of the test tube with glass rod white precipitate is produced.

Assay: Weigh accurately 2.5 g dissolve in 40ml water and titrate with carbonate free 1 M NaOH. Determine end point potentiometrically. Each ml of 0.5M sodium hydroxide $\equiv$ 0.0780g of $NaH_2PO_4.2H_2O$

Use: It is used as saline cathartic and as buffer in pharmaceutical preparations. As urinary acidifier, source of phosphorous.

(ii) Disodium Hydrogen Phosphate:

Chemical Formula: $Na_2HPO_4.12H_2O$ *Mol. Wt.* 358.14

I.P limit: It contains not less than 98.0% and not more than 101% of Na_2HPO_4 calculated with reference to the dried substance.

Properties: Colorless, odorless, crystalline powder. Soluble in water and practically in soluble in alcohol. Very efflorescent.

Preparation

1. It is prepared by reaction of orthophosphoric acid calculated quantity of sodium hydroxide.

$$2NaOH + H_3PO_4 \rightarrow Na_2HPO_4 + 2H_2O \quad \text{.....(9.16)}$$

2. From bone ashes or mineral phosphorite, which is treated with sulphuric acid

Assay: Weigh accurately 4gm of substance and dissolve in 25ml of water add 25ml 1N HCl and titrate potentiometrically with 1M NaOH to first inflection point of the pH curve (n1) continue titration until second inflection of curve is reached. The total volume of NaOH required is n2 ml. calculate percent content from the expression

1420(25-n 1)/w (100-d) d is percentage of water content.

Use: Widely used as saline cathartic. Orally as antihypercalcemic. It is a pharmaceutical aid used as buffering agent.

(iii) Sodium Potassium Tartarate:

Chemical Formula: $COONaCH(OH).CH(OH)COOK.4H_2O$

I.P limit: It contains not less than 99.0% and not more than 102% of C_4H_4KNa calculated on the anhydrous basis.

Preparation: It is prepared by boiling a solution of sodium carbonate and potassium bitartarate for sometime and allowing the reaction mixture to stand at 60°C. The solution is filtered, concentrated and crystallized.

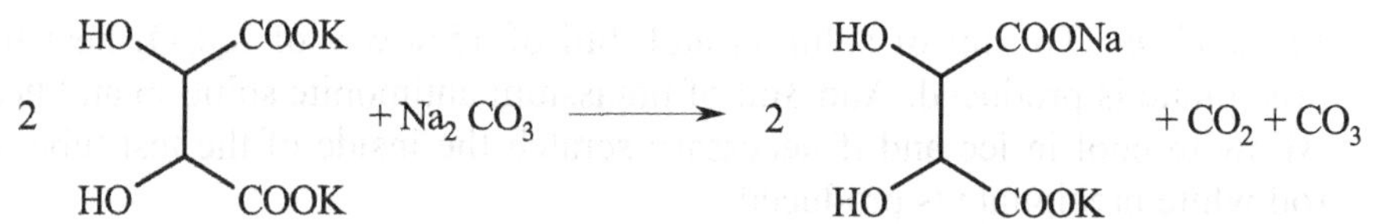

.....(9.17)

Test for Identification

For potassium: To 1ml of solution add 1ml dilute acetic acid and 1ml of 10% w/v sodium cobalt nitrite a yellow color produced.

For sodium: To 2ml of solution of 2ml of 15% w/v of K_2CO_3 heat to boil, no precipitate is produced. Add 3ml of potassium antimonite solution and heat to boil.

Allow to cool in ice and if necessary scratch the inside of the test tube with glass rod white precipitate is produced.

For tartarate

- Warm the substance with sulphuric acid charring occurs and carbon monoxide which burns with blue flame is evolved.
- To 5ml of sample solution add 1% w/v solution of ferrous solution and 0.05ml of hydrogen peroxide (10 vols) a transient yellow color is produced. After color disappears add 2M NaOH intense blue color is produced.

Assay: Weigh accurately 2g and heat until carbonized, cool and boil the residue with 50ml of water and 0.5N sulphuric acid (50ml). It is filtered washed with water and the filtrate and washing are titrated with 0.5N NaOH using methyl orange as indicator. Each ml of 0.5M sodium hydroxide ≡ 0.07056g

Uses: It is used as laxative, food additive, as stabilizer in cheese and meet products.

(iv) Magnesium Sulphate:

Chemical Formula: $MgSO_4.7H_2O$ *Mol. Wt.* 246.47

I.P. limit: It contains not less than 99.0% and not more than 100.5% of magnesium sulphate calculated with reference to dried substance.

Properties: It forms colorless prismatic crystals. It dissolves in water, is practically insoluble in alcohol. It has cooling saline bitter taste.

Preparation:

1. It can be prepared by neutralizing hot dilute sulphuric acid with magnesium or its oxides or carbonate. The solution is filtered; the filtrate is concentrated and recrystallized.

$$MgCO_3 + H_2SO_4 \rightarrow MgSO_4 + H_2O + CO_2 \quad .....(9.18)$$

$$MgO + H_2SO_4 \rightarrow MgSO_4 + H_2O \quad .....(9.19)$$

2. On commercial scale it is manufactured by reacting sulphuric with dolomite. Magnesium sulphate so formed is dissolved in the solution and the sparingly soluble calcium sulphate is deposited. The liquid is filtered the filtrate is concentrated and crystallized.

$$\underset{\text{Dolomite}}{MgCO_3CaCO_3} + 2H_2SO_4 \rightarrow MgSO_4 + CaSO_4 + 2H_2O + 2CO_2 \quad .....(9.20)$$

Test for Identification

For magnesium: To solution of sample add dilute nitric acid solution a white precipitate is produced that is redissolved by adding 1ml of 2M ammonium chloride, add 0.25M disodium hydrogen phosphate a white crystalline precipitate is produced.

For sulphate: To 5ml of sample solution add 1ml of dilute HCl and 1ml barium chloride solution white precipitate. Add 1ml of iodine solution to the suspension, the suspension remains yellow (distinction from sulphites and dithionites) but decolorizes on adding stannous chloride (distinction from iodates).

Assay: Weigh accurately about 6.3g of sample dissolve in 50ml of water, add 10ml of strong ammonia ammonium chloride solution and titrate with 0.05M disodium EDTA using 0.1g of moderate black II mixture as indicator until blue color is obtained.

Each ml of 0.05M disodium EDTA $\equiv$ 0.00602 g of $MgSO_4$

Uses: It is used as osmotic laxative, in treatment of electrolyte deficiency, in wet dressing in boils, in treatment of cholecystitis, sea sickness, hypertension etc.

(v) Calomel:

Chemical Formula: HgCl *Mol. Wt.* 236.1

I.P. limit: It contains not less than 99.0% of HgCl

Properties: A dull white heavy powder, odorless, almost tasteless. Volatilizes when strongly heated. Stable in air but gradually darkens when exposed to light. It is insoluble in water, alcohol, ether and in cold dilute acids.

Test for Identification

(i) Blackens by contact with dilute ammonia solution or with solution of alkali hydroxide.

(ii) Heat the sample in hard glass tube with an equal quantity of anhydrous sodium carbonate a sublimate of metallic mercury is obtained. Dissolve the residue in dilute HNO_3 and filter. To the filtrate add silver nitrate solution shake and allow to stand, a curdy white precipitate is obtained which is insoluble in HNO_3 but soluble, after being well washed with water, in dilute ammonium hydroxide solution from which it is reprecipitated by addition of dilute HNO_3.

Assay: Weigh accurately about 0.7g of substance and mix with 10ml water in a glass stoppered flask and add 50ml of 0. 1N I_2 and 5g KI dissolved in 10ml water. Close the flask and set aside, shaking occasionally until solution is complete. Titrate the excess of I_2 with 0.1N sodium thiosulphate using starch as indicator. Each ml of 0.1N I_2 $\equiv$ 0.02301 g of HgCl

Uses: It is used as cathartic.

(vi) Sulphur:

Sulphur is an essential element in the human organism. It is widely distributed throughout the body as sulphydryl group of cystine, disulphide linkage in proteins and sulphate salts and esters found in polysaccharides. Elemental sulphur exists in variety of allotropic forms solid or liquid, e.g. Rhombic sulphur, Monoclonic sulphur, Amorphous sulphur; plastic sulphur.

Rhombic sulphur: it is made by first shaking sulphur powder with CS_2 for sometime. The mixture is filtered and kept aside the excess of CS_2 evaporates away slowly leaving large crystals of rhombic sulphur. It is soluble in CS_2, insoluble in water and alcohol. Pale yellow crystals with faint odor and taste.

Monoclonic Sulphur: sulphur is heated in an evaporating porcelin dish until dish is almost full of molten sulphur and then cooled. A crest of solid forms over the surface, as soon as this crest is continuous, it is pierced in two points at opposite ends of diameter and the molten sulphur still left is poured out. The crest is then cut round and lifted off, it reveals long needle shaped crystals of monoclonic sulphur attached to the under surface of the crust due to the bottom of the evaporating dish. It has density of 1.96 g per cc and melts at 119 °C.

Amorphous sulphur: This is non crystalline variety of sulphur which is almost white. It is usually produced by chemical action between materials in aqueous solution as

(i) acidification of sodium thiosulphate solution

$$Na_2S_2O_3 + 2H^+ \rightarrow H_2O + SO_2 + S\downarrow + 2Na^+ \quad .....(9.21)$$

(ii) action between hydrogen sulphide and sulphur dioxide

$$2H_2S + SO_2 \rightarrow 2H_2O + S\downarrow \quad .....(9.22)$$

A yellow amorphous solid which is nearly insoluble in any solvent. It is usually very fine divided and difficult to filter. It may form a colloidal solution in water, known as milk of sulphur.

(vii) Precipitated sulphur (milk of sulphur):

Sulphur reacts with boiling solution of metal hydroxide to form mixture of metal sulphides and thiosulphate.

$$3Ca(OH)_2 + 12S \rightarrow 2CaS_5 + CaS_2O_3 + 3H_2O \quad .....(9.23)$$

Sulphur can be precipitated by the addition of HCl

$$2CaS_5 + CaS_2O_3 + 6HCl \rightarrow 3CaCl_2 + S + 3H_2O \quad .....(9.24)$$

It is very fine, pale yellow amorphous or microcrystalline powder, without odour or taste. It is soluble in carbon disulphide, insoluble in water and alcohol.

Uses: 1%, 5%, 10% and 20% precipitated sulphur ointments are used as mild irritant and antimicrobial agent in the treatment of skin diseases, e.g., as fungicide, paraceticide, keratolytic etc.

CHAPTER 10

Dental Products

10.1 Introduction

The teeth are accessory digestive organs. People use their teeth to bite and chew food, the first step in the digestion of food. The long, sharp canine teeth tear up food. The wide, flat molars grind and mash up food. While we chew food, the tongue pushes the food to the teeth and saliva helps digestion and wets the food.

A number of inorganic compounds are used in maintaining the oral and dental hygiene. Most of them are over the counter (OTC) products. Dental products include anticaries agents, polishing agents, and desensitizing agents.

10.2 Tooth Anatomy

A typical tooth has three major external regions: the crown, root, and neck.

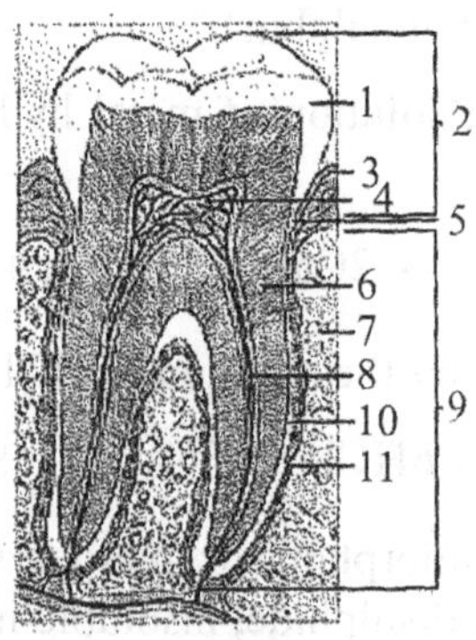

Fig. 10.1 Figure of a tooth and surrounding structures.

1. *Enamel:* Hard calcified (consists primarily of calcium phosphate and calcium carbonate) tissue covering dentin of the crown of tooth.
2. *Crown:* The crown is the visible portion of tooth above the level of the gums.
3. *Gingiva (gums):* Soft tissues overlying the crowns of un erupted teeth and encircling the necks of those that have erupted.
4. *Pulp Chamber:* The space occupied by the pulp.
5. *Neck:* The area where the crown joins the root.
6. *Dentin:* That part of the tooth that is beneath enamel and cementum.
7. *Alveolar Bone (jawbone):* The part of the jaw that surround the roots of the teeth.
8. *Root Canal:* The portion of the pulp cavity inside the root of a tooth; the chamber within the root of the tooth that contains the pulp.
9. *Root:* Embedded in the socket are one to three roots.
10. *Cementum:* Hard connective tissue covering the tooth root, giving attachment to the periodontal ligament.
11. *Periodontal Ligament:* A system of collagenous connective tissue fibers that connect the root of a tooth to its alveolus.

10.3 Anticaries Agents

Dental caries, or tooth decay, involves a gradual demineralization (softening) of the enamel and dentin. If it is not treated then microorganisms may invade the pulp, causing inflammation and infection, with subsequent death of the pulp and abscess of the alveolar bone surrounding the root's apex, requiring root canal therapy.

Dental caries (i.e. cavities) are formed by the growth and implantation of cariogenic microorganisms. Bacteria (primarily streptococcus mutans and lactobacillaceae) produce acids, mostly lactic acid that demineralize the enamel. The demineralized enamel initially appears as a white, chalky area and eventually becomes brown or yellow. Diet is another factor in the development of dental caries. Diet with a high concentration of fermentable carbohydrates increases the risk of dental caries. Masses of bacterial cells, sticky polysaccharide (produced from sucrose) and other debris adhering to teeth constitute dental plaque. Fermentable sugar such as sucrose is converted by bacterial plaque into volatile acids that destroy the hydroxyl apatite. The formation of bacterial plaque also helps the decay process by forming pockets or crevices on the tooth surface in which the food particles can stick and be decayed by the bacteria. If plaque is not removed it calcifies into calculus when calcium salt precipitates from the saliva. Brushing the teeth helps in removing the material from the tooth surface before it hardens into calculus.

Dental caries can be prevented and oral and dental hygiene can be maintained with the help of dentifrices. These are the products that enhance the removal of stain and dental plaque by the toothbrush. The most accepted approach to prevent caries includes flossing and brushing accompanied by administration of fluoride either internally or topically to the teeth.

Fluoride is anticariogenic as it replaces the hydroxyl ion in hydroxyapatite with the fluoride ion to form fluorapatite in the outer surface of the enamel. Fluorapatite hardens the enamel and makes it more acid resistant. Fluorapatite has also shown antibacterial activity. Fluoride is most beneficial up to an age of 12 or 13 because un erupted permanent teeth are mineralizing during that time.

Fluoride can be administered by two routes, orally and topically. The fluoridation of the public water supply is the most convenient and effective may of oral administration. This can be achieved by adding sodium fluoride, giving a fluoride concentration of 0.7-1.0 ppm. Fluoride can also be administered orally as sodium fluoride tablets or drops added in water or fruit juice. Fluoride in solution or in rapidly soluble salts when administered internally is readily absorbed from the gastrointestinal tract, partially deposited in the bone or developing teeth and the remainder gets excreted by the kidneys. It is not always feasible to administer fluoride internally and in post-adolescent individuals, it is not beneficial. In such cases, fluoride can be administered topically. A 2% aqueous solution of sodium fluoride is widely used topically. A freshly prepared 8% solution of stannous fluoride is also extensively used for topical application of fluoride.

Inorganic phosphate salts can also be useful in the prevention of dental caries. It has been shown that soluble salts of phosphates (e.g., calcium sucrose phosphate, sodium dihydrogen phosphate, sodium monohydrogen phosphate etc.) can cause caries reduction in men.

10.4 Polishing Agents

Dentifrices contain agents for cleaning tooth surfaces and providing polishing effect on the cleaned teeth. These agents are abrasive in nature. They are responsible for physically removing plaque and debris. Examples include dicalcium phosphate, sodium metaphosphate, calcium pyrophosphate, calcium carbonate and calcium monohydrogen phosphate. Pumice is too abrasive for daily use in a dentifrice.

10.5 Desensitizing Agents

Desensitizing agents reduce the pain in sensitive teeth caused by cold, heat or touch. These products should be non-abrasive and should not be used on a regular basis unless directed by a dentist. Examples include strontium chloride and zinc chloride.

A. Sodium Fluoride:

Chemical Formula: NaF *Mol. Wt.* 41.99

It contains not less than 98.5 percent and not more than 100.5 percent of NaF, calculated with reference to the dried substance.

Preparation: It is prepared by reacting hydrofluoric acid with sodium carbonate. Sodium fluoride being not very soluble precipitates out.

$$2HF + Na_2CO_3 \rightarrow 2NaF + H_2O + CO_2 \uparrow \quad \text{..... (10.1)}$$

The precipitate is contaminated with fluorosilicate and the acid salt. It is made alkaline to phenolphthalein with sodium carbonate and then heated to neutralize the acid salt and decompose the fluorosilicate.

$$Na_2SiF_6 + 2H_2O \rightarrow 2NaF + 4HF + SiO_2 \quad \text{..... (10.2)}$$

Identification Tests: Sodium fluoride is an ionic salt and gives reactions for fluoride ion and sodium ion.

1. Calcium chloride gives a white gelatinous precipitate of CaF_2 with fluoride ions. The precipitate dissolves in ferric chloride solution.

   $$CaCl_2 + 2NaF \rightarrow 2NaCl + CaF_2 \downarrow \quad \text{.....(10.3)}$$

   2.5g is dissolved in sufficient carbon dioxide free water without heating to make 100 ml (solution A). To 2 ml of solution A, 0.5 ml of calcium chloride solution is added. A white gelatinous precipitate is produced which dissolves on adding 5 ml of ferric chloride solution.

2. In another test for fluoride, an alizarin-zirconium lake is prepared having a red colour. This red colour changes to yellow colour of alizarin-sulphonic acid by the fluoride ions due to the formation of colourless hexafluorozirconate (IV) ion $[ZnF_6]^{2-}$.

   Add about 4mg to a mixture of 0.1 ml of alizarin red S solution and 0.1 ml of zirconyl nitrate solution and mix. The red colour changes the yellow.

3. Sodium salts react with an alkaline solution of potassium antimonite, $KsbO_3.3H_2O$ to give white precipitate of sodium antimonite $NaSbO_3.xH_2O$, which rapidly changes to the crystalline sodium pyroantimonate, $Na_2H_2Sb_2O_7.6H_2O$.

   0.1g is dissolved in 2 ml of water. To this add 2 ml of 15% w/v solution of potassium carbonate and heat to boiling. No precipitate is formed. 4 ml of a freshly prepared potassium antimonite solution is added and heat to boiling. On cooling in ice a dense white precipitate is formed.

Test for purity: It has to be tested for acidity or alkalinity; fluorosilicate; clarity and colour of solution; chloride; sulphate and loss on drying.

Acidity or Alkalinity: From the method of preparation described above, sodium fluoride may be contaminated with sodium hydrogen fluoride or sodium carbonate leading to acidity or alkalinity respectively. Titration with 0.IM sodium hydroxide or hydrochloric acid using phenolphthalein as indicator gives a limit for these impurities. 2.5g of potassium nitrate is dissolve in 40 ml of solution A and the volume is made to 50 ml with carbon dioxide-free water. Cool to 0° and then add 0.2 ml of dilute phenolphthalein solution. If the solution is colourless, not more than 1.0 ml of 0.IM sodium hydroxide is required to produce a red colour that persists for at least 15 seconds. If the solution is red, not more than 0.25 ml of 0.IM hydrochloric is required to change the colour of the solution. Keep the neutralized solution for the test for fluorosilicate.

Fluorosilicate: When the neutralized solution is heated to boiling, it results in hydrolysis of any fluorosilicate present to give free hydrofluoric acid, which is then titrated with 0.1 M sodium hydroxide.

The neutralized solution reserved in the test for acidity or alkalinity is heated to boiling and then titrated while hot with 0.IM sodium hydroxide until a red colour is produced. Not more than 1.5 ml of 0.IM sodium hydroxide is required to produce the red colour.

$$Na_2SiF_6 + 2H_2O \rightarrow 4HF + 2NaF + SiO_2 \quad ..... (10.4)$$

Loss on drying: It should not be more than 0.5%, determined on 1 g by drying in an oven at 130° for 3 hours.

Assay: It is assayed by non-aqueous titration method.

Weigh accurately about 80 mg and add a mixture of 5 ml of acetic anhydride and 20 ml of anhydrous glacial acetic acid to it. Heat to dissolve, cool, add 20 ml of dioxan and titrate with 0.1M perchloric acid using crystal violet solution as indicator until a green colour is produced.

Carry out a blank determination and make any necessary correction.

Each ml of 0.1M perchloric acid is equivalent to 0.004199 of NaF

Uses: It is used as preventive for dental caries because of its fluoride ion content.

Usual dose: 2.2 mg (equivalent to 1 mg of fluoride ion)

Application: 1.5-3.0 ppm (equivalent to 0.7-1.3 ppm of fluoride ion) in drinking water; topically, as 2% solution to the teeth.

Formulations: Sodium fluoride is administrated as solution, tablet, oral gel for systemic use or as mouth wash for local use.

B. Stannous Fluoride:

Chemical Formula: SnF_2 *Mol. Wt.* 156.69

It contains not less than 71.2 percent of stannous (Sn^{2+}) ions and not less than 22.3 percent and not more than 22.5 percent of fluoride, calculated with reference to the dried substance.

Preparation: Stannous fluoride is prepared by heating stannous oxide with gaseous hydrofluoric acid in the absence of oxygen

$$SnO + 2HF \rightarrow SnF_2 + H_2O \qquad ..... (10.5)$$

Identification Tests

1. Calcium chloride gives a white precipitate of CaF_2 with fluoride ions. Take 5 ml of solution (1 in 100) in a test tube and add 2 ml of calcium chloride solution to it. A fine, white precipitate of calcium fluoride is formed.

$$CaCl_2 + SnF_2 \rightarrow CaF_2 \downarrow + SnCl_2 \qquad ..... (10.6)$$

2. Mix 2 drops of solution (1 in 100) with 2 drops of silver nitrate solution on a spot plate. A brown-black precipitate is produced.

$$2AgNO_3 + SnF_2 \rightarrow 2AgF_2 + Sn(NO_3)_2 \qquad ..... (10.7)$$

3. One drop of solution (1 in 100) is added to 2 drops of mercuric chloride solution, there is formation of white, silky precipitate. A brown-black precipitate is formed on further addition of the solution (1 in 100).

Test for purity: It has to be tested for pH, loss on drying and water insoluble substances pH: A freshly prepared 0.4% solution has pH between 2.8 and 3.5

Loss and drying: When it is dried at 105° for 4 hours, the loss should not be more than 0.5% of its weight.

Water-Insoluble substances: About 10g of accurately weighed in taken in a 400 ml plastic beaker. To this add 20 ml of water and stir with a plastic rod for 3 minutes, or until no more solid dissolves. Then filter through a tared filtering crucible and wash thoroughly first with ammonium fluoride solution (1 in 100) and then with water. The residue is dried at 105° for 4 hours, cooled and weighed. The weight of the residue should not be more than 0.2%.

Antimony

Standard Preparation: 55.0 mg of accurately weighed antimony potassium tartarate is transferred to a 200 ml volumetric flask, dissolved in water, diluted with water to make up the volume and mixed. 5 ml of this solution is transferred to a 500 ml volumetric flask and to this 6N hydrochloric acid is added to the volume and mixed.

Test Preparation: 1.0 g of accurately weighed stannous fluoride is transferred to a 50 ml volumetric flask. To this 6N hydrochloric acid is added to the volume and mixed.

Method: 5 ml each of the standard preparation and the test preparation is taken into separate 125 ml separators. Then, 15 ml of hydrochloric acid and 1g of ceric sulphate are added and allowed to stand for 5 minutes; with occasional shaking 500 ml of hydroxylamine hydrochloride is added and shaken for 1 minute. 15 ml of isopropyl either is pippeted into the mixture, shaken for 30 seconds followed by addition of 7 ml of water and mixed. Then, cooled in water bath at room temperature for 10 minutes, shaken for 30 seconds, allowed the layers to separate and aqueous phase is discarded. 20 ml of Rhoda mine B solution is added, shaken for 30 seconds and the aqueous layer is discarded. The either layer is decanted from the top of the separator and centrifuged to obtain a clear solution. The absorbance of the either solution from the test preparation and the standard preparation are concomitantly determined at the wavelength of maximum absorbance at about 550 nm, with a suitable spectrophotometer, using water as the blank. The absorbance of the test preparation should not exceed that of the standard preparation (0.005%).

Assay: Assay for Stannous Ion: Transfer about 250 mg of accurately weighed stannous fluoride to a 500 ml conical flask. To this flask, add 300 ml of hot, recently boiled 3N hydrochloric acid. Swirl the flask to dissolve the stannous fluoride and cool to room temperature. Add 5 ml of potassium iodine and titrate with 0.IN potassium iodine-iodate using 3 ml of the starch as indicator added towards the end point.

Each ml of 0.IN potassium iodine-iodate is equivalent to 5.935 mg Sn^{2+}

Assay for Fluoride

Buffer solution: Dissolve 5.7 ml of glacial acetic acid, 58 g of sodium chloride and 4 g of (1,2-cyclohexylenedinitrilo) tetra acetic acid in 500 ml of water. Adjust the pH to 5.25 ± 0.25 with 5N Sodium hydroxide, dilute with water to 1000 ml and mix.

Standard Preparation: Dissolve the accurately weight quantity of sodium fluoride so as to obtain a solution in water containing 420 *ug* per ml (Standard preparation A). Each ml of this solution contains 190 *ug* of fluoride ion. Transfer 25.0 ml of standard preparation A to a 250 ml volumetric flask, make up the volume with water and mix (Standard preparation B). Transfer 25.0 ml of standard preparation B to a 250 ml volumetric flask, make up the volume with water and mix (Standard preparation C). This solution contains 1.9 *ug* of fluoride ion.

Assay Preparation: Transfer about 10 mg of accurately weighed stannous fluoride to a 250 ml volumetric flask. Add 50 ml of water, mix vigorously for 5 minutes, make up the volume with water and mix. Transfer 10.0 ml of this solution to a 50 ml volumetric flask, make up the volume with water and mix.

Method: Pipette 20 ml of each standard preparation and the assay preparation into separate plastic beakers. Add 20 ml of butter solution into each beaker. Concomitantly measure the potential of the solutions from the standard preparations and assay preparation using a pH meter equipped with a fluoride specific. Ion indicating electrode and a calomel reference electrode.

Plot the logarithms of the fluoride ion concentrations, in g per ml of the standard preparations versus potential, in mV. Determine the concentration, C, in μg per ml, of fluoride ion in the assay preparation from the measured potential of the assay preparation. Calculate the percentage of fluoride by the formula.

$$C/W$$

Where C is the concentration of fluoride determined in assay preparation and W is the weight of the stannous fluoride taken.

Uses: It is used as a preventive for dental caries. A freshly prepared 8% solution is used at 6 to 12 month intervals.

Formulations: It is administered as solution, gel, mouth wash or dentifrice. It is for topical use only.

10.6 Disinfectants

Disinfectants are antimicrobial agents that are applied to non-living objects to destroy microorganisms. The process is known as disinfection. Disinfectants are different from antibiotics and antiseptics in that, antibiotics, destroy microorganisms within the body, and antiseptics, destroy microorganisms on living tissue. Bacterial endospores are most resistant to disinfectants; however some bacteria and viruses also possess some tolerance. A perfect disinfectant produces complete sterilization, without causing any harm to other forms of life, is inexpensive, and non-corrosive. Unfortunately there is no ideal disinfectant. Most disinfectants are also, by their very nature, potentially harmful (even toxic) to human beings. They should be treated with appropriate care. Most of them come with safety instructions printed on the packaging, which should be read completely before using the disinfectant.

They are frequently used in hospitals and dental surgeries, to kill infectious organisms. The selection of the disinfectant to be used is made according to the particular situation. Some disinfectants kill a wide range (kill nearly all microorganisms), while others kill a smaller range of disease-causing organisms but are preferred for other properties (that may be non-corrosive, non-toxic, or inexpensive).

A. Zinc chloride:

Chemical Formula: $ZnCl_2$ *Mol. Wt.* 136.29

It contains not less than 95 percent and not more than 100.5 percent of $ZnCl_2$.

Preparation: It is prepared by heating granulated zinc with hydrochloric acid. When evolution of hydrogen ceases, the solution is filtered and evaporated to dryness

$$Zn + 2HCl \rightarrow ZnCl_2 + H_2\uparrow \quad \text{..... (10.8)}$$

Identification Tests: 1. To 2 g, add 38 ml of carbon dioxide—free water prepared from distilled water. Add 2 M hydrochloride acid drop wise until solution is complete and make up the volume to 40 ml with carbon dioxide—free water prepared from distilled water (Solution A). Solution A gives the following reactions of Zinc salts.

(i) A solution of zinc salts when treated with dilute sodium hydroxide solution forms a white precipitate of zinc hydroxide which dissolve in excess of sodium hydroxide solution (forms zincate which is soluble). The solution remains clear on addition of ammonium chloride, On addition of a few drops of sodium sulphide solution, a white precipitate of zinc sulphide is formed.

$$ZnCl_2 + 2NaOH \rightarrow 2NaCl + Zn(OH)_2\downarrow \quad \text{.....(10.9)}$$

$$Zn(OH)_2 + 2NaOH \rightarrow Na_2ZnO_2 + 2H_2O$$

$$Na_2ZnO_2 + Na_2S + 2H_2O \rightarrow 4NaOH + ZnS\downarrow \quad \text{.....(10.10)}$$

Take 5 ml of solution A and add 0.2 ml of sodium hydroxide solution to it. A white precipitate is formed. Further, add 2 ml of sodium hydroxide solution to dissolve the precipitate. The solution remains clear on addition of 10 ml of ammonium chloride solution. Add 0.1 ml of sodium sulphide solution. A flocculent white precipitate is formed.

(ii) When solution of zinc salts is acidified with dilute sulphuric acid and treated with one drop of copper sulphate solution followed by addition of amrnonium mercurithiocyanate solution, a violet precipitate is obtained due to complexation of zinc ion.

Acidify 5 ml of the solution A with dilute sulphuric acid. Add one drop of a 0.1% w/v solution of cupric sulphate and 2 ml of ammonium mercurithiocyanate sollution. A violet colour precipitate is produced.

(iii) Solution of zinc salts when treated with potassium ferrocyanide solution forms a white precipitate of zinc ferrocyanide which is insoluble in dilute hydrochloric acid.

$$2\,Zn^{2+} + \left[Fe(CN)_6\right]^{4-} \rightarrow \underset{\text{Zinc ferrocyanide}}{Zn_2\left[Fe(CN)_6\right]} \quad \text{..... (10.11)}$$

To 5 ml of solution A, add 2 ml of potassium ferrocyanide solution. A white preparation is formed which is insoluble in dilute hydrochloric acid.

A 5% w/v solution in 2M nitric acid (Solution B) gives the following reactions of chloride.

(i) An aqueous solution of chloride on acidification with dilute nitric acid followed by treatment with silver nitrate solution forms a curdy white precipitate of silver chloride which is insoluble in dilute nitric acid but soluble in dilute ammonia solution. The precipitate reappears on addition of dilute nitric acid.

$$NaCl + AgNO_3 \rightarrow AgCl\downarrow + NaNO_3 \quad \text{..... (10.12)}$$

$$AgCl + 2NH_3 \rightarrow [Ag(NH_3)_2]^+ Cl^- \quad \text{.....(10.13)}$$

$$[Ag(NH_3)_2]^+ Cl^- + HNO_3 \rightarrow AgCl\downarrow + (NH_4)_2NO_3 \quad \text{.....(10.14)}$$

Acidify 2 ml of the solution B with dilute nitric acid, add 0.5 ml of silver nitrate solution, shake and allow to stand for some time. A white curdy precipitate is produced, which in insoluble in nitric acid but soluble in dilute ammonia solution after being well washed with water. The reprecipitation occurs on addition of dilute nitric acid.

(ii) Sample containing chloride ion on treatment with potassium dichromate and sulphuric acid gives volatile chromyl chloride which reacts with diphenylcarbazide solution to produce a violet colour.

$$4Cl^- + K_2Cr_2O_7 + 3K_2SO_4 \rightarrow 2CrO_2Cl_2\,(g) + K_2SO_4 + 2SO_4^{2-} + 3H_2O \quad \text{.....(10.15)}$$

Take in test tube the quantity equivalent to 10 mg of chloride ion, add 0.2 g of potassium dichromate and 1 ml of sulphuric acid. Place a filter paper moistened with 0.1 ml of diphenylcarbazide solution over the month of the test tube. The paper turns to violet red in colour.

Test for Purity: It has to be tested for ammonium salts; alkalis and alkaline earths; oxychloride; sulphate and heavy metals.

(i) *Ammonium Salts:* Ammonium salts are detected by warming a solution of zinc chloride with sodium hydroxide. No odour of ammonia should be perceptible.

$$NH_4^+ + OH^- \rightarrow NH_3 + H_2O \quad \text{.....(10.16)}$$

To 5 ml of a 10% w/v solution, add 1M sodium hydroxide until the precipitate first formed is redissolved and then warm the solution. No odour of ammonia is perceptible.

(ii) *Alkalis:* When sample is dissolved in dilute hydrochloric acid followed by addition of ammonical ammonium sulphide, zinc gets precipitation as ZnS.

The ZnS is filtered off and an aliquot of the filtrate is acidified with sulphuric acid, evaporated to dryness and ignited. The residue obtained after ignition is weighted. The residue should not be more than 1%.

(iii) *Oxychloride:* Aqueous solution of zinc chloride is acid to litmus due to hydrolysis to hydrogen chloride and on evaporation it gives a residue containing varying proportions of the Oxychloride, Zn (OH) Cl.

$$ZnCl_2 + H_2O \rightarrow HCl + Zn(OH)Cl \qquad .....(10.17)$$

Dissolve 1.5 g in 1.5 ml of carbon dioxide-free water. To this solution, which should be clear, is added 7.5 ml of 95% ethanol. The solution may become cloudy within 10 minutes but becomes clear on the addition of 0.2 ml of 2M hydrochloride acid.

(iv) *Sulphate:* 15 ml of solution A complies with the limit test for sulphates.

Assay: It is assayed by complexometry using strong ammonia—ammoniurn chloride solution as buffer, eriochrome black T solution as indicator and titrating with 0.1 M disodium edetate.

$$\underset{}{ZnCl_2} + \underset{\text{Disodium edetat}}{C_{10}H_{14}N_2Na_2O_8} \rightarrow \underset{\text{Zinc edentate}}{C_{10}H_{14}N_2O_8Zn} + 2NaCl \qquad ..... (10.18)$$

Weight accurately about 3 g, dissolve in 125 ml of water add 3 g of ammonium chloride and make up the volume to 250 ml with water. To 25 ml of the resulting solution add 100 ml of water and 10 ml of strong ammonia-ammonium chloride solution. Titrate with 0.IM disodium edetate using eriochrome black T solution as indicator until a deep blue colour is produced.

Each ml of 0.1 M disodium edetate is equivalent to 0.01363 g of $ZnCl_2$

Uses: It is used as an antiseptic astringent to the skin and mucous membrane as a 0.5—2.0% solution. It ranks very low among disinfectants.

It is used as an active ingredient to prepare magnesia cements for dental fillings and certain mouthwashes.

It is also used as dentin desensitizer, topically as a 10% solution to the teeth.

Formulation: It is for topical use only and is administered as solution and mouthwash.

B. Zinc-Eugenol Cement

Preparation: Zinc-Eugenol cement consists of two parts and is prepared as follows

Part I. The Powder:

Zinc Acetate 0.5 g

Zinc Stearate	1 g
Zinc Oxide	70 g
Rosin	27.5 g

Rosin is powdered and mixed with about an equal weight of zinc oxide. The mixture is sifted through a sieve of not less than 100—mesh. The material retained on the sieve is re grinded with additional zinc oxide, and sifted again. The process of regrinding and sifting is repeated until all of the material passes readily through the sieve and the two mixtures are then mixed with the remainder of the zinc oxide.

Part II. The Liquid:

Eugenol	85 ml
Cottonseed Oil	15 ml

The liquids are mixed together in the proportion specified.

Zinc-Eugenol cement is prepared by mixing 10 parts of the powder with 1 part of the liquid to a thick paste immediately before use. For obtaining any desired consistency the amount of the liquid may be varied.

The Powder

Identification Tests

1. 10 parts of the powder is mixed with 1 part of the liquid and the resulting mixture is transferred to a beaker containing water at 25°. The mixture hardens in not more than 20 minutes.
2. 1 g of the powder is triturated with 10 ml of solvent hexane. The mixture is filtered and to the filtrate 10 ml of freshly prepared cupric acetate solution (1 in 200) is added, shaken for a few minutes, and allowed the liquid layers to separate. The solvent hexane layer is green.
3. The residue obtained in the assay for total zinc as zinc oxide, dissolved in a slight excess of hydrochloric acid, gives reactions for zinc.
4. 5 g of the powder is triturated with 25 ml of water, the triturate is filtered and to 10 ml of the clear filtrate, 1 ml of ferric chloride TS is added in a colour-comparison tube. A standard is prepared by adding 1 ml of ferric chloride TS to 10 ml of water in a matched colour-comparison tube. The intensity of the colour produced in the filtrate is more than that of standard colour when viewed downwards against a white surface.

Assay for Rosin: About 1 g of the accurately weighed powder is placed in a beaker, to this 50 ml of chloroform is added, and stirred for several minutes. The resultant is filtered through a tared porcelain filtering crucible and the insoluble material is transferred as completely as possible with additional portions of chloroform. The crucible is washed with chloroform, dried in an oven at 80°C, cooled, and weighed. The loss in weight of the powder taken should not be less than 300 mg.

Assay for total Zinc as Zinc oxide: Ignite the residue obtained in the assay of rosin to constant weight, cool, and weigh. The weight of the ZnO so obtained is not less than 680 mg and not more than 720 mg.

The Liquid

Identification Test: 1 ml of the liquid is shaken with 20 ml of water, filtered, and to 5 ml of the clear filtrate, 1 drop of ferric chloride TS is added. A transient pale yellow-green colour is produced.

Specific Gravity: Between 1.043 and 1.048.

Refractive Index: Between 1.528 and 1.531 at 20°

Uses: It is temporary cement with a zinc oxide eugenol base for crown and bridge procedures. It can be used in dentistry as a filling or cement material. It is used in temporary restorations, in managing dental caries as a temporary filling.

It is also used as an impression material during construction of complete dentures

Zinc-Eugenol cement has anaesthetic and antimicrobial effect due to the eugenol content and is used in painful conditions of dental pulp.

C. Pumice

Pumice is a substance of volcanic origin, produced when lava with a very high content of water and gases is thrown out of a volcano. As the gas bubbles escape from the lava, it becomes frothy. When this lava cools and hardens, it results in a very light rock material filled with tiny bubbles of gas. It consists mainly of complex silicates of aluminum, potassium, and sodium. It is very light, hard, rough, porous, greyish mass. It is odorless and stable in air

Labeling: Powdered pumice is labeled to indicate the fineness of the powder. Powdered pumice shall meet the following requirements:

"Pumice Flour" or "Superfine Pumice": not less than 97.0 percent of pumice flour or superfine pumice passes through a No.200 standard mash sieve. "Fine Pumice": not less than 95.0 percent of the fine pumice passes through a No 150 standard mash sieve and not more than 75.0 percent passes through a No.200 mash sieve.

"Coarse Pumice": not less than 95.0 percent of coarse pumice passes through a No.60 standard mash sieve and not more than 50 percent passes through a No.200 standard mash sieve.

Test for purity: It has to be tested for water soluble substances, acid soluble substances and iron.

Water-soluble substances: 10 g is boiled with 50 ml of water 30 minutes, adding water from time to time to maintain approximately the original volume, then filter; the filtrate is neutral to litmus, and one-half of this filtrate, when evaporated and dried at 105° for 1 hour, yields not more than

It has to be tested for water-soluble substances, acid 10 mg of residue (0.20%)

Acid-soluble substances: Boil 1 g of pumice with 25 ml of 3N hydrochloric acid for 30 minutes, adding water from time to time to maintains approximately the original volume, and then filter the liquid. Add 5 drops of sulphuric acid to the filtrate, evaporate to dryness, ignite, and weigh the residue; not more than 60 mg of residue is obtained (6.0%).

Iron: Acidify the remaining half of the filtrate from the test for water-soluble substances with hydrochloric acid, and add a few drops of potassium ferrocyanide TS. No blue colour is produced.

Uses: It is used as an abrasive. Finely ground pumice is added to some toothpastes and heavy-duty hand cleaners as a mild abrasive.

It is prepared in various grits and is used for finishing and polishing in dentistry. It is used in the polishing of natural teeth during a prophylaxis but is being replaced by less abrasive synthetic particles in some polishing pastes.

CHAPTER 11

Topical Agents

The term topical means pertaining to a particular spot. Topical agents are the compounds that act locally on the body surfaces i.e., skin and mucous membranes. They act mainly by physical or chemical manners. Their effects are seen primarily at the surfaces to which they are applied. Topical application of these drugs may extend to such body cavities that are open to the outside. Examples of such body cavities include oral, vaginal and colonic cavities. As they are not absorbed in systemic circulation, they are having very little distinct pharmacological properties.

The topical agents may produce variety of effects such as adsorbent, astringent, demulcent, emollient or protective. Some topical agents may exhibit antimicrobial and astringent activity. A large number of inorganic compounds are known which do have topical action of overlap type.

Categories

- Protective and adsorbents
- Antimicrobial agents
- Astringents
- Emollients
- Caustics

11.1 Protectives and Absorbents

Protectives are substances applied to protect areas of the skin which are subject to constant irritation due to moisture and/or friction, or areas that have become irritated or inflamed due to friction, allergy and the like. Many protectives are also adsorbents that adsorb moisture from the surface of the skin. Protectives are not applied to areas that are abraded and exuding fluid as the likelihood of systemic absorption is enhanced and they may mix with the exudates and form a crust which adheres to the open tissue.

The compounds of substances most suitable as protectives have the following properties:

- Insoluble (minimizes systemic absorption)
- Chemically inert (to prevent interactions with the tissue)

The adsorbent and protective action is maximized with small particle size. A fine state of sub division of particles possesses good adhering properties, enhances the soothing effect and minimizes irritation.

Topical inorganic protectives include formulations containing talc, Zinc oxide, calamine, Zinc stearate, basic aluminum carbonate and titanium dioxide.

11.1 (a) Purified Talc

Purified Talc is purified magnesium silicate corresponding approximately to the formula $Mg_6 (Si_2O_5)_4 (OH)_4$

It is naturally occurring hydrous magnesium silicate (sometimes containing a small proportion of aluminium silicate) and is called soapstone or French chalk. It is said to be the softest mineral and has more free flow property than kaolin.

Preparation: Pharmaceutically required product is purified from native talc by boiling the finely powdered talc with dilute hydrochloric acid and then washing the insoluble talc well and drying.

Description: Purified talc is a very fine, white or yellowish – white powder, which is odourless and tasteless. It is free from grittiness, readily adheres to skin and is unctuous to the touch. Its solution is neutral to litmus.

Tests for Identity: The I.P gives the following test for identification is reproduced here.

Mix 0.5 g with about 0.2g of anhydrous sodium carbonate and 2g of anhydrous potassium carbonate and heat the mixture in a platinum crucible until fusion is complete. Cool and transfer the fused mixture to a dish or beaker with the aid of about 50 ml of hot water. Add hydrochloric acid to the liquid until it ceases to cause effervescence then add 10 ml more of the acid and evaporate the mixture to dryness on a water bath. Cool, add 20 ml of water, boil and filter the mixture. Dissolve in the filtrate about 2g of ammonium chloride and add 5 ml of dilute ammonia solution. Remove by filtration any precipitate which may form and add solution of sodium phosphate to the filtrate, a white crystalline precipitate of magnesium ammonium phosphate separates.

Tests for Purity: It is tested for acid-soluble substance; reaction water soluble substances; water soluble iron salts; carbonates and loss on drying.

Acid Soluble Substances: It is performed by digesting 2g with 40 ml dilute hydrochloric acid for fifteen minutes, filtering the solution and evaporating the filtrate. The residue obtained is ignited after the addition of 0.1 ml of dilute sulphuric acid to constant weight. The residue weight not more than 20 mg.

Water-soluble Substances: It is performed by evaporating 25 ml of the filtrate obtained by boiling 10 g with 50 ml of water for 30 minutes to constant weight at 105°. The residue does not weigh more than 10 mg.

Water-soluble Iron: It is done by preparing a solution (as given below) which should comply with the limit test for iron. The test solution is prepared by boiling 5 g of purified talc with 25 ml of water for thirty minutes, replacing the water lost during evaporation and filtering the solution. The solution after the addition of 5 ml of nitric acid is diluted to 50 ml with water and used for the test.

Carbonates are tested by adding 1 g to 20 ml of dilute hydrochloric acid and noting the absence of effervescence.

Loss on Ignition: The substance loses not more than 6 percent of its weight when ignited to constant weight at 1000 °C.

Storage: Purified talc is stored in a well closed container.

Uses: It is an important pharmaceutical aid. It is used in a dusting powder under the name talcum powder. It is perfumed. It is used in a steering medium for all classes of preparations as it does not absorb and retain active principles.

11.1 (b) Zinc Oxide

Chemical Formula: ZnO *Mol. Wt.* 81.38

Zinc oxide contains not less than 99% of ZnO calculated with reference to the substance ignited to constant weight.

Preparation:

(i) Zinc oxide is manufactured on a large scale by heating metallic Zinc to bright redness in a current of air. The vapour of the metal burns to form the oxide, which is collected as a fine white powder but Zinc oxide so prepared is not used for pharmaceutical purposes.

$$2Zn + O_2 \rightarrow 2ZnO \quad .....(11.1)$$

(ii) A purer product can be obtained from Zinc sulphate. A solution of Zinc sulphate is added to a boiling solution of sodium carbonate. The precipitated basic carbonate of Zinc is collected, washed until freed from sulphate, dried and gently ignited. It loses carbon dioxide and water, leaving the Oxide.

$$ZnSO_4 + Na_2CO_3 \rightarrow ZnCO_3 + Na_2SO_4 \quad .....(11.2)$$

$$ZnCO_3 \rightarrow ZnO + CO_2 \quad .....(11.3)$$

Description: It occurs as a soft, white or faintly white, very fine powder, free from gritty ness. It is odorless and tasteless. When exposed to air, it slowly adsorbs carbon dioxide from the air.

Zinc oxide is insoluble in water and alcohol but since it is amphoteric it is soluble in solution of alkali hydroxides and in dilute mineral acids.

Tests for Identity

- On being strongly heated, Zinc oxide assumes a yellow colour, which disappears on cooling.
- A solution of Zinc oxide in dilute hydrochloric acid, after neutralization of the acid, yields reactions characteristic of Zinc (APP. I)

Tests for Purity: It is tested for alkalinity carbonate and insoluble impurities; arsenic, iron, metallic zinc and less on drying.

Alkalinity: It is tested by mixing 1g of the substance with 10 ml of hot water, adding 2 drops of solution of phenolphthalein and noting the colour. In case a red colour is produced, the amount of 0.1N hydrochloric acid required to discharge the red colour should be more than 0.3 ml.

Carbonate and Insoluble Impurities: It is carried out by mixing 2g of the substance with 10 ml of water, adding 30 ml of dilute sulphuric acid and heating the solution on a water bath with constant stirring. No effervescence should occur and the solution should remain clear and colorless.

Arsenic as tested by the general method described is not more than 8 parts per million.

Iron is tested by dissolved 0.1 g of ZnO in a mixture of 5 ml of water and 0.5 ml of iron free hydrochloric acid and diluting the solution to 40 ml with water. The solution with 6 drops of thioglycolic acid should comply with limit test for iron.

Lead is tested by dissolving 2 g of ZnO in a mixture of 20 ml of water and 5 ml of glacial acetic acid and adding 5 drops of solution of potassium chromate. The solution should remain clear. In the test, precipitate of yellow lead chromate is expected to be formed in the presence of acetic acid; Zinc chromate remains in solution.

Metallic Zinc: It is tested by dissolving 2 g of ZnO in a mixture of 30 ml of dilute hydrochloric acid and 10 ml of water to which one drop of solution of lead acetate has been added. The solution should be clear and colourless.

Assay: It is titrated acidimetrically. Zinc oxide and ammonium chloride are dissolved in an excess of standard sulphuric acid with the help of heat, if necessary. The excess of acid is titrated with standard sodium hydroxide using methyl orange as indicator.

Ammonium chloride is used for preventing the precipitation of Zinc hydroxide during the titration (due to local difference is concentration) as also at the end point. The procedure is as follows:

An accurately weighed amount, about 1.5g of ZnO is dissolved with 2.5 g of ammonium chloride in 50 ml of 1N sulphuric acid.

The excess of acid is titrated with 1N sodium hydroxide using methyl orange as indicator.

$$ZnO + H_2SO_4 \rightarrow ZnSO_4 + H_2O \quad .....(11.4)$$

Each ml of 1N sulphuric acid is equivalent to 0.04069 g of ZnO

Storage: Zinc oxide is preserved in a well closed container.

Uses: Zinc oxide is used for its mild astringent, antiseptic and protective action on the skin.

- It is widely employed in the treatment of skin diseases in the form of ointment and pastes, which are official in various pharmacopoeias.
- It is used for treating eczema, impetigo, psoriasis ring worm etc. It also used in bandages and adhesives. Dentists use it as dental cement and for temporary fillings.

11.1 (c) Zinc Gelatin

Synonym: Unna's paste

Zinc gelatin contains 15 percent w/w of Zinc oxide.

Ingredients:

Zinc Oxide	150 g
Gelatin	150 g
Glycerin	350 g
Purified water Q.S to	350 ml

Preparation: The gelatin is softened thoroughly in purified water and heated with glycerin on water bath till it is dissolved. The weight of a mixture, if necessary, is adjusted to 850 g by adding more of purified water. Zinc oxide is then incorporated and the contents stirred until about to set.

Assay: An accurately weighted amount, 4g of Zinc gelatin is warmed with 25 ml of water until the basis is dissolved. The mixture is cooled and shaken wall with 25 ml of 1N sulphuric acid 0.1g of ammonium chloride is added and the excess of acid is titrated with 1N sodium hydroxide, using solution of methyl orange as indicator.

Each ml of 1N sulphuric acid is equivalent to 0.04069 g of ZnO.

Storage: Zinc gelatin is preserved in a well closed container.

Uses: It is used as a protective and is applied in the molten state between two layers of bandage for protective purpose and for supporting varicosities and lesions of the legs. The dressing is removed after about a fortnight on being soaked with warm water.

11.1 (d) Zinc Oxide Compound Paste

Synonym: Zinc paste

Zinc oxide compound paste contains 25% of Zinc oxide.

Ingredients

Zinc oxide	250 g
Starch	250 g
White soft paraffin	500 g

The white soft paraffin is melted and the Zinc oxide and the starch are incorporated and paste stirred until cold.

Assay: An accurately weighted amount about 2g ointment is taken in a tared porcelain crucible and melted gently. The heating is continued, gradually raising the temperature, until the mass is thoroughly charred. The charred mass is ignited until the residue becomes uniformly yellow. The residue is of ZnO and ignited to a constant weight.

Uses: Zinc oxide is used for its mild astringent, antiseptic and protective action on the skin. It is widely employed in the treatment of skin diseases in the form of ointment and pastes, which are official in various pharmacopoeias. It is used for treating eczema, impetigo, psoriasis, ringworm etc. It also used in bandages and adhesives. Dentists use it as dental cement and for temporary fillings.

11.1 (e) Zinc Oxide Ointment

Synonym: Zinc ointment

Zinc oxide ointment contains 15 percent of ZnO.

Ingredients:

Zinc oxide	150 g
Simple ointment	850 g

The Zinc oxide is titrated with a portion of the simple ointment until smoothed. The remainder of simple ointment is added gradually and then mixed thoroughly.

Tests for Identity

(i) The tests are performed on being strongly heated Zinc oxide assumes a yellow colour which disappears on cooling.

(ii) A solution of Zinc oxide in dilute hydrochloric acid after neutralization of the acid yields reactions characteristic of zinc.

Tests for Purity: It is tested for calcium, magnesium and other foreign substances. The test is carried out by taking the residue obtained in the assay and adding to it 6 ml of dilute hydrochloric acid. When no effervescence should occur upon heating the mixture on a water bath for fifteen minutes, no more than trace of insoluble residue should remain. The solution is filtered and diluted to 10 ml with water and to this is then added dilute ammonia solution, until the precipitate first formed is dissolved, followed by the addition of 2 ml of equal volumes of solutions of ammonium oxalate and sodium phosphate. The solution is observed for five minutes, during which time not more than slight turbidity is produced.

Assay: An accurately weighted amount about 2g of ointment is taken in a tared porcelain crucible and melted gently. The heating is continued, gradually raising the temperature, until the mass is thoroughly charred. The charred mass is ignited until the residue becomes uniformly yellow.

The residue is of ZnO and ignited to a constant weight.

Storage: The Zinc oxide ointment is preserved in a well closed container and is not allowed to be exposed to prolonged temperature exceeding 30°C.

Uses: Zinc oxide is used for its mild astringent antiseptic and protective action on the skin. It is widely employed in the treatment of skin diseases in the form of ointment and pastes, which are official in various pharmacopoeias. It is used for treating eczema, impetigo, psoriasis, ringworm etc. It is used even in bandages and adhesives. Dentists use it as dental cement and for temporary fillings.

11.1 (f) Hydrous Zinc Oxide Ointment

Hydrous Zinc oxide ointment contains 15% ZnO.

Ingredients

Zinc oxide	150 g
Hydrous Ointment	850 g

***Preparation*:** The Zinc oxide is titrated with a portion of the hydrous ointment until smooth. The remainder of the ointment is added gradually with through mixing.

Tests for Identity

(i) On being strongly heated, Zinc oxide assumes a yellow colour, which disappears on cooling.

(ii) A solution of Zinc oxide in dilute hydrochloric acid, after neutralization of the acid, yields reactions characteristic of Zinc.

Tests for Purity: It is tested for calcium, magnesium and other foreign substances. The test is carried out by taking the residue obtained in the assay and adding to it 6 ml of dilute hydrochloric acid. When no effervescence should occur upon heating the mixture on a water bath for fifteen minutes, no more than trace of insoluble residue should remain. The solution is filtered and diluted to 10 ml with water and to this is then added dilute ammonia solution until the precipitate first formed is dissolved followed by the addition of 2 ml of equal volumes of solutions of ammonium oxalate and sodium phosphate. The solution is observed for five minutes, during which time not more than slight turbidity is produced.

Assay: An accurately weighted amount about 2g of ointment is taken in a tarred porcelain crucible and melted gently. The heating is continued, gradually raising the temperature, until the mass is thoroughly charred. The charred mass is ignited until the residue becomes uniformly yellow. The residue is of ZnO and ignited to a constant weight.

Uses: Zinc Oxide is used for its mild astringent, antiseptic and protective action on the skin. It is widely employed in the treatment of skin diseases in the form of ointment and pastes, which are official in various pharmacopoeias. It is used for treating eczema, impetigo, psoriasis, ringworm etc. It is used in bandages and adhesives. Dentists use it as dental cement and for temporary fillings.

11.1 (g) Calamine

Synonym: Prepared calamine

Calamine is a Zinc oxide with small amount of ferric oxide and contains after ignition 98% to 100.5% of Zinc oxide, ZnO.

Preparation: Calamine is prepared by thoroughly mixing Zinc oxide with ferric oxide (0.5 to 1%). It is used to be prepared earlier by roasting Zinc carbonate, which alone used to be known as calamine.

Description: Calamine is pink powder, passing thoroughly through a No. 100 mesh sieve. It is odorless and tasteless.

It is insoluble in water, but dissolves completely in mineral acids.

Tests for Identity:

(i) The filtrate, obtained after dissolving 1g of calamine in 10 ml of dilute hydrochloric acid yields reactions characteristic of Zinc.

(ii) A reddish brown colour is obtained if solution of ammonium thiocyanate is added to the filtrate, obtained after boiling 1g of calamine with 10 ml of dilute hydrochloric acid and filtering the solution.

Tests for purity: It is tested for acid insoluble substances, alkaline substances, arsenic, calcium, magnesium, lead, water soluble dyes, alcohol soluble dyes and loss on ignition.

Acid-insoluble substances are tested dyes: by dissolving 1g in 25 ml of warm dil-hydrochloric acid. The insoluble residue, if any is collected on a tared filter, washed with water, dried to constant weight at 105°, cooled and weighed. The residue on cooling should not weigh more than 20 mg.

Alkaline substances are tested by digesting 1 g with 20 ml of warm water, filtering the solution and adding 2 drops of the solution of phenolphthalein to the filtrate. The red colour, if produced, should not require more than 0.2 ml of 0.1N sulphuric acid to discharge it.

Arsenic: Not more than 8 PPm

Calcium: A weighed quantity of sample (0.5 g) is dissolved in water (10 ml) made acidic with glacial acetic acid (2.5 ml) and warmed on a water bath to effect complete dissolution.

The solution is filtered and to the filtrate (0.5 ml) are added 5N ammonia (15 ml) and a 2.5% w/v solution of ammonium oxalate (2 ml).

The solution remains clear on standing for 2 minutes.

Calcium and Magnesium: The test is performed by taking 10 ml of solution prepared in the test for calcium above. To the solution is added 2 ml of solution of sodium phosphate. The turbidity produced should not be more than slight.

Lead: A weighed quantity of the sample (2g) is dissolved in a mixture of water and glacial acetic acid (20 ml: 5 ml), filtered and to the filtrate is added solution of potassium chromate (0.25 ml). The solution remains clear for 5 minutes.

Water Soluble Dyes: A weighed quantity of the sample (2g) is shaken with water (10 ml) and filtered. The filtrate is colourless.

Ethanol Soluble Dyes: Same as water soluble dyes test. Alcohol (90%) is used instead of water. The filtrate is colorless

Loss on Ignition: The sample (2g) is ignited at temperature not less than 90°C. Not more than 2% of the sample is lost on ignition.

Storage: It is stored in a well closed container.

Assay: The assay procedure is based on the principle of volumetric titration.

A weighed quantity of the sample is dissolved in a measured volume of NH_2SO_4 and the excess of H_2SO_4 is then determined by titration with 1N NaOH using methyl orange as an indicator. Zinc sulphate ($ZnSO_4$) may react with NaOH to give Zinc hydroxide $Zn(OH)_2$

$$ZnSO_4 + 2\,NaOH \rightarrow Zn(OH)_2 \downarrow + Na_2SO_4 \quad .....(11.5)$$

The $Zn(OH)_2$ gets precipitated and it leads to a poor end point, both during the titration (by local differences in concentration) and at the finish. In order to present a poor end-point, ammonium chloride is used.

If more amount of NaOH is used, it is possible to dissolve the precipitate but the end-point is not correct because NaOH with $Zn(OH)_2$ forms sodium Zincate (ZnO_2Na_2)

$$Zn(OH)_2 + \text{Excess } 2\,NaOH \rightarrow ZnO_2\,NaOH + 2H_2O \quad .....(11.6)$$

An accurately weighed sample (1.5 g) from freshly ignited calamine is dissolved in 1N sulphuric acid (50 ml) and warmed gently if required.

The solution is filtered and the residue on filter is washed with hot water, till the washings are not acidic to litmus paper.

So the combined filtrate and washing is added ammonium chloride (2.5 g) (to prevent precipitation of Zinc hydroxide) and the solution is titrated with 1N NaOH, using solution of methyl red as indicator

$$ZnO + H_2SO_4 \rightarrow ZnSO_2 + H_2O \quad .....(11.7)$$

1 ml of 1N H_2SO_4 is equivalent to 0.04069 of ZnO

Uses: Calamine is used as an astringent and protective. It is used for soothing purposes in ointments and lotions for sunburns etc.

Dose: In lotion and ointments in different concentrations.

11.1 (h) Calamine Lotion

Ingredients:

Calamine	150 g
Zinc oxide	50 g
Bentonite	30 g
Sodium citrate	5 g
Liquefied Phenol	5 ml
Glycerin	50 ml

Rose water of commerce sufficient to produce – 1000 ml calamine lotion is prepared as per calamine where by thoroughly mixing Zinc oxide with ferric oxide (0.5 to 1%). It is prepared earlier by roasting zinc carbonate, which alone used to be known as calamine.

The preparation is included in the general pharmacy syllabus and is prepared by titrating the first three ingredients with a solution of the sodium citrate in about 700 ml of rose water to which are added liquefied phenol and glycerin and sufficient rose water to make the volume to 1000 ml.

Storage: Calamine lotion is preserved in a well closed container.

Uses: It is used as protective.

11.1 (i) Zinc Stearate

Chemical Formula: $(CH_3(CH_2)_{16}COO)_2 . Zn$

Zinc Stearate consists of Zinc stearate $(CH_3(CH_2)_{16}COO)_2$ Zn, together with variable proportion of Zinc palmitate $(CH_3(CH2)_{14}COO)_2$ Zn. It contains not less than 13% and not more than 15.5 percent of ZnO.

Preparation: Zinc stearate is prepared by adding an aqueous solution of Zinc sulphate to a solution of sodium stearate and washing the precipitate with water, until freed from sulphide and drying it.

The commercial stearic acid, used for preparing sodium stearate is always available mixed with different proportions of palmitic acid. It is prepared by hydrolysis of fats and is subjected to partial purification to separate other fatty acids of low melting points, which are produced along with the stearic acid.

In order to prepare Zinc stearate, first sodium stearate is prepared by adding gradually with constant mixing. Calculated quantity of stearic acid to a hot solution of sodium hydroxide or sodium carbonate.

$$NaOH + C_{17}H_{35}COOH \rightarrow C_{17}H_{35}COONa + H_2O \quad .....(11.8)$$

$$NaCO_3 + 2C_{17}H_{35}COOH \rightarrow 2C_{17}H_{35}COONa + H_2O + CO_2 \quad .....(11.9)$$

The above solution of sodium stearate is allowed to cool and a solution of Zinc sulphate is added to it. The precipitated Zinc stearate is collected, washed and dried.

$$\underset{\text{Sodium stearate}}{2C_{17}H_{35}COONa} + \underset{\text{Zinc sulphate}}{ZnSO_4} \rightarrow \underset{\text{Zinc stearate}}{(C_{17}H_{35}COO)_2Zn} + \underset{\text{Sodium sulphate}}{Na_2SO_4} \quad .....(11.10)$$

Description: Zinc stearate is a light, fine, white impalpable, amorphous powder. It is free from gritty ness and has faint characteristic odor. It is unctuous to touch and adheres to the skin readily. Zinc stearate is insoluble in water, alcohol and ether. A suspension is neutral to moistened litmus paper.

Tests for Identity:

(i) Zinc stearate is easily hydrolyzed by heating with dilute mineral acid to give soluble zinc salt and an insoluble oily layer of stearic acid (as also palmitic acid)

$$2(C_{17}H_{35}COO)_2Zn + H_2SO_4 \rightarrow 2C_{17}H_{35}COH + ZnO_4 \quad .....(11.11)$$

$$(C_{17}H_{35}COO)_2Zn + 2HCl \rightarrow 2C_{17}H_{35}COH + ZnCl_2 \quad .....(11.12)$$

1 g of stearic acid is added to a mixture of 25 ml of water and 5 ml of sulphuric acid and the mixture is boiled.

The solution separates into an upper oily layer of stearic acid and an aqueous lower layer containing zinc salt which after neutralization, yields the reaction characteristic of zinc.

(ii) The oily layer consisting of stearic acid in the above test is separated and placed on a filter wetted with water and washed with boiling water to free it from sulphate or chloride. The stearic acid is colleted in a small beaker and cooled. The supernatant water, if any is poured off, the acid melted, filtered while hot in a dry beaker and dried at 105°C for 20 minutes. The melted stearic acid congeals at a temperature not below 54°.

Tests for purity: It is tested for reaction, alkalis and alkaline earths and free fatty acids.

Tests for Alkalis and Alkaline Earths: It is performed by boiling 1 g of the substance with a mixture of 25 ml of water and 5 ml of hydrochloric acid, filtering the fatty acids formed and washing the precipitate with 25 ml of hot water. From the filtrate, after making it alkaline with dilute solution of ammonia, Zinc sulphide is completely precipitated by using a solution of ammonium sulphide. The precipitate is filtered off and the filtrate evaporated to dryness after the addition of 0.5 ml of sulphuric acid. The residue on being ignited to constant weight should hot weight more than 20 mg.

Test for free Fatty Acids: It is carried out by mixing 5 g of the substance with 100 ml of solvent ether, shaking the mixture for half an hour. The mixture (suspension) is filtered and 50 ml of the filtrate is evaporated to dryness. The residue weighs not more than 50 mg.

Assay: In this assay the basic concept of complexometric titration is used. As per I.P standards, the presence of Zinc oxide (ZnO) should be determined.

Method: An accurately weighed amount of about 1 g of sample is boiled with 50 ml of 0.1N sulphuric acid until the fatty acid layer is clearly separated. More water is added to maintain the original volume. It is cooled and filtered. The residue is washed with water to remove all acids and then 15 ml of strong ammonia-ammonium chloride solution is added to the combined filtrate and washings. The resulting solution is heated to about 40° and is titrated with 0.05M disodium EDTA using 0.2 ml eriochrome black T solution as indicator.

The end point is blue colour

Each ml of 0.05 M disodium EDTA is equivalent to 0.004069 g of ZnO.

Uses: It is an important pharmaceutical aid and is widely used like magnesium stearate as a lubricant in tablet makings. Zinc stearate is a mild astringent and possesses antimicrobial properties.

- It is used in water repellent ointment and in dusting powders.
- It is very popularly used in dermatological practice because of its desiccating and protective properties.
- Its use as a routine dusting powder for children should be discouraged, as its inhalation is said to cause pulmonary inflammation.

11.1 (j) Titanium Dioxide

Chemical Formula: TiO_2 *Mol. Wt.* 79.90

I.P requires it to contain not less than 8% of TiO_2 calculated with reference to the dried substance.

Description: It is a white powder without odor and taste.

Solubility: It is insoluble in water and in dilute mineral acids. It can be dissolved slowly in hot sulphuric acid.

Tests for Identity:

(i) Dissolve specified amount of sample (0.5 g) along with anhydrous sodium sulphate (5 g) in water (10 ml). To this solution add H_2SO_4 (10 ml) and boil until a clear solution is obtained. Cool the solution and add more H_2SO_4 (30 ml of 25% w/v) and dilute with water (to 100 ml). Name this solution as A. To a specified volume of solution A (5 ml) add strong H_2O_2 (0.1 ml). It should develop an orange red colour.

(ii) To a specified volume of solution A (5 ml) add one piece of granulated zinc. The solution should develop a violet-blue colour after about 45 min.

Tests for Purity: It is tested for clarity and colour of solution. Acidity or Alkalinity, water soluble matter, As, Ba, Heavy metals, Iron, Loss on drying and loss on ignition.

Assay: An accurately weighed amount of titanium dioxide about 0.3 g is transferred to a beaker to which is added 20 ml of H_2SO_4 and 8 g of ammonium sulphate. The contents are mixed and heated till white fumes appear. The heating is continued over a strong flame until solution is affected. It is cooled and carefully diluted with 100 ml of water. The diluted solution is heated gently and boiled with continuous stirring. It is cooled, filtered and washed with several quantities each of 10 ml of water. To the combined filtrate and washings, 10 ml of strong ammonia solution is added and the solution cooled and diluted to 200 ml with water.

50 ml of the resulting solution is pipetted into a flask to which 100 ml of water and 4 ml of strong H_2O_2 solution are added. 50 ml of 0.05M disodium EDTA is added to the flask which is allowed to stand for five min. The pH of the solution is adjusted to 5.0 with

NaOH solution to which 5 g of hexamine is added. The contents are titrated with 0.05M Zinc chloride using xylenol orange solution as indicator.

Each ml of 0.05M disodium EDTA = 0.003995 g of TiO_2

Storage: Titanium dioxide is stored in well closed containers. Its contact with aluminum should be avoided.

Uses: TiO_2 is used as a pharmaceutical aid and as a topical protectant in ointments and creams.

11.1 (k) Aluminium

The official compounds are:

1. Aluminium hydroxide gel, BP, U.S.P, I.P.,
2. Aluminium sulphate, I.P., U.S.P.,
3. Aluminium hydroxide tablets, I.P. and
4. Aluminium U.S.P.

Aluminium is trivalent and strongly electro positive metal. Its hydroxide, Al $(OH)_3$ is amphoteric in character with weak basic properties. Aluminum is the third most abundant element in the earths crust and is the most abundant metal. The most important are of aluminium is bauxite, Al_2O_3. H_2O, but aluminum also occurs in many alumino silicate rocks and clays.

11.1 (l) Aluminium U.S.P

It is finely divided aluminium powder, oleic or stearic acid may be present as a lubricant.

Properties: Free flowing very fine, silvery powder and free from gritty or discolored particles.

Solubility: Insoluble in water or ethanol but soluble in hydrochloric acid.

Tests for Purity: Tests for acid-insoluble substances, arsenic, alkalis and alkaline earths, heavy metals, iron and suitability.

Acid-insoluble substances: A weighed sample is dissolved (as for as possible) in hydrochloric acid, the solution is filtered and the filtrate used in the test for alkalis and alkaline earth etc. The residue is washed and the lubricant extracted from it by means of acetone. The filtrate is evaporated to dryness and the residue weighed. The residue left in the crucible is washed, dried and weighed. The combined weight of the residue must not exceed 5% of the total weight of the sample taken.

Test for Suitability: Rubbing a sample between the fingers and the powder must be smooth and unctuous, showing the absence of gritty particles.

Assay: It is based upon the complexometric titration. Aluminium is only slowly complexes with edetate. Hence excess edetate is added and the mixture is heated to ensure complete complexation. Dithizone is used as an indicator.

A weighed sample is dissolved in HCL and the solution is heated and filtered from lubricant. To an aliquot of the filtrate and washing is added a known excess of M/20 sodium edetate and acetic acid – ammonium acetate buffer and the mixture is heated to complete complexation.

Dithizone is added and the excess edetate is back titrated with M/20 Zinc sulphate.

The difference in blank and back titration is equal to the amount of edetate consumed. Amount of Al is calculated from the amount of edetate consumed.

11.1 (m) Aluminum Paste U.S.P

Aluminium paste consists of aluminium, mineral oil and Zinc oxide ointment.

Assay: Sufficient quantity of dil. HCl is added to the preparation in order to extract Al and Zinc as their chloride.

$$Al \xrightarrow{HCl} AlCl_3 \qquad .....(11.13)$$

$$Zn \xrightarrow{HCl} ZnCl_2 \qquad .....(11.14)$$

After adjustment of pH and Zn are determined in a portion of the extract by the procedure described under Aluminium, above.

Zinc is determined in another portion of the extract after addition of ammonia ammonium chloride buffer. The mixture is cooled to below 5° (in order to prevent complexation of Al) mordant black is added and the zinc is titrated with M/20 sodium edetate. The difference between the titration represents the volume of edetate equivalent to the aluminium present.

11.2 Antimicrobial Agents

Antimicrobial agents are agents that are effective against microorganisms with reference to their specific activity, antimicrobial agents may be

(i) ***Antiseptic:*** an agent that kills or inhibits the growth of micro organisms when applied to the tissues.

(ii) ***Germicidal:*** This refers specifically to agents that kill microorganisms. Depending on the various classes or organisms they may be bactericidal, fungicidal, amoebicidal etc.

(iii) ***Disinfectant:*** an agent used to destroy microorganism that cause disease in man, animals or plants commonly applied to inanimate objects (e.g. instruments, equipment, rooms etc)

Those agents which kill microorganisms are called are germicidal, while those which do not called are germicidal, while those which do not kill the microbes but function by inhibiting their growth are suffixed 'stat', specifically they are bacteriostat, fungistat etc.

The mechanism of action of inorganic antimicrobial agents is by oxidation, halogenation or protein precipitation.

Antimicrobial agents that function through oxidative mechanisms are hydrogen peroxide, metal peroxides, per magnates, halogens (i.e., chlorine and iodine) and certain oxohalogen anions. They act on the – SH groups in cysteine present in most proteins to form a disulphide bridge between two – SH groups. This alters the specific function of the protein in the microorganism. This overall change is responsible for the ultimate destruction of the micro organisms.

In organic compounds like hypochlorite

Chlorinate primary and secondary amides present in the peptide linkage between amino acids groups comprising the protein molecule. This ultimately results in the distribution of the function of the microbial protein.

Metal ions of copper, silver, zinc and aluminium due to their charge and small ionic radius precipitate microbial protein through a complexation interaction. Certain metals like mercury, arsenic and antimony show enzyme specificity forming covalent bonds. The protein precipitant property of metal cations depends upon the concentration in which used. By increasing concentrations, antimicrobial, astringent, irritant and corrosive properties are successively available.

11.2 (a) Hydrogen Peroxide Solution

Chemical Formula: H_2O_2 *Mol. Wt.* 34.016

H_2O_2 solution is an aqueous solution of hydrogen peroxide and contains not less than 6% w/v of H_2O_2 corresponding to about 20 times its volume of available oxygen. H_2O_2 solution is colourless, odorless liquid. It has slight acidic taste. It is rapidly decomposed on coming in constant with oxidisable organic matter and with certain metals and also if allowed to become alkaline.

Preparation: It can be prepared by the following methods.

(i) ***From Barium Peroxide:*** This is an original method for the industrial manufacture of hydrogen peroxide. Barium peroxide is made into a thick paste in ice cold water. A calculated quantity of dilute sulphuric acid is also cooled in ice. The barium peroxide paste is then added to the well cooled dilute acid H_2O_2 and insoluble barium sulphate is formed. $BaSO_4$ is filtered off.

$$BaO_2 + H_2SO_4 \rightarrow BaSO_4 + H_2O_2 \quad .....(11.15)$$

The yield of H_2O_2 is 10-20%.

(ii) H_2O_2 can also be prepared by decomposing barium peroxide with phosphoric acid or by passing CO_2 through a suspension of barium peroxide in water.

$$3BaO_2 + 2H_3PO_4 \rightarrow Ba_3(PO_4)_2 + 3H_2O_2 \quad .....(11.16)$$

(iii) ***From Sodium Peroxide***: By treating sodium peroxide with dilute H_2SO_4 at low temperature, sodium sulphate crystallizes and H_2O_2 is distilled under 10 mm pressure.

$$Na_2O_2 + H_2SO_4 \rightarrow Na_2SO_4 + H_2O_2 \quad .....(11.17)$$

(iv) ***Electrolysis***: Now a days is manufactured by electrolysis of 50% ice cold sulphuric acid. Per sulphuric acid is obtained first, which on distillation under reduced pressure gives H_2O_2 with 30% yield.

$$2H_2SO_4 \xrightarrow{\text{electrolysis}} \underset{\text{Persulphuric acid}}{H_2S_2O_8} + H_2 \quad .....(11.18)$$

$$H_2S_2O_8 + 2H_2O \xrightarrow[\text{Under reduced pressure}]{\text{Distilled}} 2H_2SO_4 + H_2O_2 \quad .....(11.19)$$

Description: Pure H_2O_2 is a colourless and odorless liquid. It slowly decomposes at ordinary temperature and readily when heated

$$2H_2O_2 \rightarrow 2H_2O + O_2$$

The decomposition is promoted by a catalyst like Cu, fe, Mn etc., while small quantity of acid H_2SO_4, H_3PO_4 and alcohol, if added retards the decomposition of H_2O_2. They act as negative catalysts and are therefore, used as preservatives or stabilizers in commercial preparation. Some of them are boric acid, urea, acetanilide or hexamine. H_2O_2 which is available in commerce is usually adjusted to pH 2-3. It is a strong oxidizing agent and it attacks many organic materials very violently. H_2O_2 solution is miscible with any amount of water from which it can be extracted with solvent ether.

Tests for Identity

(i) when made alkaline and heated it is decomposed with effervescence, evolving oxygen

(ii) The ethereal layer gets coloured blue, if shaken with H_2O_2 in presence of dilute H_2SO_4 and solution of potassium chromate.

Tests for Purity: It is tested for acidity, preservative, loss on evaporation, barium and stability.

Acidity: It is tested by using 0.1N sodium hydroxide using methyl red as indicator for 10 ml not less than 0.2 ml and not more than 1 ml of 0.1N NaOH is required.

***Test for*:** Preservative is performed by extracting 100 ml of H_2O_2 with a mixture of chloroform and solvent ether (in 3:2 ratio) and evaporating the extract to dryness at room temperature. For 100 ml of solution of H_2O_2 the residue does not weigh more than 50 mg.

***Test for*:** Barium depends upon treating the H_2O_2 with dilute H_2SO_4. No turbidity should be produced (Barium's presence in H_2O_2 is considered a possibility, if it is manufactured from BaO_2).

Loss on Evaporation: It is tested by evaporating on water bath in a platinum dish. It should leave not more than 0.2% of residue. Stability is tested as per the I.P test reproduced here under.

The solution should not lose more than 3% of its strength by weight. 100 ml of H_2O_2 (of 20 volume strength) is taken in a 250 ml round bottom neutral glass flask to which is attached a neutral glass condenser and the solution is refluxed for 30 minutes. The flask and condenser are then cooled, cleaned and kept exclusively for the following test; but before the test the flask and condenser are cleaned with nitric acid and thoroughly rinsed with purified water.

Test: 100 ml of the sample (H_2O_2 solution) is pipetted in the flask to which is attached the condenser by a ground-glass joint. The flask is covered up to the shoulder in a liquid bath of a mixture of glycerin and water in the ratio of 55.45 and maintained at a temperature of about 110°C. The sample is refluxed for exactly three hours and the time period is counted from the commencement of the boiling. After three hours the boiling is discontinued and the flask is disconnected and covered with a clear paper cap and cooled rapidly under running water to attain room temperature, taking care to avoid any contamination with extraneous matter. The content of H_2O_2 solution is determined by the usual assay for H_2O_2.

Assay: 10 ml of H_2O_2 is diluted to 250 ml in a volumetric flask. To 25 ml of diluted solution is added 10 ml 5N H_2SO_4 and the solution titrated with 0.1N potassium permanganate till pink colour appears and persists.

Each ml of 0.1N potassium permanganate is equivalent to 0.001701 g of H_2O_2.

Storage: H_2O_2 is preserved in a light resistant container with stopper resistant to H_2O_2. It is stored in a dark and cool place.

Uses: It is used as a germicide and deodorant. The liberation of gaseous O_2 provides an additional cleansing action on cuts and wounds. It is used for bleaching the hair. It is considered as an effective oxidizing antidote for phosphorous and cyanide poisonings. It is used for cleaning ears and removing the surgical dressings.

It is very strong oxidizing agent because it yields nascent oxygen. The medicinal as well as industrial use of H_2O_2 is based on the liberation of nascent oxygen.

11.2 (b) Sodium Perborate

Chemical Formula: $NaBO_3 . 4H_2O$ *Mol. Wt.* 153.9

Synonym: Sod. Perbor

Sodium per borate contains not less than 96.0 percent and not more than the equivalent of 103.0 percent of $NaBO_3 . 4H_2O$

Description: It is white, crystalline granules or a white powder. It is odourless and has saline taste.

Solubility: It is soluble in 40 parts of water. It is more soluble in boric, tartaric or citric acids and in glycerin. Its solubility is also increased by the presence of magnesium or ammonium sulphate.

Tests for Identity

(i) An acidified solution of the substance discharges the colour of potassium per magnate and produces a violet colour with a solution of potassium iodide.

(ii) 1 ml saturated solution is mixed with 1 ml of dilute sulphuric acid and 0.2 ml of solution of potassium dichromate and shaken with 2 ml of solvent ether on being separated; the solvent ether layer imparts a blue colour.

(iii) A saturated solution of the substance is alkaline to solution of phenolphthalein and when acidified with hydrochloric acid, it yields reactions characteristic of sodium and borate.

Tests for Purity: It is tested for alkalinity and heavy metals.

Tests for alkalinity are performed by noting the reaction. A 1% solution is alkaline to litmus solution.

Test for Heavy Metals: It is performed by dissolving 1g in 10 ml of water and 5 ml dilute hydrochloric acid. The solution is evaporated to dryness on a water bath with frequent stirring. The residue is dissolved in 10 ml of water and again evaporated to dryness on a water bath.

The residue is then dissolved in 23 ml of water and to the solution is added 2 ml of 0.1N hydrochloric acid and the test for heavy metals applied. The limit of heavy metals is 20 parts per million.

Assay: An accurately weighed about 0.3 g of substance is dissolved in 50 ml of water and to the solution is added 10 ml of dilute sulphuric acid and solution titrated with 0.1N potassium permanganate.

Each ml of 0.1N potassium permanganate is equivalent to 0.007695g of $NaBO_3 . 4H_2O$.

Storage: Sodium per borate is preserved in a tight container and stored in a cool place.

Uses: It is an oxidant and local anti infective.

11.2 (c) Potassium Permanganate

Chemical Formula: $KMnO_4$ *Mol. Wt.* 158.0

It contains not less than 99.0% $KMnO_4$.

Preparation: It is prepared by heating a solution of KOH with MnO_2 and potassium chlorate.

$$6KOH + 3MnO_2 + KClO_3 \rightarrow \underset{\text{Potassium per permanganate}}{3KMnO_4} + KCl + 3H_2O \quad .....(11.20)$$

The solution is then evaporated to a green mass of K_2MnO_4. The green mass is extracted with boiling water and a current of chlorine is passed through the solution until all the potassium manganate is converted to $2KMnO_4$.

$$2K_2MnO_4 + Cl_2 \rightarrow 2KCl + 2MnO4 \quad .....(11.21)$$

Carbon dioxide can also be passed through the solution in place of chlorine, when two-third of manganate is converted as follows:

$$3K_2MnO_4 + 2CO_2 \rightarrow 2KMnO_4 + MnO_2 + 2K_2CO_3 \quad .....(11.22)$$

The solution is evaporated to crystallization.

Description: It occurs as dark purple, slender, prismatic crystals with metallic luster. It has no odour but sweet astringent taste. It is soluble in 13 parts of water and 3.5 parts of boiling water.

It decomposes in presence of traces of organic matters.

Tests for Purity: It is tested for Cl and SO_4. Since the purple colour of $KMnO_4$ may interfere with the test for purity, the colour is destroyed by 1000 ml of N/10 $Na_2S_2O_3$ is equivalent to 1/10 I_2 or 1 ml of N/10 $Na_2S_2O_3$ is equivalent to 0.01269g of I_2.

For Potassium Iodide: To 10 ml of the diluted solution is added 20 ml of H_2O followed by 40 ml of HCl and the solution is titrated with 0.05M potassium iodate, shaking vigorously until the dark brown colour becomes light brown. Then 5 ml chloroform is added and the titration is continued till chloroform becomes colourless and the supernatant liquid is clear yellow. From the number of ml of potassium iodate required is subtracted one-quarter of the number of ml 0.1N sodium thiosulphate used in the assay for iodine.

$$KIO_3 + 6\,HCl + 2KI \rightarrow 3KCl + \underset{\text{Iodine Monochloride}}{3ICl} + 3H_2O \quad .....(11.23)$$

1000 ml 0.05M KIO_3 is equivalent to 1/10 KI or each ml of 0.05M potassium iodate is equivalent to 0.0166 g of KI or 0.01499 g of NaI, if sodium iodide has been used for preparing solution.

Storage: Lugol's solution is preserved in a well closed container, the materials of which are resistant to iodine e.g: glass, plastic etc.

Uses: It is a good source of iodine and is taken internally.

It is a germicide and fungicide and does not cause irritation to cuts like tincture of iodine.

11.2 (d) Iodine

Chemical Formula: I_2 *Mol. Wt.* 126.9

Iodine of I.P standard contains not less than 99.5 percent of I_2.

Iodine compounds are quite common in nature. Sea water also contains traces of combined iodine, which is absorbed by some specific plants and sea weeds, like laminaria digitata fucus vesiculosus. Iodine is also present in the form of sodium iodate in crude Chile Saltpetre ($NaNO_3$).

More recently iodine has been found in the brine of oil wells and this source is utilized now a days for obtaining iodine in U.S.A and Russia.

Description: Iodine is heavy bluish black, brittle and occurs in rhombic prisms or plates with a metallic luster. It has a characteristic odour and it volatilizes at ordinary temperature. It melts at 114° and at or below 700°, it gives rise to one of the heaviest vapours. At higher temperature the density of iodine decreases as it dissociates.

$$I_2 \rightleftharpoons 2I^- \qquad .....(11.24)$$

Iodine is present in thyroid glands. Its deficiency causes serious diseases, the outstanding iodine deficiency disease being goiter.

Children, who have iodine deficiency snow depressed growth and their sexual developments is retarded. The skin becomes rough and hair becomes thin. If the deficiency is severe cretinism, feeble mindedness and deafness may occur. The reproductivity in the female is impaired while fertility in the male decreases. Goiter usually occurs in children. Iodized salt is marketed which reduces the incidence of goiter. Iodine is said to be helpful for the production of thyroid hormones.

Iodine is almost insoluble in water (one part in 3,000) but us soluble in 12 parts of alcohol, 4 parts of carbon disulphide. It is more freely soluble in chloroform, carbon tetra chlorides, solvent ether and carbon and in aqueous solutions of iodides.

Tests for Identity:

(i) Gives off violet vapour upon heating

(ii) Gives a deep blue colour with solution of potassium iodide in presence of starch.

Tests for Purity: It is tested for chloride and bromide, cyanogens and non volatile matter.

Chloride and Bromide: are tested by triturating thoroughly 3.5 g with 35 ml of water and decolorizing the filtrate by a little zinc powder. To 25 ml of the decolorized filtrate is added 5 ml of dilute ammonia solution followed by 5 ml of silver nitrate and diluting the filtrate to 50 ml and acidifying it with 4 ml of nitric acid. The opalescence produced should not be more than the opalescence produced in the limit test for chlorides boiling with 95% alcohol which does not interfere with the test.

$$2KMnO_4 + 3C_2H_5OH \rightarrow 2MnO_2 + 2KOH + 2H_2O + 3CH_2CHO \quad .....(11.25)$$

The precipitated MnO_2 is removed by filtration. Test for Cl and SO_4 is carried out in the usual manner.

Assay: The assay involves an oxidation reduction type of reaction. A solution of weighed sample is titrated with standard N/10 oxalic acid. Here $KMNO_4$ is taken in the burette. Excess of dilute H_2SO_4 is added to the $H_2C_2O_4$ before commencing the titration and the temperature is maintained at 70°C throughout the titration otherwise the reaction is slow.

$$2KMnO_4 + 3H_2SO_4 \rightarrow K_2SO_4 + 2MnSO_4 + 3H_2O + 5(0)$$

$$5H_2C_2O_4 + 5(0) \rightarrow 10CO_2 + 5H_2O \quad .....(11.26)$$

1000 ml of N/10 $H_2C_2O_4$ is equivalent to 1/50 $KMnO_4$ or 1 ml of 0.1N oxalic acid is equivalent to 0.00316 g of $KMnO_4$.

Storage: It is preserved in a well closed container, while handling potassium permanganate care must be taken as dangerous explosions may take place if it is brought in contact with organic or other readily oxidized substances, either in solution or dry state.

Uses: It is used as an antiseptic in mouthwash and in cleaning ulcers or abscesses. As anti infective it is considered of immense value. It is used in the treatment of urethritis. It is capable of oxidizing some drugs and venoms and hence it is used in case of poisonings by barbiturates, chloral hydrate, many alkaloids etc. A solution of potassium permanganate destroys the poison and prevents absorption. The solution of potassium permanganate should not be left in the stomach. In veterinary practice it is very much used as an antiseptic.

Preparation: Iodine is manufactured from kelp (sea-weeds ash) which is extracted with water. The solution is concentrated when the sulphates and chlorides of sodium and potassium crystallize out, leaving freely soluble sodium and potassium iodides in solution.

From the number of ml of 0.05M KIO_3 is subtracted ½ the number of ml of 0.1N sodium thiosulphate required for assay of iodine.

1000 ml of KIO_3 is equivalent to 1/10 KI or each ml of the remainder 0.05M potassium iodate is equivalent to 0.0166 g of KI (or) 0.01499 g of NaI.

Storage: It is stored under conditions similar to aqueous iodine solution.

Uses: It is a very well known and popular antiseptic.

11.2 (e) Weak Iodine Solution

Synonym*:* Iodine Tincture.

Iodine tincture contains 2 percent w/v of iodine and 2.5 percent w/v of potassium iodine.

Ingredients

Iodine	20 g
Potassium iodide	25 g
Alcohol 50% sufficient to produce	1000 g

Potassium iodide and iodine are dissolved in sufficient alcohol (50%) and more alcohol is added to produce the required volume.

Alcohol content 45-58 percent w/v

Description and Tests for Identity: are similar to those described under aqueous iodine solution.

Assay for Iodine: 10 ml of the solution diluted with 20 ml of water is titrated with 0.1 N sodium thiosulphate.

1000 ml of N/10 $Na_2S_2O_3$ is equivalent to 1/10 I_2 or each ml of 0.1N sodium thiosulphate is equivalent to 0.01269 g of I_2.

For Potassium Iodide: To 10 ml is added 40 ml of water, 16 ml of HCl and 10 ml of solution of potassium cyanide and the solution is titrated with 0.05M potassium iodide until dark brown solution turns pale yellow. 5 ml of solution of starch is then added and the titration is continued until the liquid becomes colourless.

Cyanogens is tested by taking 5 ml of the above filtrate and adding to it few drops of solution of ferrous sulphate and 1 ml of solution of sodium hydroxide, warming the solution gently and acidifying with hydrochloric acid. No blue colour should be produced.

Non-Volatile Matter: Not more than 0.5% residue on being volatilized on water bath.

Assay: An accurately weighed amount (about 0.5g) dissolved in a solution of potassium iodide (1g in 5 ml of H_2O) is slightly acidified with dilute acetic acid (1 ml) and titrated with N/10 sodium thiosulphate using starch mucilage as indicator. Because iodine volatizes, it should be weighed in a stoppered vessel

$$2\,Na_2S_2O_3 + I_2 \rightarrow Na_2S_4O_6 + 2NaI \qquad .....(11.27)$$

1 ml of N/10 $Na_2S_2O_3$ is equivalent to 0.01269 g of I_2.

Storage: Iodine is to be preserved in a glass stoppered bottle.

Uses: Iodine is anti infective and is used as local germicide. For proper thyroid functioning iodine is to be supplied to the body to be utilized physiologically either in elemental form or in the form of I_2 ion as in sodium (or) potassium iodide.

Elementary iodine is highly toxic and the starch and sodium thiosulphate are useful antidotes.

11.2 (f) Aqueous Iodine Solution

Synonym*:* Lugol's solution.

Aqueous iodine solution contains 5 percent w/w of aqueous iodine, I_2 (limits 4.9 to 5.1) and 10 percent w/w of potassium iodine, KI (limits 9.8 to 10.8).

Ingredients

Iodine	50 g
Potassium iodide	100 g
Purified water, sufficient to produce	1000 ml

Preparation: potassium iodide and iodine are dissolved in 100 ml of H_2O and the volume is made up to 1000 ml.

Description: Lugols solution is a transparent liquid having brown colour and odour of iodine.

Tests for Identity:

(i) Diluted solution gives blue colour with the solution of starch.

(ii) The residue left after evaporation of solution and after ignition (evaporation of I_2) yields reactions characteristic of potassium and iodide.

Assay: The assay is carried out for iodine and potassium iodide. 25 ml of the solution is diluted to 100 ml with water and this dilute solution is used for the following assays.

For Iodine: To 20 ml of the diluted solution is added 10 ml of H_2O and the solution is titrated with N/10 $Na_2S_2O_3$.

The solution after the addition of sulphuric acid is decanted. The decanted mother liquid is treated with MnO_2 and warmed to collect iodine which distills over.

$$2NaI + 2H_2SO_4 + MnO_2 \rightarrow MnSO_4 + Na_2SO_4 + I_2 + 2H_2O \qquad .....(11.28)$$

Alternatively the solution containing freely soluble iodides is treated with required proportion of chlorine and the precipitated iodine is collected and purified by sublimation.

11.2 (g) Strong Iodine Solution

It contains 10 percent of I_2 and 6% of KI.

Ingredients

Iodine	100 g
Potassium iodide	60 g
Purified water	100 ml
Alcohol 50% sufficient to produce	1000 ml

Potassium iodide and iodine are dissolved in purified water and sufficient alcohol is added to produce 1000 ml.

Description and tests for identity are same as reported under aqueous iodine solution (11.2.f). Alcohol content 74 to 79 percent w/v.

Storage: It is stored under conditions as described for aqueous iodine solution.

Uses: It is used as an antiseptic. In making all the above solutions, potassium iodine may be replaced by sodium iodide.

11.2 (h) Potassium Iodide

Chemical Formula: KI *Mol. Wt.* 166.0

It contains not less than 90% of KI calculated with reference to the substance dried to constant weight at 105°C.

Preparation:

(i) Potassium iodide is prepared by treating a hot aqueous solution of potassium hydroxide with iodine in slight excess to form a mixture of potassium iodide and potassium iodate.

$$6KOH + 3I_2 \rightarrow 5KI + KIO_3 + 3H_2O \qquad .....(11.29)$$

$$KIO_3 + 3C \rightarrow KI + 3CO \qquad .....(11.30)$$

The solution after concentrating is treated with excess of charcoal powder; evaporating the mixture to dryness followed by ignition. The charcoal (carbon) reduces the iodate to iodine, utilizing thus the total iodine to obtain the potassium iodide. The product is lixiviated with water, filtered and potassium iodide is obtained by crystallization.

(ii) Potassium iodide is also prepared from ferroso ferric iodide in the same way

$$fe_3I_8 + 4K_2CO_3 \rightarrow 8KI + fe_3O_4 + 4CO_2 \qquad .....(11.31)$$

Iodide of iron i.e., ferroso ferric iodide is boiled with potassium carbonate solution and filtered. The filtrate is evaporated to dryness. The dried residue is extracted with water and recrystallized.

Description: It occurs as colourless transparent or some white opaque crystals or a white granular powder. It is odourless but has sometimes, slightly bitter taste. It is soluble in water (1g in 0.7 ml) glycerin (1g in 2 ml), alcohol (1g in 23 ml) and in acetone (1g in 75 ml).

Tests for Purity: It is tested for AS, Ca, SO_4, alkalinity, iodate, loss on drying, cyanide and heavy metals.

Assay: An accurately weighed quantity (0.5g) of potassium iodide is dissolved in water (50 ml), acidified with HCl (15ml) and treated with a solution of KCN (6 ml) maintaining the temperature at 15°C. The solution is titrated with 0.05M KIO_3 until the dark brown coloured solution becomes pale yellow. 5ml of starch solution is added and the solution titrated slowly further until the liquid becomes colourless.

$$KIO_3 + 2KI + 3KCN + 6HCl \rightarrow 6KCl + 3ICN + 3H_2O \quad .....(11.32)$$

1 ml of 0.05M potassium iodide is equivalent to 0.0166 g of KI

Note: Molar solution is used because normality varies depending on the nature of reaction, and uniform reduction of iodate to iodide is not feasible in direct titrimetic method.

Storage: It is preserved in well closed container.

Uses: It is used as an expectorant and a source of iodine. It can be used in the form of an iodized salts (one in 1,00,000 parts of salt) to prevent goiter, in places, where iodine in the diet is not in sufficient supply. It is also used in some kinds of antifungal therapy and in veterinary practice.

11.2 (i) Sodium Iodide

Chemical Formula: NaI *Mol. Wt.* 149.9

Sodium iodide contains not less than 99% of NaI calculated with reference to the substance dried to constant weight at 105°

Preparation: The methods of preparation of sodium iodide are similar to those described under potassium iodide, the difference being in the use of iodine in place of bromine. The methods are repeated here:

(i) By action of iodine on sodium hydroxide: In this process when excess of iodine is added to the solution of sodium hydroxide, sodium iodate is formed, which is reduced with carbon to sodium iodide as follows:

$$6\,NaOH + 3I_2 \rightarrow 5\,NaI + NaIO_3 + 3H_2O$$

$$NaIO_3 + 3C \rightarrow NaI + 3CO$$

(ii) By metathesis between ferrosoferric iodide and sodium carbonate: In this method iodide is allowed to react with iron filings and the ferrous iodide formed is decomposed with Na_2CO_3

$$Fe + I_2 \rightarrow FeI_2 \qquad .....(11.33)$$

$$3FeI_2 + I_2 \rightarrow FeI_2.\,2FeI_2 \qquad .....(11.34)$$

$$\underset{}{FeI_2.\,2FeI_3} + 4\,Na_2CO_3 \rightarrow \underset{\text{Sodium iodide}}{8NaI} + \underset{\text{ferrosoferric oxide}}{FeO.\,Fe_2O_3} + 4CO_2 \qquad .....(11.35)$$

The precipitated ferrosoferric oxide is removed by filtration and from the filtrate; sodium iodide is obtained and recrystallized.

Description: It occurs as colourless, odourless crystals or as a white crystalline powder. In moist air it tends to cake and then deliquesce, undergoing often decomposition and developing a brown tint. An aqueous solution of sodium iodide is neutral or faintly alkaline to litmus and on keeping it becomes yellow due to the formation of free iodine sodium iodide on being dissolved in water liberates heat due to the formation of dehydrate (NaI. $2H_2O$) (distinction from potassium iodide).

Sodium iodide is soluble in water (1g in 0.6 ml), in alcohol (1g in 2 ml) and glycerin (1g in 1 ml).

Tests for Identity: It gives reaction characteristic of sodium and iodides (Appendix I)

Tests for Purity: It is tested for arsenic, barium, heavy metals, cyanide, iodate, sulphate and loss on drying.

Assay: The assay procedure is similar to that of potassium iodide. Here potassium cyanide is used. It is added in excess to the solution of the substance containing HCl. To the acidified solution is added potassium cyanide solution maintaining the temperature at 15°C and titrating with 0.05M potassium iodate, using starch mucilage as indicator towards the end-point

$$KIO_3 + 2\,NaI + 3KCN + 6\,HCl \rightarrow 2NaCl + 4KCI + 3ICN + 3H_2O \qquad .....(11.36)$$

Each ml of 0.05M potassium iodate is equivalent to 0.01499 g of NaI.

Storage: sodium iodide is preserved in a well closed container as its contact with moisture and light help in liberation of iodine. An aqueous solution, if alkaline, is preserved better than an acidic solution.

Uses: Its uses are similar to potassium iodide for its therapeutic action easily interchanged with potassium iodide for its therapeutic action. It is used as an expectorant and source of iodine. It can be used in the form of an iodized slats (one in 1,00,000 parts of salt) to prevent goiter, where iodine in diet is not in sufficient supply. It is also used in some kinds of antifungal therapy and in veterinary practice.

Dose: 0.3 to 0.5g

11.2 (j) Silver Nitrate

Chemical Formula: $AgNO_3$ *Mol. Wt.* 169.0

Silver nitrate when powdered and dried in the dark over sulphuric acid for four hours contains not less than 99.8% of $AgNO_3$.

***Preparation*:** It is prepared by the action of hot nitric acid on silver metal. Nitric acid should be nearly concentrated.

$$Ag + 2H\,NO_3 \rightarrow AgNO_3 + NO_2 + H_2O \qquad .....(11.37)$$

The solution is evaporated to dryness and the residue is heated to expel nitric acid. Silver nitrate is crystallized.

Description: It is colourless or white crystals which are odourless, but have bitter and metallic taste. It darkens when exposed to atmospheric air due to reduction to metallic silver. It melts at 209°C and in soluble in water (0.4 parts), alcohol (30 parts) and only slightly in solvent ether.

Tests for Identity: Yields reactions for silver and nitrates.

Tests for Purity: It is tested for bismuth, copper and lead.

A solution in water is neutral to litmus. A solution of 2g in 20 ml of water should be clear and colourless.

Assay: It is assayed by titration with standard N/10 ammonium thiocyanate solution, using solution of ferric ammonium sulphate as indicator (about 0.50g is accurately weighed, dissolved in 50 ml of water and to the solution 2 ml of acetic acid is added before titration).

$$AgNO_3 + NH_4SCN \rightarrow Ag\,SCN + NH_4NO_3 \qquad .....(11.38)$$

Each ml of 1/10 N NH_4SCN is equivalent to 0.0169 g of $AgNO_3$

Storage: It is stored and preserved in a well closed container, protected from light.

Uses: It is used as an anti-infective or antibacterial agent. It is also used as a pharmaceutical aid in the preparation of silver proteins. As an antiseptic, silver nitrate has a very broad spectrum activity. It is very successfully used to manage severely burnt patients. Silver ion combines readily with protein and due to this the astringent, caustic and perhaps the germicidal properties are attributed to it. It is also used to prevent gonococcal eye infections of new borns.

11.2 (k) Mild Silver Protein

It is silver rendered colloidal by the presence of protein or in combination with it. It contains not less than 19.0% and not more than 23.0% of Ag.

Mild silver protein should be freshly prepared and dispensed in amber colour bottles.

Description: It is dark brown or almost black shining scales or granules, odourless, hygroscopic. It is affected by light. It is freely soluble in water forming a dark coloured solution; practically insoluble in alcohol, chloroform and ether.

Tests for Identity: It is identified by the following three tests.

(i) Heat 1 g of mild silver protein, charring will take place. Incinerate completely, a grayish white residue is obtained which after solution in nitric acid and neutralization yields the reactions characteristic of silver.

(ii) To a 1 percent w/v solution is added solution of ferric chloride, the dark colour is discharged and the solution becomes opalescent on standing.

(iii) To a 1 percent w/v solution is added solution of mercuric chloride, no white precipitate is produced and the liquid is not decolorized (distinction from strong silver protein)

Test for Silver Salts: This is done by shaking 1g of mild silver protein with 10 ml of alcohol (90 percent) and adding to the filtrate 2ml of dilute hydrochloride acid. No opalescence should be produced.

Distinction from Strong Silver Protein: Dissolve 1g in 10 ml of water. Add all at once, 7 g of ammonium sulphate and stir occasionally for 30 minutes. Filter through a quantitative filter paper into a 50 ml Nessler tube, returning the first portion of the filtrate to the filter, in necessary, to secure a clear filtrate and allow the filter and precipitate to drain. Add to the clear filtrate, 25 ml of a 1 percent w/v solution of Indian gum. In a second 50 Nessler tube dissolve 7g of ammonium sulphate in 10 ml of water and add to this solution 25 ml of the solution of Indian gum and 1.6 ml of 0.01N silver nitrate. To each tube add 2 ml of nitric acid, 2 ml of dilute hydrochloric acid and enough of the solution of Indian gum to make the volume of each solution 50 ml. Mix the contents of each tube thoroughly and allow to stand for five minutes. The turbidity of the mixture containing the mild silver protein is not greater than that to which no mild silver protein has been added (strong silver protein yields a much greater turbidity than the control).

Assay: Weigh accurately about 2g and ignite at first gently and afterwards strongly until all carbonaceous matter is destroyed. Dissolve the residue in 15 ml of nitric acid heat until no more nitrous fumes are evolved, dilute with water to 100 ml and titrate with 0.1N ammonium thiocyanate, using solution of ferric ammonium sulphate as indicator and shaking vigorously as the end point is neared.

Each ml of 0.1N ammonium thiocyanate is equivalent to 0.01079 g of Ag.

Storage: It is preserved in a well closed container and protected from light.

Uses: It is used as local antibacterial agent.

11.2 (I) Strong Silver Protein

Strong silver protein is a compound of silver and protein. It contains not less than 7.5 percent and not more than 8.5 percent of Ag.

Description: A brown powder, odourless, some what hygroscopic.

Solubility: Slowly soluble in water forming a dark brown solution, almost insoluble in alcohol, solvent ether and chloroform.

Tests for Identity: It complies with the test No. 1 and 2 as mentioned under mild silver protein and gives the following additional tests.

(i) on adding solution of mercuric chloride to a 1 percent w/v solution a white precipitate is formed and the liquid becomes colourless or almost colourless (distinction from mild silver protein)

(ii) To 5 ml of a 2 percent w/v solution, add 5 ml of solution of sodium hydroxide, 10 ml of water and 2 ml of a 2 percent w/v solution of copper sulphate, allow to stand for a few minutes, a violet colour is produced.

Silver Salts Test: Shake 1g with 10 ml of alcohol (90 percent) and filter. To the filtrate add 2 ml of dilute hydrochloric acid, no opalescence is produced.

Tests for foreign Protein:

(i) A 10% solution shows no deposit within ten minutes.

(ii) A 2 percent w/v solution shows no turbidity on the addition of an equal volume of solution of sodium chloride.

Assay: Weigh accurately about 2g and ignite at first gently and afterwards strongly until all carbonaceous matter is destroyed. Dissolve the residue in 10 ml of nitric acid, heat until no more nitrous fumes are evolved, dilute with water to 100 ml, and titrate with 0.1N ammonium thiocyanate using solution of ferric ammonium sulphate as indicator and shaking vigorously as the end-point is neared (each ml of 0.1N ammonium thiocyanate is equivalent to 0.01079 g of Ag)

Storage: It is preserved in a well closed container and protected from light.

Uses: It is used as a local antibacterial agent.

Note: Solution of strong silver protein should be freshly prepared and dispensed in amber coloured bottles.

11.2 (m) Mercury

Chemical Formula: Hg

Mercury contains not less than 99.5% of Hg.

Preparation: It is found in a free state. The chief ore of mercury is cinnabar (HgS). The ore is roasted first in air; sulphide is converted into sulphur oxide.

$$HgS + O_2 \rightarrow Hg + SO_2 \qquad .....(11.39)$$

The free mercury is liberated. It is either purified by volatilization or chemically dropping mercury into a column of dilute nitric acid to remove basic impurities.

Description: A shining silver white, heavy liquid, easily divisible into globules and extremely mobile, easily volatilizes on heating.

Practically insoluble in water, in alcohol and in hydrochloric acid, readily and completely soluble in nitric acid and in boiling sulphuric acid with evolution of SO_2 and formation of mercuric sulphate.

$$Hg + 2H_2SO_4 \rightarrow HgSO_4 + SO_2 + 2H_2O \qquad .....(11.40)$$

Density: 13.5 g/ml at 25°.

Assay: An accurately weighed quantity (0.49g) is dissolved in equal parts (20 ml) of water and nitric acid, heated gently until the solution is colorless. The solution is then diluted with water (150 ml) and sufficient quantity of potassium permanganate is added to produce permanent pink colour. A trace of ferrous sulphate to discharge pink colour is added and the solution is then titrated with standard 0.1N ammonium thiocyanate, using ferric ammonium sulphate as indicator. The temperature during the titration should not be allowed to exceed above 20°C.

$$3Hg + 8HNO_3 \rightarrow 3Hg\,(NO_3)_2 + 4H_2O + 2NO \qquad .....(11.41)$$

$$Hg\,(NO_3)_2 + 2NH_4SCN \rightarrow 2NH_4NO_3 + Hg\,(SCN)_2 \qquad .....(11.42)$$

Each ml 0.1N ammonium thiocyanate is equivalent to 0.01003 g of Hg

Uses: It is used as a pharmaceutical aid. Formerly metallic mercury was used as such therapeutically as a cathartic and parasiticide. But it is no more used as such because it is extremely poisonous and prolonged inhalation of even very minute amounts of mercury prove fatal. Almost all the salts of mercury with the exception of the sulphide are poisonous. Compounds of mercury now a day are used as diacritic, germicidal, antibacterial and anti infectious.

11.2 (n) Mercury with Chalk

Synonym: Grey powder

Mercury with chalk contains 33% of mercury and 66% of chalk, $CaCO_3$.

Ingredients

Mercury	33 g
Dextrose	1 g
Chalk	66 g

These ingredients are triturated together in a porcelain mortar until the mixture acquires a uniform pale colour and no metallic globules are visible when examined under a magnifying lens of 4 diameters.

Tests for Identity

(i) With dil. hydrochloric acid gives effervescences due to the production of CO_2 which turns lime water milky.

(ii) if a mixture of mercury with chalk and sodium carbonate in equal proportions is heated gently in a test tube for 10 minutes and after cooling emptied on a piece of white paper, globules of mercury separate and are seen.

(iii) On being triturated (1g) with precipitated sulphur (0.5g) in a porcelain mortar, the powder turns black due to the formation of mercuric sulphide.

(iv) It gives test for dextrose, when filtered aqueous solution is made alkaline and treated with Fehling's solution by forming a brick-red precipitate.

Test for Purity: It is tested for AS.

Assay for Mercury: An accurately weighed amount (about 1.2g) is boiled gently under a reflex condenser for five minutes with 10 ml of nitric acid and 25 ml of water and cooled. The condenser is washed with 25 ml of water and sufficient solution of potassium permanganate is added to produce a permanent pink colour and rest of the assay is same as described under mercury.

For Chalk: An accurately weighed amount (about 1g) is dissolved in 100 ml of water and 50 ml of 1N hydrochloric acid in a 250 ml conical flask. The solution is set aside till the reaction ceases and the excess of acid is titrated with 1N sodium hydroxide using solution of phenolphthalein as indicator.

$$CaCO_3 + 2HCl \rightarrow CaCl_2 + H_2O + CO_2 \quad .....(11.43)$$

Each ml of 1N HCl is equivalent to 0.05005 g of $CaCO_3$

Uses: It is used as a purgative

Dose: 60 to 300mg.

11.2 (o) Yellow Mercuric Oxide

Chemical Formula: HgO *Mol. Wt.* 216.3

It contains not less than 99.5% of HgO, calculated with reference to the substance dried at 105° for one hour.

Preparation: Since the yellow variety is affected by light, all operations are carried out in the dark, so that a uniformly orange yellow product is obtained. The reaction between a concentrated solution of mercuric chloride and a dilute solution of sodium hydroxide is brought about at room temperature by pouring the former slowly into the latter. After allowing the reaction mixture to stand for about an hour to being about complete precipitation of yellow mercuric oxide, the supernatant liquid is poured off. The precipitate is washed with water until free from alkali, drained from calico filter and dried in a dark place, at a temperature not exceeding 30°C.

$$HgCl_2 + 2\,NaOH \rightarrow Hg\,(OH)_2 + 2NaCl \quad .....(11.44)$$

$$Hg\,(OH)_2 \rightarrow HgO + H_2O \quad .....\ (11.45)$$

Description: An orange yellow, heavy amorphous powder, odourless, stable in air, but becomes discolored on exposure to light.

Practically insoluble in water and alcohol, readily soluble in dilute HCl and in dilute nitric acid forming colourless solution.

Test for Purity: It is tested for reaction, mercurous salts, Cl, loss on drying and sulphated ash.

Assay: Mercuric oxide is assayed by dissolving it in dilute nitric acid and carrying out the assay by titrating with 0.1N NH_4SCN using ferric alum as indicator as in the case of mercury.

$$HgO + 2HNO_3 \rightarrow Hg\,(NO_3)_2 + H_2O \quad .....(11.46)$$

1ml of N/10 ammonium thiocyanate is equivalent to 0.01083 g of HgO.

Uses: It is used as a local anti-infective and antibacterial and is used in ointments.

Preparation: Mercuric oxide eye ointment and oleated mercury.

Mercuric Oxide Eye Ointment

It is prepared under aseptic conditions by triturating very finely powdered yellow oxide of mercury with a small portion of the melted base until the mixture is smooth and then adding gradually sufficient quantity of the melted base to produce the required weight, trituration being continued until the eye ointment attains room temperature.

The amount of mercuric oxide, HgO is not less than 95% and not more than 105% of the stated amount of yellow mercuric oxide.

Assay: To an accurately weighed quantity, equivalent to about 0.1g of yellow mercuric oxide is added 10 ml of dilute nitric acid and 20 ml of water and the mixture shaken until mercuric oxide is dissolved. 50 ml of water is added to the solution and it is titrated with 0.1N ammonium thiocyanate at a temperature not exceeding 20° using solution of ferric alum as indicator.

Each ml of N/10 ammonium thiocyanate is equivalent to 0.01083 g of HgO.

Uses: It is used in ophthalmology, 1% ointment is used for treating mild inflammatory conditions for the treatment of Blepharitis and conjunctivitis.

A 2% HgO ointment is used for eczema and other skin infections.

11.2 (p) Oleated Mercury

Oleated mercury contains the equivalent of 20% of yellow mercuric oxide (limits 19.0 to 21.0)

Ingredients:

Yellow mercuric oxide	200g
Liquid paraffin	50g
Oleic acid	750g

The mercuric oxide is triturated with the liquid paraffin until it is thoroughly subdivided. Oleic acid is then added and contents mixed thoroughly. The mixture is heated at 50°, triturating occasionally until combination is effected on being cooled, a yellowish preparation is obtained.

Assay: An accurately weighed amount about 0.75g is dissolved in a mixture of 65 ml of benzene, 10 ml of glacial acetic acid and 25 ml of alcohol and warmed on a water bath to about 50°. Hydrogen sulphide is passed in for ten minutes, and the contents filtered through asbestos in a Gooch crucible, precipitate washed first with hot benzene and then with a little alcohol and dried to constant weight at 120°.

Each g of residue is equivalent to 0.9309 g of HgO.

Uses: It is used as an anti-infective and as a pharmaceutical necessity.

11.2 (q) Sulphur

Sulphur is an important element as it is an important component of a large number of natural and synthetic pharmaceutical aids and medicinal compounds. It is available in nature in a free state or in combined forms and geographically it is distributed in volcanic areas, especially in Sicily and United States, where it is found in free state. Iron pyrite, feS_2, Gypsum, $CaSO_4$. $2H_2O$ and $BaSO_4$ are some of the important natural sources. Sulphur is obtained by mining operations. Sulphate ores afford good sources for some valuable metals like antimony, mercury, bismuth, lead, zinc and molybdenum. India imports sulphur for its requirement, but recently iron pyrites deposits have been found.

Sulphur as such i.e., in its elementary form is of great pharmaceutical value. It is an old and proven germicide and fungicide and for the same purpose, it has been and is being used in various dosage forms like dusting powders, ointments, creams, lotions etc. In the form of soluble sulphides, it is used in many skin affections, being perhaps the most beloved material for dermatologists.

A large number of sulphur salts, which may look as of insignificant value perhaps because they are cheap and easily available like sodium bisulphate. Sodium metabisulphite find their utility in pharmaceutical preparations as antioxidants, preservatives and stabilizers. Sodium thiosulphate is an important antidote to iodine and cyanide, besides being useful in different parasitic skin diseases. The importance of sulphur and its compounds will be known when the compounds (sulpha drugs to mention an important group) are studied both in inorganic and organic pharmaceutical chemistry.

The following inorganic materials/preparations with sulphur, as such or in combination are included in the following text. They are official in I.P.

(i) Precipitated Sulphur (Precip. Sulp):

Precipitated sulphur is expected to contain 99.5% of S, calculated on the basis of anhydrous sulphur.

Preparation: A method which is of theoretic interest consists in acidifying a solution of thiosulphate with a mineral acid when the unstable thiosulpuric acid, first liberated gets rapidly decomposed to give precipitated sulphur. This method is costly for industrial manufacture but can be conveniently tried on small scale on laboratory scale. The reactions taking place are given below:

$$Na_2S_2O_3 + 2HCl \rightarrow H_2S_2O_3 + 2NaCl \quad .....(11.47)$$

$$H_2S_2O_3 \rightarrow S + SO_2 + H_2O \quad .....(11.48)$$

The I.P. itself mentions the method of preparation of precipitated sulphur under its monograph, which through unusual is reproduced here. Precipitated sulphur may be obtained by adding hydrochloric acid to a solution prepared by boiling sulphur, lime acid water. In actual process slurry of slaked lime (1 part) with water (10 parts) is prepared. To this slurry is added sublimed or powdered roll sulphur (2 parts) contained in water (20 parts) and the mixture is mixed and boiled for an hour, with occasional shaking till sulphur is dissolved. The liquid is cooled and filtered. It contains a mixture of thiosulphates and poly sulphides of calcium. It is treated with calculated quantity of hydrochloric acid to leave the supernatant liquid slightly alkaline.

The precipitated sulphur is collected on a filter, washed, until the washings are free from calcium and dried. The chemical reactions taking place during the process involved are complex, but they can be represented as under:

$$3\,Ca(OH)_2 + 12S \rightarrow 2CaS_5 + CaS_2O_3 + 3H_2O \quad .....(11.49)$$

Lime sulphur Polysulphide Thiosulphate

$$2\,CaS_5 + CaS_2O_3 + 6\,HCl \rightarrow 3\,CaCl_2 + 12S + 3H_2O \quad .....(11.50)$$

The yield of precipitated sulphur is about two-third of the theoretical yield as much of sulphur is lost as a result of escape of gases H_2S and SO_2 which are formed in the initial stages of the process.

Description: It is an odourless and tasteless pale grayish yellow or pale greenish yellow, soft powder, free from grittiness. It burns with a blue flame with the production of sulphur dioxide. It is insoluble in water and alcohol but soluble in carbon disulphide.

Tests for Identity:

(i) It melts at about 115° to a yellow mobile liquid which becomes dark and visible on further heating at about 160°

(ii) When viewed under microscope, it is seen to consist of grouped amorphous sub globular particles without any admixture of crystals.

Tests for Purity: As acidity, matter insoluble in carbon disulphide and sulphated ash.

Acidity: It is tested by thoroughly agitating 5g with 50 ml of purely boiled and cooled water and titrating with 0.1N sodium hydroxide, using solution of phenolphthalein as indicator. Not more than 0.5 ml of alkali is required.

Test for matter insoluble in carbon disulphide is done by dissolving 1g in 5 ml of carbon disulphide. Almost all the precipitated sulphur should be completely dissolved.

Sulphated ash should not be more than 0.25 percent.

Uses: It is a good scabicide and may be used as constituents in sulphur ointment. It is also used in the form of lotions or ointments in the treatment of acne. Precipitated sulphur is preffered in liquid mixtures, because its particles are lighter and thus they get easily suspended. Its ointment is also more smooth and preferred. Sulphur, being an active parasiticide is used in the treatment of many infection and skin disorders, such as ringworm infection, pediculosis, psoriasis, eczema etc.

(ii) Sublimed Sulphur (Sub. Sulph):

I.P does not prescribe any percentage content pertaining to purity. However, the limit tests ensure its percentage as not less than 99.5%, which is a requirement, in some pharmacopoeias including U.S.P.

Preparation: I.P mentions that sublimed sulphur may be obtained from native sulphur or from sulphides. In fact, sublimed sulphur is obtained when sulphur is heated and its vapours are lead to a chamber, which is suitably cooled. The vapour gets condensed and solidified and fall on the walls and on the bottom of the chamber in the form of a crystalline powder or friable masses. This yellowish powder is sublimed sulphur or flowers of sulphur. This sulphur after collection can be sieved.

Description: Sublimed sulphur occurs as a fine, yellow, slightly gritting, powder. It has a faint characteristic odour but is devoid of any taste.

It burns with a blue flame, producing sulphur dioxide. It is almost insoluble in water and alcohol but dissolves (1g in 2 ml) slowly and incompletely in carbon disulphide.

Tests for Identity: It melts to yellow mobile liquid on being heated at about 115°. The liquid becomes dark and visible on further heating at about 160° under the microscope it is seen consisting of opaque rounded amorphous particles or aggregates.

Tests for Purity: As, acidity, matter insoluble in carbon disulphide and sulphated ash.

Tests for Acidity: It is performed by thoroughly agitating 2g of sublimed sulphur with 50 ml of freshly boiled and cooled water and titrating the liquid with 0.1N sodium hydroxide, using solution of phenolphthalein as indicator. Not more than 1 ml of 0.1N sodium hydroxide is required. Acidity is due to the presence of traces of sulphurous and sulphuric acids which may get formed due to oxidation of sulphur during its conversion into sublimed sulphur.

Tests for Matter Insoluble in Carbon disulphide: It is carried out by agitating 1g of sublimed sulphur in 20 ml of carbon disulphide and allowing the liquid to stand for ten minutes. The liquid is filtered and the residue on filter is washed with carbon disulphide and dried. The residue should not weigh less than 0.2 g. i.e., not less than 20% of sublimed sulphur should be insoluble.

Sulphated Ash: It should not be more than 0.2 percent.

Uses: It is utilized for the same therapeutic uses as those mentioned under precipitated sulphur.

Dose: 1 to 4 g

(iii) Sulphur Ointment:

Sulphur ointment contains 10 percent of sulphur (limit 9.5 to 10.5 percent of S)

Preparation: Sulphur ointment, sublimed sulphur is one of the ingredients of liquorices compound powder.

Ingredients:

Sublimed sulphur finely sifted	100 g
Simple ointment prepared with white soft paraffin	900 g

Preparation: Sublimed sulphur is triturated with a portion of the simple ointment until the mixture is smooth. The remainder of the simple ointment is then added and mixed thoroughly.

Tests for Identity: The tests can only be applied by first extracting sulphur from the ointment with the help of light petroleum and then applying the same tests for identification. 20g of the ointment is refluxed with 25 ml of light petroleum in 250 ml conical flask for one hour and the petroleum layer is rejected. The residue in the flask is again refluxed with 25 ml of light petroleum for one hour.

The petroleum layer is decanted and the dried residue i.e., sulphur is used for identification tests given below:

It melts at about 115° to a yellow mobile liquid, which becomes darker and viscid on heating at about 160° C.

The substance from test above is further heated strongly. It burns strongly with a blue fume, forming sulphur dioxide, which can be recognized by its characteristic colour.

To 1mg of the residue dissolved in 2ml of hot pyridine is added 0.2 ml of solution of sodium bicarbonate and the mixture is boiled. A blue or green colour is produced.

Assay: The assay is carried out by converting sulphur into sodium thiosulphate by refluxing the ointment with a solution of sodium sulphite.

$$Na_2SO_3 + S \rightarrow Na_2S_2O_3 \qquad .....(11.51)$$

The solution after cooling is filtered and the filtrate along with the washings from the residual fat on the filter is treated with formaldehyde and acetic acid to fix sodium sulphite as formaldehyde sodium disulphite which in cold solution is un reactive with iodine.

$$\underset{\text{Sod.sulphite}}{Na_2SO_3} + \underset{\text{formaldehyde}}{HCHO} + H_2O \rightleftharpoons CH_2OH.OSO_2Na + NaOH \qquad .....(11.52)$$

The acetic acid present in the reaction mixture neutralized the sodium hydroxide.

$$CH_3COOH + NaOH \rightarrow CH_3COONa + H_2O \quad .....(11.53)$$

The sodium thiosulphate is titrated as usual with standard solution of iodine and the amount of sulphur in the ointment is estimated.

$$I_2 + 2Na_2S_2O_3 \rightarrow Na_2S_4O_6 + 2\ NaI \quad .....(11.54)$$

The assay procedure as given in I.P is outlined here under:

An accurately weighed amount about 1g of sulphur ointment is boiled with a solution of 2g of sodium sulphite in 40 ml of water under a reflux condenser until the sulphur is completely dissolved. The mixture is cooled and the aqueous solution is filtered and the residue of fat on the filter paper is washed with hot water. The residue is again cooled and again washed with hot water. The washings are added to the filtered aqueous solution to which is added 10 ml of formaldehyde solution and 6 ml of acetic acid. The solution is diluted to 150 ml with water and titrated with 0.1N iodine using solution of starch as indicator.

Each ml of 0.1N iodine is equivalent to 0.003206 g of S

Storage: The sulphur ointment is preserved in a well closed container, preferably in a cool place.

Uses: It is a good scabicide and may be used as constituent in sulphur ointment. It is also used in the form of lotions or ointments in the treatment of acne. Precipitated sulphur is preferred in liquid mixtures, because its particles are lighter and thus they get easily suspended. Its ointment is also more smooth and preferred.

Sulphur, being an active parasiticide is used in the treatment of many infection and skin disorders such as ringworm infection, pediculosis, psoriasis, eczema etc.

11.2 (r) Boron

Boron is a metal of the aluminium group. It is available in crystalline, amorphous and adamantine forms. Boron is a trivalent element metallic. Boron is used as an industrial catalyst in metallurgy to give hardness and because it absorbs neutrons in atomic reactors.

The body weight is reduced if boric acid is given orally for longer duration which is attributed to loss of water from cells and tissues. It was therefore used in obesity but now-a day it is no more used because of its toxicity. It gets accumulated because of slow rate of excretion. The sign of poisoning is seen through vomiting, loss of appetite, dryness of skin, itching and form confused state of mind. The lethal oral dose is 10 to 20g.

(i) Borax:

Chemical Formula: $Na_2B_4O_7.\ 10H_2O$ *Mol. Wt.* 381.4

Synonyms: Sodium Borate.

It occurs as sodium salt of pyroboric acid in the dried lakes of Tibet and India and contains not less than 99% and not more than 103% of $Na_2B_4O_7.10H_2O$

Preparation: Large quantities of borax are obtained from kunite $Na_2B_4O_7$. $4H_2O$ by simple crystallization from water. Borax crystallizes as decahydrate, $Na_2B_4O_7$. $10H_2O$.

Description: It is a colorless crystalline or white crystalline powder, without any odour but saline and alkaline taste. It effloresces in dry air and loses all its water of crystallization on ignition. It is soluble in 16 parts of cold water, in 1 part of boiling water, 1 part of glycerol and insoluble in alcohol. An aqueous solution of borax is alkaline due to hydrolysis.

$$Na_2B_4O_7 + 3H_2O \rightarrow 2NaBO_2 + 2H_3BO_3 \quad .....(11.55)$$

If the solution is diluted further, the sodium metaborate is further hydrolyzed giving rise to alkali and boric acid.

$$NaBO_2 + 2H_2O \rightarrow NaOH + H_3BO_3 \quad .....(11.56)$$

Tests for Identity: A mixture of ethyl alcohol with boric acid burns with a green edged flame due to the formation of ethyl borate.

$$H_3BO_3 + 3C_2H_5OH \rightarrow B(OC_2H_5)_3 + 3H_2O \quad .....(11.57)$$

Aqueous solution in water is acidic in nature. Also gives reaction for sodium.

Tests for Purity: It is tested for arsenic, heavy metals, iron chloride and sulphate.

Assay: An accurately weighed amount (3 g) of borax is dissolved in water (76 ml) and the solution is titrated with N/2 HCl using solution of methyl red as indicator.

$$2HCl + Na_2B_4O_7 \rightarrow 5H_2O = 2\,NaCl + 4H_3BO_3 \quad .....(11.58)$$

Each ml of N/2 HCl is equivalent to 0.09536 g of $Na_2B_4O_7.10H_2O$

Uses: Borax is used as a bacteriostatic and as a pharmaceutical aid. The emulsifying action of borax on oils is attributed to the formation of free alkali (NaOH) on hydrolysis.

(ii) Borax Glycerin:

Borax glycerin contains borax equivalent to 12% w/w of $Na_2B_4O_7.10H_2O$ and is prepared by triturating the powdered borax (120g) with glycerin (880g) and warming slowly with constant stirring till the solution is effected.

Assay: An accurately weighed amount of borax glycerin is diluted with water and the solution is neutralized with N/2 sulphuric acid using solution of methyl orange as indicator. To the boiled and cooled solution 20 ml of glycerin is added and it is titrated with N/2 NaOH using solution of phenolphthalein as indicator.

Uses: The borax glycerin is also used as bactriostatic.

11.2 (s) Antimony Sodium Tartarate

Chemical Formula: $C_4H_4O_7Sb.Na$ *Mol. Wt.* 308.8

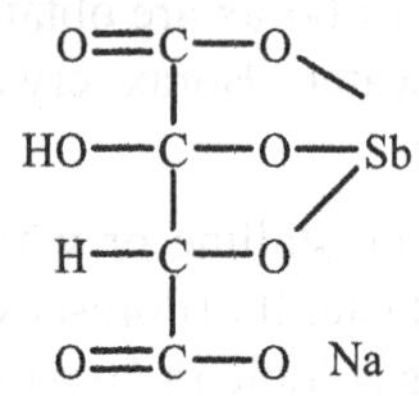

Preparation: Antimony sodium tartrate, known as tartar emetic is an example of certain closely related substance, which are formed by the interaction of an oxide of an element (antimony, Bismuth etc) which organic hydroxyl acids or their salts. Thus antimony sodium tartrate is prepared by making a paste of 5 parts of antimonious oxide and 6 parts of finely powdered sodium acid tartrate with water and setting aside the paste for a day. The paste is boiled with 40 parts of water for 15 minutes with frequent stirring. The hot liquid is filtered and the filtrate is kept aside for crystallization. The crystals of antimony sodium tartrate are collected on filter paper and dried at atmospheric temperature. The mother liquid can still be concentrated to yield another crop of tartar emetic. The I.P. requires antimony sodium tartrate to contain not less than 96% of $C_4H_4O_7Sb.Na$ calculated with reference to the substance dried to constant weight at 105°.

Description: It is a colourless and transparent or white scaly powder, which is odourless, sweetish and hygroscopic. It is freely soluble in water but practically insoluble in alcohol.

Tests for Purity: It is tested for As, acidity and alkalinity, heavy metals and loss on drying.

Acidity or alkalinity is measured by the volume of N/100 acid or alkali required to neutralize a solution of definite concentration to the green colour of bromocresol green, indicative of pH 4.5. 1 g of the substance is taken and dissolved in 50 ml of water. N/100 alkali or acid used should not be more than 20 ml for neutralization.

Loss on drying should not be more than 6 percent, when dried to constant weight at 105°C.

(The salt is hygroscopic, but the official salt is anhydrous)

Assay: An accurately weighed amount (0.5g) is dissolved in water (50 ml) and to the solution is added sodium bicarbonate (2g) and is titrated with N/10 iodine using solution of starch as indicator.

The above assay is carried out iodometrically and is an oxidation and reduction type of titration.

The following reaction which is reversible takes place. The acid which is liberated in the reaction is neutralized with sodium bicarbonate to enable the reaction to go to completion.

$$2C_4H_4O_7Sb.\,Na + 3H_2O + 2I_2 \rightleftharpoons 2Na\,HC_4H_4O_6 + Sb_2O_5 + 4HI \quad .....(11.59)$$

Factor: Each ml of N/10 I_2 is equivalent to 0.01544 g of $C_4H_4O_7$ Sb. Na

Uses: It is used as an anti schistosomal drug in schistosomiasis and as an emetic. Its emetic action is due to the irritant action on the gastrointestinal mucosa. It is given by intravenous injection only; orally it is used as a reflex expectorant.

Antimony Sodium Tartrate Injection

The injection is sterile solution of antimony sodium tartrate in water for injection and it contains not less than 95% and not more than 105% of antimony sodium tartrate. $C_4H_4O_7Sb.Na$ corresponding to the amount stated on the label. The injection is sterilized by heating in an autoclave or by filtration.

Tests for Identity: The injection is tested by evaporating 5 ml to dryness and testing the residue for reaction characteristic of sodium and antimony. Reaction characteristic of tartrate are given after antimony is removed.

Assay: It is assayed in the same way as described under antimony sodium tartrate, using an accurately measured volume equivalent to about 0.5 g of antimony sodium tartrate.

Uses: Injection is used intravenously in treatment of schistosomiasis in divided doses until the total quantity administered is 1.5g.

11.2 (t) Ammoniated Mercury

Chemical Formula: NH_2HgCl *Mol. Wt.* 252.1

Synonym: Amino chloride of mercury.

Ammoniated mercury is called amino chloride of mercury, as an amino group ($-NH_2$) replaces an atom of chlorine from mercuric chloride.

$$Cl - Hg - Cl \xrightarrow[+\,NH_2]{-\,CL} NH_2 - Hg - Cl \quad .....(11.60)$$

Preparation: Ammoniated mercury is prepared by adding a solution of mercuric chloride (3 parts) in water (60 parts) to a mixture of dilute solution of ammonia (4 parts) and water (20 parts) with constant stirring.

$$HgCl_2 + 2NH_3 \rightarrow HN_2HgCl + NH_4Cl \quad .....(11.61)$$

The precipitate of NH_2HgCl is collected on the filter paper, washed with cold water and dried at a temperature nor exceeding 30°. The washing of the precipitate is not done for a long time in an effort to remove NH_4Cl as long washing will only give yellowish product. I.P requires ammoniated mercury to contain not less than 98.0% of NH_2HgCl.

Description: Ammoniated mercury is a heavy, odourless, white, amorphous powder, which is stable in air, but darkens on exposure to light. It is insoluble in water and alcohol but soluble in warm acetic acid. It gradually decomposes in cold water and in boiling water it gets hydrolyzed to yellow basic compound: NH_2HgCl; HgO.

In boiling alkalies it gets completely decomposed as under:

$$NH_2HgCl + NaOH \rightarrow NH_3 + HgO + NaCl \quad .....(11.62)$$

This reaction is also used for identification

Tests for Identity

- on being heated it volatilizes without fusion
- on being heated with sodium hydroxide, it gives yellow mercuric oxide and ammonia
- solution in acetic acid yields reactions characteristic of mercuric salts and chlorides

Tests for Purity: It is tested for mercurous chloride, carbonates and sulphated ash.

Mercurous chloride and carbonates are tested by trituration with acetic acid followed by heating to 70° with occasional shaking. A clear solution is obtained with no effervescence.

Assay: An accurately weighed amount (0.25g) of NH_2HgCl is transferred to a stoppered flask containing water (50 ml) and potassium iodide (3g). The mixture is shaken until solution is complete. The liberated alkalis, NH_3 and KOH are titrated with N/10 HCl, using solution of methyl orange as indicator.

$$NH_2HgCl + 2KI + H_2O \rightarrow HgI_2 + NH_3 + KOH + KCl \quad .....(11.63)$$

1.0 ml of N/10 HCl is equivalent to 0.01261 g of NH_2HgCl

Uses: It is used as anti infective substance.

Ammoniated mercury ointment: It is also official in I.P. It is called white precipitate ointment and contains 2.5% of NH_2HgCl.

Ingredients:

Ammoniated mercury	25 g
Simple ointment	975 g

Preparation: Ammoniated mercury is triturated with a portion of the simple ointment to a smooth consistency and then the remainder of the simple ointment is added gradually and the preparation mixed thoroughly.

Tests for Identity: The tests (1 to 3) listed above under ammoniated mercury are carried out with the residue after completely removing the ointment base with the help of a mixture of equal volumes of solvent ether and light petroleum.

The ointment is assayed by the above acid base titration method, except that the ointment is first treated with solvent ether and light petroleum to remove the fatty and greasy materials. An accurately weighed amount of ointment representing about 0.20 g of ammoniated mercury in it (in about 5 g) is taken and assayed as above.

Each ml of 0.1N HCl is equivalent to 0.01261 g of NH_2HgCl

11.3 Astringents

Astringents are substances that precipitate surface proteins when applied topically. They have low cell penetrability and action is limited to the cell surface and interstitial spaces. Astringents are normally applied in very dilute solutions, since many are irritants or caustics in moderate to high concentrations. Most astringents are also antiseptics.

The astringent action is accompanied by contraction of the tissue and hardening and pathological transcapillary movement of plasma protein is inhibited. Local edema, inflammation and exudation are there by reduced. Mucous or other secretions may also be reduced and the affected area becomes drier.

Astringents are used therapeutically to reduce the volume of exudates from wounds and skin eruptions, to arrest hemorrhage by coagulating the blood, to check diarrhea, reduce inflammation of mucous membranes, promote healing toughen skin or decrease sweating due to their ability to constrict pores. They also possess deodorant properties due to their ability to destroy microorganisms that produced body odours.

Substances that possess astringent action include aluminium acetate, aluminium chloride, aluminium chlorohydrates, aluminium sulphate, calamine, zinc oxide, zinc sulphate, salts of manganese, bismuth, permanganates, tannins or related polyphenolic compounds.

11.3 (a) Aluminium Sulphate

Chemical Formula: $Al_2 (SO_4)_3$ *Mol. Wt.* 342.14 (anhyd)

Aluminium sulphate contains not less than 51% and not more than 59% of $Al_2 (SO_4)_3$. It contains varying amount of water of crystallization.

Preparation

- It can be prepared by dissolving aluminium hydroxide in sulphuric acid.

$$2Al(OH)_3 + 3H_2SO_4 \rightarrow Al_2(SO_4)_3 + 6H_2O \quad .....(11.64)$$

The solution is concentrated by removing water and left for crystallization.

– From pyrites shale, $Al_2O_3 \times SIO_2 + FeS_2$

$$2FeS_2 + 2H_2O + 7O_2 \rightarrow 2FeSO_4 + 2H_2SO_4 \quad .....(11.65)$$

$$Al_2O_3 \times SiO_2 + 3H_2SO_4 \rightarrow Al_2(SO_4)_3 + SiO_2 + 3H_2O \quad .....(11.66)$$

Description: It is a white, crystalline powder, shining plates or crystalline fragments, odourless, with sweet taste at first and then mildly astringent.

Solubility: It is very soluble in water, giving an acid solution, due to hydrolysis. Insoluble in alcohol.

$$Al_2(SO_4)_3 + 6H_2O \rightarrow 2AI(OH)_3 + 3H_2SO_4 \quad .....(11.67)$$

Tests for Identity: A solution (1:20) gives the reactions of aluminium and sulphates.

Reaction: It shows pH 3 to 4 when determined in a 2% w/v solution in CO_2 – free water.

Tests for Purity: It is tested for clarity and colour of solution, alkalis and alkaline earths, ammonium salts, arsenic, heavy metals and iron.

Tests for Clarity and Colour of Solution: A 5% w/v solution is clear and colourless.

Test for Alkalis and Alkaline Earths: 1g of sample is dissolved in 100 ml of water, to which methyl red solution as an indicator and enough dil. NH_3 solution to get a distinct yellow colour (pH 6.5) are added. The solution is diluted to 150 ml with water, boiled and filtered while hot, 75 ml of the filtrate is evaporated to dryness and ignited to constant weight. The wt. of the residue does not exceed 2 mg.

Test for Ammonium Salts: 1g of sample is heated with 10 ml solution of NaOH on water bath for 1 minute. The colour of ammonium is not perceptible.

Test for Arsenic: Does not exceed 3 ppm.

Test for Heavy Metals: Not more than 40 ppm, determined by dissolving 0.5g of the sample in 1 ml of dil. CH_3COOH and sufficient water to produce 25 ml.

Test for Iron: It is determined by adding potassium ferrocyanide solution to the sample solution (25 ml of a 1 in 150 solution) when no blue colour is produced immediately.

Assay: It is based upon the complexometric titrations. Disodium edetate is used as a sequestering agent. Xylenol orange is used as an indicator for metallic ions in acidic solution. It yields intensely red complexes with metals and is it self lemon yellow in acidic pH.

An accurately weighed quantity of the sample (0.6g) is dissolved in NH_4Cl (2 ml) and water (50 ml) to which is added 0.05M disodium edetate (50 ml) and neutralized to methyl red solution with 1N NaOH. The solution is heated to boiling, left on a water bath for 10 minutes and cooled rapidly. Xylenol orange mixture (50 mg) and hexamine (5g) is added and the solution is titrated with 0.05M lead nitrite. A blank determination is also carried out.

Each ml of 0.05M disodium edetate is equivalent to 0.008554 g of $Al_2 (SO_4)_3$

Storage: It is stored in a well closed container.

Use: It is used as a pharmaceutical aid (for mineral carrier for adsorbed vaccines). It is a powerful astringent. It is widely used as a local antiperspirant. It is used for water purification.

11.3 (b) Alum

Chemical Formula: $KAl (SO_4)_2 . 12H_2O$ *Mol. Wt.* 474.4

Alum is potash alum. i.e., it is the potassium aluminium sulphate. Alums are double sulphates of a univalent metal and a trivalent metal.

I.P. requires it to contain aluminium equivalent to not less than 99.5% of $KAI (SO_4)_2$.

Preparation: Alum is prepared by adding a hot concentration solution of potassium sulphate to a hot solution of aluminium sulphate.

$$Al_2 (SO_4)_3 + K_2SO_4 + 24H_2O \rightarrow 2KAI (SO_4)_2 . 12H_2O \quad \text{.....(11.68)}$$

The alum crystallizes out on cooling by slow crystallization large characteristic, regular octahedral crystals are obtained on being heated on water bath temperature alum melts in its water of crystallization. It loses the whole of its water below 20°C, leaving a white residue of anhydrous aluminium and potassium sulphate.

Description: It is available as a colourless, white powder or transparent crystalline mass. It is sweetish or astringent in taste and is freely soluble in water and insoluble in alcohol.

Tests for Identity: It gives reactions characteristic of aluminium, potassium and sulphates.

$$\underset{\text{Bauxite}}{Al_2O_3, 3H_2O} + 2\,NOAH \rightarrow \underset{\text{Sodium metaluminate}}{2\,Na\,AlO_2} + 4H_2O \quad \text{.....(11.69)}$$

$$NaAlO_2 + CO_2 + H_2O \rightarrow NaHCO_3 + Al\,(OH)_3 \quad \text{.....(11.70)}$$

$$2Al\,(OH)_3 + 3H_2SO_4 \rightarrow Al_2\,(SO_4)_3 + 3H_2O \quad \text{.....(11.71)}$$

Tests for Purity: It is treated for As, Cu, Zn, Fe and heavy metals. Potash alum is also required to comply with a test for ammonium salts.

Zinc is tested by comparison with a control and by dissolving 1 g of alum in 20 ml of water, adding 0.5ml of dilute H_2SO_4 and 2g of NH_4Cl and diluting the mixture to 20 ml of water. 1ml of solution of potassium ferricyanide is added and the solution is set aside for five minutes. Any opalescence which is produced is not greater than that produced in a control test made by adding 0.5ml dilute HCl, 2g of ammonium chloride and 1ml of solution of potassium ferricyanide to 4ml of 0.11% v/w $ZnSO_4$ and diluting the volume to 50 ml and setting aside for five minutes as in the case of test solution.

Ammonium salts are tested by comparison with a control test by dissolving 1g of alum in 1000 ml of ammonia free water. To the 10 ml of solution are added 40 ml of ammonia-free water and 2 ml of alkaline solution of potassium mercuric iodide. A colour, if produced is not deeper than in a control made by adding 2ml of alkaline solution of potassium mercuric iodide to 1 ml of dilute solution of ammonium chloride in 50 ml of ammonia-free water.

Assay: An accurately weighed amount (2g) is dissolved in water (300 ml) and to the solution is added solution of ammonium chloride (20 ml) and 5 drops of methyl red and sufficient dilute ammonia solution to produce a distinct yellow colour in the mixture. The mixture is boiled and filtered. The precipitate is washed with 25 percent w/v of solution of ammonium nitrate until it is free from chloride. The precipitate is dried to constant weight at a temperature above 120°C and the residue (Al_2O_3) is weighed.

Factor: 10 g of residue is equivalent to 9.307 g of the alum.

Uses: It is used as a pharmaceutical aid and as an astringent. It is also considered as an antiseptic and used as local wetting by barbers and rubbing alum on skin after shave, perhaps for its astringent and antiseptic action.

11.3 (c) Zinc

Chemical Formula: Zn *Mol. Wt.* 65.38

Zinc widely and quite regularly occurs in animal tissues and it amounts on an average 20-30 microgram per 1 g of fresh tissue. Some organs like genital organs are especially rich in zinc. Its presence in insulin is of interest. Most of the dietary materials contain zinc (e.g., 1 litre of milk contains 4 mg) and hence an adult receives 10-15 mg of zinc daily and this amount far exceeds the requirement. Zinc is excreted in faces.

- Bio chemically, zinc is found in association with some metalloenzymes like carbonic anhydrase, aldolase, carboxy peptidase etc. It is also found bound to RNA.

Earlier, zinc was supposed to be an essential dietary mineral and was regularly supplemented, but recently this view has changed and so far no one knows whether it is daily required and in how much quantity it is required. Food materials rich in zinc are milk, meat, fish, nuts etc.

Zinc itself has no therapeutic value, but its compounds have astringent, anti infective antiseptic and protective properties.

Zinc compounds show very mild toxicity. Acute toxicity in some cases is seen with an intake of 1g of zinc salts, 3-4g of zinc sulphate or chloride is fatal. An immediately available antidote is sodium bicarbonate. Dimercaprol is an effective antidote for zinc toxicity whose symptoms include periodic chills and fevers, malaise, coughing, headache and salivation.

The most important sources of zinc are zinc ores, zinc blende and calamine ($ZnCO_3$).

Zinc Sulphate

Chemical Formula: $ZnSO_4 . 7H_2O$ *Mol. Wt.* 287.6

Zinc sulphate contains not less than 55.6% and not more than 61% of $ZnSO_4$, corresponding to not less than 99.5% and not more than the equivalent of 102.0 percent of the hydrated salt, $ZnSO_4 . 7H_2O$

Preparation: Zinc sulphate is prepared by boiling an excess of metallic zinc with dilute sulphuric acid. The action is allowed to continue till evolution of hydrogen gas ceases.

$$Zn + H_2SO_4 \rightarrow ZnSO_4 + H_2 \quad .....(11.72)$$

The solution is filtered to separate the undissolved zinc and evaporated to crystallization.

In order to obtain purer product conforming to pharmacopeial requirement, the solution after filtration is reacted with chlorine water to oxidize ferrous sulphate impurity, if any to ferric state. The ferric salt is then precipitated as ferric hydroxide on agitating zinc carbonate or zinc oxide.

Description: zinc sulphate occurs as colourless transparent crystals, prisms or needles or as a granular, crystalline powder. It is odourless and has an astringent and metallic taste. It effloresces in dry air.

It is very soluble in water (0.6 parts) and glycerin (2.5 parts) but insoluble in alcohol. An aqueous solution of zinc sulphate is acidic to litmus due to hydrolysis of the salt and has pH of about 5.

The solution is acid to solution of phenol red and not acid to solution of methyl orange.

Tests for Identity: It yields reactions characteristics of zinc and sulphates.

Tests for Purity: It is tested for reaction for As, Cu, Zn, Fe, heavy metals, aluminium, copper, manganese and nickel, As, Fe, chloride and alkalis and alkaline earths.

Tests for Al, Cu, Mg, Mn and Ni: It is performed by adding in excess dilute solution of ammonia to the solution (1g in 20 ml of water) of the substance. The solution should remain colourless and produce no precipitate within thirty minutes.

The test for Alkalis and Alkaline Earths: It is done by dissolving 2g of the substance in about 150 ml of water contained in a 200 ml volumetric flask and precipitating the zinc completely by solution of ammonium sulphide and adding water to make 200 ml. The solution in the flask is mixed well and filtered through a dry filter rejecting the first portion of the filtrate. The subsequent 100 ml of the filtrate, after the addition of a few drops of sulphuric acid is evaporated to dryness in a tared dish and then ignited to constant weight. The residue does not exceed 5 mg.

Assay: Zinc sulphate is also estimated by using complexometric titration. The titrant is disodium EDTA and the indicator is eriochrome black T solution.

An accurately weighed amount of sample of about 0.3g is dissolved in 100 ml of H_2O to which 5 ml of ammonia – ammonium chloride solution and 0.1 ml of eriochrome black T solution are added. It is titrated with 0.05M disodium EDTA until the solution is deep blue in colour.

Each ml of 0.05M disodium EDTA $\equiv$ 0.01438 g of $ZnSO_4 . 7H_2O$

Storage: Zinc sulphate is preserved in a well closed container.

Uses: Zinc sulphate is used as an emetic and astringent. As emetic it is used as a reflex emetic. Its action is very rapid and does not cause local irritation to gastric mucosa. As an astringent it is usually found as a component in some of the astringent solution. Externally it is used as an ophthalmic astringent and its aqueous eye solution have been used in 0.25% concentration. But the eye solutions, if prepared without a borate buffer are acidic (pH 5).

Usually Gifford's borate buffer (pH 5.8 to 6.2) is used. If an alkaline solution is required, sodium citrate may be used as a sequestering agent (to keep the zinc ion from precipitating as the hydroxide)

Dose: 0.6 to 2 g.

11.4 Emollients

These are fatty substances (e.g., waxes, veg oils) which are topically applied to the skin, mucous membranes or abraded tissues.

11.5 Caustics

These substances are able to induce then destruction of tissues at the site of application. Hence, they find use to destroy warts, moles and hyperplastic tissues. They are having keratolytic action.

E.g., Potassium hydroxide, Silver nitrate, etc.

CHAPTER 12

Essential Trace Ions

12.1 Introduction

Transition Elements

3B	4B	5B	6B	7B	8B			1B	2B
Sc^{21}	Ti^{22}	V^{23}	Cr^{24}	Mn^{25}	Fe^{26}	Co^{27}	Ni^{28}	Cu^{29}	Zn^{30}
Y^{39}	Zr^{40}	Nb^{41}	Mo^{42}	Tc^{43}	Ru^{44}	Rh^{45}	Pd^{46}	Ag^{47}	Cd^{48}
La^{57}	Hf^{72}	Ta^{73}	W^{74}	Re^{75}	Os^{76}	Ir^{77}	Pt^{78}	Au^{79}	Hg^{80}
Ac^{89}	Unq^{104}	Unp^{105}	Unh^{106}						

The transition elements are those elements having a partially filled d or f sub shell in any common oxidation state. The term "transition elements" most commonly refers to the d-block transition elements. The 2B elements zinc, cadmium and mercury do not strictly meet the defining properties, but are usually included with the transition elements because of their similar properties. The f-block transition elements are sometimes known as "inner transition elements". The first row of them is called the lanthanides or rare earths. The second row consists of the actinides. All of the actinides are radioactive and those above Z = 92 are manmade in nuclear reactors or accelerators.

The general properties of the transition elements are

- They are usually high melting point metals
- They have several oxidation states
- They usually form colored compounds
- They are often paramagnetic

The transition elements include the importance metals iron, copper and silver. Iron and titanium are the most abundant transition elements. Many catalysts for industrial reactions involve transition elements.

12.2 Haematinics

Haematinics are the agents which improve the Quality of the blood increasing the hemoglobin level and the number of erythrocytes. These are used in the treatment of anemias.

(a) Ferrous Sulphate:

Chemical Formula: $FeSO_4.7H_2O$ *Mol. Wt.* 278.0

Ferrous sulphate contains not less than 97.0 percent and not more than the equivalent of 103 percent of $FeSO_4.7H_2O$

Description: Transparent, green crystals, or a pale blush-green, crystalline powder; odourless; taste is metallic and astringent. Efflorescent in dry air. On exposure to moist air, the crystals rapidly oxidize and become coated with brownish-yellow basic ferrous sulphate, when ferrous sulphate has the deteriorated it must not be used.

Preparation: It is made by adding slight excess of iron to dilute sulphuric acid.

$$H_2SO_4 + Fe \rightarrow FeSO_4 + H_2 \uparrow \quad .....(12.1)$$

When effervescence stops, the liquid is concentrated by boiling, filtered and allowed to cool and the crystals are separated and dried at air temperature.

Solubility: Soluble in 1.5 parts of water, and in 0.5 part of boiling water, practically insoluble in alcohol.

Identification: Yields the reactions characteristic of ferrous salt and of sulphates.

Tests for purity: It is tested for acidity, arsenic, copper, heavy metals, alkaline salts and alkaline earths and oxysulphate.

- *Arsenic:* Not more than 2 ppm
- *Heavy metals:* Dissolve 0.75 g in a mixture of 1 ml of dilute H_2SO_4 and 20 ml of water. Add 50 mg of hydroxylamine hydrochloride and boil for one minute. Cool, and add water to make 45 ml. To 15 ml of the solution and 4 ml of standard lead solution and dilute with water to 30 ml (A). Add to this solution and to the remaining 30 ml of ferrous sulphate solution

(B) 10 ml of solution of hydrogen sulphide. Solution B is no darker than solution A, the limit of heavy metals is 160 ppm.

- *Alkaline salts and alkaline earth*: Dissolve 1 g in 10 ml of water, oxidize by warming with a few drops of nitric acid, made alkaline with dilute ammonia solution, filter, evaporate the filtrate and ignite the residue, not more than 1 mg of residue is left.
- *Oxysulphate:* 1 g is dissolved in 2 ml of recently boiled and cooled water forms a clear solution which is not more than faintly turbid.

Assay: Weigh accurately about 1 g, and dissolve in 20 ml of dil H_2SO_4 and titrate with 0.1N potassium permanganate. Each ml of 0.1N $KMnO_4$ is equivalent to 0.0278 g of $FeSO_4.7H_2O$

Storage: Preserve in well-closed container.

Category: Haematinic

Dose: 0.2 to 0.3 g (Ferrous sulphate contains in 0.3 g about 60 mg of iron).

(b) Ferrous Fumarate:

Chemical Formula: $C_4H_2FeO_4$ *Mol. Wt.* 169.91

Description: It occurs as a reddish-orange brown fine powder with slight odour and slightly astringent taste.

Solubility: It is slightly soluble in water and very slightly in alcohol.

Preparation: It can be prepared by reaction between ferrous sulphate and sodium fumarate.

$$FeSO_4 + \begin{matrix} CHCOONa \\ || \\ CH\ COONa \end{matrix} \longrightarrow C_4H_2FeO_4 + Na_2SO_4 \quad .....(12.2)$$

Tests for identity:

- An acidic solution of the compound gives the reaction of ferrous salts.
- with resorcinol in presence of few drops of H_2SO_4 gives a deep-red semi-solid mass on heating. On dilution with water, it gives an orange yellow solution.

Tests for Purity: It is tested for arsenic, heavy metals, sulphate, ferric iron and loss on drying.

- *Arsenic:* Not more than 5 ppm.

- *Heavy metals:* Not more than 20 ppm.
- *Sulphate:* Boil a specific amount of sample (0.3g) in dil. HCl (10 ml) and water (30 ml), after cooling with ice and filtering the filtrate compiles with the limit test for sulphates.
- *Loss on drying:* Not more than 1% determined on 1g by drying at 105°C.

Assay: For the determination of ferrous iron, the compound is titrated with 0.1N ceric ammonium sulphate, using ferroin sulphate solution (1, 10-phenanthroline) as an indicator. This indicator forms a complex with iron. The complex of 1, 10-phenanthroline with ferrous iron is known as ferroin.

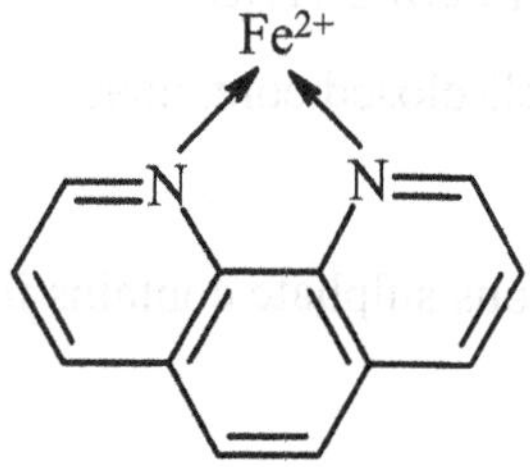

Ferroin

In the presence of reducing agent ceric ammonium sulphate undergoes reduction to the cerous state.

$$20\ Ce(SO_4) + 2FeSO_4 \rightarrow Ce_2(SO_4)_3 + Fe_2(SO_4)_3 \quad .....(12.3)$$

Accurately weight the substance (0.3 g) and dissolve in dil. H_2SO_4 (15 ml) and heat. Cool, add water (50 ml) and titrate with 0.1N ceric ammonium sulphate, using ferroin indicator.

Each ml of 0.1N ceric ammonium sulphate ≅ 0.01699 g of $C_4H_2FeO_4$

Storage: It is stored in well-closed containers.

Use: It is used as Haematinic.

Dose: 0.2-0.6 g daily.

(c) Ferrous Gluconate:

Chemical Formula:

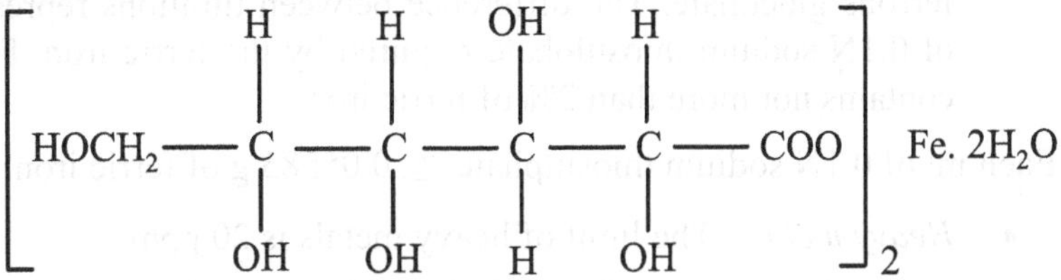

(or)

$C_{12}H_{22}O_{14}Fe.\ 2H_2O$

Mol. wt: 482.2

Ferrous gluconate contains not less than 95% of $C_{12}H_{22}O_{14}Fe.2H_2O$ calculated with reference to substance dried at 105°C for five hours.

Description: It is available in the form of yellowish grey (or) pale greenish yellow, fine powder or granules; odour slight, resembling that of burnt sugar.

Preparation: It is prepared by the reaction of ferrous sulphate and barium gluconate. The precipitate of barium sulphate is removed by filtration and the filtrate is evaporated. The reactions involved are as follows:

$[CH_2OH.(CHOH)_4.COO]_2\ Ba + FeSO_4 \rightarrow$
Barium gluconate

$[CH_2OH.(CHOH)_4.COO]_2\ Fe + BaSO_4 \downarrow$ .....(12.4)
Ferrous Gluconate

Solubility: Soluble in 10 parts of water, more readily soluble on harming, almost insoluble in alcohol.

Identification: Yields the reactions characteristic of ferrous salts.

Test for purity: It is tested for acidity, arsenic barium, ferric iron, heavy metals, chloride, sulphate, oxalic acid, dextrose and sucrose and loss on drying.

- *Test for acidity:* A 5% w/v solution is acid to solution of litmus.
- *Arsenic:* Not more than 2 ppm
- *Barium:* Dissolve 0.1g in 50 ml of water, add 5 ml of dil. H_2SO_4 and allow to stand for five minutes, no turbidity is produced.
- *Ferric iron:* Weigh accurately about 5g and transfer to a glass-stoppered flask and dissolve in 100 ml of freshly boiled and cooled water and 10 ml of hydrochloric acid and add 3g of potassium iodide. Shake well and allow

to stand for 5 minutes. Titrate any liberated iodine with 0.1N sodium thiosulphate, using solution of starch as an indicator. Repeat the experiment with same quantities of the same reagents in the same manner omitting ferrous gluconate. The difference between titrations represents the amount of 0.1N sodium thiosulphate required by the ferric iron. Ferrous gluconate contains not more than 2% of ferric iron.

Each ml of 0.1N sodium thiosulphate ≅ 0.05585 g of ferric iron.

- *Heavy metals:* The limit of heavy metals is 20 ppm.
- *Chloride:* 0.5g complies with the limit test of chlorides.
- *Sulphate:* 1g complies with limit test for sulphate.
- *Oxalic acid:* It is done by precipitating oxalate as calcium oxalate from the residue of ether extract.
- *Dextrose and sucrose:* They can be tested by reduction with Fehling's solution after removal of iron as ferrous sulphide. Both get hydrolyzed to reducing sugars during the process of removal of iron.
- *Loss on drying:* Looses not less than 7% and not more than 10% of its weight when dried at 105°C for five hours.

Assay: Weigh accurately about 1.5g and dissolve in 75 ml of water and 15 ml of dilute H_2SO_4 in a 300 ml conical flask fitted with a Bunsen valve. Add 0.25g of zinc dust, and allow to stand at room temperature for 20 min or until the solution becomes colourless. Filter the solution through a gooch crucible containing an asbestos mat coated with a thin layer of zinc dust and wash the crucible and contents with 10 ml of dilute sulphuric acid and then with 10 ml of water. Titrate the combined filtrate and washings immediately with 0.1N cerric ammonium sulphate, using solution of ferroin indicator.

Each ml of 0.1N cerric ammonium sulphate ≅ 0.04462 g of $C_{12}H_{22}O_{14}Fe.2H_20$

Dose: 0.3 to 0.6 g

Category: Haematinic

(d) Ferric ammonium citrate:

It occurs as a complex form. It contains not less than 20.5% and not more than 22.5% of Fe. This preparation of iron is preferred to ordinary iron salts because of its comparatively non-astringent nature and also because of the fact that this form can be dispended with ammonia or alkali carbonates without the precipitation of iron as ferric hydroxide.

Description: Thin, transparent, dark red scales or granules or a burnished granular powder, odourless; astringent taste. Deliquesces in air and is affected by light.

Preparation: The first step in the preparation of ferric ammonium citrate is the production of freshly precipitated ferric hydroxide by the interaction of a solution of a ferric salt with an alkali. The ferric solution should be added, with constant stirring, to the alkali and not the vice-versa

$$Fe_2(SO_4)_3 + 6\,NaOH \rightarrow 2Fe(OH)_3 + 3Na_2SO_4 \qquad .....(12.5)$$

The ferric hydroxide is collected and washed in filter press, without being dried, is stirred with enough citric acid solution to dissolve nearly the whole of it. A slight excess of ammonia is then added and the small quantity of ferric hydroxide is removed by filtration. The clear, reddish brown filtrate is evaporated to syrup. The syrup is painted finally on glass plates, dried below 40°C. The scales are scraped off.

Solubility: Freely soluble in water, almost insoluble in alcohol.

Identification:

1. Ignite gently and dissolve the residue in HCl, the solution yields the reactions characteristic of ferric slats.
2. Warm with solution of sodium hydroxide, ammonia is evolved and the solution yields the reactions characteristic of citrates.

Tests for purity: It is tested for acidity, arsenic, lead, zinc, chloride, free ferric compounds and sulphate.

- *Test of acidity:* A 5.0% w/v solution of litmus is neutral or slightly alkaline
- *Arsenic:* not more than 4 ppm
- *Lead:* Not more than 20 ppm
- *Chloride:* 0.2g dissolved in 5ml of water and boiled with 2ml of nitric acid complies with limit test for chlorides.
- *Free ferric compounds:* A 1% w/v solution gives no blue precipitate with solution of potassium ferrocyanide unless acidified with hydrochloric acid.
- *Sulphates:* 0.1g dissolved in 5ml of water and boiled with 2ml of hydrochloric acid complies with limit test for sulphates.

Assay: Weigh accurately about 0.5g, dissolve in a mixture of 15 ml of water and 1 ml of H_2SO_4 and warm until the dark brown colour becomes yellow. After cooling the solution to 15°C add, drop by drop 0.1N potassium permanganate till a pink colour persists for five seconds. Add 15 ml of HCl and 2g of KI. Allow to stand for three minutes, add about 60 ml of water and filtrate with 0.01N sodium thiosulphate, using solution of starch indicator.

Each ml of 0.1N sodium thiosulphate $\cong$ 0.005585 g of Fe.

Storage: Store in well-closed container, protected from light.

Category: Haematinic

Dose: 1 to 3 g

Ferric ammonium citrate contains in 3g about 0.6g of iron.

(e) Iron and Dextran Injection (I.P, B.P)

This compound is official in both I.P and B.P. It is sterile colloidal solution having a complex with dextrans of low molecular weight (B.P-80, specifies molecular weight of dextran between 5,000 and 7,000) in water for injection. It should not contain less than 4.75% but not more than 5.25% w/v of iron. By heating in an autoclave, it is sterilized.

Limits: The pharmacopeial product is slightly acidic (pH 5.2 to 6.5) limits are prescribed in respect of non-volatile residue, copper, zinc, chloride, heavy metals and arsenic. Besides these biological tests are prescribed for most injections absorb from injection site, pyrogens and undue toxicity.

Assay: The preparation is assayed by dissolving the ferric iron in sulphuric acid and reducing the ferric iron to ferrous iron with zinc amalgam reductor. The reduced ferrous iron is titrated with cerric ammonium sulphate solution.

Uses: This preparation is generally useful for patients for whom oral administration is not possible for some reason or other. It may also be given intravenously by infusion at a slow rate. However there may be allergic reactions to the dextran, as also contra-indications for its therapy.

Note: Another preparation called iron *dextrin injection* (or) dextriferron is intended for intravenous use and is difference from the pharmacopeial product.

Dose: Its usual dose is 1-2 ml daily and is administered by deep intramuscular injection.

12.3 Mineral Supplements

Keeping minerals in proper balance throughout the body while providing all of them in sufficient quantities needed for optimal health is complex. Balance is important to all areas of our lives and nutrition, but it is particularly crucial when it come to essential and trace minerals. It is becoming increasingly evident when studying the relationship of mineral to human health that keeping the level of mineral in balance in every tissue, fluid, cell and organ in the human body may be the key to maintaining human health. Minerals should be ionic to be readily absorbed through transfer in the small intestine. They also have unique properties that distinguish them from each other, and allow them to freely take part in biochemical communication throughout the body.

Elements found in the body in mg/kg concentrations are called *trace elements*, whereas in concentrations of μg/kg or less are *ultra-trace elements*.

Essential and trace ions are required by the body in minute concentrations for its normal functioning. Its deficiency results in impairment of certain biologic or biochemical functions, which can be corrected by the replacement of the ion, Excess concentrations may be toxic. Deficiency of trace elements can occur as a result of decreased intake as in malnutrition states, decreased absorption, conditions resulting in chronic loss, chronic disease of liver and kidney, alcoholism or as a result of genetic abnormalities. Excessive or toxic concentrations can occur as a result of accidental or deliberate ingestion of the ion-containing compound (acute toxicity), continuous use of utensils containing the element (chronic toxicity) or impaired excretion of the ion from the body.

Essential and trace mineral supplements are available orally in their salt form, as oral colloidal mineral solutions or from plant sources to enable better absorption and bioavailability.

(a) Copper

Copper plays an important role as a component of enzymes or proteins involved in redox reactions. Certain enzymes containing copper are cytochrome C oxidase, tyrosinase and lysyl oxidase.

The biochemical functions of ionic copper include pigmentation, bone development, oxygen transport, protein and nucleic acid synthesis. Deficiency characteristics of copper are disorders in pigmentation, retarded growth, anaemia in children, Wilson's disease and Menke's syndrome. An early feature of copper deficiency is neutropenia. Copper deficiency causes a microcytic hypochromic anaemia associated with low concentrations of ceruloplasmin that can be corrected by the administration of ceruloplasmin. Severe copper deficiency affects collagen maturation and blood vessel defects.

Excess copper can cause free radical production and damage. Copper overload in brain and liver can cause cirrhosis of the liver and brain lesions. Administration of dimercaprol, pencillamine, or ammonium terathiomolybdate chelates copper and increases its urinary excretion. It has been reported that pencillamine and dimercaprol treatment is associated with harmful side effects, where as ammonium molybdate blocks copper absorption and appears to preserve neurologic functions.

There are no official preparations for the administration of copper deficiency although solutions of copper salts are available in the market. *Cupric sulphate* is official as an antidote for phosphorus poisoning and is used topically as fungicide.

(b) Zinc

Zinc containing enzymes are found in every enzyme class and zinc is a co-factor for more than 300 enzymes. Enzymes containing zinc include DNA polymerase, alkaline phosphatase, alcohol dehydrogenase, carbonic anhydrase etc.

Zinc enzymes are essential to growth, haemoglobin synthesis, collagen metabolism, bone development, wound healing, and protection from free radical damage, reproductive function and to immune system. Symptoms or deficiency include skin lesions, diarrhea, male impotence, dwarfism, sensory alterations and susceptibility to infections.

Zinc itself is relatively non-toxic. It however competes for absorption with copper and iron, and high zinc intake can interfere with the transfer of copper to the circulation. Zinc toxicity has resulted from the ingestion of acid food kept in a galvanized metal container and from industrial workers inhaling zinc oxide.

Oral zinc sulphate has been suggested for wound healing and doses of 220 mg three times daily have been reported to increase healing following surgery. Zinc sulphate is official as a topical astringent.

(c) Chromium

Inadequate chromium intake may affect overall health. Numerous studies have demonstrated that taking extra chromium daily in the form of a supplement may improve glucose tolerance in people whose blood sugar levels range from slightly elevated to full-blown diabetes.

Chromium supplementation in non-insulin-dependent diabetes mellitus (NIDDM) has demonstrated improved glucose tolerance, reduces insulin concentrations, and decreased total cholesterol.

Chromium increases the effects of insulin and decreases insulin requirements. Supplementation doses of chromium did not produce toxic effects at concentrations associated with improved glucose tolerance. However, chromium supplements won't help people who have high blood sugar inspite of getting adequate dietary chromium. Excessive intake can result in toxicity characterized by renal failure and pulmonary cancer.

The adequate intake for chromium is 35 μg daily for men and 25 μg (microgram) for women. High sugar intakes, trauma and hard exercise can increase chromium excretion. Because the mineral improves insulin function, a shortfall can impair the cells ability to remove excess sugar from the blood stream.

Chromium picolate is a well absorbed and popular chromium supplement sold today. A new formulation has also been developed as a complex of chromium and the amino acid histidine, which according to its developer, Richard Anderson, is absorbed atleast 50% better than chromium picolate.

(d) Manganese

The minimum daily requirement of manganese has been estimated at 3.9 mg. Manganese functions in many metalloproteins as a non-specific cation. Manganese is associated with several enzymes including pyruvate carboxylase, mitochondrial superoxide dismutase, arginase and glucokinase. It is therefore required for the biochemical functions of growth and reproduction, oxidative phosphorylation and cholesterol metabolism. Deficiency can result in disorders in spermatogenesis, bone abnormalities and bleeding disorders.

Very high doses except by inhalation are not toxic. In patients having magnetic resonance imaging using manganese agents, deposition of manganese in brain tissue has been demonstrated which may be associated with neurologic symptoms. Excessive manganese intake can lead to chronic manganism (manganese poisoning), which in many ways is similar to Parkinson's disease. Increasingly, the use of levodopa (used in treatment of Parkinson's disease) has been successful in relieving many of the symptoms of manganism.

The rate of manganese absorption is low and is decreased by phosphate, phytate, calcium and iron. Manganese salts which were once official and used as tonics include a citrate, $Mn_3(C_6H_5O_7)_2$, a glycerphosphate, $MnC_3H_5(OH)_2PO_4$, and a hypophosphite $Mn(H_2PO_2)_2.H_2O$.

Since the effect of manganese deficiency can be minimized by substitution of other similar ions in enzymes, symptoms of its deficiency are extremely rare and there is no current therapeutic rationale for the administration of manganese.

(e) Antimony

Antimony is a generally blue-white chemical element with the symbol Sb (Latin: *stibium*). It has an atomic number of 51. As a metalloid, antimony has four allotropes. The stable form of antimony is a blue-white metalloid. Yellow and black antimony are unstable non-metals. Elemental antimony and its compounds are used in many areas; including electronics, flame proofing, paint, rubber, ceramics, enamels, drugs to treat Leishmania infection and a wide variety of alloys.

Antimony compounds may be used in medications such as antiprotozoan drugs; Antimony trioxide is the most important of the antimony compounds and is primarily used in flame-retardant formulations. The natural sulfide of antimony, stibnite, was known and used in Biblical times, as a medication and in Islamic/Pre-Islamic times as a cosmetic. Antimony has been used for the treatment of schistosomiasis.

Antimony attaches itself to sulfur atoms in certain enzymes which are used by both the parasite and human host. Small doses can kill the parasite without causing damage to the patient. Antimony and its compounds are used in several veterinary preparations like anthiomaline or lithium antimony thiomalate, which is used as a skin conditioner in ruminants.

Antimony has a nourishing or conditioning effect on keratinized tissues, at least in animals. Tartar emetic is another antimony preparation which is used as an anti-schistosomal drug. Treatments chiefly involving antimony has been called antimonials. Antimony-based drugs such as meglumine antimoniate, is also considered the drugs of choice for the treatment of leishmaniasis in domestic animals.

Antimony and many of its compounds are toxic. Clinically, antimony poisoning is very similar to arsenic poisoning. In small doses, antimony causes headache, dizziness, and depression. Larger doses cause violent and frequent vomiting, and will lead to death in a few days. Inhalation of antimony dust is harmful and in certain cases may be fatal. Prolonged skin contact may cause dermatitis. It causes damage to the kidneys and the liver. Avoid contact with eyes, skin, or clothing. Do not breathe dust. Keep in tightly closed container. Use with adequate ventilation. Wash thoroughly after handling. Antimony is incompatible with strong oxidizing agents, strong acids, halogen acids, chlorine, or fluorine. Keep away from heat.

Antimony leaches from polyethylene terephthalate (PET) bottles into liquids. While levels observed for bottled water are below drinking water guidelines, fruit juice concentrates (for which no guidelines are established) produced in the UK were found to contain up to 44.7 μg/L of antimony, well above the EU limits for tap water of 5 μg/L.

The guidelines are:

- World Health Organization: 20 μg/L
- Japan: 15 μg/L
- United States Environmental Protection Agency, Health Canada and the Ontario Ministry of Environment: 6 μg/L
- German Federal Ministry of Environment: 5 μg/L

(f) Sulphur

Sulphur us widely distributed throughout the body as sulphhydryl groups of cystine, disulfide linkages in protein and sulphate slats and esters found in mucopolysaccharides and sulpholipids. The minimum daily requirement of sulphur is 2-3 g. This requirement is met from normal plant and animal food stuffs in the diet. Hence currently there is no need for dietary supplements of sulphur.

Therapeutically sulphur has been used in the following conditions and actions.

- Cathartic action
- Parasiticide in scabies
- Stimulant in alopecia
- Fumigation
- Miscellaneous skin diseases
- Sulphides used as depilatories

(g) Iodine and Iodides

Iodine is an essential ion necessary for the biosynthesis of Triiodothyronine (T_3) and thyroxine (T_4) produced by the thyroid gland. The biochemical functions of thyroid hormones include colorigenesis and oxygen consumption through regulation of carbohydrate, lipid and protein metabolism, central nervous system activity and brain development, cardiovascular stimulation; bone and tissue growth and development, Gastrointestinal regulation and sexual maturation.

When levels of thyroid hormones are insufficient to meet the metabolic needs of the body at cellular level, hypothyroidism occurs. Symptoms of hypothyroidism include enlargement of the thyroid gland – goiter, impairment of cognition (including memory, speech and attention), Fatigue, slowing of mental and physical performance, change in personality etc. Hypothyroidism has also been associated with increased cholesterol and risk of coronary heart disease. Severe hypothyroidism results in myxedema.

Internally iodine/iodide can be administered, since iodine is reduced to iodide in the intestinal tract. For solubility reasons, the official preparations contain KI in addition to iodine which combines with iodine to form KI_3. Approximately 150 mg of iodine is absorbed in the intestine each day. The thyroid gland has very high attraction for iodine and traps it (about 70 mg/day) by active transport.

Excessive administration of iodides may result in iodism bringing about certain irritative phenomena of the skin and mucous membrane. Iodism is exhibited by coryza (head cold), rashes, headaches, conjunctivitis etc. Hence it is not recommended to give iodides in cases of acne. Gastrointestinal effects include nausea, vomiting and diarrhea.

Iodide is an essential anion necessary for the synthesis of two vital hormones produced by the thyroid gland. These hormones are triiodothyronine (T_3) and thyroxine (T_4). The following are the structures of T_3 and T_4.

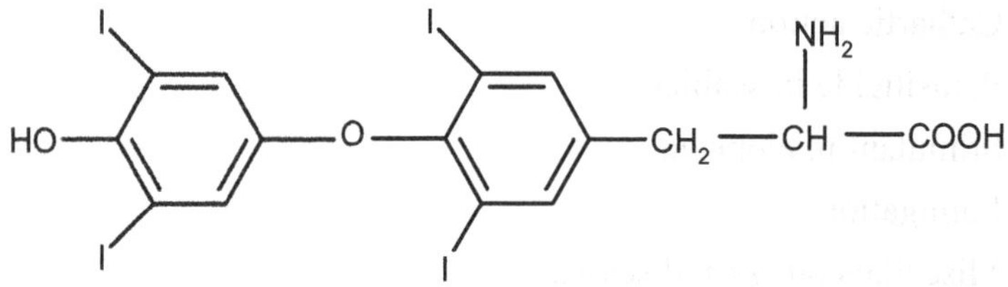

Triiodothyronine (T_3)

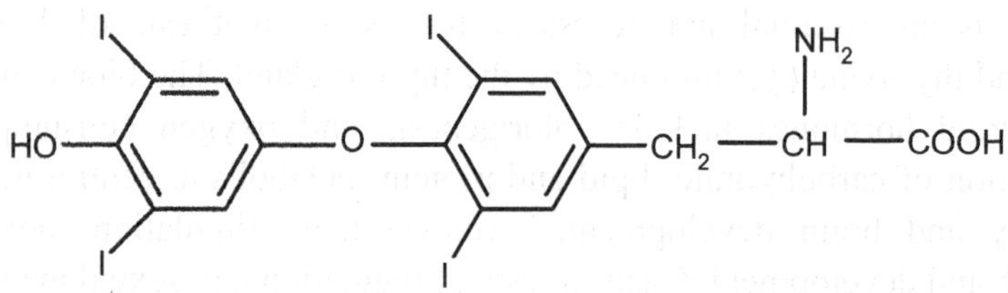

Thyroxine (T_4)

Iodine can be discussed from two standard points:

1. Its biochemical role in thyroid hormone production or formation.
2. Its pharmacological action as fibrolytic agent, expectorant and bactericidal.

The daily requirement of iodine is 140 micrograms for man and 100 micrograms for female. Lack of sufficient iodine in the diet result in an enlargement of the thyroid gland, known as simple (or) colloid goiter. It is characterized by swelling at the neck. Iodine is essential constituent of thyroid hormones.

Official iodine products

- Iodine I.P (Refer page 334)
- Aqueous iodine solution, I.P (Refer page 337)
- Strong iodine solution, I.P (Refer page 338)
- Weak iodine solution, I.P (Refer page 336)
- Sodium iodide, I.P (Refer page 339)
- Potassium iodide I.P (Refer page 338)

CHAPTER 13

Gases and Vapours

13.1 Introduction

The matter can be divided into three main categories namely gases, liquids and solids. The order of complexity being gas, liquid and solid. In general, gases are less dense than the other forms of matter and their internal friction is much less. The physical behaviour of a gas is not affected by the chemical nature of the molecules and thus all the gases respond in nearly the same way to the variables.

13.2 Gas Laws

As a consequence of study of gases, various gas laws evolved. They are:

(i) ***Boyl's law:*** It states that at constant temperature, the volume of a definite mass of gas is inversely proportional to the pressure. So the product of the pressure and volume of a definite mass of gas is constant at constant temperature

$$V \alpha \frac{1}{P}$$

or $$PV = K_1 \qquad .....(13.1)$$

where K_1 = a constant

From eq. 13.1, in certain state the volume and pressure of a given mass of gas are V_1 and P_1 and in another state V_2 and P_2 at constant temperature, then

$$P_1V_1 = P_2V_2 = K_1$$

or $$\frac{P_1}{P_2} = \frac{V_2}{V_1} \qquad .....(13.2)$$

(ii) ***Gay-Lussac's or Charles's law:*** This law states that at constant pressure, the volume of a definite mass of gas increases by the same relative fraction or amount for every degree rise is temperature. Then

$$V = V_o + (\alpha V_o)\, t$$

where V_o = Volume of a definite mass of gas at 0°C

V = Volume at any temperature t °C and same pressure

or
$$V = V_o(1 + \alpha t) \qquad \text{.....(13.3)}$$

where α = coefficient of expansion or volume change at constant pressure

$= 1/273$ per °C

At constant pressure, the volume of a definite mass of gas increases or decreases by $1/273$ of the volume for each °C raise or fall of temperature. So eq. 13.3, may be written as

$$V = V_o\left(1 + \frac{t}{273}\right) \qquad \text{.....(13.4)}$$

For V_1 and V_2 of volumes of gases at temperatures t_1 and t_2 at a given pressure, then

$$V_1 = V_o\left(1 + \frac{t_1}{273}\right) \qquad \text{.....(13.5)}$$

$$V_2 = V_o\left(1 + \frac{t_2}{273}\right) \qquad \text{.....(13.6)}$$

Dividing eq. 13.5 and 13.6

$$\frac{V_1}{V_2} = \frac{273 + t_1}{273 + t_2} \qquad \text{.....(13.7)}$$

At constant volume, the pressure of a definite mass increases by the same relative fraction or amount of every degree rise in temperature. Then eq. 13.7 can be written in terms of pressure as

$$\frac{P_1}{P_2} = \frac{273 + t_1}{273 + t_2} \qquad \text{.....(13.8)}$$

The equation 13.7 and 13.8 are variously known as the law of Dalton, the law of Gay-Lussac and the law of Charles.

If absolute temperature scale is considered i.e.,

$$T = t + 273$$

Then eq. 13.7 may be written as

$$\frac{V_1}{V_2} = \frac{T_1}{T_2} \quad .....(13.9)$$

And 13.8 may be written as

$$\frac{P_1}{P_2} = \frac{T_1}{T_2} \quad .....(13.10)$$

The eq. 13.9 can be rewritten as

$$\frac{V_1}{T_1} = \frac{V_2}{T_2}$$

i.e.,

$$\frac{V}{T} = \text{constant} \quad .....(13.11)$$

Similarly eq. 13.10 as

$$\frac{P_1}{T_1} = \frac{P_2}{T_2}$$

i.e.,

$$\frac{P}{T} = \text{constant} \quad .....(13.12)$$

Hence *the volume of a given mass of gas at a definite pressure is directly proportional to the absolute temperature. Similarly the pressure of a gas at a definite volume is directly proportional to the absolute temperature.*

(iii) ***Dalton's law of partial pressures:*** This law states that the total pressure of a mixture of gases is equal to the sum of the partial pressures of the constituent gases.

$$P = P_1 + P_2 + P_3 + \ldots \quad .....(13.13)$$

where P_1, P_2, P_3 + … = partial pressures

The partial pressure of each gas in a mixture can be defined as the pressure of the gas would exert if it alone occupied the whole volume of the mixture at the same temperature.

For e.g., the average pressure of atmosphere at sea level is 760 mm and it is composed of $\simeq$ 4 parts of N_2 and one part of O_2 by volume. According to this law the partial pressure of $N_2 = 4/5 \times 760$ mm and of $O_2 = 1/5 \times 760$ mm.

(iv) ***Graham's law of diffusion:*** The diffusion is the tendency for any substance to spread uniformly throughout the space available to it.

This law states that under the same conditions of temperature and pressure, the volume rates (viz., c.c. per second), at which different gases diffuse (or effuse) are inversely proportional to the square roots of the densities or molecular weights.

Let R_1 and R_2 be the rates of diffusion of two gases then

$$\frac{R_1}{R_2} = \sqrt{\frac{D_2}{D_1}} \qquad \text{.....(13.14)}$$

where D_1 and D_2 = Densities

Or

$$\frac{R_1}{R_2} = \sqrt{\frac{M_2}{M_1}} \qquad \text{.....(13.15)}$$

where M_1 and M_2 = Molecular weights of two gases. Densities are directly proportional to molecular weights. This law compares the rates of diffusion of different gases under the same conditions.

(v) ***Equation of state or combined gas law:*** This is the combination of results of Boyle's law and Gay-Lussac's law in an equation which will represent the relation between the pressure, volume and temperature of a given mass of gas. Such an equation is known as the combined gas law or an equation of state.

$$\frac{PV}{T} = K(\text{constant})$$

or

$$PV = KT \qquad \text{.....(13.16)}$$

13.3 Ideal Gas Equation

An ideal gas is one which obeys gas laws exactly; while a real gas (non-ideal gas) is one that obeys these laws at low pressures.

The ideal gas equation is

$$PV = nRT \qquad \text{.....(13.17)}$$

where P = Pressure of the gas

V = volume of the gas

T = Absolute temperature of the gas

R = Gas constant

n = number of moles

If one mole of gas under consideration and the eq. 13.17 will be

$$PV = RT \quad .....(13.18)$$

or

$$\frac{PV}{T} = R \quad .....(13.19)$$

where R = universal or gas constant and has the same value for all gases

$= 8.315 \times 10^7$ ergs deg^{-1} $mole^{-1}$

or

1.987 cals deg^{-1} $mole^{-1}$

The equations 13.18 and 13.19 are known as the fundamental gas equation or ideal gas law or ideal gas equation for one mole.

An ideal gas is a hypothetical gas, in our study it will be frequently converted to take the behaviour of an ideal gas as a standard and to discuss the properties of actual gases interms of their deviation from this standard.

13.4 The Kinetic Theory of Gases

This theory was developed by Kronig, Clausius, Maxwell, Bolzmann and others. This theory is capable of giving the best interpretation of the properties of an ideal gas and the reasons for the departure of real gases from ideal behaviour.

The Postulates are:

(i) Gases are considered to be made up of an exceedingly large number of minute, discrete, and perfectly elastic particles called molecules.

(ii) All molecules of any gas are supposed to be of the same mass and size but they differ in their from gas to gas.

(iii) Molecules are carelessly moving about in all directions with high velocities i.e., they are in a sort of ceaseless chaotic motion, colliding with each other and walls of the containing vessel. All molecular collisions are perfectly elastic.

(iv) These elastic impacts on the walls of the containing vessel are responsible for their phenomenon of pressure exerted by the gas. The magnitude of this pressure is dependent up on the kinetic energy of the molecules and their number.

(v) For an ideal gas, the molecules are supposed to be so small that this actual volume is practically negligible in comparison with the total volume of the gas. The molecules of an ideal gas are supposed to exert no attraction upon one another.

(vi) The average velocity of gas molecules depends on temperature.

13.5 Deviation from the Ideal Gas Laws

The gases which does not obey ideal gas equation PV = nRT, are real gases. Regnault and Amagat studies the behaviour of gases H_2, O_2, N_2, He and CO_2 under different conditions of temperature and pressure and observed that they fail obey gas equation and show certain deviations. The extent of deviation of the real gases from the ideal behaviour can be expressed through compressibility factor C_F.

$$C_F = \frac{PV}{nRT} \qquad .....(13.20)$$

For ideal gas $C_F = 1$ at all pressures and temperatures

For non-ideal gas C_F is is < 1 or C_F is > 1.

The temperature at which, no deviation is observed is called Boyle temperature and it is specific for each gas. Boyle's law could only be regarded as rough approximation. Deviations are clear from the figures which were drawn PV (in arbitrary units) Vs P (in atmospheres).

Fig. 13.1 is the graph of PV Vs P at high pressure.

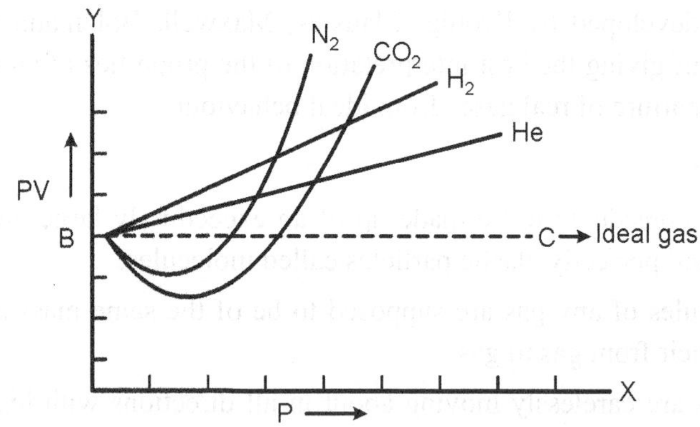

Fig. 13.1 PV Vs P for 1 mole of gas at high pressure.

From the Fig. 13.1 is clear that

- The dash line BC for ideal gas i.e., the PV for various pressures is a straight line.
- The values of H_2 and He continuously increase with increasing pressure
- The values N_2 and CO_2 decreases through a minimum and then increase above the value of ideal gas

Fig. 13.2 is the graph of PV Vs P at low pressure.

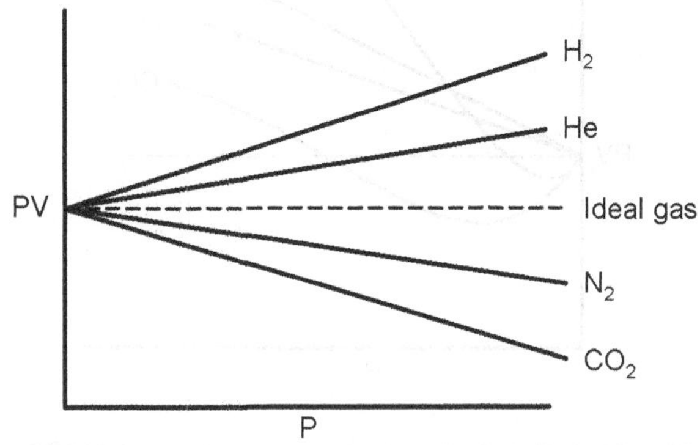

Fig. 13.2 PV Vs P at low pressure.

Fig. 13.2 shows that N_2 and CO_2 gases have lower value than that expected for the ideal gas as the values of H_2 and He are higher than the expected for ideal gas. The deviations from Boyle's law are slight and disappear as the pressure approaches zero.

Fig. 13.3 and 13.4 are the graphs of PV Vs V at low temperature and high temperature respectively.

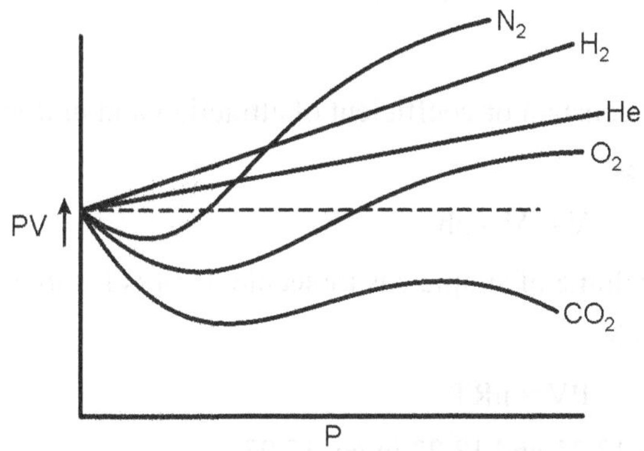

Fig. 13.3 PV Vs P at low temperature of 17°C.

The deviations from Boyle's low become less, the higher the temperature. This fact is well illustrated by the curves of CO_2 (in Fig. 13.3 and 13.4) at 17 °C and 100 °C. Hence it seems that the general nature of deviations from ideal behaviour does not depend on the nature of gas but depends on the temperature.

In general, the deviations depends both on the conditions of observation and on the nature of the gas, and that Boyle's law is more nearly obeyed, the lower the pressure, the higher the temperature, and further the gas is removed from the critical state.

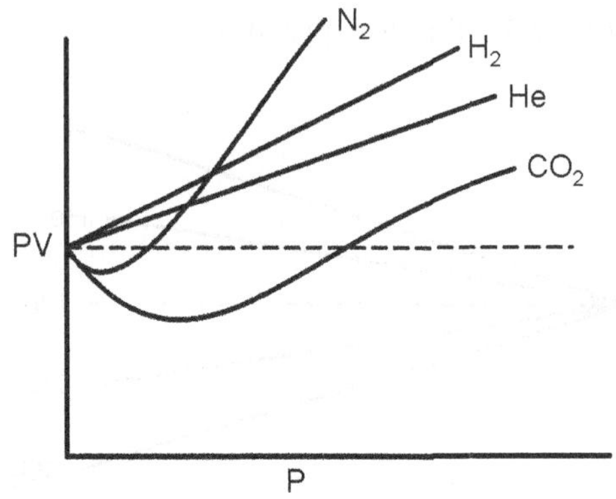

Fig. 13.4 PV Vs P at high temperature of 100°C.

13.6 Vander Waal's Equations

Vander Waal obtained a new equation after incorporation of the volume and pressure correction is the ideal gas equation.

$$P = P + a\left(\frac{n^2}{V^2}\right) \qquad \text{.....(13.21)}$$

where

a = proportionality constant or coefficient of attraction and unit is at mos.L^{2-1}, mol^{-1}

n = number of moles

$$V = V - nb \qquad \text{.....(13.22)}$$

Where b = excluded volume of simple gas molecule. Unit is L-$mole^{-1}$

The ideal gas equation is

$$PV = nRT \qquad \text{.....(13.23)}$$

on incorporation of eq.13.21 and 13.22 in eq. 13.23

$$\left[P + a\frac{n^2}{V^2}\right][V - nb] = nRT \qquad \text{.....(13.24)}$$

For one mole of gas

$$\left[P + \frac{a}{V^2}\right][V - b] = RT \qquad \text{.....(13.25)}$$

The eq. 13.25 is called Vander Waal's equation of state.

For e.g., for H_2, a = 0.244 and b = 0.02666 and for CO_2, a = 3.59 and b = 0.0371

13.7 Molecular Weights of Gases

All molecular weights are based on the arbitrarily assumed standard of 32.00 for the molecular weight of oxygen. The density of a gas is defined as the weight in grams of one liter of the gas at any given temperature and pressure. According to Avogadro's hypothesis, it can be proved at any given temperature and pressure the molecular weights of gases will be proportional to their densities and that knowing the molecular weight of the gas, that of the other can be found. The molecular weight of a gas is an important quantity required for all types of calculations.

The densities of a gas or vapour are determined at a given temperature at atmospheric pressure and at several other pressures below one atmosphere. The ratio d/P is plotted against P. If the real gases were ideal, the ratio d/p would be same at all pressures. If real gases are not ideal, the f/P changes with decreasing pressure. The plot of d/P against P is practically linear and can be extrapolated to zero pressure. At zero pressure (for the ideal gas) is

$$\left[\frac{d}{P}\right]_{P=0} = \frac{M}{RT} \qquad .....(13.26)$$

where d = density

M = molecular weight

or

$$M = RT\left[\frac{d}{P}\right]_{P=0} \qquad .....(13.27)$$

13.8 Joule-Thomson Effect

A porous plug of absorbent cotton or silk was fixed in the tube through which a stream of gas at constant pressure was passed. It was observed that the gas emerging from the plug was appreciably cooler than entering gas. This change in temperature is known as Joule-Thomson effect. It is due to a decrease of speed and hence the kinetic energy of the molecules, as energy is to be supplied to over come the molecular attractive forces when the gas expands is passing through the porous plug.

13.9 Specific Heat of Gases

The heat capacity of any substance is the quantity of heat required to raise the temperature of the substance by one degree. The heat capacity is called the specific heat of the substance when the weight of the material is one gram. The heat capacity per mole (specific heat per gram × gram molecular weight) is defined as the amount of heat required to raise the temperature of one mole of substance by 1°C.

Specific heat at constant volume C_v – the volume is kept constant while temperature and pressure allowed to raise.

Specific heat at constant pressure C_p – the pressure is kept constant and the volume is allowed to change.

The difference between C_v and C_p of gases being significantly large when compared to liquids and solids. Hence in the case of gas C_p and C_v must be taken into consideration.

In terms of kinetic theory on the problem of specific heat of gases, the energy of gas may be considered as – translational energy, vibrational energy, and rotational energy.

13.10 Critical Phenomenon in Gases

The result of a study of the pressure-volume-temperature relations of carbon dioxide Andrews discovered the essential conditions for the liquification of gases.

Andrews enclosed a definite known amount of carbon dioxide in a glass tube kept at constant temperature and measured the volumes at different pressures. The results for series temperatures (Andrews isotherms) are represented in Fig. 13.5.

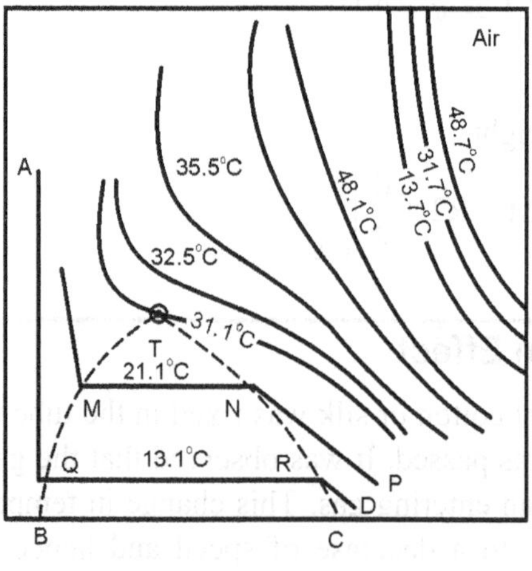

Fig. 13.5 Isotherms of carbon dioxide and air.

It will noted from the graph that:

- at 13.1°C CO_2 is entirely gaseous at low pressure and at D on increasing the pressure, the volume decreases (DR)
- at R, the liquification starts and the volume decreases rapidly and at Q the liquification completes.
- Curve DR represents gas only and along RQ both gas and liquid present in equilibrium and QA represents only liquid.

- The isotherm at 21.1°C is similar to that at 13.1°C except the horizontal portion, over which liquification takes place, is shorter.
- The boundary curve BTC is called the coexistence curve.
- as the temperature is raised, the horizontal portion of isotherms becomes less and less and finally it is reduced to a more point T, at 31.1°C
- above 31.1°C there is no indication of liquification and CO_2 can not be liquefied even at pressures of several hundred atmospheres, where as below 31.1°C a pressure of 75 atmospheres is sufficient.

Critical phenomenon: For CO_2 there is a limit of temperature above which the gas can not be liquefied, no matter what pressure is used. This phenomenon is called as critical phenomenon of gas.

Critical pressure, P_E: The pressure reduced to liquefy the gas at critical pressure.

For e.g., the P_c for O_2 is 50.1 and CO_2 72.9 atmospheres.

Critical volume, V_c: The volume occupied by 1 mole of gas at critical pressure and temperature.

For e.g., the V_c for N_2 90.1 and for O_2 74.4 C.C

Critical temperature, T_c: The temperature above which gas can not be liquefied however high the pressure may be.

For e.g., the T_c for H_2 33.2°K and for O_2 154.3 °K.

13.11 The Principle of Corresponding States

It have been observed experimentally that near the critical point all gases give the same type of isotherm as CO_2, but the actual position of the critical isotherm varies with the critical constants of the gas studied.

Let Reduced pressure $P_r = P/P_c$ or $P = P_rP_c$ .....(13.28)

Reduced volume $V_r = V/V_c$ or $V = V_rV_c$ .....(13.29)

Reduced temperature $T_R = T/T_c$ or $T = T_rT_c$ .....(13.30)

Substituting these P, V and T in Vander Waal's equation, we get

$$\left(P_rP_c + \frac{a}{V_r^2V_c^2}\right)(V_rV_c - b) = R\ T_rT_c \quad .....(13.31)$$

Substituting in eq. 13.31 the values of P_c, V_c and T_c obtained, we get

$$\left(\frac{P_r\ a}{27b^2} + \frac{a}{V_r^2.9b^2}\right)(3bV_r - b) = \frac{8aRT_r}{27bR} \quad .....(13.32)$$

Equation 13.32 is reduces to

$$\left(P_r + \frac{3}{V_r^2}\right)(3\,V_r - 1) = 8T_r \quad .....(13.33)$$

This is known as the reduced equation of state. From eq. 13.33, it can be concluded that if two substances have the same reduced temperature and pressure, they will also have the same reduces volume. This is known as the law of corresponding states and the substances are said to be in corresponding states under such conditions.

13.12 Inhalations

The drugs or chemicals which in the vapour form are inhaled or administered through the respiratory system in the body. They are inhaled by a closed mask method. Some of the inhalants acts as anesthetics and others are as respiratory stimulants.

Following are the list of inhalants

(a) Oxygen

(b) Carbon dioxide

(c) Nitrous oxide

(d) Helium

(e) Nitrogen

(a) Oxygen: *Role of oxygen:* Oxygen gas is important to all the living cells. It is very essential for normal oxidative metabolic process in the cell for the production of energy. The energy is used by cell for synthesizing adenosine triphosphate (ATP). When this ATP gets hydrolyzed energy is released. Enzymes like FADH, NADH are also known to play as important role in the reaction.

The transport of oxygen takes place through hemoglobin a constituent of blood. Oxygen combines with haemoglobin reversibly as follows

$$Hb + O_2 \rightleftharpoons HbO_2$$

where Hb = deoxyhaemoglobin and

HbO_2 = Oxyhaemoglobin

This loose combination undergoes dissociation readily to release oxygen in the medium of cell. The factors that affect the formation and dissociation of oxyhaemoglobin include temperature, electrolytes, effect of carbon dioxide, pH etc.

By inhalation during respiration, oxygenation of blood occurs in alveoli of lungs. The requirement of oxygen by body varies during different conditions when mole oxygen supply is required. It is supplied by inhalation method.

Chemical Formula: O_2 *Mol. Wt.* 32.00

Oxygen contains not less than 99.0 percent v/v of O_2. The residue contains either of argon with a trace of nitrogen or of hydrogen. For convenience it is compressed in metal cylinders.

Description: Colourless, odourless, tasteless gas.

Solubility: 1 volume dissolves in 32 volumes of water and 3.6 volumes of alcohol at 20^{0} and a pressure of 760 mm of mercury.

Preparation: It is prepared by fractionation of liquid air. It is also prepared by electrolysis of water.

Identification:

- A glowing splinter of wood bursts into flame on being plunged into the gas.
- When mixed with an equal volume of nitric oxide, red fumes are produced (distinction from nitrous oxide)

Tests for Purity: It is tested for wt. per litre, acidity and alkalinity, carbon monoxide, carbon dioxide, halogens and oxidizing substances.

Wt. per litre: At 0^{0} and a pressure of 760 mm of mercury 1.429g.

Note: Maintain the cylinder of oxygen between 23^{0} and 27^{0} for atleast six hours prior to withdrawing samples for the tests and assay, and correct the results to normal temperature and pressure.

Acidity and alkalinity: Boil 300 ml of water with 1 ml of solution of methyl red for five minutes and transfer 100 ml of this solution to each of three similar cylinders. Label the cylinders as 1, 2 and 3. While the solutions are still warm add 0.1 ml of 0.01N sulphuric acid or 0.01N hydrochloric acid to cylinder 1 and 0.2 ml of the same acid to cylinders 2 and 3, stopper cylinders 1 and 3 and pass the gas through cylinder 2; the colour in cylinder 2 is not more yellow than that in cylinder 1 and not more pink than that in cylinder 3.

Carbon monoxide: Transfer a volume of the gas equivalent to 250 ml at N.T.P to a suitable container and add 2.5 ml of dilute oxalated blood (1 part of oxalated blood to 20 parts of water) avoiding admixture with air as far as possible. Shake thoroughly for fifteen minutes and transfer to a small test tube. Add 40 mg of a mixture of equal parts of pyrogallol and tannic acid, shake the contents and set aside for forty minutes, the precipitate assume a grayish-brown colour and no red colour is seen.

***Carbon dioxide*:** Pass a volume equivalent to 1000 ml measured at normal temperature and pressure, through 100 ml of clear solution of barium hydroxide in the course of about fifteen minutes. The turbidity produced is not greater than that produced by adding 1 ml of a solution prepared by dissolving 0.1 g of sodium bicarbonate in 100 ml of freshly boiled and cooled water to 100 ml of clear solution of barium hydroxide.

***Halogens*:** Pass a volume equivalent to 2000 ml measured at normal temperature and pressure, through a mixture of 100 ml of water and 1 ml of solution of silver nitrate, no opalescence is produced.

***Oxidising substances*:** Pass a volume equivalent to 2000 ml measured at normal temperature and pressure through a freshly prepared solution of 0.5 g of soluble starch and 0.5 g of potassium iodide in 100 ml of water containing 1 drop glacial acetic acid; the colour of the liquid is not changed.

***Assay*:** Place a sufficient quantity of mercury in 100 ml calibrated nitrometer provided with a two-way stopcock and a two-way outlet, and properly connected with a balancing tube.

Connect one of the outlet tubes of the nitrometer with a gas pipette of suitable capacity. Place in the pipette a coil of copper wire which extends to the uppermost portion of the bulb and add about 125 ml of solution of ammonium chloride ammonium hydroxide. Draw the liquid (free from air bubbles) through the capillary opening connection and stopcock opening in the nitrometer by reducing the pressure in the nitrometer tube and opening the stopcock controlling the connection with the gas pipette. Then close the stopcock. Having completely filled the nitrometer, the other nitrometer exactly 100 ml of oxygen by reducing the pressure in the tube. Close the stopcock, increase the pressure on the oxygen in the nitrometer tube and open the stopcock controlling the connection with the gas pipette. Force the entire volume of gas into the pipette. Close the stopcock and rock the pipette gently, providing frequent contact of the liquid gas and copper spiral. At the end of fifteen minutes most of the gas will have been absorbed by the liquid. At this time of facilitate the absorption of the last portion of the oxygen, draw some of the liquid into the nitrometer tube, and force the residual gas back up on the surface of the liquid in the gas pipette. Again rock the pipette until no further diminution in the volume of the gas occurs. Draw the residual gas if any, into the nitrometer tube, and measure its volume. The volume of gas remaining unabsorbed does not exceed 1 ml.

Labelling: The shoulder of the metal cylinder is painted white and the remainder is painted black. The cylinder carries a label stating the name of the gas and in addition, the name of the gas or the symbol O_2 is stenciled in paint, on the shoulder.

Storage: Pressure oxygen in cylinders. Containers used for oxygen must not be treated with any toxic sleep inducing or narcosis producing compounds and must not be treated with any compound that will be irritating to the respiratory tract when the oxygen is used.

(b) Carbon dioxide:

Chemical Formula: CO_2 *Mol. Wt.* 44.01

Carbon dioxide contains not less than 9.0 percent v/v of CO_2. For convenience in the use it is compressed in metal cylinders.

Description: A heavy colourless gas, odourless; its aqueous solution is faintly acidic in taste.

Solubility: Soluble in about an equal volume of water.

Preparation: It is produced on the large scale as the by-product of the fermentation industries. It is also prepared by the action of acids on native carbonates.

Identification:

1. Extinguishes a flame
2. Pass through a solution of barium hydroxide, a white precipitate is produced, ass acetic acid, the precipitate dissolves with effervescence.

Note: Maintain the cylinder of carbon dioxide at 25° for atleast six hours prior to with drawing samples for the following tests and assay.

Test for purity: It is tested for acid and sulphur dioxide, phosphine, hydrogen sulphide and organic reducing substances and carbon monoxide.

Acid and Sulphur dioxide: Pass a volume of a gas equivalent to 1,000 ml at N.T.P through 50 ml of carbon dioxide free water. Regulate the flow so as to require fifteen minutes for the deliver of 1000 ml of the gas. The delivery tube must have an orifice about 1 mm in diameter and must extend to within 2 mm of the bottom of the vessel containing water. The vessel must be of such a diameter that a hydrostatic column of from 12 to 14 cm is produced with 50 ml of water. After the passage of the gas, transfer the liquid to a colour-comparison tube (A) and add 0.1 ml of solution of methyl orange. To similar tube (B) add 50 ml of carbon dioxide-free water, 1 ml of 0.01N Hydrochloric acid and 0.1 ml of solution of methyl orange. Compare the colours in the tubes over a white surface viewed down ward; the liquid in tube (A) shows no deeper shade of red than that in tube (B)

Phosphine, hydrogen sulphide and organic reducing substances: Pass a volume of the gas equivalent to 1000 ml at N.T.P under conditions stated in the test for acid and sulphur dioxide through a mixture of 25 ml of solution of silver ammonium nitrate and 3 ml of dilute ammonia solution. Repeat the experiment with the same quantities of the same reagents in the same manner omitting carbon dioxide; no turbidity or darkening is produced as shown by comparison with the control solution.

Carbon monoxide: Transfer a volume of the gas equivalent to 250 ml at N.T.P to a suitable container and add 25 ml of dilute oxalated blood (1 part of oxalated blood to 20 parts of water), avoiding admixture with air as far as possible. Shake thoroughly for fifteen minutes and transfer to a small test tube. Add 40 mg of a mixture of equal parts of pyrogallol and tannic acid, shake the contents and set aside for forty minutes. The precipitate assumes a greyish-brown colour and no red colour is seen.

Assay: Place sufficient quantity of mercury in 100 ml of nitrometer provided with a two way stopcock and a two way outlet and properly connected with a balancing tube. Connect one of the outlet tubes of the nitrometer with a gas pipette of suitable capacity. Place in the pipette about 125 ml of a 50 percent w/v solution of potassium hydroxide. Draw the liquid (free from air bubbles) through the capillary opening, connection, and stopcock opening in the nitrometer by reducing the pressure in the nitrometer tube and opening the stopcock controlling the connection with the gas pipette. Close the stopcock. Having completely filled the nitrometer the other stopcock opening and the other intake tube with mercury, draw into the nitrometer, by reducing the pressure in the tube, exactly 100 ml of carbon dioxide at N.T.P. Close the stopcock. Increase the pressure on the gas in the nitrometer tube and open the stopcock controlling the connection with the gas pipette. Force the entire volume of gas into the pipette. Close the stopcock and rock the pipette gently providing frequent contact of the liquid and the gas. At the end of five minutes when most of the gas has been absorbed by the liquid, facilitate the absorption of the remainder by drawing some of the liquid into nitrometer tube and forcing the residual gas back up on the surface of the liquid in the gas pipette. Again rock the pipette until no further diminution in the volume of gas occurs. Draw the residual gas; if any into the nitrometer tube and measure its volume, not more than 1 ml of gas remains.

Storage: Pressure carbon dioxide in metal cylinder

Category: Respiratory stimulant

Application: By inhalation 5 to 7.5 percent of oxygen

Labelling: The metal cylinder should be painted grey and carry a label stating the name of the gas. In addition, the name of the gas or the symbol CO_2 is stenciled in point on the shoulder of the cylinder.

(c) **Nitrous oxide:** It is used as general anaesthetic by inhalation and as an analgesic. It is stored in safe metal cylinder.

Chemical Formula: N_2O *Mol. Wt.* 44.02

It is coloured in blue. It contains not less than 99.0 percent v/v of nitrous oxide, when drawn from the gaseous phase. For convenience in use it is kept compressed in metal cylinders.

Description: A colourless gas heavier than air, odour, characteristic, taste, faintly sweetish.

Solubility: Dissolves in about 2 volumes of water, more freely soluble in alcohol, soluble in solvent ether.

Preparation: It is prepared by the action of heat on ammonium nitrate

$$NH_4NO_3 \rightarrow N_2O + 2H_2O \quad .....(13.34)$$

Identification: A glowing splinter of wood bursts into flame on being plunged into the gas.

No red fumes are produced when mixed with nitric oxide (distinction from oxygen). For the following tests place the reagent in a 100 ml cylinder which has a height of about 20 cm and is closed with a stopper containing an inlet tube and an exit tube, the inlet tube has a bore not exceeding 0.5 mm and passes to the bottom of the cylinder. A volume equivalent to 2000 ml measured at normal temperature and pressure is passed through the reagent in thirty minutes for each of the following tests.

Tests for purity: Acidity or alkalinity, halogens and hydrogen sulphide, oxidizing substances, reducing substances.

Acidity or alkalinity: Boil 300 ml of water with 1 ml of solution of methyl red for five minutes and transfer 100 ml of this solution to each of three similar cylinder. Label the cylinders as 1, 2 and 3. While the solutions are still warm add 0.1 ml of 0.01N sulphuric acid or 0.01N HCl to cylinder 1 and 0.2 ml of the same acid to cylinders 2 and 3; stopper cylinders 1 and 3 and pass the gas through cylinder 2; the colour in cylinder 2 is not more, yellow than that in cylinder 1 and not more pink than that in cylinder 3.

Halogens and hydrogen sulphide: On passing the gas through 100 ml of water containing 1 ml of solution of silver nitrate neither opalescence not darkening is produced.

Oxidising substances: On passing the gas through a freshly prepared solution of 0.5g of soluble starch and 0.5 g of potassium iodide in 100 ml of water containing 1 drop of glacial acetic, no colour is developed.

Reducing substances: On passing the gas through 100 ml of water containing 0.2 ml of 0.1N potassium permanganate, the colour is not completely discharged.

Arsine and Posphine: Through a mercuric chloride paper attach to a glass-tube as in the quantitative tests for arsenic; pass a volume equivalent in 2000 ml measured at normal temperature and pressure. No visible stain is produced.

Carbon dioxide: Not more than 50 parts per million v/v of C. For this determination the gas used is the first portion drawn from the cylinder and is taken with the cylinder in the upright position. A volume equivalent to 5000 to 10,000 ml measured at normal temperature and pressure. No visible stain is produced.

Carbonmonoixde: Not more than 50 parts per million v/v of Co. For this determination the gas used is the first portion drawn from the cylinder and is taken with the cylinder in the up right position. A volume equivalent to 5000 to 10,000 ml measured at normal temperature and pressure is passed through a purifying train comprising chromic-sulphuric acid (20 potassium hydroxide).

Phosphorus pentoxide: It is next passed through a tube containing iodine pentaoxide (previously dried at 200°) maintained at a temperature of 120° absorbing the liberated iodine in solution of potassium iodide. Sweep out the apparatus by passing through it 5,000 ml of air free from carbon monoxide. Titrate the iodine with 0.002N sodium thiosulphate and from the amount used substract the amount required in a similar experiment, in which 5,000 ml of air free from carbon monoxide is used. Each ml of 0.002N sodium thiosulphate is equivalent to 0.112 ml of CO at normal temperature and pressure.

Carbon dioxide: Pass 1,000 ml at normal temperature and pressure through 50 ml of clear solution of barium hydroxide contained in a vessel of such size and shape that the depth of the solution is from 12 to 14 cm employing a deliver tube with an orifice of approximately 1 mm diameter and extending to within 2 mm of the bottom of the vessel and regulating the flow of the nitrous oxide so as to require approximately fifteen minutes for the delivery of 1000 ml.

Any turbidity produced does not exceed that produced when 1 m of a solution of 100 mg of sodium bicarbonates in 100 ml of freshly boiled and cooled water is added to 50 ml of clear solution of barium hydroxide.

Assay: Cool a measured volume in liquid oxygen in a suitable apparatus. Not less than 99.0 percent v/v is condensed.

Labelling: The metal cylinder is painted blue and carries a label stating the name of the gas. In addition, the name of the gas or the symbol 'N_2O' is stenciled in paint on the shoulder of cylinder.

Storage: Preserve nitrous oxide in cylinders.

Category: General anaesthetic

(d) Helium: Helium is the second lightest element found in nature. This has not been included in I.P but has been official in B.P and U.S.P. It is supplied in metal cylinders under compression in 98% v/v purity. The cylinders are coloured brown. It finds medicinal uses in cases of respiratory obstruction.

(e) Nitrogen: It may be prepared by the distillation of liquid air and is supplied in cylinders painted grey with black on the neck and shoulders and the name stenciled or painted indelibly on the body of the cylinder. It finds use as a diluent for oxygen before administering to patients. It is also used in cryoscopic surgery to remove some tumours.

CHAPTER 14

Radiopharmaceuticals

14.1 Introduction

All chemical changes involve the rearrangement of atomic electrons either by means of the exchange or sharing of electrons in the outer part of the atom, the atomic nucleus remaining quite unaltered. Under special conditions, however, atomic nuclei may be altered either spontaneously as in radioactivity or artificially as in atomic transformations. Such reactions which involve changes occurring inside the nuclei of atoms and depend wholly on their structure are known as nuclear reactions. In nuclear reactions, there will be a change either in the number of protons or neutrons or both leading to atomic transformation of one element to another. These reactions involve greater energy changes than ordinary chemical reactions ($\simeq$ 100000 times energetic). Nuclear reactions involve-natural radioactivity, artificial radioactivity, artificial transmutation, nuclear fission and nuclear fusion. Nuclear instability is the cause of radioactivity.

Over four hundred radioactive isotopes have been made in the laboratory. The use of radioactive isotopes as traces has become a valuable technique in modern scientific and medical research. These discoveries have much vital effects on our lives and have found so many applications

14.2 Radioisotopes

Two or more elements possessing the same excess of protons over electrons in the nuclei but with a different total number of protons and electrons.

The isotopes (Greek same place) can also be defined as two or more elements differing in their atomic weights or radioactive properties but having identical chemical properties and thus requiring to be placed in the same place in the periodic table, since isotopes possesses the same atomic number.

In various cases of artificial radio elements, one and the same nucleus with the same mass number and atomic number disintegrates in different ways with different decay periods. These nucleides are called nuclear isomers and the phenomenon as nuclear isomerism by analogy with chemical isomerism. In 1917, Soddy suggested that such isomers might exist among the natural radio elements.

The existence of different energy levels of the nucleus of an atom represents different isomeric states these nucleides are isotopic isobars or isobaric isotopes having same decay schemes with different half lifes.

The example is

$$_{30}Zn^{68} + {}_1D^2 \rightarrow {}_{30}Zn^{69} + {}_1H^1 \quad \ldots\ldots (14.1)$$

13.8h ↑ (excited) → the excited state decays with emitting γ -rays

$$_{31}Ga^{69} \xrightarrow[\substack{7.5\,min \\ (stable)}]{-\beta} {}_{30}Zn^{69} \text{ (Ground state)}$$

So $_{30}zn^{69}$ shows nuclear isomerism

The other example is

$$\underset{\text{Excited}}{_{35}Br^{79}} \xrightarrow[4.4h]{\gamma\text{-rays}} \underset{\text{ground state}}{_{35}Br^{80}} \xrightarrow[18\,min]{-\beta} \underset{\text{Stable}}{_{36}Kr^{80}} \quad \ldots\ldots (14.2)$$

Wiezsacker Explanation **(1936):** The existence of low excited states whose angular momentum (i.e., nuclear spin) is considerable, differing by several units from that of the ground state. Such an excited state would be metastable and nucleus can remain in this state for a sufficient time as a separate nucleus and then decay. This metastable state goes over to ground state either with the emission of γ -rays and particle or emit a β-particle directly.

Two major types of isotopes are found in nature:

1. Stable isotopes maintain their elemental integrity and do not decay.
2. Unstate or radioactive isotopes decay by emission of nuclear particles, into other isotopes of the same or different elements.

***Binding Energy*:** The binding energy of a nucleus is the energy of which must be given to a nucleus to break it completely into its protons and neutrons.

The mass change is a measure of energy change according to Einstein's mass energy relation, then

$$B = ZM_H + (A - Z)\,M_N - M \text{ amu (atomic mass units)} \quad \ldots..(14.3)$$

Where B = binding energy of a nucleus

M_H = mass or proton

M_N = mass of neutron

M = mass of nucleus

A = Total nucleus

Z = number of protons or neutrons

For e.g.: $_2He^4$. Four nucleons, two protons and two neutrons. Mass of proton is 1.0073 and neutron 1.0087 amu

$$\text{Total mass} = 2 \times 1.0073 + 2 \times 1.0087 = 4.0320 \text{ amu}$$

B = mass difference of nucleons and necleous

$$= 4.0320 - 4.0028 = 0.0292 \text{ amu (atomic mass units)}$$

$$\text{In Mev} = 0.0292 \times 931 = 27.2 \text{ Mev}$$

The binding energy per nucleon, $\bar{B}$, of a nucleus is B divided by the total number of nucleons.

For e.g: In $_2He^4$ $\bar{B} = \frac{27.2}{4} = 0.8\,\text{Mev}$

Packing Fraction and Binding Energy per Nucleons

Packing fraction *f* is given by the equation

$$f = \frac{M - A}{A} \qquad \text{..... (14.4)}$$

Where A = integral mass number

M = exact atomic mass

or

$$\frac{M}{A} = 1 + f \qquad \text{.....(14.5)}$$

$$\bar{B} = \frac{B}{A} = \frac{Z}{A} M_H + \left(1 - \frac{Z}{A}\right) M_N - \frac{M}{A} \qquad \text{..... (14.6)}$$

$$= \frac{Z}{A}\left[M_H - M_N\right] + M_N - (1 + f)$$

The minimum value of *f* corresponds to maximum value of $\bar{B}$. Since *f* is rarely exceeds 10^{-3} amu the value of $\bar{B}$ is roughly constant which is a consequence of the fact that all nuclear forces are short range forces.

Units of Radioactivity

(i) ***Curie*:** Curie can be defined as that quantity of any radioactive substance which undergoes the same number of disintegrations per second as one gram of pure radium.

Curie can also be defined, as per IUPAC, as that quantity of any radioactive material which disintegrates at a rate of exactly 3.7×10^{10} per second.

So 1 curie c = 3.7×10^{10} disintegrations per second

1 milli curie mc = 3.7×10^{7} disintegrations per second

1 micro curie mc = 3.7×10^{4} disintegrations per second

(ii) ***Rutherford*:** This is another unit is defined as the amount of radioactive substance which undergoes ten disintegrating per second.

14.3 Radioactive Disintegration

Rutherford and Soddy (1903) stated that the atoms of the radio elements are assumed to be unstable and spontaneously undergoing change.

During the change, Becquerel rays are emitted and the original radioactive atom is converted into a new atom which is quite different physically and chemically from its parent. This new element may in turn also be unstable and emit a particle with the production of still another element and there may infact be a succession of transformations each accompanied by its characteristic rays until a non-radioactive end product is reached. The atomic weight and other properties of the elements produced in the radioactive disintegration will depend on the nature of the particle emitted.

For e.g:

$$ {}_{Z}X^{A} \rightarrow {}_{z.2}Y^{A-4} + {}_{2}H^{4}(\alpha) \quad \text{..... (14.7)}$$

$$ {}_{Z}P^{A} \rightarrow {}_{2+1}Q^{A} + \beta \quad \text{..... (14.8)}$$

Radioactive Constant

At any instant the rate of decay is proportional to the activity at that instant. Hence

$$\frac{I_t}{I_o} = e^{-kt} \quad \text{..... (14.9)}$$

Where

I_o = activity of the substance in the commencement

I_t = activity after time t

t = radioactive constant or the disintegration constant or decay constant of the radioactive element

The radioactive decay is first order process. So

$$-\frac{dN}{dt} = KN \qquad \text{..... (14.10)}$$

Where dN = change in no. N of atoms of a species in an infinitesimal time period

As N is decreasing, the left hand side of the eq. (14.10) is – ve. On integrating eq.10, we get

$$\log_e \frac{Nt}{N_o} = -kt \qquad \text{.....(14.11)}$$

or

$$N_t = N_o\, e^{-kt} \qquad \text{.....(14.12)}$$

Correcting to normal logarithms, we get

$$\log_{10} \frac{N_t}{N_o} = -\ 0.4343\ kt \qquad \text{..... (14.13)}$$

The radioactive constant, K, is thus defined as the fraction of the total no. of atoms present which break up per unit of time and is characteristic of element under consideration irrespective of physical condition or state of combination.

The life of any radioactive atom i.e., the length of time it can exist before it disintegrates can have all possible values from zero to infinity. It is essential to define life of the radioactive element in terms of half-life period and average life period.

Half-life Period

It is defined as the time in which one half of the amount of the radio element initially available decays.

The half life period $\left(T^{1/2}\right)$ is obtained by substituting ½ for N_t/N_o and p for t in eq. 14.13

$$\log_{10} ½ = -\ 0.4343\ KP \qquad \text{.....(14.14)}$$

$$T_{1/2} = P = \frac{0.693}{K} \qquad \text{..... (14.15)}$$

***Mean or Average life Period,* θ:** It is defined as the time within which the whole material would disintegrate if the number of atoms disintegrating per unit time were to remain the same as initially.

$$\theta = {}^{1}/_{k} \qquad \text{..... (14.16)}$$

Where k = radioactive constant

or $$p = 0.693\ \theta \qquad \text{.....(14.17)}$$

14.4 Radioactive Disintegration Particles

If the radioactive isotope disintegrates, it emits certain particles or quantities of energy that are characteristic of particular isotope involved. The major decay particles of interest are:

Alpha particles (α, $_2He^4$): The characteristics are:

- High speed helium nuclei and the velocity being nearly one tenth of velocity of light. The charge of a α-particle is $2 \times 4.802 \times 10^{-10}$ e.s.u (electrostatic units) which is same as the charge on the helium nucleus.
- The particle consists of two protons and two neutrons with an atomic mass of 4 (wt is $\simeq 6.6 \times 10^{-24}$ g) and atomic number is 2.
- The penetrating power is much less and is about 1/100 of that of β rays and 1/1000 of that of γ-rays. A thick sheet paper can stop α-rays completely.
- The ionizing power is great, 100 times that due to β-rays and 10,000 times that due to γ-rays.
- They produce fluorescence when they fall on certain substances like diamond, zinc sulphide etc.
- They have high kinetic energy so that effect the photographic plate and cause luminosity when strike a zinc sulphide plate

Alpha radiation is usually emitted only from elements having atomic numbers greater than 82. The emission of α-radiation is usually accompanied by other radiations.

For e.g:

$$_{88}Ra^{226} \rightarrow {}_{86}Ra^{222} + {}_2He^4(\alpha) \qquad \text{..... (14.18)}$$

***Beta Particles* (β^- or β^+):** Generally the β particles are negative charged species having the mass of an electron ($\sim 9.1 \times 10^{-28}$). The characteristics are:

- Have high velocity between 0.36 to 0.98 (33% to 90%) times the velocity of light. Due to this variation in velocity, beta radiation is not homogeneous.
- Charge to mass (e/m) ratio is found to be identical with that of an electron.
- The ionizing power of β-radiation is nearly 100 times that of γ-rays.
- The penetration power is 100 times higher than α-rays and travel 10 to 15 mm in water, penetrate 1 inch thickness of Al foil.

- The effect on a photographic plate is greater than that of α-particles since these produce x-rays when then incident on the photographic plate.

Beta particles, sometimes called megatrons, are emitted by unstable nuclei having neutrons in excess of protons.

For e.g:

$$_0n^1 \rightarrow {}_1p^1 + \beta^- \qquad \text{..... (14.19)}$$

$$_6C^{14} \rightarrow {}_7N^{14} + \beta^- \qquad \text{..... (14.20)}$$

(higher at. no.)

The positron β^+ is identical to the electron with the exception of having positive charge and is emitted from nuclei having proton/neutron ratio above stable limits.

$$_1P^1 \rightarrow {}_0n^1 + \beta^+ \qquad \text{..... (14.21)}$$

$$_{30}Zn^{65} \rightarrow {}_{29}Cu^{65} + \beta^+ \qquad \text{..... (14.22)}$$

(lower at. no)

Positrons are very short lived and have little importance in biological applications. With electron gives γ -rays.

$$\beta^+ + e^- \rightarrow 2\gamma \qquad \text{..... (14.23)}$$

***Gamma Rays* (γ):** These are very short electromagnetic waves and the wavelength is of the order of 10^{-10} cm and has no mass and no charge.

The characteristics are:

- The penetrating power is very high and is 100 times more than β-rays.
- The ionizing power is very low i.e., 1/100 times that of β-rays.
- They do not show any deviation in a magnetic and electric field since they do not carry any charge.
- They produce very little effect on Zns and photographic plate.
- These are diffracted indicating that γ-rays are waves.
- These interact with atoms and molecules in a particular medium to produce ions and free radicals secondarily by dislodging electrons from orbitals.

The emission of γ-rays is always accompanied by other forms of radiation. If γ-emission alone occurs, it will involve the transition of metastable state of an isotope either to a stable form or to a form of same isotope which will continue to decay by other way.

For e.g:

$$_{27}Co^{59}(n,\gamma)_{27}Co^{60(m)} \rightarrow {}_{27}Co^{60} + \gamma \rightarrow {}_{28}Ni^{60} + \beta^- + 2\gamma \quad .....(14.24)$$

m is metastable

X-rays (K-Capture)

The emission of x-rays is through a process of K-capture (orbital electron capture). The proof of its reality was obtained by L.W. Alvarez in 1938.

X-ray radiation is produced by isotopes with an unstable proton/neutron ratio but with insufficient energy to emit positron. In such condition nucleus captures an electron from the K(1 s) L(2s) etc shells.

If the nucleus captures electron from K shell i.e., 1s orbital (k-electron capture) which combines with a proton to form neutron. The rearrangement of orbital electrons with the release of energy in the form of x-rays. The intensity of x-rays falls off as the active material decays. The loss of proton in the nucleus that the isotope will decay to the element having one atomic number less.

A 330 day $_{23}V^{49}$ isotope to the element $_{22}Ti^{49}$ by K-capture.

Another example K-capture is

$$_{80}Hg^{197} \rightarrow {}_{79}Au^{197} + x - rays \quad ....(14.25)$$

Isotopes emitting γ-radiation are used frequently in biological applications. The high penetrating power of this radiation is sufficient to reach deep into tissues, and to be detected outside of the body.

The energies are expressed is units of million electron volts (Mev) and 1 Mev is equivalent to 1.6×10^{-6} ergs molecule^{-1} or 2.3×10^{7} K.cal mol^{-1}.

The properties of radioactive decay particles are given in table 14.1.

Table 14.1 Properties of radioactive decay particles.

S.No.	Radiation	Particle type	Mass	Charge	Velocity	Effect of Emission	
						At. No.	Mass No.
1.	α	$_2He^4$	6.6×10^{-24} g	+2	$0.1 \times 3 \times 10^{10}$	–2	–4
2.	β^-	electron	9.1×10^{-28} g	–1	$0.4 + 0.9 \times 3 \times 10^{10}$	+1	0
3.	β^+	positron	9.1×10^{-28} g	+1	$0.4 - 0.9 \times 3 \times 10^{10}$	–1	0
4.	γ	Photon	0	0	3×10^{10}	0	0
5.	x (K-capture)	Photon	0	0	3×10^{10}	–1	0

14.5 Kinetics of Isotopic Decay

Geiger and Nuttah found in their experiment that, in general, those materials which decay slowly emit α-particles of short range. While those which disintegrate rapidly emit more energetic particles. A relationship between the decay constant λ and the range R is given by equation.

$$\log \lambda = A + B \log R \qquad .....(14.26)$$

Where A and B = constants

As per the eq. 14.26, for elements of a particular series a plot of log λ vs log R will give a straight line, where R is the range of standard air.

Atoms of every radioactive element are constantly disintegrating (Rutherford and Soddy) to fresh radioactive products with the emission of α, β and γ-rays. The number of atoms breaking per second at any instant is proportional to the number present at that instant i.e.,

$$-\frac{dN}{dt} \alpha N \qquad ..... (14.27)$$

Where dN = number of nuclei which decay

N = nuclei present

The eq. 14.27 can be rewritten as

$$\frac{dN}{dt} = -\lambda N \qquad .....(14.28)$$

Where λ = radioactive constant

λ is a definite and specific property of a given radioelement

Then eq. 14.28 written as

$$\frac{dN}{N} = -\lambda\, dt \qquad ..... (14.29)$$

On integration

$$\log_e N = -\lambda t + C \qquad .....(14.30)$$

Where C = constant of integration

When t = 0, N = N_o, then eq. 14.30 becomes

$$\log_e N_o = -\lambda_{xo} + C \qquad .....(14.31)$$

or $C = \log N_o$

Combine eq 14.30 and 14.31

$$\log_e N = -\lambda t + \log_e N_o$$

or $$\log_e \frac{N}{N_o} = \lambda t \text{ or } \frac{N}{N_o} = e^{-\lambda t} \quad .....(14.32)$$

The number of nuclei in radioactive element decreases exponentially with time.

The number of disintegrations per second is given by

$$A = -\frac{dN}{dt} = \lambda N \quad .....(14.33)$$

Where A = activity

$$A_o = \frac{dN_o}{dt} = \lambda N_o$$

Where A = activity at t = 0

Dividing eq. 14.33 by 14.34,

$$\frac{A}{A_o} = \frac{N}{N_o}$$

$$\therefore \quad \frac{A}{A_o} = e^{-\lambda t} \quad .....(14.34)$$

Since $\frac{N}{N_o} = e^{-\lambda t}$ as per eq. 14.32

The unit of activity is the curie.

14.6 Nuclear Reactions

The nuclei of atoms interact with other nuclei or lighter particles or a photon, resulting in the formation of new nuclei and one or more lighter particles are called nuclear reactions.

According to Bohr, the nuclear reaction involving two steps.

(i) In the first step the projectile combines with nucleus to form compound nucleus. Kinetic energy of the projectile is distributed uniformly among all nucleons (protons, neutrons) in the nucleus (excited state)

Incident particle (projectile) + initial nucleus → compound nucleus .....(14.35)

(ii) In the second step the compound nucleus formed to give the final products.

Compound nucleus → product nucleus + outgoing particle .....(14.36)

For e.g.,

Examples of types of Nuclear Reactions

(i) Capture reactions

$${}_{6}C^{12} \rightarrow {}_{1}H^{1} \rightarrow {}_{7}N^{13} + \gamma \qquad .....(14.37)$$

$${}_{35}Br^{74} \rightarrow {}_{0}n^{1} \rightarrow {}_{35}Br^{75} + \gamma \qquad .....(14.38)$$

(ii) Particle–particle reactions

$${}_{7}N^{14} \rightarrow {}_{0}n^{1} \rightarrow {}_{6}C^{14} + {}_{1}H^{1} \qquad .....(14.39)$$

$${}_{5}B^{11} \rightarrow {}_{1}H^{1} \rightarrow {}_{6}C^{11} + {}_{0}n^{1} \qquad .....(14.40)$$

(iii) Fission reactions

$${}_{92}U^{235} \rightarrow {}_{0}n^{1} \rightarrow {}_{96}Ba^{141} + {}_{56}Kr^{92} + 3\,{}_{0}n^{1} + \text{energy} \qquad .....(14.41)$$

(iv) Spallation reactions

$${}_{29}Cu^{63} + {}_{2}He^{4}\,(400\text{Mev}) \rightarrow {}_{17}Cl^{37} + 14\,{}_{1}H^{1} + 16\,{}_{0}n^{1} \qquad .....(14.42)$$

(v) Fussion Reactions

$${}_{1}H^{2} + {}_{1}H^{2} \rightarrow {}_{1}H^{3} + {}_{0}n^{1} + 3.25\ \text{Mev} \qquad .....(14.43)$$

$${}_{1}H^{3} + {}_{1}H^{2} \rightarrow {}_{2}He^{4} + {}_{0}n^{1} + 17.8\ \text{Mev} \qquad .....(14.44)$$

(vi) Bombarding reactions

α, p type

$${}_{13}Al^{27} + {}_{2}He^{4} \rightarrow {}_{14}Si^{30} + {}_{1}H^{1} \qquad .....(14.45)$$

p, γ type

$${}_{7}N^{14} + {}_{1}H^{1} \rightarrow {}_{8}O^{15} + \gamma \qquad .....(14.46)$$

p , α type

$${}_{13}Al^{27} + {}_{1}H^{1} \rightarrow {}_{12}Mg^{24} + {}_{2}He^{4} \qquad .....(14.47)$$

d, p

$${}_{3}Li^{6} + {}_{1}H^{2} \rightarrow {}_{2}He^{4} + {}_{2}He^{4} \qquad .....(14.48)$$

d, p

$${}_{15}P^{31} + {}_{1}H^{2} \rightarrow {}_{15}P^{32} + {}_{1}H^{1} \qquad .....(14.49)$$

γ, x

$${}_{4}Be^{9} + \gamma \rightarrow {}_{4}Be^{8} + {}_{0}n^{1} \qquad .....(14.50)$$

γ, p

$$_{13}Al^{27} + \gamma \rightarrow \left(_{13}Al^{27}\right) \rightarrow _{11}Na^{24} + _1H^1 + _1H^1 + _0n^1 \quad .....(14.51)$$

High energy

The isotopes used in the Radiopharmaceuticals and Biological research are given in Table 14.2.

Table 14.2 Isotopes used in Radio pharmaceuticals and Biological research.

S.No	Isotope	$T_{1/2}$	Energy in Mev and type of radiation				Application
			β^-	β^-	γ, x-ray (K capture)	α	
1.	Au^{198}	2.7d	0.959	-	0.412	-	Therapeutic Diagnostic
2.	C^{14}	5700 y	0.16	-	-	-	Research
3.	Ca^{45}	165 d	0.26	-	-	-	Diagnostic
4.	Ca^{47}	4.5 d	1.1	-	1.3	-	Diagnostic
5.	CO^{57}	270 d	-	-	0.137 and 0.123	-	Diagnostic
6.	CO^{58}	71 d	-	0.47	0.81(captur e	-	Diagnostic
7.	CO^{60}	5.27 Y	0.312	-	1.17281.132	-	Therapeutic Diagnostic
8.	Cr^{51}	27.8 d	-	-	0.321 (k capture)	-	Diagnostic
9.	Cs^{131}	9.7 d	-	-	0.029 (K capture)	-	Diagnostic
10.	Cs^{137}	30 Y	0.51881.17	-	0.66	-	Research
11.	F^{18}	1.7 h	-	0.6	-	-	Diagnosis
12.	Fe^{59}	45 d	0.46280.27 1	-	131.1	-	Diagnostic
13.	H^3	12.3 Y	0.018	-	-	-	Diagnostic and Research
14.	Hg^{197}	2.7 d	-	-	0.077	-	Diagnostic
15.	Hg^{203}	46.9 d	0.21	-	0.279	-	Diagnostic
16.	I^{125}	60 d	-	-	0.027 (k capture)	-	Diagnostic Therapeutic
					0.035		
17.	I^{131}	8.8 d	0.608	-	0.722	-	Diagnostic
			0.335	-	0.637	-	Therapeutic
			0.250	-	0.364	-	Research

Table 14.2 *Contd...*

S.No	Isotope	$T_{1/2}$	Energy in Mev and type of radiation				Application
			β^-	β^-	γ, x-ray (K capture)	α	
18.	Ir^{192}	74.4 d	0.67	-	0.3280.47	-	Therapeutic
19.	K^{42}	12.4 h	2.04803.58	-	1.5	-	Research
20.	Mo^{99}	2.8 d	1.23	-	0.14	-	Source of TC^{m99}
21.	Na^{22}	2.6 Y	-	0.54	1.2 (K capture)	-	Diagnostic
22.	Na^{24}	15 h	1.39	-	1.38 & 2.75	-	Diagnostic
23.	P^{32}	14.3 d	1.71	-	-	-	Diagnostic Therapeutic Research
24.	Ra^{226}	1620 Y	-	-	0.19	4.77	Therapeutic
25.	Rb^{86}	18.8 d	1.77	-	1.08	-	Diagnostic
26.	Rn^{222}	3.8 d	-	-	-	5.5	Therapeutic
27.	S^{35}	88 d	0.167	-	-	-	Research
28.	Se^{75}	120 d	-	-	0.13680.265 0.401 (K capture)	-	Diagnostic
29.	Sr^{85}	64 d	-	-	0.513 (K capture)	-	Diagnostic
30.	Sr^{90}	28 Y	0.54	-	-	-	Therapeutic
31.	Ta^{182}	115 d	0.36, 0, 44 & 0.51	-	0.068, 1.12 and 1.22	-	Therapeutic
32.	Y^{90}	2.6 d	2.26	-	-	-	Diagnostic and Therapeutic
33.	Yb^{169}	32 d	-	-	0.0638 0.198	-	Diagnostic
34.	Zn65	245 d	-	0.32	1.11 (K capture)	-	Research

14.7 Radiation Dosimetry

Radiation dosimetry is the calculation of the absorbed dose in matter and tissue resulting from the exposure to indirectly and directly ionizing radiation. It is a scientific sub specific in the fields of health physics and medical physics that is focused on the calculation of internal and external doses from ionizing radiation.

Dose is reported in gray (Gy) for the matter or sieverts (Sv) for biological tissue, where 1 Gy or 1 Sv is equal to 1 joule per kilogram. Non-SI units are still prevalent as well, where dose is often reported in rads and dose equivalent in rems. By definition, 1 Gy = 100 rad and 1 Sv = 100 rem.

(A) ***Radiation effects on living tissue***: The distinction between absorbed dose (Gy) and dose equivalent (Sv) is based upon the biological effects of the weighting factor (denoted w_r) and tissue/organ weighting factor (W_T) have been established, which compare the relative biological effects of various types of radiation and the susceptibility of different organs.

(B) ***Organ dose weighting factors***: By definition, the weighting factor for the whole body is 1, such that Gy of radiation delivered to the whole body (i.e., an evenly distributed 1 joule of energy deposited per kilogram of body) is equal to one sievert (for photons with a radiation weighting factor of 1, see below). Therefore, the weighting factors for each organ must sum to 1 as the unit gray is defined per kilogram and is therefore a local effect. As the table below shows, 1 gray (photons) delivered to the gonads is equivalent to 0.08 Sv to the whole body –in this case, the actual energy deposited to the gonads, being small, would also be small.

Organ or tissue	W_T
Gonads	0.25
Breasts	0.15
Red Bone Marrow	0.12
Lung	0.12
Thyroid	0.03
Bone surfaces	0.03
Remainder	0.30
Whole body	1.0

(C) ***Radiation weighting Factors***: By definition, x-rays and gamma rays have a weighting factor of unity, such that 1 Gy = 1 Sv (for whole-body irradiation). Values of w_r are as high as 20 for alpha particles and neutrons, i.e., for the same absorbed dose in Gy, alpha particles are 20 times as biologically potent as X or gamma rays.

(D) ***Dose versus activity***: Radiation dose refers to the amount of energy deposited in matter and/or biological effects of radiation, and should not be confused with the unit of radioactive activity (Becquerel, Bq). Exposure to a radioactive source will give a dose which is dependent on the activity, time of exposure, energy of the radiation emitted, distance from the source and shielding. The equivalent dose is then dependent upon the weighting factors above. Dose is a measure of deposited dose, and therefore can never go down-removal of a radioactive source can only reduce the rate of increase of absorbed dose, never the total absorbed dose.

The worldwide average background dose for a human being is about 3.5 mSv per year, mostly from cosmic radiation and natural isotopes in the earth. The largest single source of radiation exposure to the general public is naturally-occurring radon gas, which comprises approximately 55% of the annual background dose. It is estimated that radon is responsible for 10% of lung cancers in the United States.

(E) ***Measuring dose*:** There are several ways of measuring doses from ionizing radiation. Workers who come in contact with radioactive substances or the NPL in the UK operate a graphite-calorimeter for absolute photon dosimetry. Graphite is used instead of water as its specific heat capacity is one-sixth that of water and therefore the temperature rises in graphite are 6 times more than the equivalent in water and measurements are more accurate.

Significant problems exist in insulating the graphite from the laboratory in order to measure the tiny temperature changes. A lethal dose of radiation to a human is approximately 10-20 Gy. This is 10-20 joules per kilogram. A 1 cm^3 piece of graphite weighing 2 grams would therefore absorb around 20-40 mJ. With a specific heat capacity of around 700 $J.kg^{-1}.K^{-1}$, this equates to a temperature rise of just 20 mK.

(F) ***Medical dosimetry*:** Medical dosimetry is the calculation of absorbed dose and optimization of dose delivery in radiation therapy. It is often performed by a professional medical dosimetrist with specialized training in the field. In order to plan the delivery of radiation therapy, the radiation produced by the sources is usually characterized with percentage depth dose curves and dose profiles measured by medical physicists.

(G) ***High dose effects*:**

Table 14.3

Dose (Rad)	Effect observed
15 - 25	Blood count changes in a group of people
50	Blood count changes in an individual
100	Vomiting (threshold)
150	Death (threshold)
320 – 360	LD 50/60 with minimal care
480 – 540	LD 50/60 with supportive medical care
1,100	LD 50/60 with intensive medical care (bone marrow transplant)

Every acute exposure will not result in death. If a group of people is exposed to a whole body penetrating radiation dose, the above effects might be observed. The LD 50/60 is the lethal dose at which 50% of those exposed to that dose will die within 60 days.

It is sometimes difficult to understand why some people die while others survive after being exposed to the same radiation dose. The main reasons are the health of the individuals at the time of the exposure and their ability to combat the incidental effects of radiation exposure, such as the increased susceptibility to infections.

Summary of Biological Response to High Doses of Radiation

Dose	Response
< 5 rad	- No immediate observable effects
~ 5 rad to 50 rad	- Slight blood changes may be detected by medical evaluations
~ 50 rad to 150 rad	- Slight blood changes will be noted and symptoms of nausea, fatigue, vomiting, etc, likely
~ 150 rad to 1,100 rad	- Severe blood changes will be noted and symptoms appear immediately. Approximately 2 weeks later, some of those exposed may die. At about 300-500 rad, up to one half of the people exposed will die within 60 days without intensive medical attention. Death is due to the destruction of the blood forming organs. Without white blood cells, infection is likely. At the lower end of the dose range, isolation, antibiotics, and transfusions may provide the bone marrow time to generate new blood cells and full recovery is possible. At the upper end of the dose range, a bone marrow transplant may be required to produce new blood cells.
~ 1,100 rad to 2,000 rad	- The probability of death increases to 100% within one to two weeks. The initial symptoms appear immediately. A few days later, things get very bad, very quickly since the gastrointestinal system is destroyed. Once the GI system ceases to function, nothing can be done, and medical care is for comfort only.
> 2,000 rad	- Death is a certainty. At doses above 5,000 rad, the central nervous system (brain and muscles) can no longer control the body functions, including breathing, blood circulation. Everything happens very quickly. Nothing can be done, and medical care is for comfort only.

As noted, there is nothing that can be done if the dose is high enough to destroy the gastrointestinal or central nervous system. That is why bone marrow transplants don't always work.

In summary, radiation can affect cells. High doses of radiation affect many cells, which can result in tissue/organ damage, which ultimately yields one of the Acute Radiation Syndromes. Even normally radio-resistant cells, such as those in the brain, cannot withstand the cell killing capability of very high radiation doses.

(H) ***Other high dose Effects*****:** Besides death, there are several other possible effects of a high radiation dose.

Effects on the skin include (reddening like sunburn), dry desquamation (peeling), and moist desquamation (blistering). Skin effects are more likely to occur with exposure to low energy gammas, X-ray, or beta radiation. Most of the energy of the radiation is deposited in the skin surface. The dose required for erythema to occur is relatively high, in excess of 300 rad. Blistering requires a dose in excess of 1,200 rad.

Hair loss, also called epilation, is similar to skin effects and can occur after acute doses of about 500 rad.

Sterility can be temporary or permanent in males, depending upon on the dose. In females, it is usually permanent, but it requires a higher dose. To produce permanent sterility, a dose in excess of 400 rad is required to the reproductive organs.

Cataracts (a clouding of the lens of the eye) appear to have a threshold of about 200 rad. Neutrons are especially effective in producing cataracts, because the eye has a high water content, which is particularly effective in stopping neutrons.

(I) ***Acute Radiation Syndrome*** **(ARS):** If enough important tissues and organs are damaged, one of the Acute Radiation Syndromes could result.

The initial signs and symptoms of the acute radiation syndrome are nausea, vomiting, fatigue and loss of appetite. Below about 150 rad, these symptoms, which are no different from those produced by a common viral infection, may be the only outward indication of radiation exposure.

As the dose increases above 150 rad, one of the three radiation syndromes begins to manifest itself, depending upon the level of the dose. These syndromes are (Table 14.4):

Table 14.4

Syndrome	Organs affected	Sensitivity
Hematopoietic	Blood forming organs	Most sensitive
Gastrointestinal	Gastrointestinal Systems	Very sensitive
Central nervous System	Brain & muscles	Least sensitive

(J) ***Effects of Exposure to Low Doses of Radiation:*** There are three general categories of effects resulting from exposure to low doses of radiation. These are:

Genetic: The effect is suffered by the offspring of the individual exposed.

Somatic: The effect is primarily suffered by the individual exposed. Since cancer is the primary result, it is sometimes called the Carcinogenic Effect.

In-Utero: Some mistakenly consider this to be a genetic consequence of radiation exposure, because the effect, suffered by a developing embryo/fetus, is seen after birth. However, this is actually a special case of the somatic effect, since the embryo/fetus is the one exposed to the radiation.

(K) ***Genetic Effects:*** The Genetic Effect involves the mutation of very specific cells, namely the sperm or egg cells. Mutations of these reproductive cells are passed to the offspring of the individual exposed.

Radiation is an example of a physical mutagenic agent. There are also many chemical agents as well as biological agents (such as viruses) that cause mutations.

One very important fact to remember is that radiation increases the spontaneous mutation rate, but does not produce any new mutations. Therefore, despite all of the hideous creatures supposedly produced by radiation in the science fiction literature and cinema, no such transformations have been observed in humans. One possible reason why genetic effects from low dose exposures have not been observed in human studies is that mutations in the reproductive cells may produce such significant changes in the fertilized egg that the result is a nonviable organism which is spontaneously resorbed or aborted during the earliest stages of fertilization.

(L) ***Somatic Effects:*** Somatic effects (carcinogenic) are, from an occupational risk perspective, the most significant since the individual exposed (usually the radiation worker) suffers the consequences (typically cancer).

Radiation is an example of a physical carcinogenic, while cigarettes are an example of a chemical cancer causing agent. Viruses are examples of biological carcinogenic agents.

Unlike genetic effects of radiation, radiation induced cancer is well documented. Many studies have been completed which directly link the induction of cancer and exposure to radiation. Some of the population studied and their associated cancers are:

Lung cancer - uranium miners

Bone cancer - radium dial painters

Thyroid cancer - therapy patients

Breast cancer - therapy patients

Skin cancer - radiologists

Leukemia - bomb survivors, in-utero exposures, radiologists, therapy patients

(M) ***In-Utero Effects*:** The in-utero effect involves the production of malformations in developing embryos.

Radiation is a physical teratogenic agent. There are many chemical agents (such as thalidomide) and many biological agents (such as the viruses which cause German measles) that can also produce malformations while the baby is still in the embryonic or fetal stage of development.

The effects from in-utero exposure can be considered a subset of the general category of somatic effects. The malformation produced does not indicate a genetic effect since it is the embryo that is exposed, not the reproductive cells of the parents.

The actual effects of exposure in-utero that will be observed will depend upon the stage of fetal development at the time of the exposure,

Table 4.5

Weeks	Post Conception	Effect
0 – 1	(Preimplantation)	Intrauterine death
2 – 7	(Organogenesis)	Developmental abnormalities growth retardation/caner
8 – 40	(fetal stage)	Same as above with lower risk plus possible functional abnormalities

14.8 Biological Effects of Radiation

Whether the source of radiation is natural or man-made, whether it is a small dose of radiation or a large dose, there will be some biological effects. Radiation causes ionizations of atoms which may affect molecules which may affect cells, which may affect tissues, which may affect organs, which may affect the whole body. Although we tend to think of biological effects in terms of the effect of radiation on living cells, in actuality, ionizing radiation, by definition, interacts only with atoms by a process called ionization. Thus, all biological damage effects begin with the consequence of radiation interactions with the atoms forming the cells. As a result, radiation effects on humans proceed from the lowest to the highest levels as noted above.

(A) ***Cellular Damage*:** Even though all subsequent biological effects can be traced back to the interaction of radiation with atoms, there are two mechanisms by which radiation ultimately affects cells. These two mechanisms are commonly called direct and indirect effects.

(B) ***Direct Effect*:** If radiation interacts with the atoms of the DNA molecule, or some other cellular component critical to the survival of the cell, it is referred to as a direct effect. Such an interaction may affect the ability of the cell to reproduce and, thus, survive. If enough atoms are affected such that the chromosomes do not replicate properly, or if there is significant alteration in the information carried by the DNA molecule, then the cell may be destroyed by "direct" interference with its life-sustaining system.

(C) ***Indirect Effect*:** If a cell is exposed to radiation, the probability of the radiation interacting with the DNA molecule is very small since these critical components make up such a small part of the cell. However, each cell, just as is the case for the human body, is mostly water. Therefore, there is a much higher probability of radiation interacting with the water that makes up most of the cell's volume.

When radiation interacts with water, it may break the bonds that hold the water molecule together, producing fragments such as hydrogen (H) and hydroxyls (OH). These fragments may recombine or may interact with other fragments or ions to form compounds, such as water, which would not harm the cell. However, they could combine to form toxic substances, such as hydrogen peroxide (H_2O_2), which can contribute to the destruction of the cell.

(D) ***Cellular Sensitivity to Radiation*:** Not all living cells are equally sensitive to radiation. Those cells which are actively reproducing are more sensitive than those which are not. This is because dividing cells require correct DNA information in order for the cell's offspring to survive. A direct interaction of radiation with an active cell could result in the death or mutation of the cell, whereas a direct interaction with the DNA of a dormant cell would have less of an effect.

As a result, living cells can be classified according to their rate of reproduction, which also indicates their relative sensitivity to radiation. This means that different cell systems have different sensitivities. Lymphocytes (white blood cells) and cells which produce blood are constantly regenerating, and are, therefore, the most sensitive. Reproductive and gastrointestinal cells are not regenerating as quickly and are less sensitive. The nerve and muscle cells are the slowest to regenerate and are the least sensitive cells.

Cells, like the human body, have a tremendous ability to repair damage. As a result, not all radiation effects are irreversible. In many instances, the cells are able to completely repair any damage and function normally.

If the damage is severe enough, the affected cell dies. In some instances, the cell is damaged but is still able to reproduce. The daughter cells, however, may be lacking in some critical life-sustaining component, and they die.

The other possible result of radiation exposure is that the cell is affected in such a way that it does not die but is simply mutated. The mutated cell reproduces and thus perpetuates the mutation. This could be the beginning of a malignant tumor.

14.8.1 Sensitivity

The rate of reproduction of the cells forming an organ system is not the only criterion determining overall sensitivity. The relative importance of the organ system to the well being of the body is also important.

One example of a very sensitive cell system is a malignant tumor. The outer layer of cells reproduces rapidly, and also has a good supply of blood and oxygen. Cells are most sensitive when they are reproducing, and the presence of oxygen increases sensitivity to radiation. Anoxic cells (cells with insufficient oxygen) tend to be inactive, such as the cells located in the interior of a tumor.

As the tumor is exposed to radiation, the outer layer of rapidly dividing cells is destroyed, causing it to "shrink" in size. If the tumor is given a massive dose to destroy it completely, the patient might die as well. Instead, the tumor is given a small dose each day, which gives the healthy tissue a chance to recover from any damage while gradually shrinking the highly sensitive tumor.

Another cell system that is composed of rapidly dividing cells with a good blood supply and lots of oxygen is the developing embryo. Therefore, the sensitivity of the developing embryo to radiation exposure is similar to that of the tumor; however, the consequences are dramatically different.

(A) ***Organ Sensitivity:*** The sensitivity of the various organs of the human body correlate with the relative sensitivity of the cells from which they are composed. For example, since the blood forming cells were one of the most sensitive cells due to their rapid regeneration rate, the blood forming organs are one of the most sensitive organs to radiation. Muscle and nerve cells were relatively insensitive to radiation, and therefore, so are the muscles and the brain.

(B) ***Whole Body Sensitivity Factors*:** Whole body sensitivity depends upon the most sensitive organs which, in turn, depend upon the most sensitive cells. As noted previously, the most sensitive organs are the blood forming organs and the gastrointestinal system.

The biological effects on the whole body from exposure to radiation will depend upon several factors. Some of these are listed above. For example, a person, already susceptible to infection, who receives a large dose of radiation, may be affected by the radiation more than a healthy person.

14.8.2 Radiation Effects

Biological effects of radiation are typically divided into two categories. The first category consists of exposure to high doses of radiation over short periods of time producing acute or short term effects. The second category represents exposure to low doses of radiation over an extended period of time producing chronic or long term effects.

High doses tend to kill cells, while low doses tend to damage or change them. High doses can kill so many cells that tissues and organs are damaged. This in turn may cause a rapid whole body response often called the Acute Radiation Syndrome (ARS).

Low doses spread out over long periods of time don't cause an immediate problem to any body organ. The effects of low doses of radiation occur at the level of the cell, and the results may not be observed for many years.

14.9 Handling and Storage of Radioactive Materials

Great care has to be taken in handling and storage of radioactive materials for protecting people and personnel who handle it from harmful radiations which the radioactive material emits.

In order to have protection from hazards of radiation, radioactive materials must be stored in an area not frequently visited by people. Shielding may be required. Thick glass or Perspex containers usually provide sufficient shielding. In order to protect from the gamma radiations, lead shielding has to be used. The storage area must be regularly checked for the radioactivity.

The working area should not get contaminated with radio active materials. If radioactive liquid has to be handled, it must be carried in trays having absorbent tissue paper so that any spillage will get absorbed by paper. Rubber gloves have to be used when working with radioactive liquids.

Pipettes operated by mouth should never be employed. Before making use of glass apparatus, it must be ensured that they have been inactive. The waste radioactive material has to be stored till the activity becomes low before its disposal.

The following precautions have to be kept in mind while handling and storage of radioactive substances.

- One should not touch radioactive emitter with hand but it should be handled by means of forceps or suitable instruments.
- Smoking, eating and drinking activities should not be done in the laboratory where the radioactive materials are handled.
- Sufficient protective clothing or shielding have to be used while handling the materials.
- Radioactive materials have to be stored in suitable labeled containers, shielded by lead bricks and preferably in a remote corner.
- Areas where radioactive materials have been stored or used should be monitored, i.e., tested for radioactivity regularly.
- Disposal of radioactive materials should be carried out with great care.

14.10 Radiopharmaceutical Preparations

(a) Iodinated [^{125}I] Albumin Injection: It is a sterile solution of albumin that has been iodinated with iodine-125 and subsequently freed from iodide [^{125}I] ion, made isotonic with blood by the addition of sodium chloride and containing a suitable antimicrobial preservative such as benzyl alcohol. It is prepared from albumin solution and contains not less than 1% of protein.

Content of iodine-125 activity stated on the label at the date stated on label: 85.0 to 115.0%

Half-life of iodine 125: 60.1 days; emission: γ -radiation and x-rays.

Characteristics: It is a clear, colourless or faintly yellow solution.

Identification

(i) The γ-ray and x-ray spectrum does not differ significantly from that of a standardized Iodine-125 solution.

(ii) On examination in an ultracentrifuge it has the sedimentation coefficient of normal human albumin

Acidity or Alkalinity: pH 6.5 to 8.5

Radiochemical purity: It is determined by paper electrophoresis using a volume containing not less than 0.5 mg of albumin applied to a strip of filter paper and a solution containing 5 g of barbitone sodium, 3.25 g of sodium acetate, 4 g of sodium octanoate and 34.2 ml of 0.1M HCl in sufficient water to produce 100 ml. Not less than 95% of radio activity on the paper occurs in a position corresponding to that which would be occupied by treating normal human albumin at the same time and in the same manner.

Assay: The activity is determined by comparison with a standardized Iodine-125 solution or by measurement in an instrument calibrated with such a solution.

Storage: Iodinated [^{125}I] albumin injection should be stored at a temperature of 2° to 8° C. All the tests described under iodinated [^{125}I] albumin are applied to iodinated [^{131}I] albumin.

Uses: Radio iodinated serum albumin preparations are used to determine the plasma volume, and simultaneously with sodium chromate Cr-51 and ferric citrate Fe-59 to determine total blood volume. These preparations are also used to study circulation time and cardiac output. They are useful diagnostic acids for localizing. Neoplasm's of the brain. They have also been used to evaluation the circulation of cerebrospinal fluid.

These preparations have the advantages of emitting low energy radiation with no β particles. Therefore radiation exposure to the patient is minimized.

Doses: The i.v. doses for blood and plasma volume studies range from 3 to 20 μc for both ^{125}I and ^{131}I and for brain, lung and liver scanning range from 200 to 500 μc.

(b) Ferric Citrate [^{59}Fe] Injection: Ferric citrate [^{59}Fe] injection is a sterile solution containing iron [^{59} Fe] in the iron [III] state, 1.0% w/v of sodium citrate and sufficient sodium chloride to make it isotonic with blood.

Content of iron-59 activity at the date stated on the label: 90.0 to 110.0%

Half-life of iron-59: 44.6 days; emission; γ-radiation.

Characteristics: It is a clear colourless or faintly orange brown solution.

Identification

(i) The γ-ray spectrum does not differ significantly from of a standardized iron-59 solution

(ii) A suitable quantity is boiled with an excess of mercury (II) sulphate solution, filtered and the filtrate boiled and 0.15 ml of dilute potassium permanganate solution added. The colour is discharged and a white precipitate is produced.

Acidity or Alkalinity: pH 6.0 to 8.0

Total iron: Complies with the limit tests for iron

Assay: The activity is determined by comparison with a standardized iron-59 solution or by measurement in an instrument calibrated with such a solution.

Labelling: The label states the content of total iron.

Uses: The preparation is used for determining the mechanism or estimating the status of erythropoiesis and of iron metabolism in normal and disease persons. The parameters of iron metabolism which can be measured, include clearance of iron from the plasma, appearance of iron in the red cells, appearance of iron in various sites and adsorption of iron from the intestine. Suitable test for these parameters and the inferences drawn from them are based on the fact that blood plasma is a medium in which iron is in continuous flux. Iron-59 has sufficient γ-radiation energy for scintillation counting of the radioactivity in various tissues associated with erythrocyte formation and destruction such as in spleen, sacrum and liver from outside the body.

Doses: The usual oral and i.v dose for ferrous citrate fe-59 is 2 to 5 μ_c and may increase to 10 μ_c. Ferric chloride Fe-59 is also used for the same purposes and at the same dosage as the above compound in sterile solution.

(c) Sodium Iodide [^{131}I] Solution: Sodium iodide [^{131}I] solution is a solution suitable for oral administration containing iodine 131 in the form of sodium iodide and containing sodium thiosulphate or other suitable reducing agent. It is suitable for oral administration.

Content of iodine: 131 activity at the date and hour stated on the label: 90.0 to 110.0%.

Half-life of iodine 131: 8.04 days; emission γ-radiation.

Characteristics: It is a clear colourless solution

Identification:

(i) The γ-ray spectrum does not differ significantly from that of a standardize I-131 solution.

(ii) In the test for purity, the distribution of radioactivity in the chromatogram is an indication of the identification of the preparation.

Radiochemical purity: Ascending paper chromatography is used to determine the radiochemical purity of Iodine-131 using a mobile phase consisting of met hand-water (75:25 v/v). The distribution of radioactivity is determined using a suitable instrument. The spot corresponding to the iodide contains not less than 95% of the total radioactivity of the chromatogram.

Assay: The activity is determined by comparison with a standardized iodine-131 solution or by measurement in an instrument calibrated with such a solution.

(d) Sodium Iodide [^{125}I] solution: All the tests described under sodium iodide [^{131}I] solution are applied to sodium iodide [^{125}I] solution.

Uses: It is the most common isotope and chemical form used as a diagnostic aid to study the function of thyroid gland and in scanning the thyroid to determine size, position and possible tumor location. The study of thyroid function involves the measurement of uptake of radioactive iodine in a 24-hour period after oral administration or i.v injection. The normal patient will take up from 15 to 45% of the administered dose in 24 hrs. If the uptake is less than 10%, the patient is hypothyroid and an uptake of over 50% indicates hyperthyroidism.

In thyroid scanning the radioactive does is about 2-3 times than that used in uptake studies.

Scanning of the gland in the neck area provides a picture of its size, shape and location as well as areas of high and low iodine concentrating ability. Most of the tissue effect of iodine-131 is due to β-radiation which will penetrate 2-3 mm into tissue.

Sodium iodide [^{131}I] is also employed to destroy thyroid tissue or to alter the function of its tissue size. The particular use of the isotope is in disease states such as hyperthyroidism, thyroid carcinoma and severe cardiac disease. Severe cardiac diseases like angina pectoris and congestive heart disease may be eased by the use of the isotope required to induce hypothyroid state in order to reduce the work load of the heart.

Scanning can be done with iodine-125 which has lower energy and hence an advantage of lower radiation exposure to the patient.

I-125 does not emit β-radiation and therefore has minimum damage potential. The shelf life of Iodine-125 preparations is longer than that of iodine-131 preparations giving them a longer shelf life in the laboratory.

Sodium iodide [^{125}I] is used in the treatment of hyperthyroidism to impair the hormone synthesizing capability of the apex of the thyroid cells. It has been demonstrated that the properties of iodine-125 are more desirable than iodine-131 in this therapy. The lower energy and shorter path length radiation from iodine-125 are advantageous in limiting its deleterious effect on cell muscles without impairing its effect on areas of hormone synthesis. The concept behind this therapy is to avoid over treatment of hyperthyroidism.

Doses: Diagnostic preparations of sodium iodide [^{131}I] or [^{125}I] are available in capsule or solution form. Oral or i.v doses for uptake or general thyroid scanning range from 5 to 50 μ_C. Meta static thyroid cancer scanning requires doses around 300 μ_c. Oral or i.v. doses for therapeutic purposes may vary. In hyperthyroidism doses range from 80 to 120 μ_c/g of gland tissue. Thyroid cancer is treated with doses ranging from 100 to 200 mc. Cardiac diseases need 10 to 25 mc, which may be repeated in 2 to 6 weeks.

(e) Sodium Phosphate [^{32} P] Injection:

Sodium phosphate [^{32}P] injection is a sterile solution of disodium and monosodium orthophosphate [^{32}P] that is made isotonic with blood by the addition of sodium chloride.

Content of phosphorus-32 at the date and hour stated on the label: 90.0 to 110.0%

Half-life of phosphorous-32: 14.3 days; emission: β-radiation

Characteristics: It is a clear colourless solution.

Identification

(i) The β-ray spectrum or the β-ray absorption curve does not differ significantly from that of a standardized phosphorus-30 solution measured under the same conditions.

(ii) In the test for radiochemical purity (see below). The distribution of the radioactivity in the chromatogram is an indication of the identification of the preparation.

Acidity or alkalinity: pH 6.0 to 8.0

Radiochemical purity: It is determined by ascending paper chromatography using a mobile phase consisting of propan-2-ol-water-10 M ammonia (75:25:0.3 v/v) with 5 g trichloroacetic acid dissolved in it. The radio active spot is located by autoradiography or by measuring the radioactivity of the whole chromatogram. Not less than 95% of the total radioactivity corresponds to the spot of orthophosphoric acid.

Assay: The activity is determined by comparison with a standardized phosphorus-32 solution or by measurement in an instrument calibrated with such a solution.

Uses: Sodium phosphate [^{32}P] is used for both diagnosis and treatment of neoplastic diseases. Since phosphate is utilized in cell metabolism, the rapidly proliferating cells as in the case of cancer show up the highest turnover of phosphate, compared with those of the non cancerous cells. The primary diagnostic use of this preparation is in the localization of intraocular and cerebral tumors. The β-radiation from phosphorus-32 has sufficient energy to penetrate in tissues of about 8 mm. It is particularly helpful in the location of eye tumors which are relatively near the surface.

Sodium phosphate [^{32}P] is therapeutically used in diseases associated with both red and white blood cells. It is used in the treatment of polycythemia, which involves increases in the number and absolute mass of RBC. The effect of radioactivity in this disease is primarily through the reduction in erythrocytes formation.

It is also used in the palliative treatment of chronic granulocytic or myelocytic leukemia that results in an increase in the number of white blood cells.

Dose: For diagnostic purposes doses range from 250 to 500 mc with counts being taken at one hour, 24 hours and 48 hours over both eyes. Therapeutic doses in polycythemia vera ranges from 2-10 mc with an average of 6 mc. i.e., the dose is 75% of this dose.

(f) Cyanocobalamin [^{57}Co] Solution

Cyanocobalamin [^{57}Co] may be prepared by the growth of appropriate microorganisms on a medium containing cobalt [II] [^{57}Co] ion.

Content of cobalt 57: activity at the date stated on the label: 90.0 to 110.0%

Half-life of cobalt 57: 271 days; emission γ -radiation.

Characteristics: It is a clear, colourless or slightly pink solution.

Identification

(i) The γ -ray and spectrum does no differ significantly from that of a standardized cobalt-57 solution.

(ii) In the radiochemical purity test, the principal peak in the radio chromatogram obtained with the test solution has a retention time similar to that of the peak in the chromatogram obtained with a reference solution.

Acidity: pH 4.0 to 6.0

Radiochemical purity: This is determined by liquid chromatography using a stationary phase of silica modified by chemically-bounded octylsilyl groups, and a mobile phase of 10% of w/v disodium hydrogen orthophosphate-methanol (73.5:26.5 v/v) adjusted to pH 3.5 with orthophosphoric acid at a detection wavelength of 351.

The percentage of cobalt-57 present as cyanobalamine is determined by normalization.

Assay: The activity is determined by comparison with a standardized cobalt-57 solution.

Storage: Cyanocobalamine [^{57}Co] solution should be stored at a temperature of 2° to 8°C and protected from light.

Uses: Cyanocobalamine [^{57}Co] solution is used in diagnostic procedures for pernicious anaemia.

It is also used to study the effect of the liver on the intestinal absorption of vitamin B_{12}.

The radioactivity from an oral dose of the solution is detectable in the urine of the normal patient and absent or at a very low level in the urine of patients with pernicious anemia since the patients lack intrinsic factor which is necessary for the proper intestinal absorption of vitamin B_{12}.

Doses: For oral and i.m. injection the range is 0.5 to 1.0 μc corresponding to 0.5 to 2.0 μg of cyanocobalamin.

(g) Cyanocobalamin [^{57}Co] Capsules:

Cyanocobalamin [^{57}Co] may be prepared by the growth of appropriate microorganisms on a medium containing cobalt (II) [^{57}Co] ion.

Content of Cobalt 57*:* activity at the date stated on the label 90.0 to 110.0%.

Half life of cobalt 57*:* 271 days; emission; γ radiation.

Characteristics: Hard gelatin capsules.

Identification:

(i) The γ-ray and spectrum does not differ significantly from that of a standardized cobalt-57 solution.

(ii) In radio chemical purity test the retention time of the principal peak in the radio chromatogram obtained with the test solution is similar to the peak in the chromatogram obtained with the reference solution.

Radio chemical purity: This is determined by liquid chromatography using the centrifuged supernatant solution of the dissolved contents of a capsule in one ml of water, a stationary phase of silica modified by chemically bonded octylsilyl groups and the mobile phase comprising 1% w/v disodium hydrogen orthophosphate methanol (73.5 : 26.5 v/v) adjusted to pH 3.5 with orthophosphoric acid and a detection wavelength of 361 mm. The percentage of cobalt-57 present as cyanocobalamine is determined by normalization.

Disintegration: Complies with the requirement of hard capsules using only one capsule.

Uniformity of content: The radioactivity of one capsule is measured, the procedure is repeated with further nine capsules and the average radio activity for capsule is determined. The radioactivity of no capsule differs by more than 10% from the average radioactivity (RSD < 3.5%).

Assay: The average of 10 or more individual results obtained in the tests for uniformity of content is used to determine average radioactivity per capsule.

Storage: Cyanocobalamine [^{57}Co] capsules should be kept in an air tight container protected from light and stored at a temperature of 2° to 8°C.

Uses: Cyanocobalamine [^{57}Co] solution is used in diagnostic procedures for pernicious anaemia. It is also used to study the effect of the liver on the intestinal absorption of vitamin B_{12}.

The radioactivity from an oral does of the solution is detectable in the urine of the normal patient and absent or at a very low level in the urine of patients with pernicious anemia, since the patients lack intrinsic factor, which is necessary for the proper intestinal absorption of vitamin B_{12}.

Doses: For oral and 1°.m injection the range is 0.5 to 1.0 μ_C corresponding to 0.5 to 2.0 μg of cyanocobalamin.

(h) Colloidal Gold [^{198}AU] Injection:

It is a sterile, apyrogenic; colloidal dispersion of gold-198 stabilized with gelatin and contains reducing agents such as glucose or ascorbic acid. Content of god-198 activity at the date and hour mentioned on the label 90.0 to 110.0%.

Half life of gold 198: 2.70 days; emission: β and γ -radiations.

Characteristics: A dark red liquid which does not deposit metal on the surface of the container.

Identification

(i) The γ -ray spectrum does not differ significantly from that of standardized gold-198 solution.

(ii) In the test for radiochemical purity, the distribution of radioactivity in the chromatogram is an indication of the preparation.

Acidity or Alkalinity: pH 4.0 to 8.0

Radiohemical purity: It is determined by using ascending paper chromatography: using a mobile phase consisting of acetone-water-hydrochloric acid (70:20:10).

The distribution of radioactivity is determined by using a suitable instrument. The spot corresponding to colloidal gold contains not less than 98% of the total radioactivity of the chromatogram.

Assay: The activity is determined by comparison with a standardized gold-198 solution or by measurement in an instrument calibrated with such a solution.

Labelling: The particle size range of the gold is stated on the label.

Uses: Gold-198 solutions are most frequently used for therapeutic purposes. They are administered by intracavitary injection into the pleural and peritoneal cavities as an aid in the management of pleural effusion (accumulation of serous fluid in pleural cavity) and ascites (accumulation of serous fluid in the peritoneal cavity). These fluid accumulations when secondary to neoplastic disease in the area can be inhibited by the effect of β-radiation in the cancerous tissue cells. Another use of this preparation is made in as a prophylactic benefit against the growth of more tumors after surgical removes of tumors from a major cavity.

The colloidal gold Au-198 solutions are also used for diagnostic scanning of liver. Liver scanning with Au-198 is used in determining the position, shape and size of the organ as well as to obtain information on the distribution of the isotope concerning the functioning of the kupffer's cells. The isotope does not enter the tumor tissue, abscesses or cysts, therefore, these appear as light areas in the liver scans.

Doses: The usual dose of golf Au-198 injection for pleural effusion is 35 to 75 mc and for ascites is 100 to 125 mc. The i.v. dose is around 300 μc with a range of 100 to 500 μc.

14.11 Radio-Opaque Contrast Media

These substances are having the ability to stop the passage of γ -rays and hence appear opaque on X-ray examination. Such compounds and their preparations are known as γ -ray contrast media.

X-ray is electromagnetic radiation of short wavelength and thus has high penetrating power. The electrons of high atomic number element can interact with X-rays. The interaction brings about interference in their passage through the medium.

In diagnostic study using X-rays the soft tissue have been permeable to passage of X -rays because the tissue are mostly composed of elements of low atomic numbers like carbon, hydrogen, oxygen and nitrogens and hence cause darkening on X-ray film. The bony structure casts shadow on film as the bones are having elements of high atomic number such as calcium and phosphorus. The cause of this bony tissue can be distinguished on an exposed X-ray film.

Inorganic compounds like barium sulphate and some bismuth compounds thus are useful as radio-opaque contrast media for diagnostic use. They cannot be categorized as drugs because they are usually devoid of any pharmacological effects.

Many organic iodinated compounds are used as radio-opaque contrast media.

These are administered either by (i) systemic i.e., orally or intravenously or (ii) (by retrograde i.e., by mechanical means backwardly for various diagnostic purposes. These compounds have been used successfully for examination of gastrointestinal tract, kidney (urography), liver (cholcystography) gall bladder and bileduct, blood vessels of heat (angiography and cardiography), bronchial tract (bronchography) and that of urethra, vagina etc.

(a) Barium sulphate:

Chemical Formula: $BaSO_4$ *Mol. Wt.* 233.4

Preparation:

(i) For pharmaceutical purposes barium sulphate is prepared by treating an aqueous solution containing barium ions with a solution containing sulphate ions.

$$Ba(OH)_2 + H_2SO_4 \rightarrow BaSO_4\downarrow + 2H_2O \quad .....(14.52)$$

$$BaCl_2 + H_2SO_4 \rightarrow BaSO_4\downarrow + 2HCl \quad .....(14.53)$$

The precipitated salt is thoroughly washed, dried and then screened

(ii) It is also prepared by the action of dilute H_2SO_4 on BaS

$$BaS + H_2SO_4 \rightarrow \underset{ppt}{BaSO_4} + H_2S \quad .....(14.54)$$

(iii) Barium sulphate is also obtained as a by-product from many industries such as the manufacture of H_2O_2 from $BaO_2.8H_2O$. However, it is not pure enough for use for a pharmaceutical agent.

(iv) $BaSO_4$ is also obtained by purifying native barium sulphate. However, the substance so obtained is not pure enough for use as a pharmaceutical agent. If $BaSO_4$ is prescribed, the title must be written in full in order to avoid any confusion with the poisonous barium sulphide and barium sulphite

Properties: It is heavy, fine, white, odourless, tasteless and bulky powder, that is free from grittiness. It is insoluble in water, organic solvents and dilute acids and alkalies. However, it is soluble in concentrated H_2SO_4 due to the formation of barium sulphate.

$$BaSO_4 + H_2SO_4 \rightarrow Ba(HSO_4)_2 \quad .....(14.55)$$

Barium sulphate has been so insoluble, that it is undergoing very few reactions. It may be solubilized with H_2SO_4 by fusing it with alkali carbonates, once it gets converted to a carbonate, it reacts with acids readily

$$2BaSO_4 + Na_2CO_3 + K_2CO_3 \rightarrow 2KNaSO_4 + 2BaCO_3\downarrow \quad .....(14.56)$$

$$BaCO_3 + 2CH_3COOH \rightarrow (CH_3\ COO)_2\ Ba + H_2O + CO_2\uparrow \quad .....(14.57)$$

Tests for identity: 0.5 g of barium sulphate is fused with 2 g of anhydrous sodium carbonate and anhydrous potassium carbonate in a crucible and the resulting fused mass is treated with hot water and filtered. The filtrate is acidified with hydrochloric acid and used for with hydrochloric acid and used for reactions which are characteristic of sulphates.

The residue obtained from the above is dissolved in hydrochloric acid and the solution is used for carrying out the characteristic flame test for barium.

Tests for Purity: It has to be tested for Arsenic, acid insoluble substances, soluble barium salts, heavy metals, phosphate, sulphide, sulphite, bulkiness and loss on drying.

As there has been no assay as such and as barium sulphate is used internally in large quantities it becomes essential to outline in detail all the I.P limits tests here under. Some of these tests, especially tests for acid insoluble substances soluble barium salts and bulkiness should be performed, observations made and recorded. These tests have been also important because they have been the main limit tests for barium sulphate compound powder i.e., Barium meal.

Reaction: The powder is boiled with water and filtered. The filtrate has been natured to litmus.

Acid soluble substances: 10 g of the sample is boiled with 100 ml water having 10 ml dilute HCl and then cooled, filtered till a clear filtrate is obtained. 50 ml of the filtrate is transferred to a tarred vessel and evaporated on water bath (adding 2 drops of HCl during evaporation). The residue is dried at 105 ^{o}C, cooled and weighed to a constant weight. The residue should not be more than 0.15% w/w. The residue is used for the test of soluble barium salt.

Soluble barium salt: The residue obtained in the acid soluble substances is digested with 10 ml of water and filter. Now 0.5 ml dilute H_2SO_4 is added to the clear filtrate and set aside for 30 minutes, no turbidity is produced.

Heavy metals: 4 g of sample is boiled with a mixture of 2 ml of glacial acetic acid and 48 ml of water for ten minutes. Now water is added to make 50 ml. The solution is filtered and the first 5 ml of filtrate is rejected, the limit of heavy metals on the next 25 ml of filtrate has been 10 parts per million.

Phosphate: 2 g of sample is treated with a mixture of 5 ml nitric acid and 5 ml water. The solution is cooled and filtered. To the filtrate, 5 ml solution of ammonium molybdate is added and set aside for one hour, no yellow precipitate is produced.

Sulphide: 10 g of sample is shaken with 100 ml of water having 10 ml of dilute HCl, and during the boiling lead paper is exposed to the escaping vapours; the paper does not darken.

Sulphite: 1 g of sample is boiled with 10 ml of water. To it 1 ml of dilute H_2SO_4 and 2 drops of a solution of potassium permanganate are added. The colour is not discharged within 15 minutes.

Bulkiness: 5 g of sample is taken in 50 ml stoppered graduated cylinder. The 50 ml graduation mark should be between 8.5 to 11.5 cm above the 5 ml mark. After adding 30 ml of water the cylinder is shaken for 3 minutes and then water is added to produce 50 ml. It is set aside for 15 minutes. The upper level of the barium sulphate should not be below the 20 ml graduation mark.

Loss on drying: It loses not more than 2.0 percent of its weight when it is dried to constant weight at 105°.

Storage: Barium sulphate is stored in a well closed container.

Uses: Barium sulphate is a diagnostic drug which is used medicinally in X-ray examination. It is administered by enema before X-ray examination in the form of barium meal to make the intestinal tract opaque to X-rays so that it could be photographed. As the amount administered is large (60 to 250g) it becomes important that barium sulphate used in x-ray examination should be IP standard.

Doses: The oral dose in suitable suspension has been generally 300 g (60 – 450g) while rectal dose has been usually 360 g (200 – 600g).

B. Barium Sulphate for Suspension:

Synonym: Barium meal or shadow meal.

It is a dry mixture of barium sulphate having suitable colour, flavour, preservative and suspending or dispersing agent. It contains not less than 90% w/v of barium sulphate.

Composition: It is having the following ingredients.

Barium sulphate	1000 g
Saccharin solution	0.25 g
Vanillin	0.10 g

***Preparation*:** Barium meal is prepared by mixing Saccharin and Vanillin with barium sulphate and is given to the patient immediately. Saccharin is a sweetening agent and vanillin is a flavouring agent.

***Action and Use*:** It is used as a barium meal.

***Assay*:** Barium sulphate for suspension is arranged by fusion of the known weight of compound with sodium carbonate and potassium carbonate at 1000° for fifteen minutes. It is cooled, suspended in water and decanted. The residue is made to wash with 2% sodium carbonate solution (until free from sulphate). Dilute hydrochloric acid is added followed by ammonium acetate and potassium dichromate and urea. The suspension is digested at 80-85° for 16 hours and then filtered through sintered glass crucible. The residue is washed with potassium dichromate solution followed by water and the contents are weighed after drying at 105 °C.

(C) Bismuth Compounds

Bismuth nitrate and bismuth carbonate in 30-60 g as suspension in water were used to be employed for examination of alimentary tract. These compounds have been replaced by other better compounds.

CHAPTER 15

Miscellaneous Pharmaceutical Agents

15.1 Expectorants and Emetics

Expectorants: Expectorants are the drugs that help in removing sputum from the respiratory tract either by increasing the fluidity of sputum or increasing the volume of fluids that have to be expelled from the respiratory tract by coughing. The action and mechanism of expectorants have been questioned and remain controversial.

Examples of inorganic expectorants are ammonium chloride, potassium iodide and sodium iodide. It is advisable to give the doses of expectorants that could be tolerated by the patient along with flavours, sweeteners and with antitussives (cough suppressants).

Emetics: Sometimes emetics in low doses have been used in cough preparations. These are the drugs which give rise to forced emesis or vomiting by which the contents of the stomach get expelled through the oral cavity. The emetics act either by local irritation of gastric mucosa or directly on the chemo receptor trigger zone (CTZ) in the medulla.

The following are the examples of official expectorants and emetics (I.P. 1966).

(i) Ammonium chloride

(ii) Potassium iodide

(iii) Sodium Iodide

(iv) Antimony potassium Tartarate

(v) Zinc sulphate

Ammonium Chloride, I.P. (Refer page 275)

Ammonium Chloride Tablets, I.P

Ammonium chloride tablets may be prepared by process of I.P 1966. They are enteric coated. The weight of ammonium chloride NH_4Cl, in each tablet of average weight, as determined by the assay described below, is not less than 95.0 percent and not more than 105.0 percent of the stated amount of ammonium chloride.

Identification: A filtered solution of the tablets equivalent to 10 percent w/v solution of ammonium chloride, yields the reactions characteristic of ammonium salts and of chlorides.

Disintegration time: Maximum time, two hours and thirty minutes, when tested as for enteric coated tablets.

Assay**:** Weigh 20 tablets and reduce to a fine powder. Weigh accurately a portion of the powdered tablets equivalent to about 0.2 g of ammonium chloride, transfer to a beaker and add 40 ml of water, stir the suspension intermittently during thirty minutes, filter and wash until the last washing yields a clear solution when 2 drops of solution of silver nitrate and 1 ml of dilute nitric acid are added 5 ml of the wash liquid. Combine the filtrate and washings and transfer to a 500 ml glass stoppered flask add 50 ml of 0.1N silver nitrate, 3 ml of nitric acid and 5 ml of nitro benzene, shake vigorously, add 2 ml of solution of ferric ammonium sulphate and titrate the excess of silver nitrate with 0.1N ammonium thiocyanate. Each ml of 0.1N silver nitrate is equivalent to 0.005349 g of NH_4Cl. Calculate the weight of ammonium chloride NH_4Cl in each tablet of average weight.

Category: Expectorant, Diuretic, Systemic acidifier.

Dose: Ammonium chloride 3 to 6 g daily in divided doses.

Usual Strength: 0.5 g

Potassium Iodide I.P (Refer page 338)

Sodium Iodide I.P. (Refer page 339)

Antimony Potassium Tartarate:

Chemical Formula: $C_4H_4KO_7Sb$ *Mol. Wt.* 333.93

Synonym: Tartar emetic **Mol. wt: 333.93**

It contains 99.0 to 103 percent of $C_4H_4KO_7Sb.\ 2H_2O$

Preparation: It is obtained by mixing 5 parts of antimony trioxide Sb_2O_3 with 6 parts of potassium acid tartrate (6 parts) in a fine paste. Then this paste is kept aside for a day. It is then boiled with water for fifteen minutes with constantly stirring. The liquid is then filtered hot. Now the filtrate is left for crystallization. The crystals are colleted on a filter. They are dried at atmospheric temperature.

$$2KHC_4H_4O_6 + Sb_2O_3 \rightarrow 2K\,(SbO)\,C_4H_4O_6 + H_2O \qquad .....(15.1)$$

Properties: It occurs as colourless crystals, odourless having a sweetish taste on exposure to air, crystals effloresce. It is soluble in water but insoluble in alcohol. Its solubility is 1 in 12 of water at 25°C and 1 in 3 of boiling water and 1 in 15 of glycerin.

Identification: It gives reactions which are characteristic of potassium and antimony.

Tests for Purity: It is to be tested for As, Pb acidity or alkalinity and loss on drying. Acidity or alkalinity may be measured by the volume of N/100 acid or alkali which is required to neutralize a solution of definite concentration to the green colour of bromo cresol green, indicative of pH 4.5. One gram of the substance is taken and is then dissolved in 50 ml of water, N/100 acid or alkali used should not be more than 20 ml for neutralization.

Loss on drying should not be more than 6.0 percent when it is dried to constant weight at 105°C (The salt is hygroscopic, but the official salt has been anhydrous).

Assay: An accurately weighed amount (0.5 g) is dissolved in water (50 ml). To this solution 2g of sodium bicarbonate is added. Now it is titrated with N/10 iodine, using solution of starch as indicator.

The above type of titration is iodometric. The reaction is oxidation reduction type of reaction. The following reaction is reversible and the acid which gets liberated in the reaction gets neutralized with sodium bicarbonate to enable the reaction to go to completion.

$$2C_4H_4O_7SbK + 3H_2O + 2I_2 \rightarrow 2KHC_4H_4O_6 + Sb_2O_6 + 4HI \qquad .....(15.2)$$

Each ml of N/10 I_2 = 0.01544 g of $C_4H_4O_7Sb$.

Uses: Like antimony sodium tartarate it finds use to treat schistosomiasis and is administered by I.V. injection. It is acting as an emetic (hence the name tartar emetic) because of its irritant action on the gastric mucosa.

Zinc Sulphate :

Chemical Formula: $ZnSO_4.\ 7H_2O$ *Mol. Wt.* 287.6

Zinc sulphate contains not less than 55.6 percent and not more than 61.0 percent of $ZnSO_4$ corresponding to not less than 99.5 percent and not more than the equivalent of 102.0 percent of the hydrated slat, $ZnSO_4.7H_2O$.

Description: Colourless, transparent crystals or a crystalline powder, odourless, taste astringent and metallic. It is efflorescent in dry air.

Solubility: Soluble in 0.6 part of water, practically insoluble in alcohol, soluble in 2.5 parts of glycerin.

Preparation: It is prepared by the reactions of metallic zinc with dilute sulphuric acid.

$$Zn + H_2SO_4 \rightarrow ZnSO_4 + H_2 \qquad .....(15.3)$$

The liquid is filtered from undissolved metal and evaporated to crystallization.

Identification: Yields the reactions characteristic of zinc and of sulphates.

Tests for Purity: It is tested for reaction, aluminium, copper, magnesium, manganese and nickel, arsenic, iron, chloride and alkalis and alkaline earths.

Reaction: A 5 percent w/v solution is acid to solution of phenol red but is not acid to solution of methyl orange.

Aluminium, copper, magnesium, manganese and nickel: Dissolve 1 g in 20 ml of water and dilute ammonia solution in excess and allow to stand. The solution remains colourless and no precipitate is produced within thirty minutes.

Arsenic: Not more than 10 parts per million.

Iron: 0.14 g complies with the limit test for iron, using 6 drops of thioglycollic acid.

Chloride: 1g complies with the limit test for chlorides.

Alkalis and Alkaline Earths: Dissolve 2 g in about 150 ml of water contained in a 200 ml volumetric flask, precipitate the zinc completely by means of solution of ammonium sulphide and add water to make 200 ml, mix well and filter through a dry filter rejecting the first portion of the filtrate. To 100 ml of the subsequent filtrate add a few drops of sulphuric acid, evaporate to dryness in a tared dish and ignite, the weight of the residue does not exceed 5 mg.

Assay: Weigh accurately about 1g, and dissolve in about 100 ml of water. Heat the solution to about 90° and add solution of sodium carbonate to precipitate all of the zinc, taking care to avoid a large excess of sodium carbonate. Boil for about five minutes and set aside to allow the precipitate to subside. Collect the precipitate in a tared Gooch crucible and wash with hot water until free from alkali. Dry the residue, ignite and weigh. Each gram of residue equivalent to 1.984 g of $ZnSO_4$.

Storage: preserve zinc sulphate in a well closed container.

Category: Emetic and Astringent

Dose: 0.6 to 2 g.

15.2 Poisons and Antidotes

Antidotes are substances which neutralize the effect of poisons.

A poison is generally defined as a compound that when administered, inhaled or swallowed is capable of acting deleteriously on the body even in relatively small quantities and by a chemical action can cause death or disability. The diagnosis of poisoning is often difficult. It is based on the history, physical examination (or) pathological changes and laboratory procedures. The investigation includes personal, occupational and family history. Physical examination and pathological changes may be quite characteristic for certain types of poisoning. In many instances, however laboratory procedures are essential to make a definitive diagnosis.

In many cases in which no traces of poisoning has been found although circumstantially it was almost quite certain that poison was the cause or illness of death, following explanations need to be observed:

- The case may have been one of disease only.
- Poison may have been eliminated by vomiting or other means or neutralized or metabolized.
- The analysis has been performed in a faulty manner.

In general the treatment of the patient in case of poisoning consists of:

- Removal of the unabsorbed poison from the body.
- Use of antidotes.
- Elimination of poison absorbed into the system by excretion and
- Symptomatic treatment.

The use of antidotes, which is vividly described forms an important remedial measure in the management of poison cases because antidotes are used not only for the prevention of absorption of the poison but also for inactivating the poison or opposing its action following absorption.

Antidotes must counteract the effect of poison by way of interfering or neutralizing. We must admit considerable ignorance as to precise mode of action of many poisons.

It is possible to conceive of complete antagonism but in practice we find that it is usually limited. Antidotes usually cause such alteration of the surface of or enzyme system of a cell or prevent a particular activity so that poison may be unable to act upon. The inactivation or neutralization may be either by direct chemical action or it may be a mechanical or physiological nature. Thus the antidotes are divided into three classes-mechanical, chemical and physiological.

(i) Mechanical Antidotes

These render the poison inert by mechanical action or prevent their absorption. For example, finely powdered activated charcoal in a dose of 4-8 g acts mechanically by adsorbing and retaining within its poles organic and also to a lesser degree mineral poisons and thus prevents the absorption from the stomach.

One gram activated charcoal absorbs, 1800 mg mercuric chloride, 1000 mg of sulphanilamide, 950 mg strychnine, 800 mg morphine hydrochloride, 700 mg atropine sulphate; 700 mg nicotine 550 mg salicylic acid, 400 mg phenol, 300 mg barbiturates, 300 mg alcohol, 35 mg potassium cyanide.

Demulcents such as fats, oils, egg albumin, milk, starch, milk of magnesia, aluminium hydroxide gel; etc prevents the action of the poison by forming a coating on the mucous membrane of the stomach. However fats and oils should not be used for oil soluble poisons such as kerosene, phosphorous, organo-phosphorus compounds like hexaethyl tetra phosphate (HETP), tetraethyl pyrophosphate (TEPP), Malathion, DDT, Phenol, turpentine, acetone, carbon tetrachloride, etc. However demulcents do not inactivate the poison already, impregnated in the stomach wall. Bulky foods such as bread loaf, ispaghula husk, etc act as a mechanical antidote to glass by imprisoning its particles within its meshes and thus prevent damage being affected by the sharp glass particles.

(ii) Chemical Antidotes

These are the substance which counteracts the action of poison by forming harmless or insoluble compounds or by oxidizing poison when brought into contact with them. For example:

- Common salt decomposes silver nitrate by direct chemical action forming the insoluble silver chloride
- Albumin precipitates mercuric chloride
- Dialyzed iron is used to neutralize arsenic
- Copper sulphate is used to precipitate phosphorus.
- Sodium sulphate is used to precipitate lead
- Alkalies neutralize mineral acids by direct chemical reaction. It is safer to use a little weak solution of an alkali hydroxide, magnesia or ammonia than the respective bicarbonates because of the possible rise of rupturing the stomach owing to the liberated CO_2.
- Acids neutralize alkalies by direct chemical action. However, only those substances which are by themselves harmless should be selected, e.g., vinegar, lemon juice, canned fruit juice etc.
- Potassium permanganate is a chemical antidote owing to its oxidizing properties. Its 1:5000 solution introduced by means of a stomach tube is used in poisoning for opium and its derivatives and other oxidisable substances like phosphorous, hydrocyanic acid, barbiturates, strychnine, atropine and other alkaloids. When it reacts with the poison in the stomach it loses its pink colour. The wash must be continued till the solution coming out of the

stomach is of the same pink colour as that of the solution put in. However, a caution is warranted in the use of potassium permanganate because being a powerful oxidizing agent it is likely to further damage the gastric mucosa, there by affecting the recovery of the patient.

- Tannic acid 4% or tannin in the form of a strong tea is used to precipitate certain metals and alkaloids and some other organic substances. Its efficacy is increased by alkalies and is attenuated by alcohol and acids. It precipitates apomorphine, cinchona, strychnine, nicotine, cocaine, aconite, pilocarpine, lead, aluminium, silver, cobalt, copper, mercury, nickel and zinc.

  As discussed above the use of chemical antidotes was first visualized to serve the purpose of inactivation of the poison in the stomach but later developments lead to the use of systemically active chemical antidotes. The most important examples are being the chelating agents. Also protective agents against radiation are now viewed as antidotes against free radicals produced by radiation.

Chelating Agents

These are soluble organic compounds having the peculiar property of fixing metallic ions into their molecular structure, thus inactivating these metallic ions.

The chelating agents form stable, soluble, non ionisable complexes with calcium and certain heavy metals.

B.A.L: (Dimercaprol, 2, 3-mercapto propanol)

It is effectively used as an antidote in arsenic, mercury and gold. Its usefulness is not certain in antimony and bismuth poisoning. Dimercaprol should not be used in the treatment of poisoning caused by cadmium, iron or selenium. In the case of cadmium, it has been shown that dimercaprol complexes are more toxic than the metal itself. Many heavy metals have a great affinity to sulfhydryl (SH) radicals and combine with them in tissues and deprive the body of the use of respiratory enzyme of tissue cells. Dimercaprol has two unsaturated sulfhydryl groups which combine with the metal and thus prevent union of an arsenic with the – SH groups of the respiratory system. The compounds formed by the heavy metal and dimercaprol is relatively stable which is carried out into the tissue fluids particularly plasma and is excreted into the urine.

Adverse effects of dimercaprol include pain at the site of injection, weakness, nausea, salivation, elevation of blood pressure, coma and convulsions.

Dimercaprol is a potentially dangerous drug. Even doses of less than 100 mg/kg can cause swift death in animals. It is believed that in these high concentrations the drug inhibits metals containing enzymes that are essential in cellular respiration. Large doses may cause vascular collapse as a consequence of capillary damage.

B.A.L should not be used when liver is damaged. Alkalization of the urine is claimed to protect against dissociation of the dimercaprol metal complex and nephrotoxicity.

Dimercaprol (BAL) is available in peanut oil for injection, 100mg/ml. Dosage varies from 3 to 5 mg/kg intramuscularly, 4 hourly for 2 days, 6 hourly on the third day and then 12 hourly for 7 days.

Sodium Calcium Edetate: The calcium chelate of the disodium salt of ethylenediamine tetra-acetic acid ($CaNa_2$ EDTA) has revolutionized the treatment of lead poisoning. Despite the high stability of this chelate, calcium is displaced from it by lead forming the lead chelate which is excreted in urine leaving behind a harmless amount of calcium. A few other heavy metals like zinc, chromium, copper, cadmium, manganese and nickel can also get exchanged with calcium of sodium calcium edetate.

Disodium edetate is quite dangerous when injected intravenously since it chelates calcium forming sodium calcium edetate in the body inducing hypocalcaemia due to which tetany may occur that may be fatal. This action can be used in hypocalcaemia and in digitalis poisoning by the experts.

Calcium disodium edetate is administered by intravenous drip for 2 or 3 days in the treatment of lead poisoning. Oral administration of calcium disodium edetate for the treatment of lead poisoning is not effective and may even by hazardous.

In contrast to dimercaprol, calcium disodium edetate is not metabolized in the body and is excreted as soluble metal complex in the urine.

Adverse effects of calcium disodium edetate include renal damage, hyper sensitivity reactions and transient bone marrow depression. Sodium calcium edetate given i.v 1.0g 12 hourly (over 1 hr) usually in 5 days courses, repeated as necessary after 7 day interval. It is available as 5 ml ampoule of a 20% solution (1 g) to be diluted for intravenous infusion.

Penicillamine: It is a hydrolysis product of penicillin. It has a stable – SH group. It is a chelating agent having maximum efficiency for the mobilization and excretion of heavy metals.

Penicillamine is available in capsules containing 250 mg. Penicillamine is given in a dose of 30 mg/kg body weight up to a total of 2 g/day in four divided doses orally for 7 days. 1 to 3 g can be given in slow normal saline drip daily for 2 to 4 days.

Desferrioxamine: It is a very potent and specific iron chelating agent. It is used in the treatment of iron poisoning and for the removal of iron from the body. The drug has great affinity for iron in the ferric form.

In acute iron poisoning desferrioxamine 8 to 12 g is administered by gastric tube to absorb iron in the stomach. In addition 1 to 2g of the drug may be injected intramuscularly or intravenously to bind absorbed iron to be repeated 12 hourly if necessary. Desferrioxamine mesylate is available as powder for injection 500 mg.

Universal Antidote

This antidote is a combination of physical and chemical antidote. It is an antidote which can be used in cases where the nature of the poison swallowed is not definitely known or where it is suspected that two or more poisons have been taken.

The universal antidote has the following composition.

(i)	Powdered animal charcoal	2 parts
(ii)	Tannic acid (or strong tea)	1 part
(iii)	Magnesium oxide	1 part

A table spoonful is mixed in glassful of water and given by mouth. It may be repeated once or twice. Charcoal adsorbs alkaloids, tannic acid precipitates alkaloids, glycosides and many of the metals and magnesia neutralizes acid without gas formation.

(iii) Physiological Antidotes:

These are also called as antagonists. These act on the tissue of the body and produce symptoms exactly opposite to those caused by the poison acting on them or the enzyme. They are used after some of the poison is absorbed into the circulation. Their use is somewhat limited and not without danger. These agents act on the principle of antagonism utilizing where possible, the power which different drugs possess of interfering with another's action up on the enzymes, tissue cells or opposing nerve systems.

Most of the known physiological antidotes are partial in their action. Atropine and Physostigmine are two real physiological antidotes as both of them affect nerve endings and produce opposite effects on the heart rate, state of the pupil and glandular secretory activity. I.V. atropine sulphate is also used to antagonize the peripheral action of organo phosphorus compounds. Other examples are atropine and morphine, barbiturates and picrotoxin or amphetamine, strychnine and barbiturates, digitalis and aconite, cyanides and amyl nitrite, naloxone against morphine and morphine like drugs and pralidoxime against anti cholinesterase.

In practice most emergency treatment of acute poisoning is symptomatic and not specific to the poison because as indicated earlier the diagnosis of poisoning is often difficult. In clinical practice a selective antidote (antagonist) is available for less than 2% episodes. However, where there is selective antidote, its use is vital.

Official Antidotes

The following are the examples of the official antidotes (I.P 1966)

- Sodium nitrite (Refer page 217)
- Sodium thiosulphate (Refer page 215)
- Sodium thiosulphate injection (Refer page 293)
- Activated charcoal (Refer page 293)
- Light kaolin (Refer page 292)
- Copper sulphate

Copper Sulphate:

Chemical Formula: $CuSO_4 . 5H_2O$ *Mol. Wt.* 249.7

Copper sulphate contains not less than 98.5 percent and not more than the equivalent of 101.0 percent of $CuSO_4.5H_2O$.

Description: Blue triclinic prisms or a blue crystalline powder.

Solubility: Soluble in water, very soluble in boiling water, almost insoluble in alcohol, very slowly soluble in glycerin.

Identification: Yields the reactions characteristic of copper and of sulphates.

Preparation: It is formed when concentrated sulphuric acid is heated with metallic copper

$$Cu + 2H_2SO_4 \rightarrow CuSO_4 + SO_2 \uparrow + 2H_2O \quad .....(15.4)$$

Tests for Purity: It is tested for acidity and clarity of solution, arsenic, iron, lead and zinc.

Acidity and Clarity of Solution: 1g, dissolved in 20 ml of water forms clear blue solution, which becomes green on the addition of 0.1 ml of solution of methyl orange.

Arsenic: Not more than 8 parts per million.

Iron: To 5g, add 25 ml of water and 2 ml of nitric acid boil and cool. Add excess of ammonia solution, filter and wash the residue with 20 percent v/v of dilute ammonia solution. Dissolve the residue if any, on the filter with 2 ml of charcoal acid, diluted with 10 ml of water, to the acid solutions add dilute ammonia solution till the precipitation is complete, filter and wash the residue after ignition weighs not more than 7 mg.

Lead and Zinc: Dissolve 1 g in 10 ml of water, add 1 g of citric acid and 10 ml of dilute ammonia solution. Add solution of potassium cyanide so that the blue colour is completely discharged, then add 1 drop of solution of sodium sulphide; no opalescence or not more than a slight darkening is produced.

Assay: Weigh accurately about 1 g and dissolve in 50 ml of water add 3 g of potassium iodide, 5 ml of acetic acid and titrate the liberated iodine with 0.1N sodium thiosulphate, using solution of starch as indicator. Continue titration till a faint blue colour remains, add 2g of potassium thiocyanate stir well and continue the titration until the blue color disappears. Each ml of 0.0N sodium thiosulphate is equivalent to 0.02497g of $CusO_4$. $5H_2O$.

Category: Astringent and antidote.

15.3 Sedatives

Any substance which brings about mild depression of the central nervous system and calming a person is known as sedative. A slightly higher dose of a sedative may cause sleep (hypnotic). Several sedatives and hypnotics find use as anti-epileptic or anti-convulsants. Among the inorganic compounds only bromides were employed for this purpose. When bromides were taken for a long period, a condition called "bromism" is developed. Due to the introduction of many safer and less toxic and more specific organic anticonvulsants and sedatives, bromides are not used at present. However, due to its historical importance, I.P has retained potassium bromide.

Potassium Bromide:

Chemical Formula: KBr *Mol. Wt.* 119.0

Potassium bromide contains not less than 98.5 percent of KBr, calculated with reference to the substance dried to constant weight at 105°.

Description: Colourless, transparent or opaque crystals or a white, granular powder, odourless, taste saline and faintly bitter.

Solubility: Soluble in 1.5 parts of water, in 250 parts of alcohol (90 percent) and in 5 parts of glycerin.

Preparation: Potassium bromide and potassium bromate are formed when bromine, in slight excess, is added to concentrated solution of potassium hydroxide. The pale yellow solution is evaporated to dryness and the residue is heated with charcoal to reduce the bromate:

$$6KOH + 3Br_2 \rightarrow KBrO_3 + 5KBr + 3H_2O \quad .....(15.5)$$

$$KBrO_3 + 3C \rightarrow KBr + 5CO \quad .....(15.6)$$

The product is extracted with water, filtered and the filtrate is evaporated to crystallization.

Identification: yields the reactions characteristic of potassium and of bromides.

Tests for Purity: It is tested for arsenic, barium, iron, heavy metals, sodium chloride, bromate, sulphate, alkali and loss on drying.

Arsenic: Not more than 2 parts per million.

Barium: Dissolve 0.5g in 10 ml of water and add 1 ml of dilute sulphuric acid, no turbidity is produced within five minutes.

Iron**:** 0.5g complies with the limit test for iron, Heavy metals. Dissolve 2g in 10 ml of water, add 2 ml of dilute acetic acid and water to make 25 ml; the limit of heavy metals is 10 parts per million.

Heavy Metals: Dissolve 2g in 10 ml of water, add 2 ml of dilute acetic acid and water to make 25 ml, the limit of heavy metals is 10 parts per million.

Sodium: Complies with the test for sodium describes under potassium chloride.

Chloride: Dissolve 1 g in 75 ml of water and 25 ml of nitric acid in a 500 ml distillation flask fitted with a bung carrying a thermometer and a tapered air inlet tube adjusted so that it is above the surface of the liquid. Pass a gentle stream of air, heat to 105°, lower the inlet tube into the liquid and continue to heat for one minute, maintaining the temperature at 105°. Remove the source of heat and pass a brisk stream of air for ten minutes. Add 5 ml of 0.1N silver nitrate and 5 drops of nitrobenzene and shake. Titrate the excess of silver nitrate with 0.1N ammonium thiocyante, using solution of ferric ammonium sulphate as indicator, and shaking vigorously near the end point, not less than 3.7 ml of 0.1N ammonium thiocyanate is required.

Bromate: Powder 1 g and add 1 ml of dilute sulphuric acid, no yellow colour is produced immediately.

Sulphate: 2 g complies with the limit test for sulphates.

Alkali: Dissolve 5 g in 50 ml of freshly boiled and cooled water and add 0.2 ml of 0.1N sulphuric acid, no colour is produced on the addition of a drop of solution of phenolphthalein.

Loss on Drying: Loses not more than 1.0 percent of its weight when dried to constant weight at 105°.

Assay: Weigh accurately about 0.4 g and dissolve in 40 ml of water and 5 ml of nitric acid. Add 50 ml of 0.1N silver nitrate and 5 ml of nitrobenzene and shake. Titrate with 0.1N ammonium thiocyanate, using solution of ferric ammonium as indicator and shaking vigorously as the end point of approached. Correct for the amount of chloride present, as determined by the test for chloride. Each ml of 0.1N silver nitrate is equivalent to 0.0119 g of KBr.

Storage**:** Preserve potassium bromide in a well closed container.

Category**:** sedative.

Dose**:** 0.3 to 1.2 g.

15.4 Diagnostic Agents

Diagnostic agents are used as aids in clinical diagnosis of diseases. They serve the physician by permitting examination of the body and to diagnose abnormalities and impairment of organ functions. These agents are not having any therapeutic effect in the body. The diagnostic agents are two categories, namely:

(i) Radio paques and

(ii) Drugs for testing functioning of different organs.

Radio paques are the agents which have the ability to absorb x-rays as they pass through the body. They cast a contrasting shadow on x-ray film. The constant medium is administered into the body through a suitable route for visualization.

Barium Sulphate I.P (Refer Page no.427)

Barium Sulphate Compound Powder I.P (Refer Page no.429)

15.5 Anti-Neoplastic Agents

Cancer or neoplastic disease may be degraded as a family of related disorders. A common feature in different types of cancer is an abnormal and uncontrolled cell division at a rate greater than that of most normal body cells. In cancer, mass of new tissue grows independently and invades surrounding structures. This new tissue has no physiological use.

Cisplatin is the widely used inorganic anti-cancer agent.

Cisplatin: Chemically it is cis-diammine-dichloro platinum (CDDP) and the molecular formula is $(NH_3)_2 PtCl_2$. It exists in the form of two isomers.

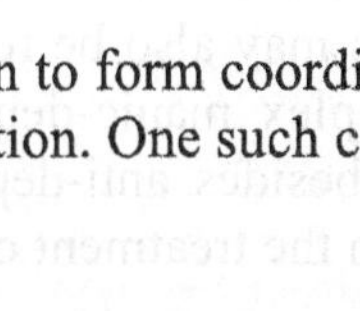

(Cis- platin)

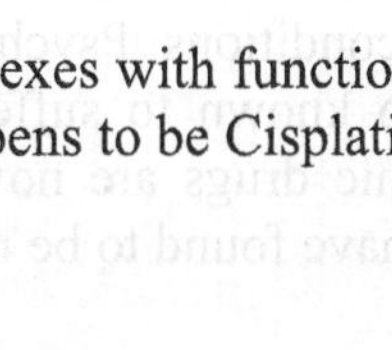

(Trans - platin)

Platinum is known to form coordinated complexes with functional group attached to it in a planar configuration. One such complex happens to be Cisplatin.

Cis-platin occurs as a white powder, soluble in water (1 in 100) and in dimethyl form amide (1 in 42). Solutions deteriorate and should be used within 20 hrs of preparation and have to be protected from light. Generally administered by intravenous injection.

Like gold compounds, platinum compounds are also toxic. Toxic symptoms are associated with kidneys, bone marrow, ears and nervous system.

Its used should be under medical care.

Uses: Cisplatin has been proved to be effective in testicular and ovarian tumors for which it is currently used.

Note: cisplatin has been found to be incompatable with meta sulphites (commonly used as anti-oxidants) and aluminium metal (used in tubing, joints, caps, lining etc.)

15.6 Anti-Thyroid Agents

Thyroid hormones regulate growth and development. Deficiency of thyroid hormones results in hypothyroidism or when particularly severe, in myxoedema, a condition marked by slowing down of all metabolic processes of the body. Excessive secretion of thyroid hormones results in hyperthyroidism. This cause Grave's disease characterized by thyrotoxicosis and opthalmopathy.

Anti-thyroid drugs are the compounds which are capable of interfering directly or indirectly with the synthesis of thyroid hormones and are of clinical value for the control of hyperthyroidism. These drugs are able to limit the uptake of inorganic iodide and reducing the inorganic iodide already present in the gland in order to treat hyperthyroidism.

Potassium perchlorate is widely used inorganic anti thyroid drug.

Potassium perchlorate:

Chemical Formula: $KClO_4$ *Mol. Wt.* 138.5

Potassium per chlorate occurs as a white crystalline powder, soluble in water. When other organic drugs are not tolerated by the patient, it can be employed as antithyroid drug. It prevents the uptake of inorganic iodide already accumulated in the thyroid gland. For this reason, it can also find use in hastening elimination of radioactive iodide.

15.7 Anti-Depressants

These are the drugs which are used to relieve depressed state of mind which may be caused due to psychic conditions. Psychic conditions may also be responsible for mania. Many patients are also known to suffer from complex manic-depressive states. Anti-psychotic and anti-manic drugs are now available besides anti-depressant drugs. Only compounds of lithium have found to be most stable in the treatment of manic conditions.

Lithium Carbonate:

Chemical Formula: $LiCO_3$ *Mol. Wt.* 73.89

Physical Properties: It occurs as a white crystalline powder, odourless but having slight alkaline taste. It is sparingly soluble in boiling water and insoluble in alcohol.

Purity: Its purity should be 99.5% and free from arsenic, calcium, magnesium, chlorides, heavy metals, aluminium and iron, potassium, sodium, sulphate and moisture.

Identification: It shows characteristic reactions of lithium and carbonate ions.

The margin of safety with lithium is narrow because the therapeutic dose and toxic dose are very close to each other. For this reason it finds use continuously and under dose observation in manic depressed state.

Contraindications: As lithium toxicity becomes worse by depletion of sodium, diuretics are contraindicated in lithium therapy.

Sodium bicarbonate (antacid) tends to increase urine excretion of lithium, there by reducing effectively plasma levels.

Each gram of lithium carbonate represents 27 millimoles of lithium. Optimum therapeutic effects could be obtained with plasma levels of 0.6 to 1.2 millimoles/1. A plasma level exceeding 1.5 millimol//l is regarded to be a toxic level. Hence plasma levels of lithium could be closely monitored during the lithium therapy.

15.8 Surgical Aids

These are the materials and chemicals which are not therapeutic agents used to assist in surgery, in recovery, treatment or cure of a diseased condition a tissue/an organ.

As surgical aids come in direct contact with the skin/other parts of the body, they should be prepared, processed and controlled carefully. Only plaster of Paris is official in I.P.

Plaster of Paris: (Calcium sulphate semihydrate)

Chemical Formula: $CaSO_4.\ 2H_2O$ *Mol. Wt.* 145.2

Synoym: Gypsum, Calcium sulphate semihydrate.
Preparation: Calcium sulphate can be prepared by adding dilute. H_2SO_4 to a solution of calcium chloride.

$$CaCl_2 + H_2SO_4 + 2H_2O \rightarrow CaSO_4.\ 2H_2O + 2HCl \quad .....(15.7)$$

Calcium sulphate being sparingly soluble in water gas precipitated and is filtered.

Description: It is a white, hygroscopic powder, which is odourless and also tasteless.

Solubility: Sparingly soluble in water, solubility decreases sharply with rise of temperature. It is insoluble in alcohol but dissolves in dilute HCl.

Tests for Identity: It gives tests for calcium and sulphate ions.

Tests for Purity: It is tested for acidity / alkalinity, acid insoluble matter, loss on drying and setting properties.

Stage: Plaster of Paris is preserved in a sound, clean, dry and water-tight container.

Uses: It is a surgical aid and is used in making casts in plaster of paris bandages. Also used as diluent in manufacture of compressed tablets.

Part – II

Practical Lab Manual

1. INTRODUCTION

The Practical Pharmaceutical inorganic chemistry consists of

- Qualitative Analysis
- Quantitative Analysis
- Analysis of Pharmaceutical Dosage forms
- Limit tests
- Preparation

Modern medicines for human use are required to meet exact standards. The standards are related to their

- Quality
- Safety
- Purity

Chemical purity may be described as complete freedom from foreign matter.

For pharmaceutical use the standards demanded of chemicals are determined by number of factors. British, European, United States and Indian Pharmacopoeias each have their own standards and distinctive formats in official monographs like:

- Definition of the nature the material
- Statement of minimum standard of purity (assay)
- Description of its physical characteristics
- Tests for use in verifying the identity
- Limit tests
- Official quantitative procedure for determination of the active ingredients.
- Physical constants
- Packing and storage conditions
- Dosage
- Cautionary notices

Any salt consists of two radicals i.e., basic radical and acidic radical.

For e.g., NaCl when dissolved in water dissociate into Na^+ and Cl^-. Here Na^+ is basic radical (cation) and Cl^- is acidic radical (anion). In nature many salts occur in admixture. Qualitative analysis may be carried out on various scales (Table P.1).

Table P.1

S.No	Class	Volume for Analysis	Quantity of the substance
1.	Macro analysis	20 ml	100 ml to 500 mg or more
2.	Semi micro analysis	1 ml	10 mg to 50 mg
3.	Micro analysis	0-1 ml	1 mg to 5 mg
4.	Sub micro or ultra micro	0.05 ml	< 1 mg
5.	Spot test analysis	Few drops	1 μg
		$1\ \mu g = 0.001\ mg = 10^{-6}\ g$	

2. Apparatus

Apparatus used in Qualitative Analysis

2.1 Apparatus for Macro Analysis

- Test tubes – 15 × 2 cm (25 ml)
- Beakers – 50, 100, 250 ml
- Conical flasks – 50, 100, 250 ml
- Stirring rods – 20 cm
- Wash bottle – 500 ml
- Filter papers
- Centrifuge tubes
- Kipps apparatus (for H_2S)
- Test tube brushes
- Bunsen burners
- Spatulas
- Tripod stand and wire gauge
- Porcelain crucible
- Watch glass
- Droppers

2.2 Apparatus for Semi- Micro Analysis

- Test tubes – 75 × 10 mm (4 ml)
- Beakers – 250 ml (Griffin form) 5 ml, 10 ml
- Conical flasks – 10 ml, 25 ml
- Water bath
- Wash bottle
- Stirring rods
- Flat bottomed flask – 50 ml and 100 ml with rubber bulb

- Centrifuge tubes – 3 ml
- Semi micro boiling tubes – 65 × 25 mm (20 ml)
- Gas absorption pipette –75 × 10 ml
- Reagent droppers
- Capillary droppers
- Porcelain crucibles – 3, 6 and 8 ml
- Measuring cylinder – 5 ml
- Watch glass – 3.5 cm diameter
- Cobalt glass – 3 × 3 cm
- Microscope slides
- Platinum wire
- Forceps
- Semi-micro spatula
- Semi-micro test-tube holder
- Semi-micro test-tube brush
- Spot plate
- Tripod stand and wire gauge
- Retort stand with iron ring (7.5 cm)
- Semi- micro burner
- Bunsen burner
- Measuring cylinder – 5 ml

2.3 Apparatus for Micro Analysis/spot Tests

- Micro beakers – 5 ml
- Micro crucibles – porcelain, silica or platinum (0.5 to 2 ml)
- Micro conical flasks – 5 to 10 ml
- Micro test tubes – 40–50 × 8 mm
- Micro volumetric flasks – 1 ml, 2 ml and 5 ml
- Micro nickel or platinum spatula (flattened at one end)
- Micro burner
- Micro agate pestle and mortar
- Micro centrifuge tubes – 0.5, 1.0 and 2.0 ml

- Capillary droppers
- Transfer capillary pipette
- Magnifying lenses – 5X and 10X
- Platinum spoon – 0.5 to 1.0 ml

2.4 Cleaning of Apparatus

Cleaning of apparatus is important for any analysis. Water can be used to clean. If apparatus is dirty or greasy then it is cleaned with Chromo Sulphuric acid (100 g of potassium dichromate in one litre of sulphuric acid). After cleaning with acid thoroughly washed with tap water and distilled water.

2.5 The Apparatus used in Quantitative Analysis

The Apparatus used commonly in Quantitative Analysis are:

- Wash bottles
- *Desiccators*: Its purpose is to provide a storage place of low moisture content for dried samples and crucibles. The bottom of desiccator is charged with anhydrous $CaCl_2$, conc. H_2SO_4, P_2O_5, anhydrous Magnesium Perchlorate (dehydrate) or Ca_2SO_4 semi hydrate (drierite).

Fig. P.1 Desiccator.

- Porcelain, silica, platinum crucibles to ignite precipitates
- Pipettes (Fig. 5.1) The pipettes are designed to deliver a fixed volume. The usual capacities are 5, 10, 25, and 50 ml. The certain amount of liquid should remain in the tip. This is never blown. The tolerances, permitted by British Standard Institute(BSI) for various capacity pipettes are given in Table P.2

Table P.2

Total capacity (ml)	1	2	5	10	20	25	50	100
Tolerance ± ml	0.01	0.01	0.02	0.02	0.02	0.03	0.04	0.06
Delivery time (sec)	5-10	7-15	10-20	15-25	15-30	20-35	25-40	30-50

Presently advanced automatic pipettes without sucking are available.

- Burettes (Fig. 5.1) The burettes are used to deliver variable volumes. These are long tubes provided with a stop cock one end and is generally graduated in units of 1/10 milli litre. The stop cock is prepared, instead of glass, with poly tetra fluroethylene PTFE or Teflon. The tolerances permitted by BSI for various capacity burettes are given in Table P.3.

Table P.3

Total Capacity (ml)	1	2	5	10	25	50	100
Tolerance ± ml	0.01	0.01	0.02	0.02	0.04	0.06	0.1

Presently advanced burettes with different graduations are available.

- Volumetric flasks (Fig. 5.1) The volumetric flasks are designed to contain a given volume when filled so that the meniscus in tangent to the ring, which has been etched on the neck of the flask, the eye being at the level of the ring. The lower meniscus is to be taken for clear and colourless solutions and the upper meniscus is to be taken for clear and coloured solutions. The tolerance permitted by BSI for various calibrated flasks are given in Table P.4.

Table P.4

Total Capacity (ml)	5	10	25	50	100	100	250	1000
Tolerance ± ml	0.02	0.02	0.03	0.04	0.06	0.10	0.15	0.20

- Conical Flasks (Fig. 5.1)
- Measuring jars (Fig. 5.1)
- *Filtering crucibles*: Used in filtering precipitates by suction and the precipitate may be dried at moderate heat. These are:
  - (i) Gooch crucibles (Fig. 5.11) – 30 ml
  - (ii) Seals crucible
  - (iii) Glass filtering crucible
- Crucible holders
- *Weighing bottles*: Used for weighing samples exactly for the preparation of standard solutions.

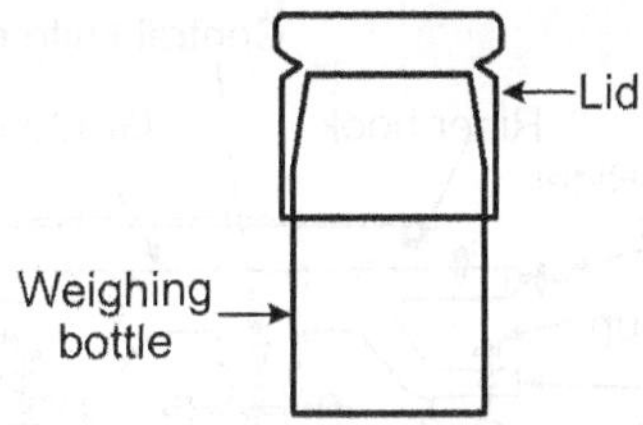

Fig. P.2 Weighing bottle.

- Glass funnels - various diameters and capacity

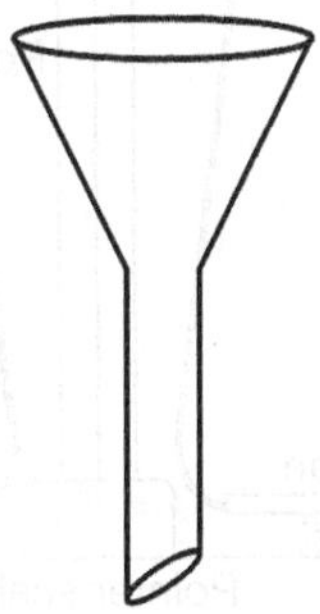

Fig. P.3 Glass funnel.

- Stirring rods of various lengths – the glass rods are fine polished at both ends.
- Rubber policeman – one of the glass rods is used as rubber policeman. A solid piece of soft rubber is attached on one end of the glass rod.

  This is useful to loosen precipitates that may adhere to the inner walls.

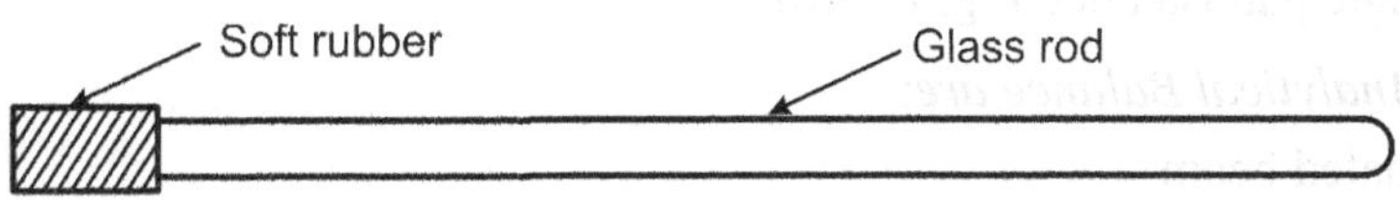

Fig. P.4 Rubber Policeman.

- Rough balance
- Analytical balance

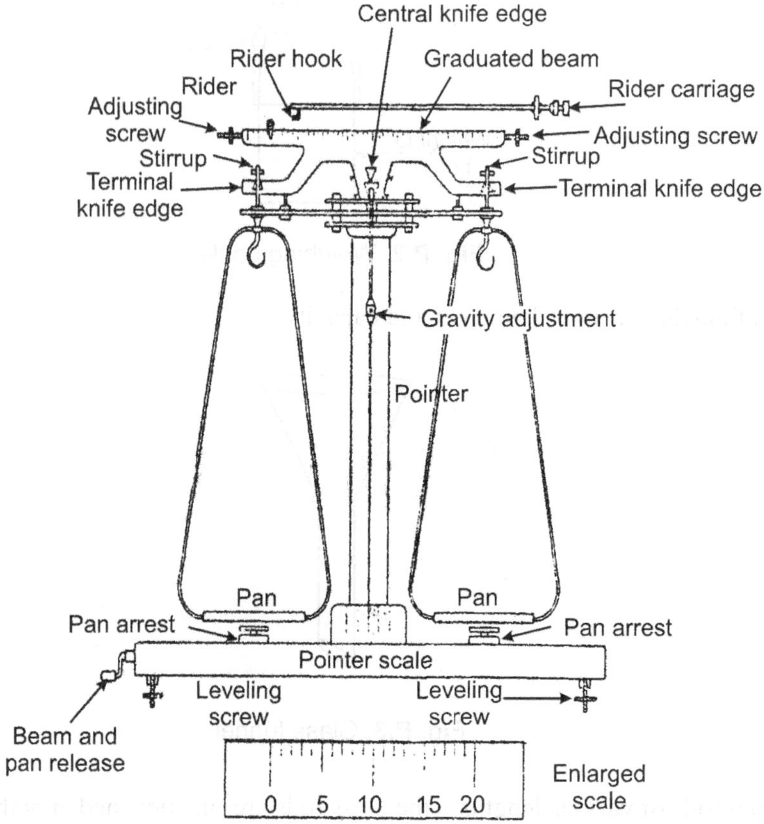

Fig. P.5(a) Analytical balance.

Presently so many very sensitive electrical single pan balances are available.

Mettler single pan Balance Fig. P. 5(b)

The parts of Analytical Balance are:

(i) Graduated beam

(ii) Centre knife-edge

(iii) Supporting stirrups for pass

(iv) Pan

(v) Rider hook

(vi) Rider rod

(vii) Beam release knob

(viii) Push-button pan arrest

(ix) Pointer

(x) Pointer scale

(xi) Adjusting screws for points

(xii) Balance case

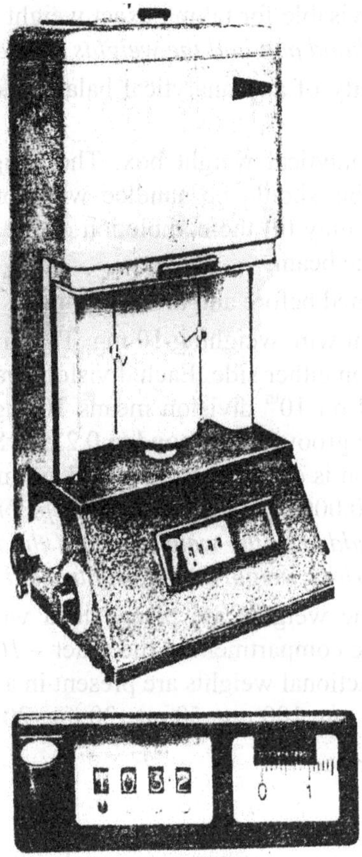

Fig. P.5(b)

Rules in Weighing

(i) The balance case door is kept closed when the balance is not in use in order to protect it from dust and fumes. It is closed in the final adjustment of weights and while swings are being observed, to avoid drafts.

(ii) Neither weights nor objects shall be placed upon or removed from the balance pan unless the beam is raised off the knife edges and the pan arrests are in place.

(iii) The beam support and pan arrests must be raised and lowered gently to avoid damage to the agate knife-edges.

(iv) No object to be weighed unless it is at room temperature. Heated objects placed in the balance, can cause air currents, which gives an erratic results.

(v) No powder or crystalline material must ever be placed directly on the balance pan. Use a suitable container like weighing bottle, watch glass etc. Use of non-stick papers is not advisable for taking exact weight of the sample. *The sample is to be kept on the left hand pan and the weights on the right hand pan.*

(vi) The maximum capacity of any analytical balance is 200 g, so never overload a balance.

(vii) Use only standard analytical weight box. The weights should be checked and calibrated. The weights should be handled with forceps only. Use the weights from one weight box only till the completion of the weighing. Place the rider on the zero position of the beam.

(viii) Balance must be cleaned before and after weighing.

(ix) *Rider*: It is a thin bent wire weighing 10 mg. The graduated beam is divided into 0-10 equal divisions on either side. Each division again divided into five grooves. When rider is placed on 10th division means 10 mg i.e., full weight of rider is added in the pan. One groove correspond to 0.2 mg. So 10 divisions is equal to 10 mg (0.01 g), 1 division is equal to 1 mg (0.001 g) and 1 small division or groove is equal to 0.2 mg (0.0002 g). *The rider is kept on the right hand side of the beam, the weight is added to the weights and if the rider is kept on the left hand side of the beam, the rider weight is to be subtracted from the total weight.*

(x) In the weight box the weights are placed in a wooden box. The weights are present in the separate compartment in the order – *100 g, 50 g, 20 g, 20 g, 10 g, 5 g, 2g, 2g, 1 g*. The fractional weights are present in a separate box and the order is 500 mg, 200 mg, 200 mg, 100 mg, 50 mg, 20 mg, 20 mg, 10 mg and rider.

All the weights must be calibrated

Standard Temperature

The capacity of glass vessel varies with the temperature and it is therefore necessary to define the temperature at which its capacity intended to be correct. A temperature of $20^{O}C$ has been universally accepted. The other temperature $27^{O}C$ is accepted by BSI for tropical climates. US Bureau of Standards accepts $25^{O}C$. The temperature corrections for 1000 ml of volumetric glass flask for the expansion of borosilicate glass at $20^{O}C$ are given in Table P.5.

Table P.5

Temperature (^{O}C)	5	10	15	20	25
Correction (ml)	–0.15	–0.01	–0.05	+10.05	+0.10

3. REAGENTS

3.1 Introduction

Quality of the reagents is important for accuracy and in quantitative analysis. The student must see the label on the bottle before use. Important ones are:

- Chemical Name of the substance
- Chemical formula
- Molecular weight
- Melting point of solid and boiling point if liquid
- Purity
- Specification and nature of possible impurities
- Grade
- Make

Regarding grade there are different ones depending on the purity and their usage. *The grades are*:

- Commercial grade – used for when high purity is not required.
- Chemically pure (C.P grade) – specifications are not mentioned and are not useful for analytical work. These are almost equal to commercial grade.
- Laboratory reagent grade (LR grade) – Purity is labeled. These can be used for the secondary standard chemicals in analytical work.
- Analytical reagent grade or guaranteed reagent grade (A.R or G.R) – These are extremely pure. The purity and nature of impurities are labelled. These chemicals are used for preparation of primary standard solutions in analytical work.
- Pharmacopoeial grade chemicals – These chemicals should confirm to tolerances set by pharmacopoeias.
- Spectroscopic grade chemicals – There are extremely pure used in spectroscopic instrumentation work.

The reagents in the lab handled properly. Care must be taken while handling acids, solvents, poisonous chemicals. After usage properly close the bottles and place it in the original position in the lab.

3.2 Preparation of Reagent Solutions in Qualitative Analysis

***Acetic acid* (2M):** Dilute 114 ml glacial acetic acid with water to 1 liter.

***Alizarin*:** (saturated solution in ethanol): To 2g alizarin, $C_{14}H_8O_4$, add 10 ml 965 ethanol and shake. Use the clear solution for the tests.

***Alizarin red S* (2%):** Dissolve 2g alizarin red S (sodium alizarin sulphonate), $C_{14}H_7O_4$. $SO_3Na.H_2O$, in 100 ml water.

***Ammonia Solution* (Concentrated)**

Ammonia Solution: To 500 ml water add 500ml concentrated ammonia solution and mix.

***Ammonium Molybdate* (0.025M):** Dilute 1 ml of 0.25M ammonium molybdate with water to 10ml.

***Ammonium nitrate* (2%):** Dissolve 20g ammonium nitrate, NH_4NO_3, in water and dilute to 1 liter.

Ammonium polysulphide: To 1 liter ammonium sulphide solution (M) add 32g sulphur, and heat gently until the latter dissolve completely and a yellow solution is formed. The formula of the reagent is $(NH_4)_2S_x$, where x is approx. 2.

Ammonium Thiocyanate (0.1M): Dissolve 7.61g ammonium thiocyanate, NH_4SCN, in water and dilute to 1 liter.

***Aqua regia*:** To 3 volumes of concentrated hydrochloric acid add 1 volume of concentrated nitric acid. Mix and use immediately. (The solution does not keep at all).

Barium Chloride (0.25M): Dissolve 61.1g barium chloride dehydrate, $BaCl_2$ in water and dilute to 1 liter.

***Benzoin α-oxime* (cupron)** (5% in alcohol): Dissolve 5 g benzoin α-oxime, $C_6H_5CH(OH)$. C (NOH) C_6H_5 in 96% ethanol and dilute with the solvent to 100 ml.

Bromine water (saturated): Shake 4 g (or ml) liquid bromine with 100 ml water. Ensure that a slight excess of undissolved bromine is left at the bottom of the mixture. The solution keeps for 1 week. When handling bromine, exercise utmost care.

Chromosulphuric acid (concentrated): To 100g potassium or sodium dichromate ($K_2Cr_2O_7$ or $Na_2Cr_2O_7$) and 1 litre concentrated sulphuric acid. Stir the mixture occasionally and keep in a stopped vessel. Because of its strong oxidizing and dehydrating properties, it is used for cleaning glassware. Handle with greatest care.

***Copper* (II) *sulphate* (0.25M):** Dissolve 62.42g copper sulphate pentahydrate, $CuSO_4$. $5H_2O$ in water and dilute to 1 litre.

Diethyl Ether

***Disodium hydrogen phosphate* (0.033M):** Dissolve 12g disodium hydrogen ortho-phosphate dodecahydrate, $Na_2HPO_4.12H_2O$ or 6g disodium hydrogen ortho-phosphate dihydrate, $Na_2HPO_4.2H_2O$ to 1 litre.

Ethanol

Hydrochloric acid (Concentrated)

***Hydrochloric acid* (6M):** To 500 ml water add 500ml Conc. HCl

***Hydrochloric acid* (3M):** To 500 ml water add 265ml Conc. HCl and dilute to 1 litre.

***Hydrochloric acid* (0.5M):** Dilute 4.5 ml concentrated HCl with water to 100 ml

***Hydrogen peroxide* (Conc. 30%)**

Hydrogen peroxide (3%)

H_2S gas

***Iron* (III) *Chloride* (0.5M):** Dissolve 135.2g iron (III) chloride hexahydrate, $FeCl_3.6H_2O$ in water add a few milliliters concentrated hydrochloric acid if necessary, and dilute to 1 litre.

***Lead acetate* (0.25M):** Dissolve 95g lead acetate trihydrate $Pb(CH_3COO)_2.3H_2O$ in a mixture of 500 ml water and 10 ml glacial acetic acid and dilute to 1 litre with water.

Lead acetate (0.0025M)

***Magnesium nitrate reagent* (Ammonical)** (0.5M for magnesium): Dissolve 128 g magnesium nitrate hexahydrate, $Mg(NO_3)_2$ $6H_2O$ and 160g ammonium nitrate, NH_4NO_3 in water and add 50 ml of conc. ammonia and dilute to 1 litre.

***Nessler's reagent*:** Dissolve 10 g potassium iodide in 10 ml water (solution A). Dissolve 6 g mercury(II) chloride in 100 ml water (solution B). Dissolve 45g potassium hydroxide in water and dilute to 80 ml (solution C). Add solution **B** to solution **A** drop wise until a slight permanent precipitate is formed, then add solution **C**. Mix and dilute with water to 200 ml. Allow to stand overnight and decant the clear solution, which should be used for the test.

Nitric acid (Concentrated)

***Nitric acid* (2M):** To 128 ml concentrated nitric acid into 500 ml water, dilute to 1 litre.

Phosphoric acid (concentrated)

***Phosphoric acid* (1M):** Dilute 63.7 ml conc. phosphoric acid with water to 1 litre.

Potassium chromate **(0.1M):** Dissolve 19.4g potassium chromate, K_2CrO_4 and dilute the solution to 1 litre.

Potassium hexacyanocobaltate **(III) reagent (4%):** Dissolve 4g potassium hexacyanocobaltate (III) K_3 [Co (CN) $_6$] and lg potassium chlorate, $KClO_3$, in water and dilute to 100 ml.

Potassium hexacyanoferrate **(II)** (Potassium Ferro cyanide) (0.025M): Dissolve 10.5g potassium hexacyanoferrate (II) trihydrate, K_4 [Fe (CN) $_6$]$3H_2O$ in water and dilute to 1 litre.

Potassium hexacyanoferrate **(III)** (Potassium ferricyanide) (0.022M): Dissolve 10.98 g potassium hexacyanoferrate (III), K_3 [Fe (CN) $_6$] in water and dilute to 1 litre.

Potassium periodate **(saturated):** To 0.1g potassium periodate, KIO_4, add 20 ml water and heat on a water bath until dissolution. Allow to cool and use the clear supernatant liquid.

Potassium permanganate **(0.02M):** Dissolve 3.16g potassium permanganate solution with water to 1 litre.

Potassium permanganate **(0.004M):** Dilute 4ml 0.02M $KMnO_4$ solution with water to 20 ml.

Potassium thiocyanate **(10%):** Dissolve l0g in 100 ml water.

Silver nitrate **(0.1M):** Dissolve 16.99g of $AgNO_3$ in water and dilute to 1 litre. Keep the solution in dark bottle.

Sodium hydroxide **(Concentrated):** To 5g solid sodium hydroxide, NaOH, add 5 ml water and mix.

Sodium Hydroxide **(2M):** To 80g solid sodium hydroxide, NaOH, add 80 ml water.

***Sodium nitroprusside reagent*:** 0.5g sodium nitroprusside dehydrate

Na_2 [Fe (CN) $_5$ NO].$2H_2O$ in 5ml water

Sodium peroxide (0.5M): Dissolve 3.9g sodium peroxide Na_2O_2 in water and dilute to 100 ml. The solution must be prepared freshly.

Sodium tetraborate **(O.1M):** Dissolve 38.lg sodium tetra borate decahydrate (borax)

$Na_2B_4O_7$. $10H_2O$

Sulphuric acid (Concentrated)

Sulphuric acid **(1M):** Add 55.4ml concentrated sulphuric acid slowly to 800 ml cold water with constant stirring.

Tartaric acid **(1M):** Dissolve 15g tartaric acid, $C_4H_6O_6$, in water and dilute to 100 ml

***Tin (II) chloride (saturated)*:** Shake 2.5g tin (II) chloride dihydrate, $SnCl_2.2H_2O$ in 5 ml conc. HCl. Allow the solid to settle and use the clear solution for the tests.

***Zirconyl chloride* (0.1%):** Dissolve 0.lg zirconyl chloride octahydrate, $ZrOCl_2\ 8H_2O$ in 20 ml conc. HCl and dilute with water to 100 ml.

3.3 Solid Reagents

Ammonium carbonate $(NH_4)_2\ CO_3$

Ammonium chloride, NH_4Cl

Ammonium sodium hydrogen phosphate, $NH_4NaHPO_4.4H_2O$

Ammonium thiocyanate, NH_4SCN

Copper turnings, Cu

Dimethylglyoxime, $CH_3\ C\ (NOH)\ C\ (NOH).CH_3$

Lead acetate, $(CH_3COO)_2\ Pb.3H_2O$

Litmus paper

Manganese dioxide, MnO_2

Oxalic acid, $(COOH)_2.\ 2H_2O$

Potassium dichromate, $K_2Cr_2O_7$

Resorcinol, 1, 3-$C_6H_4\ (OH)_2$

Sodium carbonate, Na_2CO_3

Sodium chloride, NaCl

Sodium hexanitritocobaltate (III), $Na_3\ [Co(NO_2)_6]$

Sodium nitroprusside, $Na_2\ [Fe\ (CN)_5\ NO].\ 2H_2O$

Sodium peroxide, Na_2O_2

Sodium tetraborate, (Borax) $Na_2B_4O_7.10H_2O$

4. Inorganic Qualitative Analysis

(Qualitative tests for Anions and Cations/Semi-Micro Qualitative Analysis)

The analysis of the given sample for its ion or radical starts with preliminary tests followed by systematic method. The tests are:

Dry tests – Applicable for solid materials

Wet tests – Applicable for solutions

4.1 Dry Tests

4.1.1 Action of Heat

The substance is taken in a dry test tube and heat on a Bunsen flame. The observation and inferences are given in Table P.6.

Table P.6

Observation	Inference
(a) On sublimation	
(i) white deposit	May be NH_4^+ salts, Hg_2Cl_2 or $HgCl_2$, Al_2O_3, Sb_2O_3, SeO_2
(ii) Yellow deposit	May be sulphur, As_2S_3, Hg_2I_2, HgI_2
(iii) Black	May be Hg_2S, HgS
(iv) Violet	May be Iodide

(b) *Colour change*

(i) White — May be removal of water of crystallization. *e.g.*, $CuSO_4.5H_2O$ (blue)

(ii) Charring — May be decomposition of organic ions *e.g.*, tartarate

(c) *Produce gases with characteristic smell*

(i) CO_2 — May be carbonates

(ii) SO_2 — May be sulphites, sulphates and sulphides

(iii) Cl_2 — May be certain chlorides *e.g.*, $MgCl_2$

(iv)	NO_2	May be nitrates
(v)	CO	May be oxalates, tartarates
(vi)	Cyanogen	Cyanides
(vii)	O_2	May be chlorates, perchlorates, bromates and iodates
(viii)	NH_3	May be ammonium salts

4.1.2 Flame Test

Compounds of certain metals chlorides are volatized in a non-luminous Bunsen flame and give important characteristic colours in the flame.

Let us know the different zones of the Bunsen burner flame (Fig. P.6)

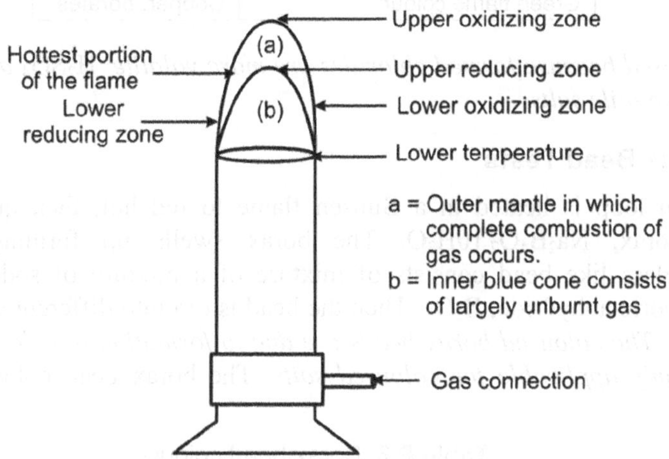

Fig. P.6 Bunsen Burner flame zones.

Different zones employed for:

(i) For testing volatile substances to determine whether they impart any colour to the flame,

(ii) For testing fusibility of substances,

(iii) For oxidation of substances in bead tests,

(iv) For all oxidation purposes,

(v) For reducing oxide in combustions to the metal,

(vi) For the reducing fused beads.

Procedure: To the salt or mixture taken on a watch glass add little concentrated hydrochloric acid and made into a paste. A thin platinum wire about 5 cm long and 0.03 to 0.05 mm diameter, fused into the end of glass rod is employed (where platinum wire is not available in the lab, the clean glass rod is using). The wire is cleaned thoroughly in HCl. The wire is dipped into the paste and then introduce into the lower oxidizing flame and the colour imparted to the flame is observed (Table P.7).

Table P.7 Flame tests.

Observation	Inference
Golden yellow flame colour	Sodium
Violet (lilac) flame colour	Potassium
Brick red flame colour	Calcium
Crimson flame colour	Strontium
Yellowish Green flame colour	Barium
Green flame colour	Copper, borates

HCl *is only used because formed chlorides are more volatile.* H_2SO_4 *and* HNO_3 *do not produce the volatile salts.*

4.1.3 Borax Bead Tests

The platinum loop is heated in a Bunsen flame to red hot, then quickly placed in powdered borax, $Na_2B_4O_7.10H_2O$. The borax swells up forming a colourless, transparent glass like bead consists of mixture of a mixture of sodium metaborate, $NaBO_2$ and boric anhydride, B_2O_3. Then the bead is dip into different salts and observe bead colours. *The coloured borax beads are due to formation of coloured borates and this test is only applicable for coloured salts.* The borax bead colours are given in Table P.8.

Table P.8 Borax bead colours

S.No	Oxidizing flame		Reducing flame		Metal
	Hot	Cold	Hot	Cold	
1.	Green	Blue	Colourless	Opaque	Copper
2.	Yellowish brown	Yellow	Green	Green	Iron
3.	Yellow	Green	green	green	Chromium
4.	violet	Amethyst	colourless	colourless	Manganese
5.	Blue	Blue	Blue	Blue	Cobalt
6.	Violet	Reddish brown	Gray	Gray	Nickel
7.	Yellow	Colourless	Brown	Brown	Molybdenum
8.	Rose-violet	Rose-violet	Red	Violet	Gold

The others of less important are microcosmic or phosphate bead test. The chemical used is sodium ammonium hydrogen phosphate tetra hydrate Na (NH_4) HPO_4. $4H_2O$ and sodium carbonate bead test.

4.2 Wet Tests

For the analysis of cations the preparation of clear and transparent solution is essential.

Preparation of Original Solution

All salts/mixtures are not soluble in one solvent. So it is necessary to use either water or acid for testing solubility.

Procedure: To the small quantity of the substance add the solvents in the order given below i.e., the substance is dissolved in cold distilled water then no other test is done and if not dissolved try next one like that.

Substance + distilled water

Substance + hot distilled water

Substance + dil. HCl

Substance + Conc. HCl

Substance + dil. HNO_3

Substance + Conc. HNO_3

Substance + aqua regia (3 parts conc. HCl and 1 part Conc. HNO_3)

After knowing the substance soluble in which solvent, we can prepare the original solution of the mixture or salt in that solvent. From the colour of the original solution we can get some information of salts (Table P.9).

Table P.9

S.No.	Observation	Inference
1.	Pink	May be CO, Mn
2.	Purple	May be permanganate
3.	Orange red	May be dichromate
4.	Blue	May be copper
5.	Green	May be Ni^{2+}, Fe^{2+}, Cr^{3+}
6.	Yellow	May be Fe^{3+}, chromates, ferricyanides

***Hints*:**

(i) For preparation conc. H_2SO_4 cannot be used as solvent since it forms sulphate precipitates with many cat ions like Pb^{2+}, Ba^{2+}, Ca^{2+}, Sr^{2+} etc. The other reason is, it is an oxidizing agent and it oxidizes H_2S to sulphur.

(ii) If conc. HCl is used as solvent, the excess acid must be evaporated and then dilute with distilled water.

If dil/conc. HNO_3 is used as solvent, the solution is evaporated to dryness and then extract with dil. HCl.

4.3 Tests for Anions (Acid-Radicals)

More than one salt is mixed, then it is called as mixture. Various anions will come across in the mixture/salt. Generally mixtures are given for semi-micro analysis.

4.3.1 Classification of Anions

The common anions can be classified as:

Class 1 - Carbonate$\left(CO_3^{2-}\right)$, Bicarbonate (HCO_3^-) Sulphide (S^{2-}), Sulphite$\left(SO_3^{2-}\right)$, Nitrite $\left(NO_2^-\right)$, Acetate (CH_3COO^-)

Reagent – Dil. Sulphuric acid.

Class 2 - Chloride (Cl^-), Bromide (Br^-) Iodide (I^-), Nitrate$\left(NO_3^-\right)$, Oxalate$\left(COO\right)_2^{2-}$, Tartarate $(C_4H_4O_6)^{2-}$

Reagent – Conc. Sulphuric acid.

Class 3 - Sulphate$\left(SO_4^{2-}\right)$, Phosphate$\left(PO_4^{3-}\right)$, Borate$\left(BO_3^{3-}, B_4O_7^{2-}, BO_2^-\right)$, Chromate$\left(CrO_4^{2-}\right)$, Arsenate AsO_4^{3-}

Reagent – separate reagents for each one.

4.3.2 Systematic Analysis of Anions in a Mixture

The analysis of anions should be done in a systematic way. First test for class 1 anions. If they are absent in the given salt/mixture, proceed for the class 2 and so on; otherwise the erratic results are obtained

Class 1 Anions – to the mixture 0.01 g to 0.1 g taken in a dry test tube, add dil. HCl or dil. H_2SO_4 and observe the reaction (Table P.10)

Table P.10

S.No.	Observation	Inference
1.	Colourless CO_2 gas is evolved which is identified by turning lime water milky or turbid after passing into it.	May be carbonate CO_3^{2-}
2.	Colourless SO_2 gas is evolved with suffocating odour of burning sulphur	May sulphite, SO_3^{2-}
3.	Colourless H_2S gas is evolved with rotten egg odour and blackening of filter paper moistened with lead acetate solution.	May be sulphide S^{2-}
4.	Brown fumes of NO_2 and the solution becomes pale blue (due to free HNO_2 or N_2O_3)	May be nitrite NO_2^-
5.	Smell of vinegar (acetic acid smell)	May be acetate CH_3COO^-

After identifying anions from the above test, the anions should be confirmed by doing confirmatory tests since many anions react with one reagent so the identification is not sufficient.

Sodium Carbonate Extract

For doing confirmatory test, sodium carbonate extract of the mixture is to be prepared because if the metal ions (cations) also present, they can also give some positive tests along with anions.

***Preparation*:** To the mixture in a beaker add excess sodium carbonate and sufficient water, and then boil. All the anions react with sodium and because they are soluble either into solution and all the cations react with carbonate to precipitate as metal carbonate. Filter the solution. The filtrate is called as sodium carbonate extract which is to be taken for doing confirmatory tests except for carbonate.

The extract is always neutralized with appropriate dilute acid to eliminate excess carbonate in the form of CO_2 gas.

Confirmatory Tests

(i) *Carbonates*: To the carbonates add barium chloride solution. A white precipitate of $BaCO_3$ is formed. The precipitate is soluble in mineral acid or carbonic acid.

(ii) *Sulphites*:

(a) To the sodium carbonate extract add $K_2Cr_2O_7$ and dil. H_2SO_4

A green colouration owing to the formation of Cr (III) ions. Sulphite is confirmed.

(b) To the sodium carbonate extract add $KMnO_4$ and dil. H_2SO_4. The solution is decolourised due to the formation of Mn (II) ions.

(iii) *Sulphides*

(a) To the sodium carbonate extract add sodium nitro prusside solution, $Na_2 [Fe (CN)_5 NO]$. If transient purple colouration is observed, the sulphide is confirmed.

$$S^{2-} + [Fe (CN)_5 NO]^{2-} \rightarrow [Fe (CN)_5 NOS]^{4-} \quad ....(P.1)$$

(b) To the sodium carbonate extract add silver nitrate solution. If black precipitate of Ag_2S is formed which is insoluble in cold dil HNO_3 but soluble in hot acid, then sulphide is confirmed.

(iv) *Nitrites*

Fe SO_4 *Test*- To the extract solution add carefully a saturated solution of $FeSO_4$ acidified with dil. acetic or sulphuric acid, a brown ring due to the compound $[Fe NO] SO_4$ is formed at the junction of two liquids, then NO_2^- is confirmed.

$$NO_2^- + CH_3COOH \rightarrow HNO_2 + CH_3COO^- \quad .....(P.2)$$

$$3 HNO_2 \rightarrow H_2O + HNO_3 + 2 NO \quad .....(P.3)$$

$$Fe^{2+} + SO_4^{2-} + NO \rightarrow [Fe NO] SO_4 \quad .....(P.4)$$

Griess-Ilosvay test – To the extract add sulphanilic acid-1-naphthylamine reagent, if red colour is observed nitrite is confirmed.

(v) *Acetates*:

(a) The extract is treated with 1 ml of conc. H_2SO_4 and 2-3 ml of ethanol in a test tube and heat. Ethyl acetate with pleasant and fruity odour is evolved. Acetate is confirmed.

$$CH_3COONa + H_2SO_4 \rightarrow CH_3COOH + NaHSO_4 \quad .....(P.5)$$

$$CH_3COOH + C_2H_5OH \rightarrow CH_3COOC_2H_5 + H_2O \quad .....(P.6)$$

(b) To the extract ferric chloride solution is added, deep red colouration is observed due to formation of complex ion $[Fe_3 (OH)_2 (CH_3COO)_6]^+$. On boiling this brownish red precipitate of basic Fe (III) acetate is formed. Acetate is confirmed.

Class 2 Anions: To the mixture/salt 0.01 g to 0.1 g taken in a dry test tube add conc. H_2SO_4 and observe the reaction in cold and hot condition (Table P.11)

Table P.11

S.No.	Observation	Inference
1.	HCl gas is evolved with pungent odour and it produces white fumes on introducing glass rod moistened with ammonia solution	May be chloride Cl^-
2.	Reddish brown bromine vapour along with HBr is evolved and the solution becomes reddish-brown and the reaction is accelerated on heating.	May be bromide, Br^-
3.	Violet vapour of iodine is evolved	May be iodide I^-
4.	Heat the mixture with conc. H_2SO_4 and copper turnings. Reddish-brown fumes of NO_2 evolved and solution acquires blue colour owing to the formation of Cu (II) ions.	May be nitrate NO_3^-
5.	CO_2 and CO are evolved and the CO is identified by burning at the mouth of test tube which gives blue flame.	May be oxalate, $(COO)_2^{2-}$
6.	Charring occurs. Smell of burnt sugar due to CO_2 and SO_2 gases.	May be tartarate $C_4H_4O_6^{2-}$

Confirmatory Tests

(i) *Chlorides*:

(a) To the sodium carbonate extract solution in a test tube add manganese dioxide and conc. H_2SO_4; the mixture is gently warmed. Chlorine gas is evolved which is identified by its yellowish green colour and suffocating odour.

$$MnO(OH)_2 + 2\,H_2SO_4 + 2Cl^- \rightarrow Mn^{2+} + Cl_2 + 2\,SO_4^{2-} + 3H_2O \quad .....(P.7)$$

Chloride is confirmed

(b) *Chromyl chloride test*: To the extract solution in a boiling test tube add excess $K_2Cr_2O_7$ and conc. H_2SO_4 and boil the mixture. Deep red vapours of chromyl chloride gas, CrO_2Cl_2 are formed. The vapours are passed into a test tube containing 2M NaOH solution. CrO_2Cl_2 reacts with Na OH to form sodium chromate and the resulting solution is yellow in colour. To this solution if lead acetate solution is added yellow precipitate of lead chromate will form. Chloride is confirmed.

(c) To the sodium carbonate extract add little dil. HNO_3 for neutralization till the effervescence is ceased, then add $AgNO_3$ solution. A white curdy precipitate of AgCl is formed, insoluble in water and dil. HNO_3 but soluble in ammonia. Chloride is confirmed.

(ii) *Bromide*:

(a) To the extract solution add little dil HNO_3 for neutralization and then add $AgNO_3$ solution. Curdy pale yellow precipitate of Ag Br is formed which is sparingly soluble in dilute ammonia but readily soluble in conc. ammonia. Bromide is confirmed.

(b) To the extract solution add chlorine water and chloroform. The chlorine liberates Br_2 if bromide is present. The Br_2 is dissolved in chloroform and the organic layer gets orange colour.

Bromide is confirmed. This test is to be done if chloride, bromide and iodides are present.

(iii) *Iodide*:

(a) To the extract solution add little dil HNO_3 for neutralization and then add $AgNO_3$ solution. Curdy yellow precipitate of AgI very slightly soluble in conc. ammonia and insoluble in HNO_3.

Iodide is confirmed.

(b) To the extract solution add chlorine water and chloroform. The organic layer becomes violet since chlorine water liberates I_2 which is dissolved.

Iodide is confirmed.

(iv) *Nitrates*:

Brown ring test: The extract is neutralised with dil. H_2SO_4. Add 3 ml of freshly prepared ferrous sulphate solution in a test tube. Pour 3.5 ml of conc. H_2SO_4 slowly down the side of the test tube so that acid forms a layer beneath the mixture. At the junction of the liquid, a brown ring is formed $[FeNO]SO_4$. Nitrate is confirmed.

(v) *Oxalate*:

(a) MnO_2 test: To the extract add Mn O_2 and dil. H_2SO_4. The red colour complex of $[Mn(C_2O_4)_3]^{3-}$ will form.
Oxalate is confirmed.

(b) Aniline blue test: To the extract solution add phosphoric acid and diphenyl amine and boil the mixture. Aniline blue will form. Other salts of organic acids like acetate, tartarates, citrates etc. do not give this text.
Oxalate is confirmed.

(c) The extract solution is acidified with dil. HCl and then add $CaCl_2$ solution. White precipitate of calcium oxalate is precipitated. Dissolve the precipitate in dil. HCl or dil. HNO_3. Now add few drops of acidified $KMnO_4$ ($KMnO_4$ + dil. H_2SO_4) and heat. The permanganate colour is decolourised.

Oxalate is confirmed.

(vi) *Tartarates*:

(a)Fenton's *test:* The extract is neutralised with dil. H_2SO_4 and then add 1 drop of $FeSO_4$ solution + 2–3 drops of H_2O_2 + NaOH solution. A deep violet or blue colour appear due to formation of dihydroxy malic acid CO_2H. C(OH) = C(OH).CO_2H. Tartarate is confirmed.

(b) *Silver Mirror Test*: The extract solution is neutralised with dil. HNO_3 in a boiling test tube and add excess of $AgNO_3$ solution and if any precipitate is formed which is filtered off. Very dil. ammonia (0.02M) is then added until the precipitate is formed is nearly re-dissolved. The solution is filtered and the filtrate is collected in a clean test tube, and then it is placed in a beaker of boiling water. A brilliant mirror is formed on the sides of the tube after few minutes due to deposition of silver on the sides. Tartarate is confirmed.

Class 3 ***Anions***: These anxious are identified by the precipitation in a suitable reagent. The tests and inferences are given in Table P.12

Table P.12

S.No	Test	Inference
1.	To the mixture add $BaCl_2$ solution. White precipitate of $BaSO_4$ is formed.	May be sulphate, SO_4^{2-}
2.	To the mixture add conc. HNO_3 and ammonium molybdate reagent. A canary yellow precipitate of ammonium phosphomolybdate, (NH_4) [P Mo_{12} O_{40}]is formed in cold	May be phosphate, PO_4^{3-}
3.	To the mixture add conc. HNO_3 and ammonium molybdate. The solution on boiling form yellow crystalline precipitate of ammonium arsenomolybdate, (NH_4) [Ar $MO_{12}O_{40}$]	May be arsenate, ASO_4^{3-}
4.	To the mixture add silver nitrate solution. Brownish-red precipitate of silver chromate, Ag_2 CrO_4 informed	May be chromate, CrO_4^{2-}
5.	To the mixture add 1 ml of conc. H_2SO_4 and 5 ml of methanol or ethanol in a small porcelain dish and the alcohol is ignited. The later will burn with green edged frame due to the formation of methyl or ethyl borate.	May be borate, BO_3^{3-}

Confirmatory Tests:

(i) *Sulphates*:

(a) The exact is neutralised with dil. HCl and then add $BaCl_2$ solution. White precipitate of $BaSO_4$ will form. Sulphate is confirmed.

(b) The extract is neutralised with dil acetic acid and then add lead acetate solution. A white precipitate of $PbSO_4$ will form. The precipitate is soluble in hot conc. H_2SO_4 and in ammonium acetate solution. Sulphate is confirmed.

(ii) *Phosphates*:

(a) Ammonium molybdate test: The extract is neutralized with dil. HNO_3 and then add conc. HNO_3 and 2-3 ml of ammonium molybdate reagent is added, canary yellow precipitate of ammonium phosphomolybdate is formed in cold. Phosphate is confirmed.

(b) Cobalt nitrate test: The sample when heated on charcoal and then moistened with a few drops of Co $(NO)_2$ soluton, give a blue mass of Na $CoPO_4$.

(iii) *Arsenates*:

(a) The extract is neutralized with dil. HNO_3 and then add $AgNO_3$ solution. Brownish red precipitate of silver arsenate Ag_3AsO_4 will form. Arsenate is confirmed.

(b) Ammonium molybdate test: The extract is neutralized with dil. HNO_3. Add conc. HNO_3 and ammonium molybdate reagent, then boil the solution. Yellow crystalline precipitate of ammonium arsenomolybdate will form. Arsenate is confirmed.

(iv) *Chromates*:

(a) The extract is neutralized with dil. HNO_3 and add $AgNO_3$ solution, brownish red precipitate of Ag_2CrO_4 will form. Chromate is confirmed.

(b) The extract is neutralized with dil. HCl and add $BaCl_2$ solution, pale yellow precipitate of barium chromate will form. Chromate is confirmed.

(v) *Borates*:

To the sample add powdered CaF_2 and a little conc. H_2SO_4 on a watch glass and made a paste. The paste is taken on a platinum loop or on the end of glass rod and bring very close to the edge of bunsen flame. The flame colours green due to formation of BF_3. Cu and Ba salts can not interfere in this test. Borate is confirmed.

Hint: By performing preliminary test and the confirmatory test only, report the concerned anion present in the salt. Do the analysis in a systematic way otherwise wrong results will obtain For e.g., to the salt if you add $BaCl_2$ solution, if white precipitate is obtained, don't report as sulphate. It can also give white precipitation with oxalate, tartarate etc. So it is essential to confirm first there are no class 1 anions, then do test for class 2 and finally class 3.

4.3.3 Interfering Anions and their Elimination

Some of the acid radicals like oxalate, tartarate, phosphate, chromate etc., can be called as interfering acid radicals since they interfere the systematic analysis of cations and they must be eliminated before analysis.

e.g., Barium oxalate, calcium phosphate, magnesium phosphate etc. precipitated in ammonical medium but they are soluble in acid medium. In the third group after adding the group reagent NH_4Cl and NH_4 OH, they will precipitate. Instead of only third group cat ions, the other Vth group, VIth group cations also precipitate. So it is not possible to analyze only IIIrd group cation because of interference of the above acid radicals.

So some of the acid radicals must be eliminated from the given mixture before analyzing systematically the cations.

Elimination

(i) **Oxalate** $C_2O_4^{2-}$: Take 2 g of mixture in a porcelain dish and heat it strongly for few minutes. Now add 3-4 ml of conc.HNO_3 and evaporate to dryness. The oxalate is eliminated as CO_2.

$$C_2O_4^{2-} \rightarrow 2\,CO_2 \qquad \text{.....(P.8)}$$

To the residue add conc. HCl and dilute with water to prepare original solution.

(ii) **Tartarate,** $C_4H_4\,O_6^{2-}$: Take 2g of mixture is a porcelain dish and heat for few minutes. Add few drops of conc. HNO_3 and evaporate to dryness.

$$H_2C_4H_4O_6 \rightarrow CO + 2C + CO_2 + 3H_2O \qquad \text{.....(P.9)}$$

To the residue add conc. HCl and dilute with water to make original solution. This solution is to be filtered because carbon also there. The filtrate is the original solution.

(iii) **Phosphate,** PO_4^{3-} : All phosphates are soluble in acid medium. So before going to III group analysis, the phosphate is to be eliminated. After the second group, the solution is boiled to remove H_2S. To about 10 ml of solution add 1 g of solid NH_4Cl, then add zirconyl nitrate till the precipitation is complete. Filter the precipitation of zirconyl phosphate and use the filtrate for analysis.

$$ZrO\,(NO_3)_2 + Na_2HPO_4 \rightarrow ZrOHPO_4 + 2NaNO_3 \qquad \text{.....(P.10)}$$

(iv) **Chromate**, CrO_4^{2-} : Take the mixture in a boiling test tube and add few ml of conc. HCl and boil the mixture until the solution becomes green *i.e.,* chromate is converted to cation chromium.

Hint: Though chromium cation is not there in mixture, Cr^{3+} come in third group. So in the report if chromate is given, the cations other than Cr^{3+} is to be given.

(v) **Fluoride**, F – Less commonly given in the analysis. To the mixture add conc. HCl and evaporate to dryness.

$$NaF + HCl \rightarrow HF + NaCl \qquad .....(P.11)$$

(vi) Borate, $BO_3^{\ 3-}$ - Same method as above

In the case of V and VI the residue in dissolved is conc. HCl and dilute with water to make original solution.

4.4 Tests for Cations (Basic Radicals)

Some of the metals or ions are present in nature abundantly like Al, Fe, Ni, Pb etc., and they are called as common cat ions and some are present in very small quantities like U, Th, Mo, Ti, w etc., and they are called as less common cat ions.

4.4.1 Classification of Common Cations

The cations are classified into different groups. They are:

Group I – Lead (II), Mercury (I) and Silver.

Group reagent – dil. HCl (2M)

Precipitated as – $PbCl_2$, white

Hg_2Cl_2, white

AgCl, white

Group II – Mercury (II), Lead (IV), Bismuth (III), Copper (II), Cadmium (II), Arsenic (III and V) Antimony (III and V), Tin (II and IV).

Group reagent – dil. HCl and H_2S gas.

Precipitated as – HgS- black, PbS- black

Bi_2S_3 – brown, CuS – Chocolate brown

CdS – yellow, As_2S_3 – yellow, As_2S_5 – yellow

Sb_2S_3 – orange, Sb_2S_5 – orange,

SnS-brown, SnS_2 – yellow.

The cations of this group are subdivided into II A (copper group) and II B (arsenic group). These can be separated from one another by adding ammonium polysulphide. II A group precipitates are insoluble while II B precipitates are soluble. By filtering they can be separated.

II A cations are – Hg (II), Pb (II), Bi (III), Cu (II), Cd (II)

II B cations are – As (III & V), Sb (III & V), Sn (II & IV).

Hint: *All the precipitates are coloured and from the colour, one can guess the probable cation.*

Group III – Iron (III), Aluminum (III) Chromium (III).

Group reagent – solid ammonium chloride and ammonia solution.

Precipitated as – $Fe(OH)_3$ – reddish brown gelatinous precipitate

$Al(OH)_3$ – white gelatinous precipitate

$Cr(OH)_3$ – bluish green gelatinous precipitate.

Group IV – Nickel (II), Cobalt (II), Zinc (II), Manganese (II)

Group reagent – Solid ammonium chloride, ammonia solution and then pass H_2 S gas.

Precipitated as – NiS – black CoS – black, ZnS – white, MnS – buff (flesh colour)

Group V – Barium (II), Strontium (II) and Calcium (II)

Group reagent – Solid ammonium chloride, ammonia solution and 1M ammonium carbonate solution.

Precipitated as – $BaCO_3$ – white precipitate, $SrCO_3$ – white precipitate, $CaCO_3$ – white precipitate

Group VI – Magnesium (II), Sodium (1), Potassium (1), Ammonium (1)

Group reagent – No specific group reagent. To be tested separately.

Hint: *The Na^+ generally not given in the mixture because it is a common cation.*

4.4.2 Separation of Cations into Groups

Prepare the original solution for the given sample (mixture) as described in 4.2. The prepared original solution after elimination of interfering anions, separate the cations into groups by following the Table P.13

To the solution add a few drops of 2M HCl in cold solution. If a precipitation forms, add 2M HCl further till no further precipitation takes place and filter the solution.

Table P.13

<table>
<tr><th>Residue</th><th colspan="5">Filtrate – Dilute filtrate to 0.3M HCl and add 1 ml 3% H_2O_2. Heat nearly to boil and pass H_2S gas and filter</th></tr>
<tr><td rowspan="4">If precipitation formed group I cation is present</td><td rowspan="4">Residue If precipitate formed group II cation is present
(Eliminate group II cations by repeating the above process)</td><td colspan="4">Filtrate – Boil till all H_2 S is to be removed. Add 3-4 ml of conc. HNO_3 to oxidise Fe^{2+} to Fe^{3+} and boil to remove organic acid. Now add solid NH_4Cl and 2M NH_3 solution. Filter (If phosphate anion is there it is eliminated before adding reagent).</td></tr>
<tr><td rowspan="3">Residue If precipitate formed group III cation is present.
(Repeat the process till all the group III cation eliminated)</td><td colspan="3">Filtrate – Add 2-3 ml of 2M NH_4OH and pass H_2S gas. Filter</td></tr>
<tr><td rowspan="2">Residue If precipitate formed group IV cation is present.
(Repeat the process till group IV cation is eliminated)</td><td colspan="2">Filtrate – Boil off H_2S gas, then add NH_4Cl, NH_4 OH and, 1M (NH_4) CO_3. Heat the solution to 50-60 °C and filter.</td></tr>
<tr><td>Residue If precipitate formed group V cation is present</td><td>Filtrate contains group VI cations if present</td></tr>
</table>

After knowing from Table P.13, which group cation is present, then in particular group which cation is present in that group by following separation and identification of groups from the given below Tables.

4.4.3 Separation and Identification of Group I Cations

The precipitation may contain $PbCl_2$, Hg_2Cl_2 and $AgCl_2$. Wash the precipitation on filter with 2M HCl and cold water. Transfer the precipitation to a test tube and boil with 5-10 ml of water. Filter the hot solution and follow Table P.14.

Table P.14

<table>
<tr>
<td colspan="2">Residue
May contain Hg_2Cl_2 and AgCl.
Pour 3-4 ml warm dil. NH_3 solution over the precipitate and collect the filtrate.</td>
<td rowspan="2">Filtrate
May contain $PbCl_2$. To one portion add 0.1M $K_2Cr_2O_7$ solution. Yellow precipitate of $PbCrO_4$ which is insoluble in dil. acetic acid.
To one more portion add 0.1 M KI. Yellow precipitate soluble in hot water.
To another portion add 1M H_2SO_4. White precipitate of $PbSO_4$ soluble in ammonium acetate.
Lead is confirmed</td>
</tr>
<tr>
<td>Residue
If black consists of Hg (NH_2) Cl
Mercury is confirmed</td>
<td>Filtrate
May contain $[Ag(NH_3)_2]^+$
1. Acidify one part with 2M HCl. White precipitate of AgCl formed.
2. Add few drops of 0.1M KI solution. Yellow precipitate of AgI formed
Silver is confirmed</td>
</tr>
</table>

4.4.4 Separation and Identification of Group II Cations

The first group cation if present, remove by precipitation from the original solution completely.

Then to this solution **not fresh original solution** add the group two reagent i.e. dil. HCl and H_2S gas.

If precipitate is formed, it may consists sulphides of group II A cations (HgS, PbS, Bi_2S_3, CuS and CdS) and of IIB cations (As_2S_3, As_2S_5, Sb_2S_3, Sb_2S_5, SnS and SnS_2). Transfer the precipitation to a porcelain dish, add about 5 ml of yellow ammonium sulphide, $(NH_4)_2S_x$ solution, heat to 50 -60° C and filter.

Residue–May contains IIA group sulphides. Wash the precipitate with dil. ammonium sulphide and then with 2% NH_4NO_3 solution. Reject all washings and proceed for II A group analysis (Table P.15)

Filtrate–May contain thiosalts of II B group- $(NH_4)AsS_4$, $(NH_4)_2SbS_4$ and $(NH_4)_2SnS_3$. Just acidify the precipitate with conc. HCl and warm. A yellow or orange precipitate indicates presence of II B and proceed for analysis (Table P.16)

(a) *Separation of Group* II *A cations*-Transfer the precipitate to a beaker, add 5-10 ml HNO_3, boil gently for 2-3 minutes, filter and wash with a little water.

Table P.15

Residue	**Filtrate**-may contain nitrates of Pb, Bi, Cu, Cd. Add 1M H_2SO_4 and filter		
May contain HgS. Dissolve in NaOCl + HCl. Boil of excess Cl_2. white precipitate turning grey or black. **Hg^{2+} present**	**Residue** White precipitate of $PbSO_4$. Dissolve the precipitate in ammonium acetate. Add 2M acetic acid and 0.1M K_2CrO_4. Yellow precipitate of $PbCrO_4$. **Pb^{4+} is present**	**Filtrate**- May contain nitrates and sulphates of Bi, Cn and Cd. Add conc. NH_4OH in excess and filter. **Residue** White precipitate of Bi $(OH)_3$ formed. Dissolve in 2M HCl and pour into Na_2 $[Sn(OH)_4]$. Black precipitate formed . **Bi^{3+} is present.**	**Filtrate** may contain [Cu $(NH_3)_4]^{2+}$ and [Cd $(NH_3)_4]^{2+}$. 1. If the complex is deep blue, copper is present. 2. To [Cu $(NH_3)_4]^{2+}$ add 2M acetic acid and K_4 [Fe $(CN)_6$]. Chocolate brown precipitate formed. **Cu^{2+} is present** 3. To [Cd $(NH_3)_4]^{2+}$ add 1M KCN in excess. Pass H_2S gas. Yellow precipitate of Cd is formed. **Cd^{2+} is present**

(b) Separation of II B cations – Transfer the filtrate obtained after adding yellow ammonium sulphide, to a conical flask, add 5-10 ml conc. HCl and boil gently for five minutes. Dilute with water and pass H_2S gas and filter.

TableP.16

Residue-	**Filtrate**-May contains sb^{3+} and Sn^{4+}. Boil to expel H_2S gas, Divide into two parts.
May contain As_2S_3, As_2S_5. Dissolve the precipitate in 2M NH_3 solution add 3-4 ml H_2O_2 (to oxidise arsenite to arsenate). Add $AgNO_3$ solution. Brownish red precipitate of $AgAsO_4$. **As is present**	1. To one part add 2M NH_3 and 1-2 g of oxalic acid (marking agent). Boil and pass H_2S gas. Orange red precipitation of Sb_2 S_3. **Sb is present** 2. To another part add 2 m NH_3 solution and clean iron wire (to reduce Sn^4 to sn^{2+}). Filter the solution into $HgCl_2$. white precipitate of Hg_2Cl_2 or grey precipitate of Hg. **Sn is present**

4.4.5 Separation and Identification of Group III Cations

After removing IInd group cations in the solution, then add IIIrd group reagent to the same solution (don't take fresh original solution). The precipitation may contain Fe $(OH)_3$, $Cr(OH)_3$ and $Al(OH)_3$ and proceed for analysis (Table P.17). Transfer the precipitate into a beaker, add 5 ml of 2M NaOH and 5 ml of H_2O_2. Boil gently for 2-3 minutes and filter.

Table P.17

Residue – Dissolve in 2M HCl and add 0.1M NH_4SCN solution. Deep red colouration. If added $[K_4Fe\ (CN)_6]$ solution gives premium blue precipitate. **Fe is present**	**Filtrate**- May contain Cr^{3+} and Al^{3+}. Divide into two parts. 1. To one part add acetic acid and 0.25 M lead acetate solution. Yellow precipitate of $PbCrO_4$ **Cr is present** 2. To another portion add 2M HCl and 2M NH_4OH until alkaline. Heat to boiling. White gelatinous precipitate of $Al\ (OH)_3$. **Al is present**

4.4.6 Separation and Identification of Group IV Cations

To the solution obtained after the analysis of group III cations, add IV group reagent. The precipitate may contain CoS, NiS, MnS and ZnS. Analysis by following the Table P.18.

Transfer the precipitate to a beaker, add 5 ml of water + 5 ml 2M HCl, stir well and filter.

Table P.18

<table>
<tr><td rowspan="2">Residue –May contain CoS and NiS. Dissolve the precipitate in a mixture of 1.5 ml 1M NaOCl + 0.5 ml 2M HCl. Then add 1 ml 2 M HCl and boil. Divide the solution into two parts.
1. To one part add 1 ml amyl alcohol +2 g NH_4SCN. Shake well. Amyl alcohol layer is blue.
Co is present
2. To another part add 2 ml NH_4Cl + 2M NH_3 solution. Now add excess dimethyl glyoxime reagent. Scarlet red precipitate
Ni is present</td><td colspan="2">Filtrate – May contain Zn and Mn. Boil off H_2S gas and add excess 2M NaOH + 1 ml 3% H_2O_2 and boil</td></tr>
<tr><td>Residue – Largely Mn $(OH)_2$. Dissolve the precipitation in 5 ml of 8 M HNO_3 + 3% H_2O_2 + 0.05 g $NaBiO_3$. Purple solution of permanganate formed.
Mn is present</td><td>Filtrate-contain $[Zn(OH)_4]^{2-}$
1. Add 2M acetic acid and pass H_2S.
White precipitate of 2ns.
2. Add 1M H_2SO_4 + 0.5 ml of 0.1M cobalt acetate solution +0.5 ml of $(NH_4)_2\ [Hg(SCN)_4]$. Pale blue precipitate.
Zn is present.</td></tr>
</table>

4.4.7 Separation and Identification of Group V Cations

To the remaining solution after the analysis of group IV, add group V reagent. The precipitate may contain $BaCO_3$, $SrCO_3$ and $CaCO_3$. Analysis by following Table P.19 Dissolve the precipitation in 5 ml hot 2M acetic acid and add 0.1M K_2CrO_4 and filter.

Table P.19

<table>
<tr>
<td rowspan="2">Residue – Yellow precipitate of $BaCrO_4$
Flame test - green
Ba is present</td>
<td colspan="2">Filtrate-May contains Sr^{2+} and Ca^{2+} Add 2 ml $(NH_4) SO_4$ +0.2g of sodium thiosulphate. Heat the solution and filter</td>
</tr>
<tr>
<td>Residue
$SrSO_4$ precipitate
Flame test–crimson
Sr is present</td>
<td>Filtrate-May contains Ca^{2+}. Add 2M acetic acid and $(NH_4)_2C_2O_4$ and warm. White precipitate of CaC_2O_4.Flame test–brick red.
Ca is present</td>
</tr>
</table>

4.4.8 Identification of Group VI Cations

The remaining solution after V group analysis may be tested for VI group cations – Mg^{2+}, Na^{+}, K^{+}, NH_4^+

(i) *Magnesium*:

(a) Add to the solution NH_4Cl + NH_4OH + Na_2HPO_4. White precipitation of magnesium ammonium phosphate.

(b) Add to the solution NH_4Cl + ammonical 8-hydroxy quinoline reagent and heat the solution. Pale yellow precipitate of magnesium oxinate.

Mg is present

(ii) Sodium- To the solution add little magnesium uranyl acetate

Reagent-yellow crystalline precipitate

Flame test-persistent yellow flame

Na^+ is present

(iii) *Potassium*

To the solution add sodium cobalt nitrate solution

Yellow precipitate of $K_3 [Co (NO_2)_6]$

Flame test – violet or lilac red flame

K^+ is present

(iv) *Ammonium*

To the solution add Nessler's reagent (alkaline solution of potassium tetra iodo mercurate (II))

Brown precipitation or brown/ yellow colouration of basic mercury (II) amido iodide.

NH_4^+ is present

Finally give report as

The given mixture contains

1. Anions - X^-, Y^- etc.
2. Cations – A^+, B^{2+} etc.

5. Inorganic Quantitative Analysis

This is again divided to volumetry (titrimetry) and gravimetry for macroamounts. The theory and principles are discussed in part I. Here only discussed the experimental part. The titrations are classified as:

- Acid-base titrations
- Redox titrations
- Complexometric titrations
- Precipitation titrations

5.1 Acid-Base Titrations

Experiments

5.1 (a) Preparation and Standardisation of 0.1M HCl

Aim: To standardise 0.1M HCl using standard solution of primary standard anhydrous sodium carbonate.

Apparatus – Burette, 25 ml pipette, conical flask, volumetric flask, beaker, measuring jar, glazed white tile etc.

Reagents:

0.1M HCl – about 2.4 ml of pure HCl acid is taken in a 250 ml volumetric flask and dilute to 250 ml to get 0.1M HCl.

0.1M sodium carbonate- Analytical reagent grade (99.9%) sodium carbonate is dehydrated at 260°-270°C for half an hour and cool in a dessicator. Prepare 0.1M by weighing 2.6498 g of dehydrated Na_2CO_3 and dissolve in water and dilute to 250 ml in a volumetric flask.

Methyl orange indicator: Dissolve 0.18 g in 100 ml distilled water.

Principle and Reaction: It is an acid base titration of strong acid Vs weak base. 0.1M Na_2CO_3 is used as a base and is titrated with 0.1M HCl using methyl orange indicator. The colour change is orange to red at a P^H of about 4.

$$Na_2CO_3 + 2HCl \rightarrow 2\,NaCl + H_2O + CO_2 \qquad .....(P.12)$$

***Procedure*:** pipette out 25 ml of standard 0.1M Na_2CO_3 into a conical flask. Add 2-3 drops of methyl orange indicator. Fill the burette with 0.1M HCl and titrate with Na_2CO_3 solution by adding drop wise and by continuously stirring the conical flask till the orange colour changes to red. Note the volume of HCl.

Burette Readings

S.No	Volume of Na_2CO_3 in ml	Burette Readings		Volume of HCl
		Initial	Final	
1	25	0	24.8	24.8
2	25	24.8	49.6	24.8
3	25	0	24.9	24.9

The volume of HCl needed for neutralizing 25 ml of Na_2CO_3 is equal to 24.8.

Calculations

The molarity of HCl is

$$V_1M_1 = V_2\,M_2$$

$$24.8 \times M_1 = 0.1 \times 25$$

$$M_1 = 0.1 \times 25/\,24.8$$

$$M_1 = 0.1008.$$

0.053 g of Na_2CO_3= 1 ml 1M HCl.

Report: The molarity of HCl is 0.1008

Note: In writing Record follow the pattern given above for all the experiments.

5.1 (b) Preparation and Standardisation of 0.1N Sulphuric Acid

Aim: To standardise 0.1N H_2SO_4 solution using standard solution of anhydrous sodium carbonate.

Reagents:

0.1N H_2SO_4 solution: About 0.75 ml of pure conc. H_2SO_4 acid is taken in a 250 ml volumetric flask and dilute slowly with distilled water to get 0.1N H_2SO_4.

0.1N Na_2CO_3 solution: Prepare 0.1N Na_2CO_3 by dissolving exactly 1.3249g of pure Na_2CO_3 in a 250 ml volumetric flask with distilled water.

Methyl orange indicator: dissolve 0.18 g in 100 ml distilled water.

Sometimes methyl orange (0.18) and indigo carmine (0.25g) in 100 ml also used as indicator for better results.

***Principle and Reaction*:** It is an acid base titration of strong acid Vs weak base O IM Na_2CO_3 is used as base and titrate with Standard 0.1N H_2SO_4 using methyl orange (orange to red) or methyl orange + indigo carmine (green to magenta) as on indicator.

$$Na_2CO_3 + H_2SO_4 \rightarrow Na_2SO_4 + H_2O + CO_2 \qquad \text{.... (P.13)}$$

Procedure: Pipette out 25 ml 0.1N Na_2CO_3 solution into a conical flask. Add 2-3 drops of methyl orange or mixed indicator. Rinse the burette into 0.1N H_2SO_4 and fill the burette with 0.1N H_2SO_4 and adjust to zero and titrate by adding drop wise and by continuously swirling the contents of the conical flask till the orange colour changes to red or faintly pink. If mixed indicator, the colour changes from green to magenta (natural gray).Tabulate burette readings.

Calculations:

The volume of H_2SO_4 needed for neutralising 25 ml of Na_2CO_3 is supposed to be X ml:

Then the normality of H_2SO_4 is

$$V_1 N_1 = V_2 N_2$$

$$X \times N_1 = 25 \text{ X } 0.1$$

$$N_1 = \frac{25 \text{ X } 0.1}{\text{X ml}}$$

$$0.053 \text{ g } Na_2CO_3 \equiv 1 \text{ ml of } 0.5N\ H_2SO_4.$$

$$9.106 \text{ g } Na_2CO_3 \equiv 1 \text{ ml of } 1.0M\ H_2SO_4$$

5.1 (c) Preparation and Standardization of 0.1N Sodium Hydroxide

Aim: To standardise in 0.1N NaOH using either standard potassium hydrogen phthalate (procedure A) or standard oxalic acid (procedure B).

Reagents:

0.1N NaOH solution: about 1g of NaOH is weighed roughly on watch glass (because it is highly hygroscopic) and dissolved in distilled water and pour into the 250 ml volumetric flask and dilute

0.1N potassium hydrogen phthalate (eq. wt 204.22): Weigh out accurately about 5.1g of A.R product and dissolve it in distilled water making it up to 250 ml in a volumetric flask.

0.1N oxalic acid (eq. wt. 63.034) - weigh out accurately 1.575g of A.R oxalic acid into a 250 ml volumetric flask and make it with distilled water.

Indicator: Phenolphthalein indicator solution

***Principle and Reaction*:** Both procedures are acid base titration of strong base Vs weak acid. Either standard potassium hydrogen phthalate or oxalic acid is titrated with NaOH solution using phenolphthalein indicator till the end point i.e., colorless to pink colour.

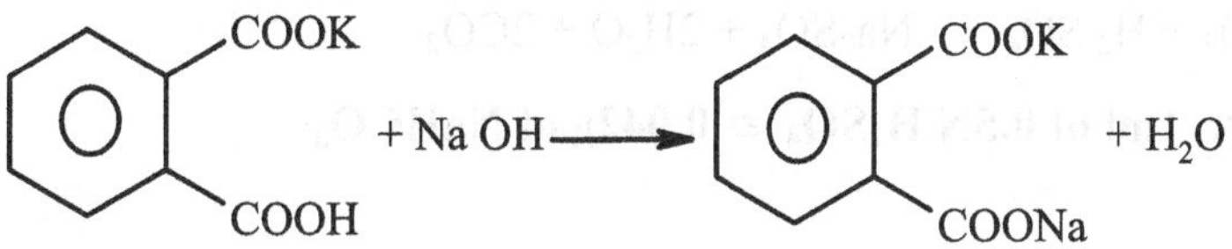

..... (P.14)

$$H_2C_2O_4 + 2\,NaOH \rightarrow Na_2C_2O_4 + 2\,H_2O \qquad .....(P.15)$$

Procedure A: Pipette out 25 ml of 0.1N potassium hydrogen Phthalate solution into a conical flask and add 2-3 drops of phenolphthalein indicator solution. Taken in a burette sodium hydroxide solution and titrate till the permanent pink colour is obtained. Repeat the titration till the concordant values are obtained.

0.2042g of K-H phthalate ≡ 1 ml of 1M NaOH.

Procedure B: Pipette out 25 ml 0.1N oxalic acid solution into a conical flask and add 2-3 drops of phenolphthalein indicator and titrate with 0.1N NaOH solution from the burette till the permanent pink colour is obtained. Tabulate burette readings.

Calculations:

. Calculate the normality of NaOH by using

$$V_1 N_1 = V_2 N_2$$

Where V_1 = Volume of NaOH say X

N_1 = Normality of NaOH.

V_2 = Volume of Pot. hydrogen phthalate or oxalic acid (25ml)

N_2 = Normality of 0.1N

$$N_1 = \frac{25 \times 0.1}{X}$$

5.1 (d) Assay of Sodium Bicarbonate

Aim: To carry out the assay of the given sample of sodium bicarbonate

Reagents:

Sodium bicarbonate, 0.5N H_2SO_4, 0.5N Na_2CO_3 solution, Methyl orange Indicator

Principle and Reaction: This is an acid-base titration (strong acid weak base). First standardise the sulphuric acid by titrating against Na_2CO_3. Then by using standard H_2SO_4 carry the assay of sodium bicarbonate by titration H_2SO_4 against $NaHCO_3$ using methyl orange indicator.

$$2NaHCO_3 + H_2SO_4 \rightarrow Na_2SO_4 + 2H_2O + 2CO_2 \quad .....(P.16)$$

Equivalent factor: 1ml of 0.5N H_2SO_4 ≡ 0.042g of $NaHCO_3$

Procedure:

Standardisation of H_2SO_4: Take 25 ml portion of 0.5N Na_2CO_3 in a conical flask and add 2-3 drops of methyl orange indicator. Pale yellow or orange in colour. The H_2SO_4 is taken in a burette and titrate against Na_2CO_3 solution till the colour changes to red. Then calculate exact normality of H_2SO_4.

***Assay of* $NaHCO_3$:** Accurately weigh 2.6 g of $NaHCO_3$ and transfer to 250 ml volumetric flask and add distilled water and make it up to mark. Pipette out 25 ml portion and add 2-3 drops of. Methyl orange indicator. Another way is to dissolve 1 g of $NaHCO_3$ in 25 ml of water in conical flask and titrate. The solution turns yellow titrate this solution with 0.5N H_2SO_4 from the burette till the solution turns red. Note the burette readings.

Calculations: I[st] method

Calculate the normality to sodium bicarbonate by using

$$V_1 N_1 = V_2 N_2$$

Where V_1 = volume of $NaHCO_3$, 25 ml

N_1 = Normality of $NaHCO_3$

V_2 = Volume of H_2SO_4, X ml

$N_2 = 0.5$

$$N_1 = \frac{X \times 0.5}{25} = Y$$

$$\text{Amount of } NaHCO_3 = \frac{\text{Normality} \times \text{Eq.Wt.}}{4}$$

$$\text{Present is 250 ml} \quad = \frac{Y \times 107}{4} = Z$$

Percentage purity

2.6 g of $NaHCO_3$ contains Z g of $NaCO_3$

100g of $NaHCO_3$?

$$\% \text{ Purity of } NaHCO_3 = \frac{Z \times 100}{2.6 \ \text{wt. of the sample}}$$

IInd Method:

Equivalent factor is

1 ml of 0.5N $H_2SO_4 \equiv$ 0.042 g of $NaHCO_3$

$$\% \text{ Purity of } NaHCO_3 = \frac{\text{vol. } H_2SO_4 \times \text{Equivalent factor} \times 100 \times \text{exact N of } H_2SO_4}{\text{Wt. of Na } HCO_3 \text{ in g} \times \text{N of } H_2SO_4 \text{ expected}}$$

Note: Actual normality is the normality given in procedure and expected normality is obtained in standardisation.

5.1 (e) Assay of Boric Acid H_3BO_3

Aim: To carry out assay of the given sample of Boric acid.

Reagents:

Boric acid, Glycerol, Phenolphthalein indicator solution, 0.1N NaOH solution and 0.1N potassium hydrogen phthalate.

Principle and Reaction:

This is an acid-base titration. First standardise 0.1N NaOH by titration against 0.1N standard potassium hydrogen phthalate. The boric acid is a very weak acid so it cannot be titrated directly against NaOH. First it is treated with glycerol (glycerol boric acid) or better mannitol (mannite) to make strong acid and can be titrated with NaOH

H [boric acid complex] + NaOH $\rightarrow$ Na [boric acid complex] + H_2O .....(P.17)

Equivalent factor

1 ml 1N NaOH $\equiv$ 0.06184 g H_3BO_3.

Procedure:

Standardisation of 0.1N NaOH: Standardise 0.1N NaOH against standard KH phthalate using phenolphthalein indicator as described in Experiment C.

***Assay of Boric acid*:** weigh out accurately 1g of boric acid into conical flask and dissolve in 25 ml water and 50 ml of glycerol. Take the standard 0.1N NaOH in burette, titrate against boric acid + glycerol by using phenolphthalein indicator till the end point i.e., colourless to permanent pink. Note the burette readings and repeat the titration till the concordant values obtained.

***Calculation*:**

$$\% \text{ Purity of } H_3BO_3 = \frac{\text{vol. of NaOH} \times 0.06183 \times 100 \times \text{actual normality of NaOH}}{\text{Wt.g } H_3BO_3 \text{ in g} \times \text{expected normality of NaOH}}$$

5.1 (f) Assay of Borax $Na_2B_4O_7$. $10H_2O$

Aim: To carry out the assay of the given sample of borax.

Reagent: Borax, 0.5N HCl, Methyl red and Methyl orange indicator solution. 0.5N Na_2CO_3 solution.

Principle and Reaction: Borax is alkaline. It is a titration of weak base vs. strong acid. First standardise HCl by titrating with standard sodium carbonate. Then the borax solution is to be titrated with standard HCl using methyl red indicator.

$$Na_2B_4O_7\ 10\ H_2O + 2\ HCl \rightarrow 4\ H_3BO_3 + 2\ NaCl + 5\ H_2O \qquad .....(P.18)$$

Equivalent factor

$$1 \text{ m } 1N\ HCl \equiv 0.10065 \text{ g. } Na_2B_4O_7$$

Procedure:

Standardisation of HCl: Standardise by following the procedure described under a.

***Assay of borax*:** Weigh out accurately 1g of borax into a conical flask and dissolve it in 25 ml distilled water. Add 2-3 drops of methyl red indicator. The colour of the solution is yellow.

Titrate with standard HCl from the burette till the yellow colour turns red. Repeat the experiment and record burette readings.

5.1 (g) Assay of Ammonium Chloride NH_4Cl (Formal Titration)

Aim: To carry out the assay of the given sample of ammonium chloride.

Reagents: NH_4Cl, 0.1N NaOH, phenolphthalein indicator, Formaldehyde solution, 0.1N potassium hydrogen phthalate.

***Principle and Reaction*:** First standardise the 0.1N NaOH. To the NH_4Cl if formaldehyde is added, HCl is liberated which is then titrated with 0.1N NaOH.

$$4\ NH_4Cl + 6\ HCHO \rightarrow (CH_2)_6\ H_4 + 4\ HCl + 6\ H_2O \qquad .....(P.19)$$

Equivalent factor

$$1\ ml\ 1N\ NaOH \equiv 0.05349\ g\ NH_4Cl.$$

Procedure:

Standardisation of 0.1N NaOH: Standardise 0.1N NaOH by following the procedure described in the experiment (C). Assay of NH_4Cl – weigh out accurately 1 g of NH_4Cl into conical flask and dissolve it in a mixture of 20 ml distilled water + 5 ml formaldehyde. Titrate the content against standard 0.1N NaOH using phenolphthalein indicator. Repeat the experiment and record the readings.

Calculation:

$$\%\ \text{purity of } NH_4Cl = \frac{\text{vol. of NaOH} \times \text{Eq. Factor} \times 100 \times \text{Actual N of NaOH}}{\text{Wt. of } NH_4Cl \times \text{expected N of NaOH}}$$

5.1(h) Assay of Zinc oxide ZnO

Aim: To carry out the assay of the given samples of zinc oxide.

Reagents:

Zinc oxide, Standard HCl, Standard NaOH.

***Principle and Reaction*:**

Zinc oxide is basic, so it can be assayed by acidimetry – alkalimetry. Zinc oxide reacts with acid slowly. So back titration method is useful

$$ZnO + 2HCl \rightarrow ZnCl_2 + H_2O \qquad .....(P.20)$$

$$HCl + NaOH \rightarrow NaCl + H_2O \qquad .....(P.21)$$

Eq. factor: 1 ml of 1N HCl $\equiv$ 0.04069 g of ZnO

Procedure:

Prepare 1N HCl and standardise

Prepare 1N NaOH and standardise

Weigh out accurately 1.0 g of ZnO to a conical flask and dissolve with 2.5 g NH_4Cl in 50 ml of 1N HCl by heating gently. When the solution is completely cooled, add few drops of methyl orange indicator and titrate the excess HCl with 1N NaOH solution till the colour changes from red to yellow.

Calculation:

Calculate the volume of HCl consumed by the sample using $N_1V_1 = N_2V_2$

$$\%\text{Purity of ZnO} = \text{(ii)} \frac{\text{Vol. of HCl} \times 0.04069 \times 100}{\text{Wt. of ZnO}}$$

5.2 Non–aqueous Titrations

5.2 (a) Assay of Sodium Benzoate, $C_7H_5O_2Na$

Aim: To carrying out the assay of the given sample of sodium benzoate.

Reagents:

0.1N Perchloric acid: Slowly add 2.2 ml of perchloric acid to 225 ml glacial acetic acid taken in a 250 ml volumetric flask with continuous mixing and add about 8 ml acetic anhydride and adjust the volume up to the mark with glacial acetic acid.

0.1 N standard potassium hydrogen phthalate

Crystal violet indicator 0.5 % in glacial acetic acid.

Sodium benzoate

0.2 % 1 – naphtholbenzoin indicator 0.2 % in glacial acetic acid.

Principle and Reactions:

It is based on non-aqueous titration using standard perchloric acid

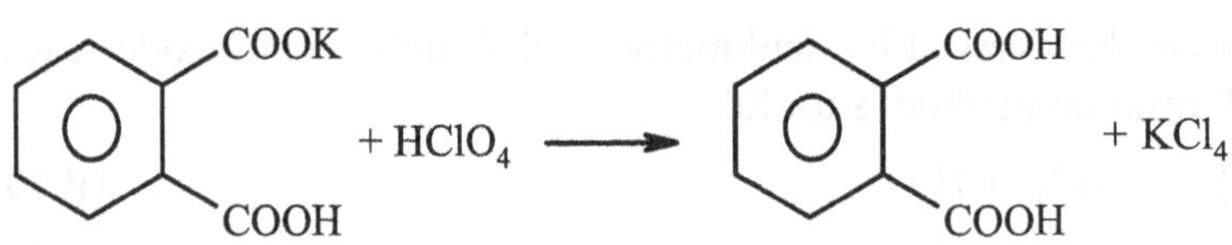

.....(P.22)

$$C_6H_5COONa + HClO_4 \rightarrow C_6H_5COOH + NaClO_4 \quad \text{.....(P.23)}$$

Equivalent factor

$$\text{1 ml of 1N } HClO_4 \equiv 0.1441 \text{ of g } C_7H_5O_2Na$$

Procedure:

Standardisation of 0.1N perchloric acid: Accurately weigh potassium hydrogen phthalate (0.5 g) into a conical flask and dissolve in 25 ml glacial acetic acid. Add few drops of crystal violet indicator and the colour of the solution is violet and titrates against perchloric acid till the colour turns blue-green, Note the reading and repeat the experiment and tabulate the burette readings.

Then calculate the normality of perchloric acid

Assay of sodium benzoate: weigh out 0.25 g of the given sample into a conical flask, dissolve anhydrous glacial acetic acid (if necessary warm) and titrate cool solution with standard 0.1N perchloric acid solution using 0.05 ml 1-naphtholbenzoin solution till the yellow solution turn to green. Repeat the experiment.

Calculations:

$$\% \text{ Purity of } C_7H_5O_2Na = \frac{\text{Vol. of HCl} \times \text{Eq. factor} \times 100 \times \text{Actual N of } HClO_4}{\text{Wt. of } C_7\ H_5\ O_2Na \times \text{expected N of } HClO_4}$$

5.2 (b) Assay of Sodium Acetate $CH_3COO\ Na.\ 3H_2O$

Aim: To carry out assay of the given sample of sodium acetate

Reagents:

0.1N perchloric acid, 0.1N Potassium hydrogen phthalate

Crystal violet indicator, given sample of sodium acetate, 1-naphthal benzein

Principle and reaction: It is based on non-aqueous titration using perchloric acid.

$$CH_3COONa + HClO_4 \rightarrow CH_3COOH + NaClO_4 \qquad \text{.....(P.24)}$$

Equivalent factor

$$1\text{ml of 1N } HClO_4 \equiv 0.3161 \text{ of } CH_3COONa.\ 3H_2O.$$

Procedure:

Standardisation of 0.1N perchloric acid: Follow the same procedure as described in experiment. (5.2 a)

Assay of Sodium Acetate: Accurately weigh out 0.25 g of the given sample of sodium acetate into a conical flask and dissolve in 25 ml of glacial acetic acid.

Add 1 drop of 1-naptha benzein and titrate against standard perchloric acid till the solution colour term from yellow to green.

Calculations:

$$\% \text{ Purity of } CH_3COONa = \frac{\text{Vol. of } HClO_4 \times \text{Eq. factor} \times 100 \times \text{Actual N of } HClO_4}{\text{Wt. of } CH_3\ COO\ Na \times \text{expected N of } HClO_4}$$

5.2 (c) Assay of Nitrazepam, $C_{15}H_{11}N_3O_3$

Aim: To carry out the assay of the given sample of Nitrazepam

Reagents: 0.1N perchloric acid, Nile blue A indicator (1% in glacial acetic acid), 0.1N KH phthalate, phenolphthalein indicator, Nitrazepam, acetic anhydride.

Principle and Reaction:

It is based on non aqueous titration using standard perchloric acid.

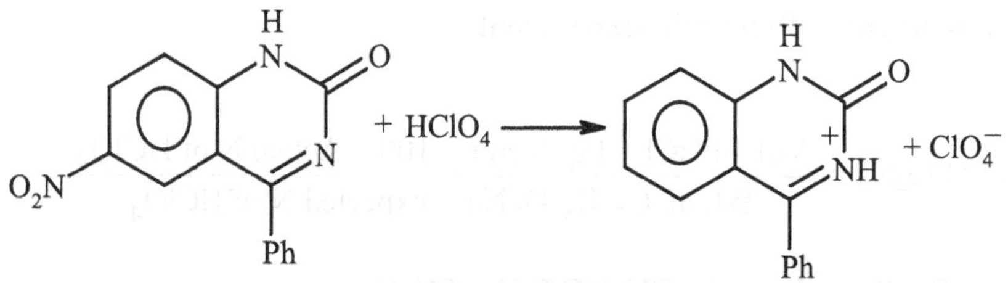

.....(P.25)

Eq. factor: 1ml 1N $HClO_4 \equiv$ 0.281 g of $C_{15}H_{11}N_3O_3$

Procedure:

Standardisation of perchloric acid: Follow the same procedure as described in experiment (5.2. a)

Assay of Nitrazepam: weigh out accurately by 0.25 g into a conical flask and dissolve in 25 ml acetic anhydride. Add 1-2 drops of Nile blue A (1% in glacial acetic acid) and titrate with perchloric acid till the colour change from blue to blue-green. Record the burette reading and repeat the experiment so as to get concordant values.

Calculation:

$$\% \text{ Purity of Nitrazepam} = \frac{\text{Vol. of } HClO_4 \times \text{Eq. factor} \times 100 \times \text{Actual N of } HClO_4}{\text{Wt. Nitrazepam} \times \text{expected } \mu \text{ of } HClO_4}$$

Note: Same method can be used for the determination of compounds, in non-aqueous media and titration with $HClO_4$, like – Adrenaline, Chlordiazepoxide, Codeine, Diazepam, Ethionamide, Metronidazole, Pyrimethamine, Chlorihexidine Acetate etc.

The reaction of Adrenaline with **$HClO_4$:** Adrenaline as an example of hormone secreted from medulla of adrenal gland

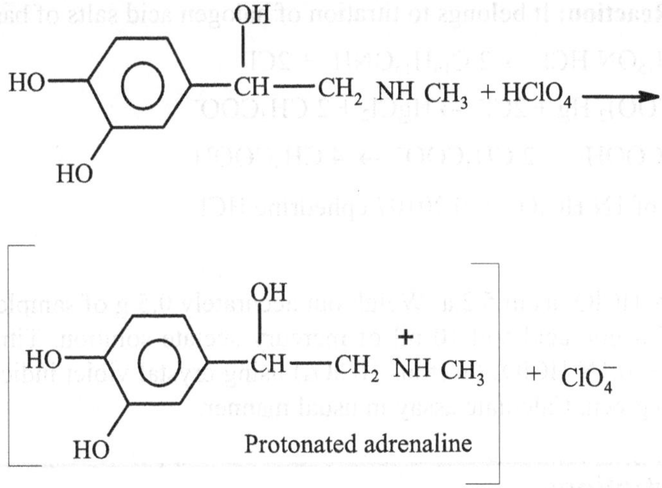

..... (P.26)

Eq. factor: 1 ml of 1N $HClO_4$ ≡ 0.1832g of adrenaline

Procedure of assay is same as above

The reaction of Metronidazole with $HClO_4$: Metronidazole is an example of anti amoebic drug.

$$HClO_4 + CH_3COOH \rightarrow \underset{\text{omium ion}}{CH_3\ COOH_2^+} + ClO_4^-$$

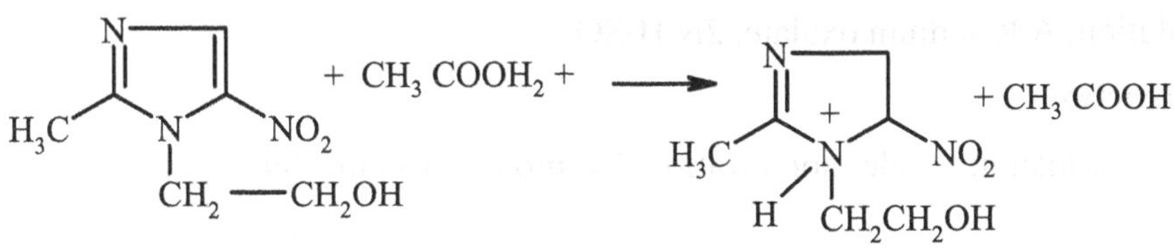

..... (P.27)

Eq. Factor:

1 ml of 1N $HClO_4$ ≡ 0.1712 g metronidazole

Procedure of assay is same as above but only 0.1N perchloric acid mixed with acetic acid.

5.2 (d) Assay of Ephedrine Hydrochloride $C_{10}H_{15}N\ HCl$

Note: The same method applicable to chlordiazepoxide HCl, chlorpromazine HCl, Lignocaine HCl.

Aim: To carry out assay of the sample of Ephedrine HCl

Principle and Reaction: It belongs to titration of halogen acid salts of bases under NAT

$$2\ C_{10}H_{15}ON\ HCl \rightarrow 2\ C_{10}H_{15}ONH^+ + 2Cl^- \quad .....(P.28)$$

$$(CH_3COO)_2\ Hg + 2Cl^- \rightarrow HgCl_2 + 2\ CH_3COO^- \quad .....(P.29)$$

$$2\ CH_3COOH_2^+ + 2\ CH_3COO^- \rightarrow 4\ CH_3COOH \quad .....(P.30)$$

Eq. factor: 1 ml of 1N $HClO_4$ ≡ 0.20107 ephedrine HCl

Procedure:

Standardise 0.1N $HClO_4$ as in 5.2.a. Weigh out accurately 0.5 g of sample and dissolve in 25 ml of glacial acetic acid and 10 ml of mercury acetate solution. Titrate the contents with acetic acid + 0.1N $HClO_4$ (acetous $HClO_4$) using crystal violet indicator till the blue colour change to green. Calculate assay in usual manner.

5.3 Redox Titrations

Theory of oxidation- reduction titrations, redox indicators were discussed in part I of this book. The experiments only discussed here.

5.3 (a) Preparation and Standardisation of 0.1N Potassium Permanganate $KMnO_4$ (Permanganometry)

Eq.Wt = 31.606 (Mol.Wt/5)

Aim: To prepare 0.1N $KMnO_4$ solution and standardize by titrating with sodium oxalate

Reagents:

0.1N $KMnO_4$ solution, A.R sodium oxalate, 2N H_2SO_4

Principle and Reactions:

This titration is the oxidation – reduction titration. **No indicator is needed.**

Procedure:

Preparation of potassium permanganate solution: Weigh about 3.2 g of A.R potassium permanganate, dissolve in a 200 ml of distilled water and dilute to 1000 ml in a beaker. Mix the solution thoroughly and keep it for two to three days, so that all organic matter in water will be oxidised and MnO_2 is completely settled. Filter the clear solution through G_4 crucible. Keep the solution in a glass-stoppered bottle.

Weigh out accurately 1.6 g of AR sodium oxalate (dried to 105 -110°C) into a 250 ml volumetric flask and make up to mark.

Fill the burette with prepared $KMnO_4$ solution. Pipette out 25 ml of sodium oxalate into conical flask and add 150 ml of 2N H_2SO_4. Titrate against $KMnO_4$ rapidly at room temperature until the first pink colour appears and allow the solution to standard until it becomes colourless. Warm the solution to 50-60°C and continue the titration to the permanent pink colour. *The end point is colourless to permanent pink colour*: Record the burette reading. Repeat the experiment till the concordant values obtained and the result are tabulated.

Calculation:

Normality of $Na_2C_2O_4$

Eq.Wt of $Na_2C_2O_4$ = Mol.Wt/2 = 134/2 = 67

1.675 in 250 ml is 0.1N

1.6 in 250 ml ?

So that normality of prepared $Na_2C_2O_4$ = 0.09552

Normality of $KMnO_4$, use the formula

$V_1N_1 = V_2N_2$

V_1 = 25 ml $Na_2C_2O_4$

N_2 = 0.09552 of $Na_2C_2O_4$

V_2 = X ml of $KMnO_4$, (burette reading)

$$N_2 = \frac{25 \times 0.09552}{X}$$

From equivalent factor can also calculated.

5.3 (b) Preparation and Standardisation of 0.1N Iodine I_2 (Iodometry)

Eq. Wt. = 126.9

Aim: To prepare 0.1N iodine solution and standardisation with arsenic trioxide

Reagents:

Iodine, arsenic trioxide, starch indicator.

Principle and Reaction:

It is based on oxidation- reduction titrations. I_2 act as oxidizing agent in an acid or in a neutral solution, it converts As_2O_3 to As_2O_5.

$$I_2 + 2e^- \rightarrow 2I^- \quad .....(P.31)$$

Eq. Wt. of AS_2O_3 = 197.8/ 4 = 49.46

Procedure:

Preparation of I_2 solution: Iodine solution can be prepared by dissolving the iodine in conc. KI solution.

Weigh out accurately 0.05 g of dried As_2O_3 into a 250 ml conical flask. Dissolve it in 10 ml 1N NaOH solution by warning if necessary. Dilute with 20 ml of water, add two drops of methyl orange solution and acidify by adding dil HCl until yellow colour is changed to pink. Add 1g of sodium bicarbonate, dilute with 25 ml water. Titrate with I_2 solution using starch as indicator till permanent blue colour is obtained. Repeat the experiment and record the result is a tabular form.

Burette readings table:

Calculation:

$$\text{Normality of iodine} = \frac{\text{Wt. of } As_2O_3 \times 1000}{49.46 \times \text{Volume of iodine solution}}$$

5.3 (c) Preparation and Standardisation of 0.1N Sodium Thiosulphate

$Na_2S_2O_3 . 5H_2O$ Mol Wt. 248.19.

Aim: preparation of 0.1N or N/10 sodium thiosulphate (hypo) solution and standardisation against potassium iodate.

Reagents:

Sodium thiosulphate, potassium iodate, starch indicator, 2N H_2SO_4, potassium iodate

Principle and Reactions:

This process involve oxidation-reduction reactions. Sodium thiosulphate is titrated against potassium iodate.

$$KIO_3 + 5KI + 3\,H_2SO_4 \rightarrow +\,6\,K_2SO_4 + 3\,H_2O\;3\,I_2 \quad \text{.....(P.32)}$$

$$I_2 + 2\,Na_2S_2O_3 \rightarrow Na_2S_4O_6 + 2\,NaI \quad \text{.....(P.33)}$$

Procedure:

Prepare 0.1N $Na_2S_2O_3$ solution by dissolving about 24.82 g of crystalled salt in hot water. Add to the solution 0.1 g Na_2CO_3, cool (for storing the solution) and transfer to 1 litre flask and make up to mark.

Weigh about 0.15 g of dry A.R potassium iodate and transfer to iodine flask. Dissolve 25 ml distilled water, add 2g of potassium iodide and 5 ml of 2N sulphuric acid. Titrate the liberated I_2 with sodium thiosulphate solution. When colour becomes pale yellow, add 2ml starch indicator and continue the titration till the colour changes from blue to colour less. Repeat the experiment, with similar quantities of KIO_3 and record the burette readings

Calculation:

$$\text{Normality of } Na_2S_2O_3 = \frac{\text{Wt of } KIO_3 \times 1000}{35.67 \times \text{Volume of } Na_2S_2O_3}$$

5.3 (d) Assay of Ferrous Sulphate $FeSO_4.7H_2O$

Mol. Wt. 278.01 Eq. Wt. = 278.01

Aim: To carry out the assay of the given samples of ferrous sulphate by titration with 0.1N $KMnO_4$

Apparatus: Titrimetry apparatus.

Reagents: Fe SO_4, 0.1N $KMnO_4$ solution, dil H_2SO_4

Principle and Reactions:

This determination depends on redox reaction. $FeSO_4$ is oxidized to $Fe_2(SO_4)_3$ by $KMnO_4$ in presence of dil. H_2SO_4. The end point is colourless to permanent pink colour.

$$2\,KMnO_4 + 3H_2SO_4 \rightarrow K_2SO_4 + 2\,MnSO_4 + 3\,H_2O + 5(O)$$

$$10\,FeSO_4 + 5\,H_2SO_4 + 5(O) \rightarrow 5\,Fe_2\,(SO_4)_3 + 5\,H_2O$$

$$2\,KMnO_4 + 10\,FeSO_4 + 8H_2\,SO_4 \rightarrow KSO_4 + 2\,MnSO_4 + 8H_2O + 5\,Fe_2\,(SO_4)_3 \quad .....(P.34)$$

***Equivalent factor*:** 1 ml of 1N $KMnO_4 \equiv$ 0.2789 g $FeSO_4$

Procedure:

Standardisation of 0.1N $KMnO_4$ – standardise the $KMnO_4$ solution by following the procedure described in experiment (a).

Assay of $FeSO_4$: weigh accurately 1 g of given $FeSO_4$.

Dissolve it in a conical flask with 40 ml of distilled water and 10 ml 2N H_2SO_4. Titrate the contents with standard $KMnO_4$ solution by continuous swirling until a permanent pink colour of permanganate colour persists. Repeat the experiment with two or three $FeSO_4$ samples. Record the burette readings and tabulate.

Burette readings table:

Calculations:

$$\%\text{ of } FeSO_4.7H_2O = \frac{\text{Vol. of } KMnO_4 \times N \times 0.0278 \times 100}{0.1 \times \text{Weight of sample}}$$

or

$$= \frac{\text{Vol. of } KMnO_4 \times \text{Eq. Factor} \times 100 \times \text{actual N of } KMnO_4}{\text{Wt. of } FeSO_4 \times \text{expected N of } KMnO_4}$$

5.3 (e) Assay of Ferrous Ammonium Salt (Mohr's Salt)

$FeSO_4 (NH_4)_2SO_4.6H_2O$

Mol. Wt: 392.16

Aim: To carry out the assay of Mohr's salt in the given sample.

Reagents: Mohr's salt, 0.1N $KMnO_4$, 2N H_2SO_4.

Principle ad Reactions: This is oxidation- reduction reaction. In presence of dil. H_2SO_4, $KMnO_4$ oxidises $FeSO_4$ of Mohr's salt to ferric sulphate. $KMnO_4$ acts as self indicator.

$$FeSO_4 (NH_4)_2SO_4.\ 6H_2O \rightarrow FeSO_4 + (NH_4)_2\ SO_4 + 6H_2O$$

$$2\ KMnO_4 + 3\ H_2SO_4 \rightarrow K_2SO_4 + 2\ MnSO_4 + 3H_2O + 5\ (O)$$

$$10\ FeSO_4 + 5\ H_2SO_4 + 5(O) \rightarrow 5\ Fe\ (SO_4)_3 + 5\ H_2O$$

$$2\ KMnO_4 + 5\ H_2SO_4 + 10\ FeSO_4 \rightarrow K_2SO_4 + 2\ MnSO_4 + 8\ H_2O + 5\ Fe_2\ (SO_4)_3$$

.....(P.35)

Equivalent factor: 1 ml of 1N $KMnO_4$ $\equiv$ 0.3921 g of Mohr's salt.

Procedure:

Standardisation of 0.1N (N/10) $KMnO_4$: Prepare and standardise $KMnO_4$ solution following procedure described in experiment 5.3(a).

Assay of Mohr's salt: Weigh accurately 1g of given Mohr's salt into a conical flask and dissolve it in 10 ml 1N H_2SO_4 and 40 ml distilled water. Titrate the contents with standard 0.1N $KMnO_4$ till the permanent pink colour obtained. Repeat the experiment with two more samples and note the burette readings.

Burette readings:

Calculations:

$$\%\ \text{of Mohr's salt} = \frac{\text{Vol. of } KMnO_4 \times N \times 0.0392 \times 100}{0.1 \times \text{Weight of sampel}}$$

or

$$= \frac{\text{Vol. of } KMnO_4 \times \text{Eq. factor} \times 100 \times \text{ actual N of } KMnO_4}{\text{Wt of sample} \times \text{ expected N of } KMnO_4}$$

5.3 (f) Assay of Copper Sulphate $CuSO_4.5H_2O$

Mol Wt. 249.7.

Aim: To carry out the assay of given samples of copper sulphate.

Reagents: Standard 0.1N $Na_2S_2O_3$ solution, copper sulphate, starch, potassium iodide.

Principle and Reaction

This is a redox titration (iodometry). In this $CuSO_4$ in acidic medium reacts with KI and forms cuprous iodide in two steps and releases free iodine which is titrated with hypo using starch as indicator.

$$2\ CuSO_4 + 4KI \rightarrow 2\ CuI_2 + 2\ K_2SO_4 \qquad \text{.....(P.36)}$$

$$2\ CuI_2 \rightarrow Cu_2I_2 + I_2 \qquad \text{.....(P.37)}$$

$$I_2 + 2\ Na_2S_2O_3 \rightarrow \underset{\text{sodium tetrathionate}}{Na_2S_4O_6} + 2\ Na\ I \qquad \text{......(P.38)}$$

Equivalent factor: 1 ml of 1N $Na_2S_2O_3 \equiv 0.2497$ g of $CuSO_4$

Procedure:

Standardisation of 0.1N hypo-Standardise the hypo solution by following the procedure described in experiment 5.3. (C).

Assay of Copper sulphate: Weigh accurately 1 g of the given sample of copper sulphate and transfer to iodine flask and dissolve in 50 ml of water. Now add 3g of KI ad 5 ml acetic acid. Titrate the liberated I_2 with standard hypo solution till the solution attains pale yellow colour is obtained and add 1 ml starch solution. Now continue the titration with vigorous shaking until the blue colour disappears. The end point is white or fleshy colour.

Note: starch indicator does not add in the beginning of the titration since starch adsorbs some liberated I_2 which gives errors. Repeat the experiment with two more $CuSO_4$ samples.

After the end points to check add 2 g of potassium thiocyanate, if blue colour appears, continue the titration. This ensures the completion of reaction and release of any adsorbed I_2. Tabulate the results.

Calculation:

$$\%\ \text{of}\ CuSO_4 = \frac{\text{Vol. of}\ Na_2S_2O_3 \times N \times 0.0249 \times 100}{0.1 \times \text{Wt. of Sample}}$$

$$= \frac{\text{Vol. of}\ KMnO_4 \times \text{Eq. factor} \times 100 \times \text{actual N of}\ Na_2S_2O_3}{\text{Wt. of}\ CuSO_4 \times \text{expected N of}\ Na_2S_2O_3}$$

5.3 (g) Assay of the Dimercaprol, $C_3H_8OS_2$

Aim: To carry out the assay of Dimercaprol iodometrically

Reagents: 0.1N iodine solution, Dimercaprol, 0.1N HCl, starch indicator.

Principle and Reaction: The determination depends on the oxidation of thiol group with iodine.

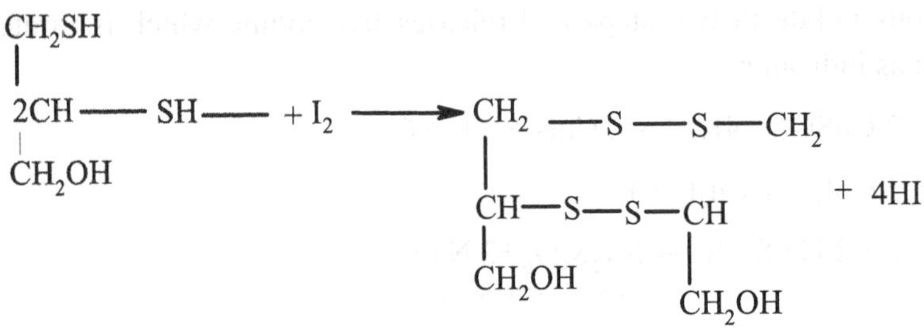

..... (P.39)

Eq. Factor: 1 ml of 0.5 M iodine≡ 0.00621 g of $C_3 H_8OS_2$

Procedure: standardise I_2 solution following the procedure of experiment (b). Weigh out accurately 0.2 g $C_3H_8OS_2$ into a conical flask, add 40 ml 0f.1N HCl and titrate with standard 0.1N Iodine solution using starch indicator. Repeat the experiment with two more samples.

Calculation:

$$\% \text{ of } C_3 H_8 OS_2 = = \frac{\text{Vol. of } I_2 \times \text{Eq. factor} \times 100 \times \text{actual N of } I_2}{\text{Wt. of } C_3H_8 \ OS_2 \times \text{expected N of } I_2}$$

5.3 (h) Assay of Ascorbic Acid, $C_6H_8O_6$

Mol. Wt. 176.1

Aim: To carry out the array of the given sample of ascorbic acid.

Reagents: Standard I_2 solution, ascorbic acid, 2N H_2SO_4, starch indicator.

Principle and Reactions: The ascorbic acid is oxidised to dihydroascorbic with acid I_2 solution.

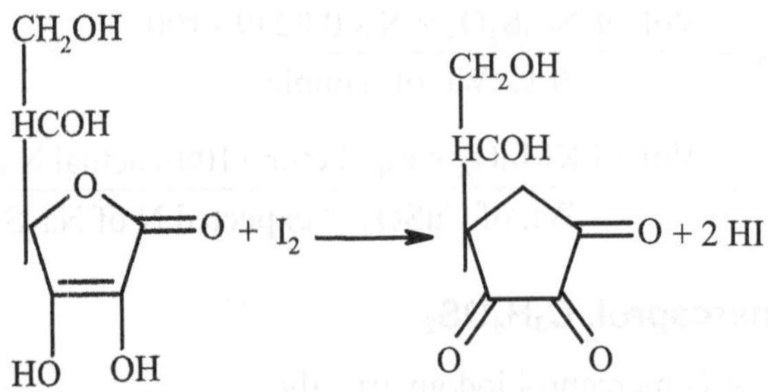

.....(P.40)

Equivalent factor: 1 ml of 0.5 M I_2 ≡ 0.0881 g $C_6H_8O_6$

Procedure:

Standardisation of iodine

Assay: Weigh out accurately 0.2 g of ascorbic acid into a conical flask. Dissolve in a mixture of 80 ml distilled water and 10 ml of 2N H_2SO_4. Titrate the contents of the flask with 0.1N iodine (follow experiment No.5.3. b) using starch solution as indicator.

5.3 (i) Determination of Iodine Value of Oils

Aim: To determine the iodine value of oils by using iodine monochloride.

Reagents: CCl_4, ICl , potassium iodide, 0.1N $Na_2S_2O_3$.

Principle: Iodine monochloride is attached to unsaturated oil to the double bonds. Excess ICl is converted to I_2 by the addition of KI. The liberated I_2 is titrated with 0.1N $Na_2S_2O_3$ in the usual way.

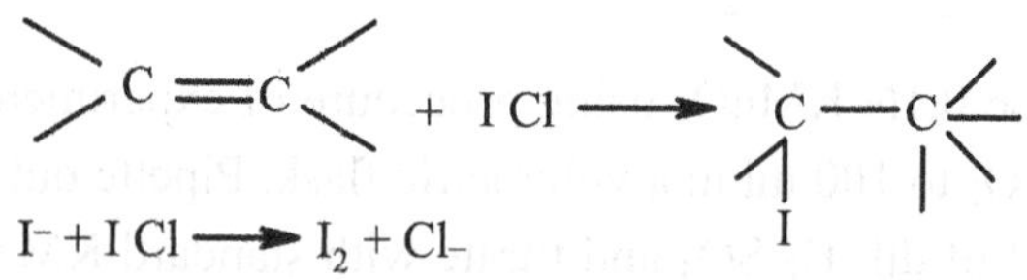

Eq. Factor $I + ICl \rightarrow I_2 + Cl^-$ .....(P.41)

1 ml 0.1N $Na_2S_2O_3 \equiv$ 0.0126 g of I_2

Procedure: Weight 20 ml of the given sample oil and place in a iodine flask. Add 10 ml of carbon tetrachloride to dissolve the oil. Run 20 ml iodine monochloride solution form automatic pipette and place the stopper of the flask and keep the flask in dark to complete the reaction at 20° C for 30 min. Now add 15 ml of 10% KI solute to the contents and back titrate the excess I_2 with standard 0.1 N $Na_2S_2O_3$ (follow expect 5.3. (c)) using starch indicator. Carry out the blank titration using all the reagents except oil sample.

Calculation

$$\text{Iodine Value} = \frac{(b-a)\times 0.01269\times 100\times \text{Normality}}{\text{Wt. of the Sample in g.}}$$

where a = test burette readings

b = blank burette reading, normality of standard $Na_2 S_2O_3$.

5.3 (j) Assay of Hydrogen Peroxide, H_2O_2

Mol. Wt. 17.01

Aim: To carry out Assay of the given $H_2 O_2$ sample.

Reagents: H_2O_2, 0.1N $KMnO_4$,

Principle and Reactions: It is redox titration. H_2O_2 and is titrated against $KMnO_4$. H_2O_2 is available in four strengths -10 volume H_2O_2, 20 volulme H_2O_2, 40 volume H_2O_2 and 100 volume H_2O_2. A 10 volume H_2O_2 when fully decomposed by heating gives 10 times its vol. of oxygen measured at N.T.P (0 °C and 760 mm)

A 10 volume H_2O_2 means 3% H_2O_2 solution

$$2H_2O_2 \rightarrow 2\,H_2O + O_2 \qquad \text{.......(P.42)}$$

$$2\,KMnO_4 + 6H + 5H_2O_2 \rightarrow 2\,Mn^{2+}\,8\,H_2O + 5\,(O) \qquad \text{.......(P.43)}$$

$$\therefore H_2O_2 \equiv 2\,K\,MnO_4 = 10\,e$$

$$\text{Eq. Wt of } H_2O_2 = \frac{\text{Mol.Wt}}{2} = \frac{34.02}{2} = 17.1$$

Procedure: Standardise 0.1N $KMnO_4$ using procedure of experiment 5.3 a. Dilute 10 ml of given sample of $H_2\,O_2$ to 100 ml in a volumetric flask. Pipette out 25 ml solution into a conical flask, add 5 ml of dil. $H_2\,SO_4$ and titrate with standard $KMnO_4$ until a faint pink colour persists. Repeat with two more 25 ml portions. Tabulate the results.

Calculations:

$$\%\text{ of } H_2O_2 = \frac{\text{Vol. of K } MnO_4 \times N \times 0.001701 \times 4 \times 100}{0.1 \times 10}$$

5.4 Complexometric Titrations

The theory of complexometric titrations and indicators are given in the part I of this Book.

The experiments are given below:

5.4 (a) Preparation and Standardisation of 0.1 M EDTA

Mol. Wt. 372.24

Analytical reagent grade available can be used as primary standard. Dry the EDTA salt at about 80° C for one hour. Dissolve exactly 37.224 of EDTA in 1 litre deionised or redistilled water to get 0.1MEDTA solution. To check further it can be standardise by titrating against Zn as in procedure given below.

Aim: To prepare and standardize 0.05M EDTA solution.

Reagents:

0.1M EDTA *solution*: Weigh out accurately 9.306 g of EDTA salt into a 250 ml volumetric flask and dissolve it in deionised water (redistilled water) and make up to mark.

***Zinc ion solution*:** Weigh out accurately 1.6345 g of A.R.granulate pure Zn or 2.0345 g of dry A.R zinc oxide into a beaker and dissolve in dil. HCl by warming, add few drops of bromine water. Boil to remove the excess bromine. Cool the solution and transfer the contents to a 250 ml volumetric flask and make up to mark with redistilled water.

Buffer solution (pH 10): To 17.5 g of NH_4Cl add 142 ml of conc. ammonia solution and dilute to 250 ml with distilled water.

Mordant Black II (Eriochrone Black T or Solochrome Black T) Indicator: Dissolve 0.2 g of the dye stuff in 15 ml of treithanolamine and 5 ml of absolute ethanol.

***Note*:** The redistilled water or deionised water is to be used in all EDTA titrations since no other ions are present in this.

***Principle and Reaction*:** It is a complexometric titration. This complex is quite stable at pH 10. The indicator forms complex with Zn to give wine red colour and at the end point Zn – EDTA complex formed and indicator is reduced so the colour is blue at 10 pH. The end point is wire red to blue

$$Zn^{2+} + EDTA \rightarrow Zn\,EDTA \qquad \text{.....(P.44)}$$

Equivalent factor: 1ml of 0.1M EDTA ≡ 0.638 mg of Zn

Procedure: Pipette out 25 ml of zinc solution into a conical flask, neutralize by 2N sodium hydroxide solution. Dilute with redistilled water and add excess buffer solution to dissolve the precipitate and to get pH 10. Add 2-3 drops of indicator and titrate against EDTA till the solution turns blue. Note the burette readings, repeat the experiment.

Calculations:

$$\text{Molarity of EDTA} = \frac{\text{wt of } z_n \text{ in 25 ml} \times 1000}{65.38 \times \text{vol.of EDTA}}$$

or by using

$$M_1V_2 = M_2V_2$$

5.4 (b) Assay of Calcium Lactate

Aim: To carry out the assay of the given sample of calcium lactate

Reagents :

0.1 M EDTA solution, calcium lactate, buffer solution (P^H-10), Eriochrome black –T indicator, 0.1M $MgSO_4$ solution.

Principle: The calcium lactate solution titrated with 0.1M EDTA solution. The volume consumed depends on the calcium lactate present in the sample

$$EDTA + Ca^{2+} \rightarrow Ca\,EDTA \qquad .....(P.45)$$

Equivalent factor: 0.1M EDTA = 0.02182 g of calcium lactate

Procedure:

Standardise 0.1M EDTA solution by following the procedure described in 5.4a.

Weigh accurately 0.2 g of calcium lactate and transfer to a conical flask. Dissolve it in 50 ml warm redistilled water. Add 5 ml of 0.05M $MgSO_4$ solution, 10 ml of pH 10 buffer solution and 2-3 drops of Eriochrome black T indicator. Titrate the contents with 0.1M EDTA solution till the colour of the solution turns from red to blue.

Titrate the blank solution without calcium lactate with EDTA solution. Repeat the titrations.

Calculation:

Actual volume of EDTA = vol. of EDTA with sample – vol. of blank

$$\%\text{ of calcium lactate} = \frac{\text{vol. of EDTA (actual)} \times \text{eq. factor} \times 100\ \text{actual M of EDTA}}{\text{wt. of calcium lactate} \times \text{expected M of EDTA}}$$

5.4 (c) Assay of Calcium Gluconate, $C_{12}H_{22}CaO_{14}.\ H_2O$

Aim: To carry out the assay of the given sample of calcium gluconate.

Reagents: calcium gluconate, 0.1M EDTA solution, pH 10 buffer solution, 0.05M $MgSO_4$ solution, Eriochrome Black T indicator solution.

Principle and Reaction: It is a complexometric titration. The vol. of EDTA consumed depends on the calcium gluconate present in the sample.

$$Ca^{2+} + EDTA \rightarrow Ca\,EDTA\ \text{complex} \qquad .....(P.46)$$

Equivalent factor

1 ml of 0.1M EDTA = 0.04484 of calcium gluconate

Procedure:

Standardise 0.1M EDTA by following the procedure described in 5.4.a. weigh accurately 0.8 g of calcium gluconate and dissolve in 100 ml redistilled water containing 5 ml of dil. HCl in a conical flask. Now add 10 ml of buffer solution, 0.05M Mg SO_4 solution and 2-3 drops of Erichrome black T. Titrate the contents of the flask with 0.1M EDTA till the colour changes from wine red to blue. Blank titration (without the sample) also perform Tabulate the results.

Calculation:

Actual vol. of EDTA = vol. of EDTA for sample – vol. EDTA for blank

$$\% \text{ of calcium gluconate} = \frac{\text{vol. of EDTA} \times \text{Eq factor} \times 100 \times \text{actual M of EDTA}}{\text{wt. of calcium gluconate} \times \text{expected M of EDTA}}$$

5.4 (d) Assay of Magnesium Sulphate, $MgSO_4 . 7H_2O$

Mol. Wt. 246.47

Aim: To carry out the assay of the given sample of magnesium sulphate

Reagents: 0.1M EDTA solution, $MgSO_4$, pH- 10 buffer solution, Mordent Black II (Erichrome Black T or Solochrome Black T) indicator.

Principle and Reaction: it is a complexometric titration. EDTA forms complex with Mg at pH 10

$$Mg^{2+} + EDTA \rightarrow Mg\ EDTA \qquad \text{.....(P.47)}$$

Equivalent factor

$$1 \text{ ml of } 0.1\text{M EDTA} \equiv 0.1203 \text{ g of } Mg\ SO_4$$

Procedure: Standardise 0.1M EDTA (experiment 5.4.a). Weigh accurately 0.5 g of Mg SO_4 and dissolve in 50 ml of water in a conical flask. Add 10 ml pH 10 buffer and 2-3 drops of indicator solution and titrate the contents of the flask with 0.1M EDTA till the colour changes from wine red to blue. Repeat the experiment.

Calculation:

$$\% \text{ of } MgSO_4 = \frac{\text{vol. of EDTA} \times \text{Eq. factor} \times 100 \times \text{actual M of EDTA}}{\text{wt. of } Mg\ SO_4 \times \text{expected M of EDTA}}$$

5.4 (e) Assay of Calcium Carbonate, $CaCO_3$

Mol.wt. 100.09

Aim: To carry out the assay of the given sample of calcium carbonate.

Reagents: 0.1M EDTA solution, calcium carbonate, conc. HCl, 10% NaOH solution, cal con/ solo chrome dark blue / Mordant Black 1I indicator

Principle and Reaction: It is a complexometric titration. Ca^{2+} forms complex with EDTA in 1:1 ratio. The indicator is cal con.

$$Ca^{2+} + EDTA \rightarrow Ca\ EDTA \qquad \text{.....(5.48)}$$

***Equivalent factor*:** 1 ml 0.1M EDTA $\equiv$ 0.01008 g Ca CO_3

Procedure: Standardise 0.1M EDTA solution (experiment 5.4.a) weigh out accurately 0.1 g of Ca CO_3 into a conical flask and dissolve in 3 ml conc. HCl. Add 10 ml of water and boil to remove the CO_2 and dilute to 50 ml. Now add about 10 ml of 10% Na OH (to adjust pH to 12.3) and 0.1 g of cal con. Titrate the contents with 0.1M EDTA solution till the end point i.e., the solution colour changes from pink to blue colour. Repeat the titration with two more samples and tabulate readings

Calculations:

$$\% \text{ of } CaCO_3 = \frac{\text{vol. of EDTA} \times \text{M EDTA} \times 0.01008 \times 100}{\text{wt. of Ca } CO_3 \times 0.1}$$

$$\text{or} \quad = \frac{\text{vol. of EDTA} \times \text{Eq. factor} \times 100 \times \text{actual M of EDTA}}{\text{wt of } CaCO_3 \times \text{expected M of EDTA}}$$

5.5 Precipitation Titrations (Argentometry)

The theory of precipitation titrations and about indicators discussed in part I. Here only experiments are described.

5.5 (a) Preparation and Standardisation of 0.1N Silver Nitrate (Mohr's of Method)

Eq. wt of $AGNO_3$ = 169.87

Aim: To prepare 0.1N silver nitrate solution and then its standardisation.

Apparatus:

Reagents:

0.1N $AgNO_3$: Dry some A.R grade silver nitrate (99.9% pure) at 120°C for 2hr and allow to cool in a desiccator. Weigh out accurately 4.248 g of $AgNO_3$ into a breaker dissolve in water and transfer to a 250 ml volumetric flask and make up to mark.

0.1N NaCl: Dissolve accurately weighed 1.4615 g of AR NaCl salt in a 250 ml volumetric flask with distilled water and make up to mark.

Potassium chromate indicator: 5% K_2CrO_4 solution (i.e., 5 g in 100 ml water).

Principle and Reaction: This is precipitation titration. In this $AgNO_3$ is used, so these are named as argentometric titrations. When titrating, $AgNO_3$ react with NaCl to give AgCl precipitation. As soon as the chloride ion is completed, $AgNO_3$ react with K_2CrO_4 indicator to give brick red precipitation.

$$AgNO_3 + NaCl \rightarrow \underset{\text{white precipitate}}{AgCl} + NaNO_3 \quad \text{.....(P.49)}$$

$$2AgNO_3 + K_2CrO_4 \rightarrow \underset{\text{Brick red}}{Ag_2CrO_4} + 2KNO_3 \quad \text{.....(P.50)}$$

Procedure: Pipette out 25 ml of NaCl solution into a conical flask and add 1 ml of indicator solution. Fill the burette with 0.1N silver nitrate solution. Titrate the contents of the conical flask with silver nitrate from the burette till the white precipitation turn to faint reddish colour persists. Note the burette reading. Repeat the experiment two to three times.

Calculate the exact normality of silver nitrate using the formula $N_1V_1 = N_2V_2$.

5.5 (b) Assay of Sodium Chloride, NaCl (Fajans Method)

*Mol.Wt.*58.45

Aim: To carry cut assay of the given sample of NaCl.

Reagents: 0.1N $AgNO_3$ solution, NaCl sample, fluorescein adsorption indicator -0.2 g in 100 ml of water (0.2%)

Principle and Reaction: This is also argentometric method. This is Fajans method. The indicator is fluorescein. As the end point is approaching, the AgCl formed coagulates appreciably and the local development of pink colour up on the addition of drop of Ag NO_3 solution becomes more and more pronounced. At the end point the precipitation suddenly assumes a pink or red colour.

$$AgNO_3 + NaCl \rightarrow AgCl + NaNO_3 \quad \text{.... (P.51)}$$

Equivalent factor: 0.1 N $AgNO_3 \equiv$ 0.05845 g NaCl.

Procedure: Standardise 0.1N $AgNO_3$ following experiment 5.5.a. Weigh out accurately 0.25 g of given sample of NaCl into a conical flask, dissolve in 50 ml distilled water and add 10 drops fluoroscein indicator. Titrate the contents with silver nitrate till the pink or red colour persists. Record the burette readings. Repeat with two more samples of NaCl. Tabulate burette readings

Calculation:

$$\%\text{ of NaCl} = \frac{\text{vol. of Ag NO}_3 \times \text{Eq. factor} \times 100 \times \text{actual N of AgNO}_3}{\text{wt. of NaCl.} \times \text{expected N of AgNO}_3}$$

5.5 (c) Assay of Potassium Chloride, (Volhord's Method)

Mol.Wt. 74.55

Aim: To carry out the assay of given sample of KCl by Volhord's method

Reagents: 0.1N $AgNO_3$, KCl, 1N NH_4SCN solution (1.9203g NH_4SCN in 25 ml water), ferric ammonium sulphate indicator (ferric alum), $NH_4Fe(SO_4)_2.12\ H_2O$ – 40 g of alum in 100 ml of water + few drops Conc. HNO_3.

Principle and Reaction: In this method the KCl solution is first treated with excess $AgNO_3$ solution and then the unreacted $AgNO_3$ is titrated with standard ammonium thiocyanate solution. In this formed AgCl is more soluble than silver thiocyanate, therefore it reacts with ammonium thiocyanate. Therefore AgCl precipitation must be prevented from reacting with thiocyanate by using nitrobenzene which forms protective film by around AgCl. When all the silver ions reacted (back titration), thiocyanate ions reacts with indicator to give deep red colour due to formation of ferric ferrothiocyanate.

$$KCl + AgNO_3 \rightarrow AgCl + KNO_3 \quad .....(P.52)$$

$$AgNO_3 + NH_4SCN \rightarrow AgSCN + NH_4NO_3 \quad .....(P.53)$$

$$2\ Fe^{3+} + 6\ NH_4SCN \rightarrow Fe\ [Fe\ (SCN)_6] \quad .....\ (P.54)$$

Procedure: Standardise 0.1N $AgNO_3$ by following 5.5.a. Weigh out accurately 0.1 g of the given sample of KCl into a conical flask and dissolve in 20 ml of water. Pipette out 25 ml of standard 0.1N $AgNO_3$ into this solution and add 1 ml conc. HNO_3 and shake. Add 5 ml nitrobenzene, 5 ml ferric alum indicator and titrate the contents with 0.1N NH_4SCN with shaking until the reddish colour persists. Repeat with 2 more samples of KCl and tabulate readings

Calculation:

Wt of KCl = W g

Vol. of $NH_4SCN = V_1$

Normality $NH_4SCN = N_1$

Normality of $AgNO_3 = N_2$

the volume $AgNO_3$ consumed = V_2

$$V_2 = \frac{N_1V_1}{N_2}$$

∴ $(25 - V_2)$ ml of $AgNO_3$ is consumed by KCl

$$\%\ of\ KCl = \frac{(25 - V_2) \times 0.07455 \times 100}{wt.\ of\ KCl}$$

5.6 Gravimetric Methods

The theory of gravimetric analysis discussed in part-I of this book

5.6 (a) Estimation of Nickel

Aim: To determine the amount of nickel present in the given sample.

Apparatus: Beaker, watch glass, glass rod, policeman, G_4 crucible, pipette, measuring jar

Principle and Reaction: Nickel can be determined as nickel dimethyl glyoximate, Ni (C_4 $O_7O_2N_2)_2$. Ni can be determined in presence of Fe (III) using citrate or tartarate masking agent. Ni can also be determined in presence of Co by using sufficient reagent.

$$Ni^{2+} + 2\,H_2DMG \rightarrow \underset{\text{scarlet red precipation}}{Ni(DMG)\downarrow} + 2H^+ \qquad .....(P.55)$$

Reagents:

1. Nickel ammonium sulphate – weigh out about 0.2 g of given nickel ammonium sulphate (or other salts) into a beaker and dissolve it in 100 ml.
2. Dimethyl glyoxime –1g of dimethyl glyoxime in 100 ml alcohol or n-propanol.
3. Dilute ammonia (1:1)
4. Wash solution – Hot water

Procedure: First clean the sintered glass crucible G_4 (gooch crucible) with conc. nitric acid, then with distilled water and dry at 120°C to constant weight. To the nickel solutions in beaker (1) add 1 ml methyl red. Now add dil. ammonia to neutralize with stirring till the colour of the solution changes to yellow. Then add few drops of acetic acid to make the solution acidic till the colour changes from yellow to red. Heat the contents of the beaker to about 70-80°C. Then add with constant stirring, dimethyl glyoxime solution (2) about 25 ml. Then add dil. ammonia drop wise till the solution become alkaline i.e. the solution colour is yellow. Digest the precipitate on a steam both for an hour and cool the solution. Filter the precipitate through a weighed sintered glass crucible; wash the precipitate with hot water till the washings are free from Cl^- and SO_4^{2-} ions. Dry the precipitate to constant weight at 110-120°C. Weigh as Ni (C_4 $H_7O_2N_2)_2$.

Calculation:

1. wt. of empty crucible = X_1 g
2. wt. of crucible + precipitation = X_2 g
3. wt. of the precipitation = $(X_2 - X_1)$ g
4. The gravimetric factor = 58.71 / 288.7 (At Wt. of Ni / formula wt of precipitate) = 0.2034
5. Amount of nickel = wt. of the precipitation × gravimetric factor

$$= (X_2 - X_1) \times 0.2034 \text{ g}$$

Note: Same Apparatus can be used for other Gravimetric methods in which Sintered Glass Crucible usage is there

5.6 (b) Estimation of Lead

Aim: To determine the lead present in the given sample by precipitating as lead chromate.

Apparatus: Same as used in above experiment

Principle and Reaction: Lead forms precipitation with chromate ion as lead chromate. This method is more accurate than Pb SO_4 method.

$$PbNO_3 + K_2CrO_4 \rightarrow \underset{\text{yellow precipation}}{PbCrO_4 \downarrow} + 2KNO_3 \quad \text{.....(P.56)}$$

Reagents:

1. Lead nitrate solution- weigh out 0.2 g of lead nitrate in a breaker and dissolve it in 100 ml of distilled water.
2. Dilute acetic acid
3. Potassium chromate – 4 % solution
4. Wash solution – dil. solution 4 % sodium acetate (4g in 100 ml)

Procedure: Clean and dry the sintered glass crucible at 120°C to constant weight. To the solution of lead nitrate (1) add dil. acetic acid till it is acidic. Heat to boiling and add 10 ml of 4 % K_2CrO_4 solution (3) slowly with constant stirring in slight excess. Boil gently for 5-10 minutes to settle the precipitate. The supernatant liquid must be yellow in colour. Filter through weighed G_4 sintered glass crucible. Wash the precipitate thoroughly with hot 4% sodium acetate solution, then with hot water till washings are colourless. Dry the precipitate in hot air over at 120°C.

Calculation:

Wt. of the empty crucible $= X_1$ g

Wt. of the crucible + precipitate $= X_2$ g

Wt. of the precipitate $= (X_2 - X_1)$ g

The gravimetric factor of lead $= 207.21 / 323.2$ Pb/ $PbCrO_4$

$= 0.6411$

The amount of lead present $= (X_2 - X_1)\text{ g} \times 0.6411$.

5.6 (c) Estimation of Barium

Aim: To determine the amount of barium present in the sample by precipitating as $BaSO_4$.

Apparatus: Beaker, watch glass, funnel, whatmaan filter paper no 42 (ash less paper), glass rod, policeman, silica crucible with lid.

Principle and Reaction: Barium reacts with sulphate ion to form white precipitation of Barium sulphate.

$$BaCl_2 + H_2SO_4 \rightarrow \underset{\text{white precipitation}}{BaSO_4 \downarrow} + 2HCl \qquad \text{.....(P.57)}$$

Reagents:

1. $BaCl_2$ solution
2. 1N H_2SO_4 solution
3. Conc. HCl
4. Wash solution-hot water

Procedure: Clean and heat the silica crucible by keeping on a clay pipe triangle using Bunsen burner. Cool in a desiccator and weigh and again keep in a desiccator.

Weigh out accurately 0.15 g of $BaCl_2$ into a beaker. Dissolve it in 100 ml distilled water and add 1 ml conc. HCl. Stir the solution and heat to boiling. Take 1N H_2SO_4 solution in another small beaker and add very slowly with constant stirring till the precipitation is completed. Allow the precipitate to settle. Test the supernatant liquid for completeness of precipitation again with one or two drops of H_2SO_4 solution. Digest the precipitate on hot water bath for about an hour. Filter the precipitate through whatmaan filter paper no 42 in a funnel. Wash the precipitation with hot water until the washings are free from sulphate and chloride ions. Dry the precipitate and fold the edges of filter paper to make it a cone and place it in a weighed silica crucible with tongs and ignite the crucible to a constant weight on a Bunsen burner (ignite first slowly and then strongly). Cool the crucible in desiccator and weigh as $BaSO_4$.

Calculation:

Wt. of the empty crucible = X_1 g

Wt. of the crucible + precipitate = X_2 g

Wt. of the precipitate of $BaSO_4$ = $(X_2 - X_1)$ g

The gravimetric factor of Ba = Ba wt / $BaSO_4$ wt = 137.36 / 233.42

= 0.58845

The amount of lead present = $(X_2 - X_1) \times 0.58845$

6. Analysis Methods of Pharmaceutical Drug Forms

6.1 Aqueous Acid-Base Analysis of Pharmaceutical Drug Forms

Name	Method	Indicator
(a) Thiamine hydrochloride	Titration with 0.1M NaOH	Thymol blue
(b) Detn. of aldehydes and ketones in essential oil like lemon oil	Reacts with hydroxyl amine hydrochloride and liberate HCl which titrated with 0.1M NaOH.	Methyl orange
(c) Aspirin	Titration with 0.1M NaOH	Phenolphthalein
(d) Benzoic acid, citric acid, nicotinic acid	Neutralization with ethanol and titrate with NaOH	Phenol red

6.2 Non-Aqueous Analysis of Pharmaceutical Drug Forms

(few example given)

Drug form	Medium and Method	Indicator
(a) Adrenaline 0.3 g	Glacial acetic acid, titration with 0.1N $HClO_4$	Crystal violet
(b) Amantadine cap	Glacial acetic acid + 10 ml mercury acetate. Titration against 0.1N $HClO_4$ and blank titration	Crystal violet

Table *Contd...*

(c) Bisacodyl 0.5 g	Glacial acetic acid. Titration with 0.1N $HClO_4$ and blank titration	1 naphthol benzein
(d) Chlorpromazine HCl 0.6 g	Acetone + mercuric acetate. Titration with 0.1N $HClO_4$ and blank titration	Methyl orange in acetone
(e) Codeine phosphate 0.4 g	Glacial acetic acid. Titration with 0.1N $HClO_4$ and blank titration	Crystal violet
(f) Dehydroemetine HCl 0.04 g	Glacial acetic acid + Hg acetate. Titration with 0.1N $HClO_4$ and blank titration	Crystal violet
(g) Ephedrine tab – 0.15 g	Glacial acetic acid + Hg acetate. Titration with 0.1N $HClO_4$ and blank	Crystal violet
(h) Imipramine HCl – 0.3 g	Chloroform + Hg acetate. Titration with $HClO_4$ and blank titration	Methanol yellow
(i) Levodopa – 0.6 g	Glacial acetic acid + Hg acetate. Titration with 0.1N $HClO_4$ and blank titration	Crystal violet
(j) Methoxamine HCl – 0.5 g	Glacial acetic acid + Hg acetate. Titration with 0.1N $HClO_4$ and blank titration	Crystal violet
(k) Nicotinamide – 0.25 g	Glacial acetic acid + acetic anhydride 0.1N $HClO_4$ titration and blank determination	Crystal violet
(l) Phenindamine tartarate tab – 0.2 g, 0.4 g	Glacial acetic acid + chloroform. Titration with 0.1N $HClO_4$ and blank determination	Orcet blue B
(m) Quinidine sulphate tab – 0.4 g	Acetic anhydride. Titration with $HClO_4$ and blank determination	Crystal violet
(n) Thiabendazole – 0.16 g	Glacial acetic acid + acetic anhydride. Titration with 0.1N $HClO_4$ and blank determination	Crystal violet

6.3 Precipitation Titration Analysis of Pharmaceutical Drugs

Drug form	Method	Indicator
(a) Aminophylline injection	Argentometric titration by Volhards method	Ferric alum
(b) Intraperitoneal dialysis fluid (or chloride)	Argentometric titration by Mohr's method	Potassium chromate
(c) Phenyl mercuric acetate	Argentometric titration by Volhard's method	Ferric alum
(d) Sodium Chloride and dextrose injection	Argentometric titration by Mohr's method	Potassium chromate
(e) Thiomersal	Argentometric titration by Volhards method	Ferric alum

6.4 Complexometric Titration Analysis of Pharmaceutical Drug Forms

Drug Form	Method and Buffer	Indicator
(a) Tricalcium Phosphate	Titration with 0.1M EDTA. NH_4 OH + NH_4Cl buffer	Mordant black II
(b) Magnesium trisilicate	Titration with 0.1M EDTA. NH_4 OH + NH_4Cl buffer	Mordant black II
(c) Zinc stereate	Titration with 0.1M EDTA Hexamine buffer	Xylenol orange
(d) Zinc undecylenate	Titration with 0.1M EDTA. Hexamine buffer	Xylenol orange
(e) Magnesium stereate	Titration with 0.1M EDTA Ammonium chloride buffer	Mordant black II

6.5 Redox Titration Analysis of Pharmaceutical Drug Forms

Drug form	Method	Indicator
(a) Analgin tablets	Titration with 0.1N iodine	Starch
(b) Dimercaprol injection	Titration with 0.1N iodine	Starch
(c) Ferrous fumarate	Titration with 0.1N ceric ammonium sulfate	Ferroin
(d) Methyl paraben	Titration with 0.1N sodium thiosulphate iodometrically	Starch
(e) Benzyl Penicillin Potassium	Titration with 0.1N mercury nitrate	Diphenyl carbazone + bromophenol blue
(f) Povidone iodine	Titration with 0.1N sodium thiosulphate iodometrically	Starch

7. Limit Tests

7.1 Introduction

The limit tests for impurity are actually the test of purity. These help to maintain the uniformity in composition and purity of substances. All the pharmacopoeias (I.P., B.P., and U. S. P.) specifies the general limit tests for the harmful impurity and harmless impurities. The limit tests are of three types.

(i) Tests in which no visible reaction

(ii) Tests in which comparison is performed with standard concentration of impurity.

(iii) Quantitative determination.

The test used as a limit test for particular impurity must be selective and highly sensitive. Some examples of impurities which are toxic are arsenic, lead, iron and heavy metals. Some other harmless impurities are acid radicals like chloride, sulphate.

Some of the experiments for limit tests are discussed for

7.2 Limit Test for Chloride

Aim: To perform the limit test for chlorides

Apparatus: Nessler's cylinders (Fig 7.1) 50 ml capacity 2 nos, beaker, glass rod etc.

Reagents: $NaHCO_3$ (which does not have chlorides), 5 % w/v silver nitrate solution, 0.01N HCl, dil. HNO_3.

Principle: The chloride impurity in $NaHCO_3$ is precipitated with $AgNO_3$ solution in presence of dil. HNO_3. The opalescence so obtained with opalescence obtained in the known quantity of chloride ions.

$$Cl^- + AgNO_3 \rightarrow \underset{\text{opalescence}}{AgCl \downarrow} + NO_3 \quad \text{.....(P.58)}$$

Procedure: Dissolve the sample of $NaHCO_3$ in a Nessler glass cylinder with 10 ml distilled water. Add 10 ml dil. HNO_3, dilute with redistilled water to 50 ml. Now add 1 ml Ag NO_3 solution. Stir the contents with glass road and kept it aside to get the opalescence which should not exceed the standard. Take for standard solution or dilute 5 ml of 0.0824 % w/v solution of NaCl in 100 ml. 1 ml of 0.01N HCl in a nessler glass cylinder which is of same capacity of the above i.e., 50 ml. Add 10 ml dil. HNO_3 and dilute to 50 ml with redistilled water. Add 1 ml of $AgNO_3$ solution and stir the contents with glass rod and set aside to get the opalescence.

Compare the opalescence produced by two solutions.

Observation: Opalescence produced by sample solution is less than that of standard opalescence.

Result: The sample complies with I.P limit for chloride

Note: The other samples can be tested for chloride limit tests by the same procedure are – $CaCO_3$, calcium gluconate, sodium citrate, magnesium trisilicate, ferrous gluconate.

7.3 Limit Test for Sulphate

Aim: To perform limit test for sulphates

Apparatus: 2 Nos 50 ml nessler's cylinders, glass rod etc.

Reagents:

Standard solution – 1ml of 0.1089 % w/v solution of K_2SO_4, sample of NaCl (or boric acid, $CaCO_3$, $AlCl_3$, ferrous gluconate, sodium citrate, magnesium trisilicate)

$BaSO_4$ reagent – (Mix 10 ml of 25 % w/v solution of $BaCl_2$ and 0.00181 % K_2SO_4 + 30 % ethanol and make the solution to 100 ml. Always it must be prepared freshly), 0.01N H_2SO_4 solution, dil. HCl acid.

Principle: This test is mainly designed to control sulphate impurity in inorganic substances. To sulphate impurity if $BaCl_2$ solution is added in presence of HCl acid, $BaSO_4$ precipitate. The turbidity so obtained compared with standard.

$$SO_4^{2-} + BaCl_2 \rightarrow BaSO_4 \downarrow 2\,Cl^- \quad \text{.....(P.59)}$$

Procedure: Dissolve the specified quantity of substance of NaCl in a nessler's cylinder in 10 ml of redistilled water. Add 2 ml of dil. HCl and dilute to 45 ml with water. Add 5 ml of $BaSO_4$ reagent. Stir with glass rod and set aside for 5 minutes.

Take 2 ml of standard solution in nessler cylinder and add 2 ml of dil. HCl. Dilute the contents to 45 ml and add 5 ml of $BaSO_4$ reagent. Stir the contents and keep aside for 5 minutes.

Compare the turbidity of the two solutions.

Observation: Turbidity produced by sample solution is less than that of standard.

Result: The sample complies with I. P limit for sulphate.

7.4 Limit Test for Heavy Metals

Aim: To perform limit test for heavy metals in the given sample.

Apparatus: Same as above

Reagents: NaCl solution, acetic acid, ammonia solution, H_2S solution, standard lead solution.

***Principle*:** This test is designed to control the metallic impurities. The heavy metal impurities are converted to their sulphides by sulphide ions imparting brown colour to the solution. The most commonly found heavy metals in pharmaceutical substances are – Ag, As, Bi, Cr, Cu, Fe, Ag, Ni, Sb, Sn, Cd, and Zn. They react with H_2S to produce corresponding metal sulphides.

$$Pb^{2+} + H_2S \rightarrow PbS + 2H^+ \quad .....(P.60)$$

$$Cu^{2+} + H_2S \rightarrow CuS + 2H^+ \quad .....(P.61)$$

Procedure

Standard Solution: Pipette out 2 ml of standard lead solution into a 50 ml Nessler's cylinder and dilute with water to 25 ml. Adjust the pH of the solution to 3-4 by addition of acetic acid or ammonia solution. Dilute to 35 ml. Add 10 ml H_2S solution and make up to 50 ml with water. Mix thoroughly and keep it aside.

Sample Solution: Dissolve specified quantity of NaCl in 10 ml of water in a Nessler cylinder. Adjust the P^H of the solution to 3-4 by adding acetic acid or ammonia and dilute to 35 ml. Add 10 ml of H_2S solution and make the solution to 50 ml. Mix well and keep it aside for 5 minutes.

View both standard and sample downwards over a white surface.

***Observation*:** The colour produced by sample solution is less than that of standard.

***Result*:** The sample complies with I. P. limit for lead

Note: Samples can be tested for heavy metals limit test are Borax, Barium sulphate, calcium carbonate, Dextrose, Magnesium Sulphate.

Lead limit test can be performed with dithizone solution in chloroform in presence of ammonium cyanide solution. The shade of violet colour produced by sample is compared with the colour obtained with standard lead solution.

7.5 Limit Test for Iron

Aim: To perform limit test for iron in the given sample.

Reagents:

Standard iron solution Dissolve 0.173g ferric ammonium sulphate in 10 ml of 0. 1N H_2SO_4. Transfer to 1000 ml volumetric flask and make it up to mark with distilled water. 1 ml contains 0.02 mg of iron.

Citric acid solution 20% w/v in distilled water, thioglycollic acid solution, ammonia buffer solution.

***Principle*:** The test is based on the formation of purple colour by the reaction of ferrous iron with thioglycollic acid to form ferrous thioglycollate in presence of ammonium citrate (citric acid + ammonia buffer). Thioglycollic acid reduces iron (III) to iron (II) and also reacts.

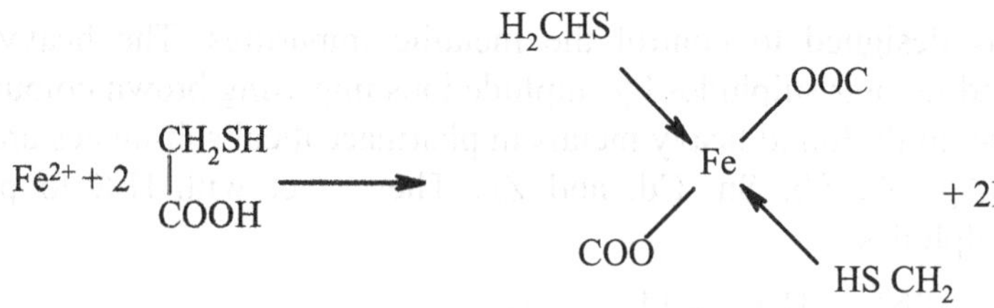

.....(P.62)

***Procedure*:**

Standard solution: To 2 ml of standard iron solution in Nessler's cylinder add 40 ml of water. Add 2 ml of 20% w/v solution of citric acid (iron free) and 2 drops of thioglycollic acid. Make the solution alkaline with ammonia solution and dilute to 50 ml with water. Allow to stand for five minutes.

Sample solution: Dissolve the specified quantity of the sample (NH_4Cl or $MgCO_3$ or NaCl or ZnO or $MgSO_4$) in water or prepare a solution as specified in I. P. in a Nessler cylinder. Make the volume to 40 ml with water. Add 2 ml of 20% w/v solution of citric acid (iron free) and 2 drops thioglycollic acid. Make alkaline with iron free ammonia solution and dilute to 50 ml with water. Keep aside for 5 minutes.

Compare the colours of two solutions by viewing transversely.

***Observation*:** The intensity of colour produced by sample is less than that of standard colour.

***Result*:** The sample complies with I.P limit for iron.

7.6 Limit Test for Arsenic

Aim: To perform limit test in the given sample.

Apparatus: Gut zeit apparatus (shown in Fig. P.7), cotton wool.

Reagents:

Standard arsenic solution: Dissolve 0.132 g of arsenic trioxide in 50 ml conc. HCl and make the solution to 100 ml with water 1 ml contains 0.0001 g of arsenic.

Lead acetate solution, standard HCl solution, Hg_2Cl_2 paper, HCl, Zn.

***Principle* and *Reaction*:** Arsenic is a physiologically harmful or toxic impurity. The arsenic in As (V) state is reduced to As (III) by nascent hydrogen in presence of potassium iodide solution. The gas is purified by passing through a glass tube containing cotton wool plug moistened with lead acetate solution. Formed arsine when it comes in contact with $HgCl_2$ paper it produces brown stain due to formation of arsenides.

$$HgCl_2 + 2\ AsH_3 \rightarrow Hg\ (AsH_2)_2 + 2\ HCl \quad .....(P.63)$$

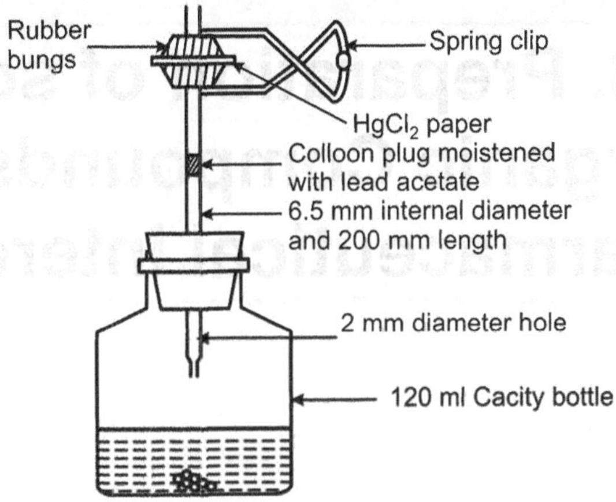

Fig. P.7 Apparatus for As Limit Test.

Procedure

Standard stain: To 50 ml water, 10 ml of stannated HCl and 1 ml of dil. Arsenic standard solution is placed in arsenic apparatus. The glass tube is lightly packed with cotton wool which was preciously moistened with lead acetate solution. A piece of $HgCl_2$ paper placed on top of the lower bung and the other bung is placed in such a manner that the borings of the two bungs meet to form true tube interrupted by a diapharams of $HgCl_2$ paper. The reaction is allowed to take 45 minutes.

Test Stain: NaCl dissolved in 50 ml of water is kept in the arsenic apparatus. Compare the stain produced with that of standard.

Observation: The stain produced by test solution is less deeper than standard stain.

Result: The sample complies with I.P. limit for arsenic.

Note: The samples for As test are – Aluminum hydroxide gel, NH_4Cl, Boric acid, Dextrose, $MgSO_4$, calcium gluconate.

8. Preparation of some Inorganic Compounds of Pharmaceutical Interest

8.1 Preparation of Ferrous Sulphate (Green Vitriol)

Chemical Formula: $FeSO_4.7\ H_2O$

Mol.Wt: 278.0

Materials required – iron filings, dil. H_2SO_4.

Reaction: $Fe + H_2SO_4 \rightarrow FeSO_4 + H_2 \uparrow$ ...... (P.64)

Procedure: Take 25 ml of dil. H_2SO_4 in a beaker and boil. Add iron filings in small portions with gradual heating until effervescence of H_2 ceases. Filter the solution and allowed to cool. The crystals are separated. The crystals are dried and recrystallise again by dissolving in water and boiling.

Uses: Heamatinic. It is used in preparation of oral iron formulation, it is used in condition of anemia.

8.2 Preparation of Copper Sulphate (Blue Vitriol)

Chemical Formula: $CuSO_4.\ 5\ H_2O$

Mol.Wt: 249.6

Materials required – copper turnings, conc. H_2SO_4 and conc. HNO_3.

Reaction:

$$Cu + 2H_2SO_4 \xrightarrow{HNO_3} CuSO_4 + SO_2 \uparrow + 2H_2O \quad .....(P.65)$$

Procedure: In a beaker take 4 ml of conc. H_2SO_4 add 25 ml of distilled water slowly. Then add about 5 ml. conc. HNO_3 and 0.5 g of copper turnings. Heat gently until the evolution of SO_2 gas ceases. Evaporate the solution to dryness. Dissolve the residue in

25 ml of water. Filter and evaporate the filtrate to crystallization. Recrystallize from water. Blue crystals of $CuSO_4$. 5 H_2O will be obtained. Report the yield.

***Uses*:** $CuSO_4.5H_2O$ is no longer official in pharmacopoeias. It is an example of astringent and fungicide.

8.3 Preparation of Magnesium Oxide

Chemical Formula: MgO

Mol.Wt: 40.32

Materials required – Light magnesium carbonate.

***Reaction*:**

$$3\ MgCO_3\ Mg\ (OH)_2.\ 3\ H_2O \rightarrow 4\ MgO + 3CO_2 + 4H_2O \quad .....(P.66)$$

***Procedure*:** Take basic magnesium carbonate in a porcelain dish and heat strongly until no CO_2 is evolved. Collect the white powder of MgO.

***Use*:** It is potent antacid and mild laxative

8.4 Preparation of Potash Alum

Chemical Formula: $K_2SO_4.Al_2\ (SO_4)_3\ .24\ H_2O$

Mol.Wt: 916.2

Materials required – Potassium sulphate, aluminium sulphate.

***Reaction*:**

$$K_2SO_4 + Al_2\ (SO_4)_3.18\ H_2O + 6\ H_2O \rightarrow K_2O_4Al_2(SO_4)_3.24\ H_2O \quad .....(P.67)$$

***Procedure*:** About 6 g of $Al_2\ (SO_4)_3$ dissolve in sufficient water (50 ml) and about 1 g of K_2SO_4 dissolve in water (10 ml) in a separate beaker. Heat both the solutions and mix hot solution in a separate beaker and boil again to reduce the volume. Cool the beaker and keep it aside. Alum crystals separate. Filter and dry.

***Use*:** Alum is an example of antiseptic and astringent.

8.5 Preparation of Boric Acid

Chemical Formula: H_3BO_3

Mol.Wt: 61.82

Materials required – Borax, dil. H_2SO_4

Reaction:

$$Na_2B_4O_7 + H_2SO_4 + 5\,H_2O \rightarrow Na_2SO_4 + 4\,H_3BO_3 \qquad .....(P.68)$$

Procedure: Dissolve about 30 g of borax in a beaker in 50 ml distilled water. Boil the solution. Now add dil. H_2SO_4 (6 ml in 60 ml of water) slowly to the boiled solution with stirring. Hot liquid is filtered and kept aside for crystallization. Dry the boric acid formed.

Use: It is an example of topical agent (antiseptic).

8.6 Preparation of Yellow Mercuric Oxide

Chemical Formula: HgO

Mol.Wt: 216.61

Materials required – Mercuric chloride, 10% sodium hydroxide solution.

Reaction:

$$HgCl_2 + 2NaOH \rightarrow Hg\,(OH)_2 + 2NaCl \qquad .....(P.69)$$

$$Hg(OH)_2 \rightarrow HgO + H_2O \qquad .....(P.70)$$

Procedure: In a beaker take 5 g of $HgCl_2$ and dissolve it in 50 ml of water. Add to this 50 ml of 10% NaOH solution at room temperature and stir well. Allow to stand for 1 hr in the dark (light sensitive). Decant the supernatant liquid. Wash the precipitate with water till it is free from alkali. Dry at room temperature in the dark. (See not to exceed 30°C).

Use: Yellow HgO is an example of antiseptic (in ophthalmic preparations).

8.7 Preparation of Ferrous Ammonium Sulphate (Mohr's Salt)

Chemical Formula: $FeSO_4\,(NH_4)_2\,SO_4.\,6H_2O$

Mol.Wt: 392.16

Materials required – Iron fillings, dil. H_2SO_4, ammonium sulphate, alcohol.

Reaction:

$$Fe + H_2SO_4 \rightarrow FeSO_4 + H_2 \uparrow$$

$$FeSO_4 + (NH_4)_2\,SO_4 + H_2O \rightarrow FeSO_4\,(NH_4)_2\,SO_4\,6H_2O \qquad .....(P.71)$$

Procedure: To about 6g of iron fillings add prepared dil. H_2SO_4 i.e., 10 ml of conc. H_2 SO_4 in 50 ml water, and heat the contents of the beaker. After the evolution of H_2 gas ceases, add about 1.5g $(NH_4)_2SO_4$ solution slowly with stirring. Then concentrate the total solution by evaporation on a sand bath. The solution is filtered while hot and allowed to crystallize by cooling. Add 2ml of alcohol to increase the yield of mohr's salt. The mohr's salt is recrystallised from water.

***Use*:** It is an example of antiseptic.

8.8 Preparation of Ammonium Chloride

Chemical Formula: NH_4Cl

Mol.Wt: 53. 46

Materials required: Conc. HCl, ammonia solution, litmus paper.

***Reaction*:**

$$NH_4OH + HCl \rightarrow NH_4Cl + H_2O \quad \text{.....(P.72)}$$

***Procedure*:** Take 20 ml of con. HCl in a beaker. Add ammonium solution slowly with stirring until red litmus turns blue by keeping in a fuming cup board. Keep the beaker on water bath and evaporate the solution till crystals appear, cool the beaker to room temperature. Filter the contents. The residue is NH_4Cl crystals. Dry the crystals by keeping in between filter papers.

***Use*:** It is used as expectorant, diuretic and systemic acidifier.

***Note*:** In quantitative preparations exact quantities of the reactants to be taken and also will have to report yield of the prepared product.

Books Referred

1. Indian Pharmacopoeia.
2. The United States Pharmacopoeia.
3. A.H. Beckett, & J.B.Stenlake, Practical Pharmaceutical chemistry.
4. L.M. Atherden, Bentley and Driver's Textbook of Pharmaceutical Chemistry.
5. J.D.Lee, Concise Inorganic Chemistry.
6. Remington's Pharmaceutical Sciences.
7. J.H.Block, E.Roche, T.O.Soine, & C.O.Wilson, Inorganic Medicinal and Pharmaceutical Chemistry.
8. Vogel's qualitative inorganic analysis.
9. L.A. Discher, *Modern Inorganic Pharmaceutical Chemistry.*
10. Dr.A.V.Kasthure, Dr. S. G. Wadodkar, *Pharmaceutical Inorganic Chemistry.*
11. Day Underwood, Qualitative Analysis.
12. S.M.Khopkar, Basic concepts of Analytical Chemistry.
13. Kaza Somasekhar Rao & K.N.K.Vani, Practical Inorganic Chemistry.
14. Skoog, Fundamentals of Analytical Chemistry.
15. W.L.Jolly, Modern Inorganic Chemistry.
16. Vogel's quantitative inorganic analysis.
17. A. H. Backett and J. B. Stenloke, Practical Pharmaceutical Chemistry.
18. G. Devala Rao, Practical Pharmaceutical Analysis.

Index

B

C

D

E

F

G

H

I

J

K

L

M

N

O

P

Q

R

S

T